Models of Disequilibrium and Shortage in Centrally Planned Economies

INTERNATIONAL STUDIES IN ECONOMIC MODELLING

Series Editor

Homa Motamen

Executive Director
Canadian Imperial Bank of Commerce
Securities Europe Ltd
London

Economic Modelling in the OECD Countries
H. Motamen
Modelling the Labour Market
M. Beenstock
Input-Output Analysis
M. Ciaschini

In preparation

Economic Models of Latin America
D. E. Hojman
Economic Models of Trade Unions
P. Garonna

Models of Disequilibrium and Shortage in Centrally Planned Economies

Edited by

Christopher Davis

Centre for Russian and East European Studies
University of Birmingham, UK

and

Wojciech Charemza

Department of Economics
University of Leicester, UK and
University of Gdansk, Poland

London
CHAPMAN AND HALL
New York

First published in 1989 by
Chapman and Hall Ltd
11 New Fetter Lane, London EC4P 4EE
Published in the USA by
Chapman and Hall
29 West 35th Street, New York NY 10001

Typeset in 10/12pt Sabon by
Thomson Press (India) Ltd
Printed in Great Britain by
J. W. Arrowsmith Ltd, Briston

ISBN 0 412 28420 0

British Library Cataloguing in Publication Data

Models of disequilibrium and shortage in centrally planned economies.—(International studies in economic modelling).
1. Planned economies. Econometric models
I. Davis, C. (Christopher), II. Charemza, W. (Wojciech) III. Series
330.12′4′0724

ISBN 0–412–28420–0

Library of Congress Cataloging in Publication Data

Models of disequilibrium and shortage in centrally planned economies/edited by Christopher Davis and Wojciech Charemza.
p. cm.—(International studies in economic modelling)
The volume resulted from the Conference on Modelling of Disequilibrium and Shortage in Centrally Planned Economies, held in Apr. 1987 at the University of Birmingham, Birmingham, England.
Bibliography: p.
Includes index.
ISBN 0–412–28420–0
1. Central planning—Poland—Econometric models—Congresses. 2. Poland—Economic policy—1981—Econometric models—Congresses. 3. Central planning—Hungary—Econometric models—Congresses. 4 Hungary—Economic policy—1968—Econometric models—Congresses. 5. Central planning—Czechoslovakia—Econometric models—Congresses. 6. Czechoslovakia—Economic policy—1965—Econometric models—Congresses. 7. Equilibrium (Economics)—Congresses. 8. Scarcity—Congresses. I. Davis, Christopher, 1948-. II. Charemza, Wojciech. III. Conference on Modelling of Disequilibrium and Shortage in Centrally Planned Economies (1967: University of Birmingham) IV. Series.
HC270.28.M63 1989 88-31494
338.9438—dc 19 CIP

Contents

Contributors

Josef C. Brada	Department of Economics, Arizona State University, USA
Wojciech Charemza	Department of Economics, University of Leicester, UK and University of Gdansk, Poland
Christopher Davis	Centre for Russian and East European Studies, University of Birmingham, UK
Vladimir Dlouhý	Institute for Forecasting, Czechoslovak Academy of Sciences, Czechoslovakia
Miroslaw Gronicki	Department of Statistics, University of Gdansk, Poland
Irena Grosfeld	Centre d'Economie Quantitative et Comparative, France
Paul Hare	Department of Economics, Heriot-Watt University, UK
Katalin Hulyak	Institute of Economic Planning, Hungary
Zsuzsa Kapitány	Institute of Economics, Hungarian Academy of Sciences, Hungary
David M. Kemme	Department of Economics, Wichita State University, USA
Arthur E. King	Department of Economics, Lehigh University, USA

János Kornai	Institute of Economics, Hungarian Academy of Sciences, Hungary and Harvard University, USA
Mária Lackó	Institute of Economics, Hungarian Academy of Sciences, Hungary
Domenico Mario Nuti	European University Institute, Italy
Richard Portes	Centre for Economic Policy Research and Birkbeck College, UK
Richard E. Quandt	Department of Economics, Princeton University, USA
Judit Szabó	Institute of Economics, Hungarian Academy of Sciences, Hungary
Aleksander Welfe	Institute of Econometrics and Statistics, University of Lodz, Poland
Wladyslaw Welfe	Institute of Econometrics and Statistics, University of Lodz, Poland

Acknowledgement

This volume resulted from a conference supported by the Joint Committee on Eastern Europe, of the American Council of Learned Societies and the Social Science Research Council, with funds provided by the Ford Foundation.

Introduction to the Series

There has been a growing dependence in the past two decades on modelling as a tool for better understanding of the behaviour of economic systems, and as an aid in policy and decision making. Given the current state of the art globally, the introduction of a series such as this can be seen as a timely development. This series will provide a forum for volumes on both the theoretical and applied aspects of the subject.

International Studies in Economic Modelling is designed to present comprehensive volumes on modelling work in various areas of the economic discipline. In this respect one of the fundamental objectives is to provide a medium for ongoing review of the progression of the field.

There is no doubt that economic modelling will figure prominently in the affairs of government and in the running of the private sector, in efforts to achieve a more rational and efficient handling of economic affairs. By formally structuring an economic system, it is possible to simulate and investigate the effect of changes on the system. This in turn leads to a growing appreciation of the relevance of modelling techniques. Our aim is to provide sufficient space for authors to write authoritative handbooks, giving basic facts with an overview of the current economic models in specific areas and publish a useful series which will be consulted and used as an accessible source of reference.

The question may arise in some readers' minds as to the role of this series *vis-à-vis* other existing publications. At present, no other book series possesses the characteristics of *International Studies in Economic Modelling* and as such cannot fill the gap that will be bridged by it. Those journals which focus on this area do not present an exhaustive and comprehensive overview of a particular subject and all the developments in the field. Other journals which may contain economic modelling papers are not sufficiently broad to publish volumes on all aspects of modelling in a specific area, which this series is designed to cover.

A variety of topics will be included encompassing areas of both micro and macroeconomics, as well as the methodological aspects of model construction. Naturally, we are open to suggestions from all readers of, and contributors to, the series regarding its approach and content.

Finally, I would like to thank all those who have helped the launch of this series. The encouraging response received from authors who have contributed the forthcoming volumes and from the subscribers to the series has indicated the need for such a publication.

Homa Motamen
London
December 1987

Preface

The centrally planned economies (CPEs) of the Soviet Union and Eastern Europe have experienced severe imbalances in domestic and external markets over the past several decades. As a result, they have been chronically afflicted by problems such as excess demand, repressed inflation, deficits of commodities, queues, waiting lists, and forced savings. Economists have responded to these phenomena by developing appropriate theoretical and empirical models of CPEs. Of particular note have been the pioneering studies of Richard Portes on disequilibrium econometric models and János Kornai on the shortage economy. Each approach has attracted followers who have produced numerous, innovative macro- and microeconomic models of Poland, Czechoslovakia, the German Democratic Republic, Hungary, and the USSR.

These models have proved to be of considerable value in the analysis of the causes, consequences and remedies of disequilibrium phenomena. Inevitably, the new research has also generated controversies both between and within the schools of shortage and disequilibrium modelling, concerning the fundamental nature of the socialist economy, theoretical concepts and definitions, the specification of models, estimation techniques, interpretation of empirical findings, and policy recommendations. Furthermore, the research effort has been energetic but incomplete, so many gaps exist in the field.

The objectives of this volume on models of disequilibrium and shortage in CPEs are to provide comprehensive and coherent surveys of past work on this subject, identify key intellectual controversies, and fill some gaps in research. The book is intended for mainstream economists interested in disequilibrium phenomena and models as well as for specialists on the CPEs.

The book is divided into three parts. Part One examines the foundations of disequilibrium and shortage models. In Chapter 1 the editors review the history of disequilibrium phenomena in CPEs and the development of related models. Richard Portes surveys the field of disequilibrium econometric models in Chapter 2 and Paul Hare makes an equivalent assessment of the shortage model in Chapter 3. These are followed by evaluations of the chronic excess demand hypothesis by David Kemme and of repressed inflation by Domenico Mario Nuti. In Chapter 6

Richard Quandt assesses the state of disequilibrium econometrics for CPEs. Part Two of the volume contains surveys of existing macroeconomic disequilibrium models of the CPEs by experts from the three countries that have been most studied: Vladimir Dlouhý on Czechoslovakia, Wladyslaw Welfe on Poland, and Katalin Hulyak on Hungary. Part Three is devoted to sectoral shortage and disequilibrium models. Mária Lackó reviews sectoral shortage models in Hungary in Chapter 10. Next, Wojciech Charemza evaluates past disequilibrium models of consumption and presents an original model for Poland that incorporates second economy activities. Novel approaches to disequilibrium analysis are also adopted by Aleksander Welfe in his study of consumption and savings in Poland and by Josef Brada and Arthur King in their investigation of Soviet foreign trade. Disequilibrium models of investment in Poland are evaluated by Irena Grosfeld in Chapter 14. The next two chapters examine the same commodity market, for automobiles, using different modelling approaches. Zsuzsa Kapitány, János Kornai and Judit Szabó assess the automobile market in Hungary using a shortage model whereas Wojciech Charemza, Miroslaw Gronicki and Richard Quandt make use of a disequilibrium model to analyse the Polish car market. The sectoral studies are rounded off by the shortage model analysis of the socialist medical system by Christopher Davis. The final chapter by the editors assesses the contributions made in the book and outlines a future programme of research in the area of the modelling of disequilibrium and shortage in CPEs.

The editors designed the volume in 1986 and arranged contributions from leading analysts of disequilibrium and shortage in CPEs from both Western countries (USA, Britain, France and Italy) and Eastern Europe (Poland, Hungary and Czechoslovakia). The initial drafts of the chapters were presented and criticized at a working conference in Birmingham, UK in April 1987. (Unfortunately János Kornai, the founder of the 'shortage school', could not attend due to conflicting commitments.) The revised chapters were completed by early 1988.

The editors of the volume would like to express their gratitude to a number of institutions and people for their support. The individual and collaborative research of the editors in the field of modelling of disequilibrium and shortage in CPEs has been funded since September 1986 by grant no. B00232171 of the Economic and Social Research Council. The Centre for Russian and East European Studies of the University of Birmingham has provided an excellent institutional base for the editors.

The April 1987 Conference on Modelling of Disequilibrium and Shortage in Centrally Planned Economies was generously supported by a variety of organizations. The editors would like to give special thanks to the Joint Committee on Eastern Europe of the American Council of Learned Societies and the Social Science Research Council for the major grant, based upon funds provided by the Ford Foundation, that made the conference possible. Supplemental assistance was provided by the BUAS–NASEES Joint Research and Development Committee, International Research and Exchanges Board (IREX), the University of Birmingham, the British Council, the British Academy, and the Economic and Social Research Council.

The editors would like to express their appreciation to the numerous people who participated in or helped organize the conference. A strong team of specialists discussed and criticized the volume participants' papers. These included Ron Amann, John Burkett, Martin Cave, Irwin Collier, Julian Cooper, Robert W. Davies, Michael Ellman, Wojciech Maciejewski, Mark Schaffer, Karoly Attila Soós, Michael Spagat, and Thomas Wolf. Our ESRC project colleague David Young not only fully participated in the conference but also provided substantial organizational assistance. We are especially grateful for his contribution. Additional help with the conference was provided by Sandra Cumberland, Patricia Smith, Melanie Ilic, David Moon and by the programme designer Antonia Cunliffe Davis.

Three of the chapters of this book have been published earlier: Chapter 5 by D. M. Nuti in *Contributions to Political Economy* (1986); Chapter 15 by Z. Kapitány, J. Kornai and J. Szabó in *Soviet Studies* (1984); and Chapter 16 by W. Charemza, M. Gronicki and R. E. Quandt in *European Economic Review* (1988). The editors would like to thank the journals for granting permission to reprint these articles.

The editors are also most appreciative of the encouragement provided by the Editor of the Series on International Studies in Economic Modelling, Homa Motamen. Several people have been of considerable help in producing the book, particularly Phillip Read, Elizabeth Johnston and all the editorial and production staff at Chapman and Hall. Thanks are also due to Ceinwen Procter, the copy-editor.

Finally, the editors would like to thank their wives, Antonia Cunliffe Davis and Krystyna Charemza, for their valuable support of project research, conference organization, and book preparation.

PART ONE

The foundations of disequilibrium and shortage models of centrally planned economies

1

Introduction to models of disequilibrium and shortage in centrally planned economies

CHRISTOPHER DAVIS
AND
WOJCIECH CHAREMZA

The analysis and modelling of disequilibrium and shortage phenomena in economic systems have attracted substantial interest in recent years. Concern about the adequacy of the conceptualization of equilibrium and its dynamics has stimulated advances in the elaboration of theories and models that explain the behaviour of economies outside the conventional equilibrium state, such as fixed-price equilibrium theory and the theory of the shortage economy. Developments in econometric estimation techniques and hypothesis testing have also contributed to the intensification of quantitative research on disequilibrium issues. Finally, work on disequilibrium and shortage problems has been stimulated by the fact that since the mid-1970s most economies have functioned for prolonged periods in disequilibrium states – Keynesian unemployment in market economies and repressed inflation or chronic shortages in socialist countries – and by the demands of policy-makers for appropriate solutions to economic imbalances.

One subset of theories and models of shortage and disequilibrium is comprised of those developed for the centrally planned economies (CPEs). Although the CPE models have some universal elements, it is generally recognized that they possess unique features that are functions of particular institutional arrangements and behavioural patterns as well as of their focus on issues such as unsatisfied consumer demand, queues, waiting lists, chronic shortages and forced savings.

International Studies in Economic Modelling Series.
Models of Disequilibrium and Shortage in Centrally Planned Economies.
Edited by C. Davis and W. Charemza.
Published in 1989 by Chapman and Hall Ltd, 11 New Fetter Lane, London EC4P 4EE.

Examination of the CPE models can therefore provide both a contrast with the more prevalent market economy variants and insights into disequilibrium phenomena that exist, but are neglected, in the West.

The study of these CPE models can also assist in understanding the actual problems of disequilibrium and pervasive shortages in socialist economies and the prospects for their solution. In the 1980s the Soviet and East European economies have experienced at least occasional severe domestic and external imbalances, which have adversely affected economic performance indicators and the behaviour of planners, managers, workers and consumers. Interesting and increasingly open debates have been carried out by economists in these countries on the causes and consequences of disequilibria. To restore balance in their economies socialist governments have resorted to policies such as retail price increases, wage control, restriction of investment, and reduction in imports. Furthermore, CPEs have introduced reforms, ranging from modest to radical in nature, in efforts to correct micro- and macroeconomic deficiencies.

This introduction to models of disequilibrium and shortage in CPEs covers three related topics. On the assumption that not all readers are experts on the socialist economies, section 1.1 surveys the history of the CPEs of the Soviet Union and Eastern Europe to show that problems with disequilibrium have been recurrent and important in these economic systems. This is followed in section 1.2 by a brief review of the development of models of disequilibrium and shortage in CPEs. Finally, section 1.3 identifies the main controversies and neglected issues in the field and compares the features of disequilibrium and shortage models of CPEs.

1.1 AN HISTORICAL SURVEY OF SHORTAGE AND DISEQUILIBRIUM IN THE CENTRALLY PLANNED ECONOMIES

The centrally planned economies of the Soviet Union and Eastern Europe appear to have been afflicted, with varying degrees of intensity, by imbalances in internal and external markets and shortages of goods and services throughout their histories. It is important to appreciate this historical record in order to place in perspective the problems of the socialist economies in the 1980s, to assess the debate in the disequilibrium econometric literature stimulated by Portes and his colleagues (see Chapters 2 and 4 of this volume) concerning the excess demand hypothesis, and to evaluate the arguments of Kornai (1980) that current agents' expectations and institutional behaviour are products of past experiences in a chronic shortage environment. Unfortunately, a concise disequilibrium-oriented history of the socialist economies is not yet available. Although specialists in the comparative economics field are familiar with this topic, many readers of this volume may not be. Accordingly, this section presents a summary history of the CPEs that focuses on disequilibrium and shortage phenomena. It should be stressed that this is intended to be a preliminary, subjective guide rather than a definitive statement.

Substantial additional research is required by economic historians on disequilibrium phenomena in all periods of every socialist country to clarify their importance and effects.

Marxist theorists have argued that the socialist economy would be superior to other economic systems with respect to structure of output, income distribution and efficiency (e.g. Mandel, 1968; Dobb, 1970). However, the actual performance of CPEs has been quite different from the *ex ante* predictions of its advocates. Of relevance to this book is the fact that for much of their history the Soviet Union and East European countries have had pervasive disequilibria in markets and chronic shortages of labour and commodities. This has been due to a combination of factors, including objective conditions, specific institutional arrangements, and policy choices.

The first socialist economy developed in the USSR following the October 1917 revolution. At that time Russia had many of the characteristics of a less developed country and its economic system had been severely strained by World War I and the revolutionary process. During 1918–21 a civil war was fought on Soviet territory that destroyed a substantial share of productive capacity. To cope with this situation the Bolshevik government adopted the policies of War Communism, which included state ownership of industry and trade, government requisitioning of agricultural products, and administrative allocation of resources. Although this programme assisted in the war effort it also caused hyperinflation, acute shortages of producer and consumer goods, and a transition to barter exchange (Malle, 1985).

The New Economic Policy (NEP) that governed the Soviet economy in the 1920s resulted in a re-monetization of market transactions, some restoration of economic balance, and recovery of output to pre-war levels. Despite these achievements, the behaviour of markets for industrial goods was distorted by imperfect competition resulting from state domination and administrative restrictions. The unstable nature of exchange between town and country was the cause of economic disruption and political tension. The 'scissors crisis' of 1923 over the growing differential between industrial and agricultural prices, in favour of the former, raised fears of an eventual decline in the supply of agricultural products to the cities and of insufficient rural demand for industrial goods. In its attempt to close the scissors the government reduced or fixed the prices of industrial products. Since the output capacity of industry grew slowly relative to the population's income, an excess demand situation developed that generated widespread shortages of commodities (called the Goods Famine) and an expansion of the second economy (Dobb, 1948; Gregory and Stuart, 1981). Furthermore, throughout the NEP period there was a chronic shortage of housing and social services, such as medical care (Davis, 1983). In contrast, unemployment was high in the twenties. So the economy was often in a classical unemployment disequilibrium state, with excess demand for goods and excess supply of labour (see Malinvaud (1977) for a discussion of disequilibrium states).

During 1924–28 theoreticians and Bolshevik Party leaders engaged in debates on future industrialization policy and approaches to planning, the outcomes of which had profound consequences for the economy in general and disequilibrium phenomena in particular (Erlich, 1960; Carr and Davies, 1969). The main issue was how the economy should develop given a rate of industrial growth lower than that believed to be essential by the politicians, and the intensification of repressed inflation and commodity shortages. In brief, one faction of the Bolshevik Party (later called the Right Opposition) supported a balanced growth strategy that would have involved gradual expansion of state and collective ownership, reliance on markets for distribution, a modest rate of capital accumulation, distribution of investment between sectors to promote balanced development, substantial improvement of efficiency, and imports of foreign technology. This strategy was linked to an approach to planning that was predominantly genetic in nature (Erlich, 1960), which meant that it recognized the complexity of interrelationships in the economy and the importance of devising consistent plans. In contrast, their opponents (the Left Opposition) argued for accelerated, unbalanced growth. This would have involved a drastic increase in capital accumulation financed by forced savings (primitive socialist accumulation), concentration of investment in heavy industry, expansion of state control in industry and agriculture, subordination of the market to the plan, autarky in foreign trade, and severe constraints on consumption. Adherents of this programme believed that planning should have more of a teleological character, which meant that it should reflect political objectives and priorities, dominate market forces, and attempt to overcome constraints imposed by existing input-output relationships. Although they did not explicitly advocate ignoring considerations of equilibrium and financial stability, their strategy implied that this would be done (Dobb, 1948; Carr, 1958; Erlich, 1960). See also Charemza and Kiraly (1988) for an assessment of the genetic–teleological debate in the context of post-war CMEA country planning).

In the late 1920s Stalin gained control of the Bolshevik Party apparatus and initiated a radical transformation of the Soviet economy. A system of mandatory central planning was introduced and the adopted First Five Year Plan initially expressed in more cautious form the unbalanced growth strategy (Carr and Davies, 1969; Zaleski, 1971). In 1929–30 the plans were revised in a manner that made them extremely ambitious and generated pervasive shortages throughout the economy. This situation contributed to the decision to collectivize the agricultural sector in order to facilitate the extraction of food and materials needed to support the rapid industrialization drive (Davies, 1980, 1988). Within a short period of time these developments produced a shift in the disequilibrium regime to one of repressed inflation (or chronic shortage).

The pervasive, chronic disequilibrium in markets and shortages of goods during the 1930s were in part caused by conscious policy. The rapid industrialization strategy did not just ignore economic balance, but actively promoted disequilibrium, supposedly to facilitate attainment of its objectives. In the late twenties and early thirties Stalin and other top Bolshevik leaders (e.g. Mikoyan and Rykov) criticized the 'bourgeois theory' of equilibrium and argued that excess demand in markets acted as a stimulus to industry (Davies, 1988, pp. 165, 465 and

478). For example, at the XVI Party Congress in 1930, Stalin claimed that the repressed inflation disequilibrium state was a positive characteristic of socialism:

> Here in the USSR the growth of consumption (purchasing power) always goes ahead of the growth of production, driving it forward, but over there, with the capitalists, the growth of consumption (purchasing power) of the masses never catches up with the growth of production and always lags behind it, thus condemning production to crises. (Stalin, 1949, pp. 322–3)

Unrealistic assumptions and unexpected developments contributed as well to disequilibria in the Soviet economy. Economic plans intentionally established output targets that were overambitious relative to available resources and justified them by extravagant expectations concerning labour and capital productivity, efficiency of the investment process, and successes in the agricultural sector (Zaleski, 1971, 1980). The unanticipated phenomena included the massive destruction of livestock by the peasantry following collectivization, the extent of the chaos in industry (e.g. in 1930) caused by the inconsistent plans, the build-up of inflationary pressure in the system, the bad weather that contributed to the famine of 1932–33, and the deterioration of the international situation in the late thirties which necessitated an acceleration of arms production in the USSR.

The economic system that emerged in the 1930s institutionalized disequilibrium phenomena. Economic agents expected to encounter imbalances and shortages and made appropriate adjustments. They gradually began to respond primarily to quantity signals rather than to prices. Central planners initiated the transition to the repressed inflation regime by their overambitious plans and determined its local features through the setting of sectoral priority. But once subordinate units adapted to the new circumstances, the planners had to cope with consciously distorted information from below, and formal fulfilment of targets in quantity terms at the expense of quality (Ellman, 1979; Zaleski, 1980). Managers in industrial enterprises learned to behave in a manner suitable to their operating conditions, which included inconsistent plans imposed from above, a seller's market for output, an incentive structure that rewarded plan fulfilment in volume terms and inhibited technological innovation, shortages of inputs, supply-linked bottlenecks in production, and non-binding financial constraints. Their responses included a drive to continually increase the quantity of output, neglect of product quality, hoarding of inputs, hiding reserves from central authorities, energetic struggle for more investment and other inputs, and acquisition of supplies in second economy markets (Berliner, 1957).

By the early 1930s widespread shortages existed in labour markets in the USSR. However, the ability of workers to take advantage of market conditions in wage bargaining through the manipulation of labour supply was restricted by political pressures, administrative measures, and the struggle to maintain a subsistence living standard. Although there was substantial labour turnover as workers searched for better conditions, periods of unemployment were necessarily short.

Since real wages were relatively inflexible and a supply response was unfeasible, the workers' reaction to excess demand was expressed in poor work discipline, low productivity and diversion of work time to leisure and second economy activities (Chapman, 1963; Zaleski, 1980).

The authorities did try to maintain a financial balance in the economy in the early 1930s by coordinating plans and budgets, reducing costs, and stabilizing the currency. However, these efforts were only partially successful. In the consumer sphere, for example, disposable income usually exceeded planned levels whereas the output of consumer goods and services consistently fell below target (Zaleski, 1980). Although official prices were regularly adjusted to market conditions (e.g. in 1935), these changes were insufficient to prevent the development of considerable excess demand in official retail markets and widespread deficits of goods and services. Private activity partially compensated for these deficiencies. The collective farm (*kolkhoz*) markets and second economy operated throughout the Stalinist period. Serious deficiencies existed in areas of collective consumption as well, such as provision of medical care, that caused shortages, queues, rationing, and unequal distribution (Davis, unpublished).

World War II gave the Soviet economy a catastrophic shock. Extensive areas of the most developed territory in the USSR were occupied by enemy forces, a substantial amount of capital stock was destroyed, supplies of all types of industrial and agricultural intermediate goods were disrupted, and over 20 000 000 Soviet citizens perished from war-related causes. The Soviet Union was able to respond successfully to the threat to its survival. In the economic realm this involved further centralization of control, establishment of defence production as the highest priority task, reliance on short-term physical plans to determine output and allocate resources, imposition of strict labour discipline, and rationing of consumer goods and services (Zaleski, 1980; Harrison, 1985). As a result of these conditions and policies there was turmoil in the economy and severe shortages of most goods and services during the years 1941–45.

After the war the more extreme controls that had been imposed on the Soviet economy were relaxed and many of the features of the late 1930s economic system and the old behavioural patterns of planners, managers and workers were re-established. Due to the pressures of reconstruction and the Cold War, the defence and heavy industry sectors maintained their high priority during 1945–53. Investment in consumer goods industries, agriculture, and social services was drastically limited. A confiscatory currency reform was introduced to bring the population's demand in balance with the modest supplies of consumer goods and services and to reappropriate illegal earnings. Other demand suppression measures included compulsory bond sales, wage controls, and restraints on social security transfer payments (Nove, 1969). This effort resulted in some reduction in excess demand in retail markets. In the immediate post-Stalin period, investment priorities shifted in favour of agriculture, housing, and light industries. This resulted in an increase in the supply of consumer goods and services. But excess demand remained in retail markets due to the simultaneous increase in the population's income due to reforms of wages and the social security system.

After World War II socialist economies emerged in Poland, East Germany, Czechoslovakia, Hungary, Romania and Bulgaria (the socialist economic system of Yugoslavia is not examined in this volume). Although there was considerable diversity in national cultures, initial levels of development, and the scale of war damage, the East European countries adopted fairly uniform economic systems that were derived from the existing Soviet model of state ownership, central planning, and unbalanced growth. As a result, during the decade 1945–55 the East European economies exhibited advantages and defects that were similar to those of the Soviet economy. For example, Kornai's study of planning and management in Hungarian industry in the 1950s described many of the same phenomena that Berliner independently identified in his research on the Soviet factory (Berliner, 1957; Kornai, 1959). Other parallel characteristics of relevance to this volume were the imbalances and deficits of inputs in the production spheres of the East European economies, chronic shortages in retail markets, and widespread second economy activity.

By the end of the 1950s it was apparent that the traditional CPE was effective in promoting rapid development of basic industries, production of high priority goods (e.g. heavy machinery), and extensive growth. But its chronic disequilibrium created numerous problems with respect to efficiency, product quality, and intensive growth. Over the next two decades the socialist countries attempted to define more unique development paths, to reform the planning, supply and management systems (e.g. the USSR in 1957 and 1965, Hungary in 1968), and to realign priorities in favour of consumption.

Although some success was achieved by these reform efforts, subsequent investigations by both national and foreign experts revealed that up to the late seventies it was modest in scale (Libura, 1979; Schroeder, 1979; Hewett, 1981; Balassa, 1983; Kornai, 1986b). In particular, during 1955–75 disequilibrium existed some of the time in the industrial, agricultural and foreign trade sectors of the socialist economies. There is more uncertainty about the situation in retail markets. Most analysts believe that consumer markets were characterized by chronic excess demand (Nuti, Chapter 5). However, Portes and others have argued that there was not sustained repressed inflation in markets for consumer goods and services in East European countries during this period (Portes, Chapter 2).

Disequilibrium and shortage phenomena in CPEs probably assumed greater significance in the 1980s and certainly were studied more carefully. Since the empirical chapters in this book are primarily devoted to the analysis of problems in the contemporary period a detailed survey of recent development is not made here. Instead only a few general comments are made about external and internal causes of this imbalance and instability.

In the early 1980s the East European countries were adversely affected by foreign economic trends, such as: growing difficulties in servicing the hard-currency debt they had accumulated in the seventies (often to finance imports for unproductive projects) due to the increase of real interest rates; poor performance in export markets (as a result of uncompetitive products and the recession in the OECD region); and a credit squeeze by Western institutions that reflected both the world debt crisis and a re-evaluation of country risk in the socialist bloc (Vanous, 1985; Balassa and Tyson, 1985). This situation was made worse by the simultaneous deterioration of East Europe's terms-of-trade with the USSR as a result of the lagged impact of the 1979–80 rise in the world price of oil on intra-CMEA transferable rouble prices. The uniform responses in Eastern Europe were to drastically reduce imports, promote exports and restrain domestic demand. Investment was cut back and attempts were made to hold down the growth in consumption. This combination of external disturbances and domestic policies stimulated substantial disequilibrium in some of the East European markets, shortages of commodities, and short-term deterioration in economic performance indicators. It should be noted that the economic crisis in Poland started earlier, had somewhat different causes, and produced the most extreme collapse of output, imbalances and shortages (Fallenbuch, 1986; Kazmer, 1986; Beksiak, 1987).

Serious economic problems developed in the late seventies in the Soviet Union, during what General Secretary Gorbachev now refers to as the period of stagnation, and were initially due to domestic factors. Among these were: unusually bad weather; poor harvests; depletion of easily accessible resource bases; ageing of the capital stock; a drastic reduction in the rate of growth of investment; and slackening of labour discipline and wage control (Levine, 1983b). These difficulties caused numerous bottlenecks in the production sphere (e.g. rail transport), shortages of industrial commodities such as metals, repressed inflation in official retail markets, forced savings, and expansion of the second economy (Grossman, 1977, 1979). Substantial problems also arose in welfare areas, such as medical care (Davis, 1987). Disequilibrium in domestic Soviet markets was intensified in the 1980s by foreign trade developments: tighter Western restrictions on technological exports to the USSR; the depreciation of the dollar; the drop in the world price of oil; failures of Third World countries to repay hard-currency debts to the Soviet Union; the decline in the Soviet share of world exports of industrial goods; and the forced reduction in industrial imports from Western nations (McIntyre, 1987).

In the second half of the 1980s party, government, and academic specialists in the socialist countries have carried out wide-ranging critical assessments of their economic systems, identified key deficiencies, and introduced a new set of relatively radical economic reform measures. Much attention has focused on problems of disequilibrium and heated debates have taken place over their causes (e.g. incorrect relative prices and lack of financial discipline) and appropriate remedial policies.

1.2 MODELS OF DISEQUILIBRIUM AND SHORTAGE IN CENTRALLY PLANNED ECONOMIES

Although centrally planned economies were afflicted by imbalances and shortages from their inceptions, relatively few theoretical or empirical analyses were made of these phenomena before the mid-1950s. Most analysts in the East ignored or minimized the importance of the problem of pervasive disequilibria. Others followed the Stalinist line discussed above, that treated excess demand as a beneficial stimulus to industry. But some insightful studies were made at this time in the West of related topics such as excess demand in retail markets (Dobb, 1948), scarcity of investment resources (Grossman, 1953), inflation (Bergson, 1953; Holzman, 1955), and shortages in industrial wholesale markets (Granick, 1954).

Over the next decade there was an increase in the number of investigations in both socialist and Western countries that focused on or gave prominence to disequilibrium phenomena in CPEs. These primarily concerned specific activities or institutions: planning (Levine, 1959; Montias, 1959; Hunter, 1961); industry (Berliner, 1957; Kornai, 1959; Jasny, 1961); monetary system (Holzman, 1960); and consumption (Pawlowski, 1960; Kolupa, 1961; Chapman, 1963). However, nothing was produced at that time that could be classified as a shortage or disequilibrium model.

During 1965–75 there were important developments in economic theory, econometric modelling, and estimation techniques of relevance to disequilibrium analysis. Among the major theoretical contributions were: Clower (1965), Benassy (1973), and Dreze (1975) on the effective demand concept; Leijonhufvud (1968) and Kornai (1971) on quantity adjustment processes in markets and non-Walrasian equilibrium; Barro and Grossman (1971, 1974) and Benassy (1975) on disequilibrium regimes and the nature of fix-price equilibrium. Major advances were made as well in all components of quantitative economics: computational capabilities, data collection, mathematical modelling, and econometric estimation and hypothesis testing. Fair and Jaffee (1972) developed the first disequilibrium econometric model with the minimum condition. In subsequent years further progress in disequilibrium econometrics was made by Amemiya (1974), Goldfeld and Quandt (1975), and Maddala and Nelson (1975).

Despite this innovative research, no serious attempts were made during 1965–75 to adopt disequilibrium approaches to modelling CPEs. In East and West much effort was expended on the creation and application of models of the optimal functioning socialist economy (see the reviews of Ellman, 1973 and Sutela, 1984) and of input–output relations (Treml, Kostinsky and Gallik, 1973; Cave and Hare, 1981). In some respects, both of these models have equilibrium orientations. This is true of the former, on an abstract level at least, because it attempts through central planning to determine patterns of resource allocations and production of commodities that will satisfy optimality criteria which in some circumstances are attainable through equilibrium. The input–output model, which portrays a stable pattern of interdependence between industries, is based on the assumption of fixed-proportions in the production process (which is incompatible with prevalent practices in CPEs of shortage-related forced substitution), and is commonly used to evaluate the consistency of central plans (Baumol, 1972, pp. 503–5). The early econometric models of the CPEs also were based on the concepts of the *ex ante* consistency of plans and the temporary and localized nature of disequilibrium (Maciejewski, 1976; Shapiro and Halabuk, 1976).

Interesting research of relevance to disequilibrium was carried out at this time in Eastern Europe: Kornai (1971) made detailed criticisms of concepts of equilibrium and outlined ideas that subsequently evolved into the theory of the shortage economy; Beksiak and Libura (1974) produced theoretical analyses of the causes and consequences of imbalances and deficits of goods in socialist economies; Welfe (1973) discussed the significance of excess demand in econometric models of CPEs; Dyba (1973) made a pioneering, if modestly successful, attempt to formulate a model of consumption markets in a disequilibrium environment; and the impact of the inflationary gap on growth was examined by Beksiak (1966), Klacek and Klaus (1970), and Bajt (1971). Western empirical work of relevance to disequilibrium in CPEs included Levine (1966), Becker (1969), Bronsen and Severin (1970), various articles in Treml and Hardt (1972), Culbertson and Amacher (1972), Bronson and Severin (1973), Bush (1973), Feshbach and Rapawy (1973), Katsenelinboigen (1975), and Schroeder (1975).

Mainstream research on disequilibrium issues continued to expand the frontiers of analysis in the period 1976–86. Several major studies of the theory of disequilibrium were published such as Barro and Grossman (1976) and Benassy (1982). Other important theoretical contributions were made by Grandmont (1977), Drazen (1980), Silvestre (1982), Cuddington, Johansson and Loefgren (1984), and Laffont (1985). Substantial progress was also achieved in econometric theory, modelling and estimation through the work of Quandt (1978, 1982), Ito (1980), Gourieroux, Laffont and Monfort (1980) and Maddala (1986).

In the second half of the 1970s the growing instability of the socialist economies, the flourishing of disequilibrium analysis of market economies, and other theoretical developments (e.g. Kornai (1980)) combined to produce a rapid expansion, if not an explosion, of research in both East and West into the modelling and analysis of recurrent imbalances of demand and supply in markets and chronic shortages of commodities and labour in CPEs. The new literature was extensive and diverse in basic approaches, models and analytical techniques. In the interest of simplicity, however, it can be divided into three categories: (1) shortage economy models; (2) disequilibrium econometric models: (3) other approaches to disequilibrium analysis.

One approach to modelling disequilibrium in CPEs was that based upon the theory of the shortage economy. Following the introduction of the basic model by Kornai (1980) there was considerable elaboration of it by him and others. By the mid-1980s there were over eighty Hungarian and English-language studies of shortage models. The chapters of this volume by P. Hare, D. Kemme, M. Lackó, and C. Davis provide detailed, critical surveys of their characteristics and applications. Reviews of the book *Economics of Shortage* by Wagener (1982), Levine (1983a), and Marrese and Mitchell (1984) also contain numerous insights.

In brief, Kornai (1971, 1980) argued that the traditional socialist economy is characterized by non-price control mechanisms, paternalistic relations between superiors and subordinates, largely autonomous behaviour of lower level economic units, sellers' markets, soft budget constraints, and shortages of most goods and services (see Hare, Chapter 3 in this volume for a full discussion of the shortage model). The chronic and pervasive shortage environment affects the behaviour of all economic institutions in ways that both intensify and reproduce shortage. Kornai argues that due to the unique features of the resource constrained economic system the primary cause of disequilibrium is the quantity-driven behaviour of economic institutions, especially in the industrial sector. Prices play only a secondary role in explaining developments in the shortage economy.

The second category was comprised of disequilibrium econometric models of CPEs. These were based on Western theoretical and econometric approaches discussed above, but adapted to take into account the institutional characteristics and data constraints of the socialist countries. During 1975–86 over sixty studies were produced in this area by economists in the West (USA, Britain and France) and in Eastern Europe (Poland, Hungary and Czechoslovakia). Given the magnitude of this literature, only a selective overview is presented here. Detailed surveys can be found in the chapters of this volume by R. Portes (Chapter 2), R.E. Quandt (Chapter 6), V. Dlouhý (Chapter 7), W. Welfe (Chapter 8), K. Hulyak (Chapter 9), I. Grosfeld (Chapter 14), and W. Charemza (Chapter 11).

The disequilibrium econometric models of CPEs were diverse in nature and can be subdivided into two groups according to their approach: (1) chronic (known) excess demand (or disequilibrium indicator) models and (2) testable excess demand models. The disequilibrium indicator modelling approach assumed the existence of excess demand and sought to represent it by an observable synthetic indicator. An early application of this idea was made in the West by Green and Higgins (1977) in the consumption block of their econometric model of the Soviet Union. However, most work after that was carried out in Eastern Europe. These models were adapted from Fair and Jaffee (1972) and in their simplest form consisted of supply, demand, and excess demand equations plus the minimum condition. This generated a condensed form model with observable variables that could be estimated using standard econometric techniques. The first example of such a model was Charemza and Gierusz (1978). The theory of disequilibrium indicator modelling was simultaneously developed in Welfe (1978) and Charemza (1981) and applied, for example, by Hulyak, 1985, Welfe, 1985, Romanski and Welfe (1986), and Charemza and Gronicki (1988).

The testable excess demand models were derived from the theoretical work of Barro and Grossman (1976) and the econometrics of Maddala and Nelson (1975) and Goldfeld and Quandt (1975). As the name suggests, these models make use of demand and supply equations plus the minimum condition, but without any assumption concerning excess demand (see Chapter 6 by R.E. Quandt in this volume). The first application to the analysis of CPEs was made by Portes and Winter (1980) in their study of consumption markets in Eastern Europe. In subsequent years, Portes and his colleagues refined their econometric models and analysed a variety of issues in CPEs (Portes *et al.*, 1987 and Chapter 2 in this volume). Several testable excess demand models were also developed and applied in Czechoslovakia (Dlouhý, 1981 and Chapter 7 in this volume). Finally, Burkett

(1988) made a revision of the disequilibrium model by replacing the minimum condition with the assumption that the quantity transacted is lower than the minimum of demand and supply, even in an excess demand environment, due to the existence of unsaleable supplies (slacks).

The disequilibrium econometric model by Howard (1976b, 1979) of consumption and labour markets in the USSR is difficult to categorize. On the one hand, it was derived from Barro and Grossman (1976) and therefore has properties similar to those in the testable excess demand group. On the other hand, it assumed excess demand in official retail markets and examined the spillover effects in the collective farm market. Taking these various factors into account it probably makes most sense to include it with the chronic excess demand econometric disequilibrium models.

The third group includes theoretical or econometric investigations of disequilibrium and shortage phenomena of CPEs that do not clearly fall into the shortage or disequilibrium modelling categories. Given the residual nature of this definition, it should not be expected that the approaches adopted by these scholars were uniform. Indeed, most of these contributions were highly original, although they often complemented work being carried out on more conventional shortage and disequilibrium models. Due to the focus of this book and space constraints an attempt is made here only to identify and summarize these other diverse studies.

Soviet economists began to pay more attention to the modelling of disequilibrium in CPEs in the second half of the 1970s. Articles of high theoretical standards were produced by Polterovich (1980, 1982, 1986) and others. Braverman and Levin (1981) produced the first Soviet book on disequilibrium models, which was mathematical in nature. Other theoretical contributions were made in Czechoslovakia by Toms (1983), who employed a disequilibrium approach and the labour theory of value to explain price formation when supply and demand are out of balance. These models are not included in this book due to their abstract character.

Several scholars made contributions through their criticism of the shortage model. Soós (1984) and Gomulka (1985) criticized J. Kornai's concepts of firm behaviour and the soft budget constraint and offered alternative explanations for the origins and reproduction of shortage, respectively the structural inelasticity of supply and the differential between budget softness and price flexibility. Their critiques were extended by the Romanian scholar Daianu (1987) in his notable survey of disequilibrium and shortage theory.

Other investigations identified the investment process as a major source of disequilibrium in CPEs. Bauer (1978, 1981) offered numerous insights into the causes and consequences of investment cycles in socialist economies. Additional important disequilibrium-related examinations of investment were Soos (1976), Simon (1980), Winiecki (1982), and Podkaminer (1985). The role played by central authorities and planners in the generation of disequilibrium in socialist economies was examined by a number of specialists. Ellman (1979) provided a comprehensive, comparative assessment of the contribution of economic planning deficiencies to imbalances and instability in CPEs and Podkaminer (1986b) analysed planners' responsibility for causing disequilibrium by establishing incorrect relative prices and exacerbating their effects through inappropriate policies. In the case of consumption, Collier (1986) attempted to estimate the gap between notional and effective purchasing power of households in a quantity constrained economy (the GDR).

The issue of inflation (open, hidden and repressed) and its relationship to market imbalances was thoroughly re-examined in the period after 1975 by Howard (1976a), Pickersgill (1977), Adam (1979), Pindak (1983), Brus and Laski (1985), and Nuti (1982 and Chapter 5 in this volume). Numerous studies were made of the second economy in CPEs, which was shown to be substantial in scale and an expression of disequilibrium in retail and wholesale markets (Grossman, 1977, 1979, unpublished; Starzec, 1983; Duchene, 1984; Gaertner and Wenig, 1985; Collier, 1985; and Davis, 1988a). A final work that should be mentioned is Birman (1983), which analysed several issues related to market imbalances, forced savings, and deficits of goods in the socialist economy.

1.3 UNRESOLVED CONTROVERSIES AND NEGLECTED TOPICS

Although considerable progress has been achieved in analysing CPEs using disequilibrium and shortage models, section 1.1 indicated that the actual phenomena under investigation are complex and section 1.2 revealed that these models are relatively new. Given these factors it is not surprising that in 1987–88, when this book was being organized, there were numerous unresolved controversies and neglected topics in the field. This section identifies the main deficiencies of shortage and disequilibrium models and then compares their features.

1.3.1 The shortage model of CPEs

Research on the shortage model was dominated by J. Kornai from its initial formulation in 1980 through to 1986. Over the years his ideas evolved in response to continuing study of related problems and to outside criticism. For example, Kornai (1986a) presented a revision of the original soft budget constraint concept. But there were relatively few intra-group disputes. Instead, the adherents of the shortage economy theory, primarily based in Hungary, concentrated on developing the basic model and extending the range of its applications. Some of the new work is surveyed by M. Lackó in Chapter 10 of this volume.

There were, however, some controversies, besides those involving disequilibrium model analysts that are examined in section 1.3.3. Soos (1984) criticized Kornai's explanation of managerial motivation, inadequate discussion of the influence of the directive system of planning, formulation of the soft budget constraint, and neglect of the structural inelasticity of supply. Gomulka (1985) also attacked the concept of the soft budget constraint and suggested that shortages were partially due to the differentials between budget and price responsiveness. Finally, Schaffer (1986) disputed Kornai's argument that budget softness reduces technological innovation. He attempted to show that, on the contrary, in certain circumstances a soft budget constraint could promote innovation.

Since the shortage economy theory was intended to provide a comprehensive explanation of the behaviour of CPEs but has been under development for a relatively short period of time, there are some noticeable gaps in its coverage of institutions and processes. First of all, the representation of hierarchical relations appears to be insufficiently developed. Further work could be done on modelling the institutional behaviour and interactions of central authorities, planners, ministry officials, and enterprise managers. Secondly, the model needs to be extended to institutions consciously neglected in Kornai (1980) such as service producing units, domestic and foreign trade organizations, and farms. Thirdly, shortage model studies did not analyse in any detail the role of the priorities of central authorities in determining sectoral shortage intensity (Davis, Chapter 17 in this volume). Fourthly, the second economy was largely ignored by shortage modellers (Davis, 1988a). This was a surprising omission given the obvious link between commodity shortages and black markets.

One final point is that the studies of J. Kornai concentrate on explaining the reproduction of shortage in existing economic systems rather than the historical evolution of the shortage economy (Hare, Chapter 3 in this volume). However, in order to understand the shortage model it is probably best to view it as a description of an economic system that has experienced a transition to a chronic shortage state, such as the Soviet Union in the 1930s and Eastern Europe in 1945–53. The impact of this transition on institutional behaviour is discussed briefly in section 1.1. The initial analysis of available evidence suggests that a plausible sequence of events can be defined that would generate a shortage economy. But additional detailed historical analysis is required to evaluate the validity of this proposition.

1.3.2 Disequilibrium econometric models of CPEs

The disequilibrium econometric models attracted substantial criticism and generated numerous controversies in the period to 1986. This reflected the innovative nature of their theoretical contributions, the fact that their theory and estimation techniques were familiar to Western mainstream economists and specialists on CPEs, and the challenges presented to conventional opinion by the empirical results. Numerous criticisms were made of the early disequilibrium modelling work of Howard (1976b) by Katz (1979) and Nissanke (1979) and of Green and Higgins (1977) by Portes (1978). The chronic excess demand disequilibrium models were East European in origin and received disproportionately little attention or criticism (but see the comments of Charemza, Chapter 11 in this volume). Thus the main controversies concerned the pioneering testable excess demand econometric models of the CPEs.

Most of the criticisms of the models produced by R. Portes and his collaborators were presented in papers by Winiecki (1985), Kemme and Winiecki (1985), Stupnicki (1985) and Podkaminer (1986a). As a result of these interventions controversies arose concerning the models' treatment of planners, supply, consumers, labour, money, and foreign trade. Given the function of this introduction these are listed below without substantive comment.

In the case of planners, criticism focused on: the combining of planners and enterprises into one institution, thereby ignoring their interaction; the claim that planners attempt to maximize government expenditure rather than the growth in output; and the treatment of consumption as an element in the planners' utility function rather than as a constraint. On the supply side, the models were said to ignore investment and capital stock formation and to offer a deficient explanation of inventories, especially of unsaleable goods. The models' coverage of household behaviour was criticized for treating stability of saving as an indicator of equilibrium (since in the same period extra cash might be flowing out of income into the second economy due to shortages in official retail markets) and for an unrealistic depiction of consumers' expectations about future constraints on consumption. The labour component of the models was attacked for arguing that labour supply responded to changes in real wages and that workers in CPEs could decrease the quantity of labour supplied in response to excess demand (instead of reducing labour quality or increasing on-the-job leisure and second economy time). Several aspects of the modelling of the monetary sector attracted critical comment, such as: ignoring the distinction between household money and enterprise deposits; lack of demand functions for money balances of either households or firms; inadequate description of money supply; and neglect of the issue of price stability. Finally, the foreign trade sectors of the models were said to ignore both the effects of exports on the domestic economy and the usage of imports as a planners' control variable.

These criticisms did not, of course, go unanswered. R. Portes and others showed that some complaints were invalid or the result of misunderstandings whereas others represented suggestions for a future research programme (Portes, Chapter 2 in this volume). Nevertheless, in 1988 there were numerous unresolved controversies and gaps in coverage associated with the testable excess demand econometric disequilibrium models of CPEs.

1.3.3 Contrasts between shortage and disequilibrium models

Although the disequilibrium and shortage models examine many of the same phenomena in CPEs, they exhibit considerable differences in underlying theory, choice of explanatory variables, rigour in formulation of relationships, and reliance on econometric techniques. This diversity has generated numerous controversies between adherents of the alternative models, some of which appear to be based upon misunderstandings of alternative approaches by analysts.

One of the objectives of this book is to bring modellers of disequilibrium and shortage closer together by clarifying the nature of their models and identifying

Table 1.1 A comparison of the features of disequilibrium and shortage models

Feature	*Disequilibrium (testable excess demand) model*	*Shortage model*
Concept of aggregate demand	Accepts	Rejects
Concept of excess demand	Accepts	Rejects
Concept of repressed inflation	Generally accepts	Rejects
Origin of disequilibrium (shortage)	Planners' errors, shocks and incorrect relative prices	Producers' non-price behaviour and soft budget constraint
Temporal nature of disequilibrium (shortage)	Unstable, regime shifts, short-term	Stable, constant regime, long-term
Role of prices	Important	Minimal
Changes needed to correct disequilibrium (shortage)	Change planners' behaviour and relative prices	Change firms' behaviour, harden budget constraints, and planners control by physical norms
Concepts of minimum condition	Accepts with reservations	Rejects
Disequilibrium (shortage) adjustment mechanisms	Various types of disequilibrium indicators (e.g. plan adjustment)	Control by norms
Linkage of macro and micro agents	Aggregation of quantity-constrained micro agents	Linked through paternalism, control by norms mechanism, expectations
Output determination	Plans through non-price mechanisms	Enterprise managerial decisions
Micro-economic analysis of behaviour in disequilibrium	Little	Well developed
Objective of planners (firm)	Maximize utility in most cases	Quantity and expansion drives
Transmission of disequilibrium from producer to consumer markets	Does not directly portray	Portrays
Adjustment of labour supply to disequilibrium (shortage) in consumer markets	Non-positive (but subject to dispute)	Non-negative
Existence of excess of demand (shortage)	Testable	Assumed chronic
Rigorous formulation and testing of hypotheses	Yes	No
Specification and estimation of econometric models	Yes	Limited

common aspects of their research programmes. Accordingly it seems appropriate to outline in this introduction some of the apparently contrasting features of the disequilibrium (testable excess demand) and shortage models. For economy of presentation this comparison is made in Table 1.1 without additional discussion. The table is partially based upon past evaluations of Kemme and Winiecki (1985) and Stupnicki (1985) as well as on material presented by Hare in Chapter 3 of this volume. It is hoped that this comparison will provide the reader with a framework for interpreting the materials presented in subsequent chapters concerning controversies between disequilibrium and shortage modellers. A concluding assessment of these contrasts is made in Chapter 18.

ACKNOWLEDGEMENTS

The preparation of this chapter was supported by a grant from the Economic and Social Research Council (B00232171) and, in the case of Christopher Davis, the PEW Charitable Trusts.

The authors would like to thank Robert W. Davies and Karoly Attila Soos for their helpful comments on an early draft of this chapter.

REFERENCES AND FURTHER READING

Adam, J. (1979) *Wage Control and Inflation in the Soviet Bloc Countries*, Macmillan London.

Amemiya, T. (1974) A note on the Fair and Jaffee model, *Econometrica*, **42**, 759–62.

Bajt, A. (1971) Investment cycles in European socialist economies: a review article, *Journal of Economic Literature*, **9**, 53–63.

Balassa, B. (1983) Reforming the New Economic Mechanism in Hungary, *Journal of Comparative Economics*, 7, (3) 253–76.

Balassa, B. and Tyson, L. (1985) Policy responses to external shocks in Hungary and Yugoslavia 1974–76 and 1979–81, in U.S. Congress, Joint Economic Committee *East European Economies: Slow Growth in the 1980's*, USGPO, Washington DC.

Barro, R.J. and Grossman, H.I. (1971) 'A general disequilibrium model of income and employment', *American Economic Review*, **61**, 82–93.

Barro, R.J. and Grossman, H.I. (1974) Suppressed inflation and the supply multiplier, *Review of Economic Studies*, **41**, pp. 97–104.

Barro, R.J. and Grossman, H.I. (1976) *Money, Employment and Inflation*, Cambridge University Press, Cambridge.

Bauer, T. (1978) Investment cycles in planned economies, *Acta Oeconomica*, **21** (3), 243–60.

Bauer, T. (1981) *Tervgazdasag, beruhazas, ciklusok*, Kozgazdasagi es jogi Konyvkiado, Budapest.

Baumol, W.J. (1972) *Economic Theory and Operations Analysis* (Third edn), Prentice-Hall, Inc., Englewood Cliffs, NJ.

Becker, A. (1969) *Soviet National Income 1958–1964*, University of California Press, Berkeley.

Beksiak, J. (1966) Inflacja w gospodarce socjalistycznej, *Ekonomista*, 91–107.

Beksiak, J. (1987) Economic crisis in Poland, Seminar paper no. 396, Institute for International Economic Studies, Stockholm.

Beksiak, J. and Libura, U. (1974) *Rownowaga Ekonomiczna w Socjalizmie*, PWN, Warsaw.

Benassy, J.P. (1973) *Disequilibrium Theory*, PhD Dissertation, University of California at Berkeley, Berkeley.

Benassy, J.P. (1975) Neo-Keynesian disequilibrium theory in a monetary economy, *Review of Economic Studies*, **42**, 503–24.

Benassy, J.P. (1982) *The Economics of Market Disequilibrium*, Academic Press, New York.

Bergson, A. (1953) *Soviet National Income and Product in 1937*, Columbia University Press, New York.

Berliner, J. (1957) *Factory and Manager in the USSR*, Harvard University Press, Cambridge.

Birman, I. (1983) *Ekonomika nedostach*, Chalidze, New York.

Braverman, E.M. and Levin, M.J. (1981) *Neravnovesnye Modeli Ekonomischeskikh Sistem*, Nauka, Moscow.

Bronson, D.W. and Severin, B.S. (1970) Consumer welfare, in *Economic Performance and Military Burden in the Soviet Union*, US Congress, Joint Economic Committee, USGPO, Washington DC.

Bronson, D.W. and Severin, B.S. (1973) Soviet consumer welfare: the Brezhnev era, in *Soviet Economic Prospects for the Seventies*, US Congress, Joint Economic Committee USGPO, Washington DC.

Brus, W. and Laski, K. (1985) Repressed inflation and the second economy under socialism, in W. Gaertner and A. Wenig (eds) *The Economics of the Shadow Economy*, Springer Verlag, Berlin/New York.

Burkett, J. (1988) Slack and shortage and discouraged consumers in Eastern Europe: estimates based on smoothing by aggregation. *Review of Economic Studies*, **55**, 493–506.

Bush, K. (1973) Soviet inflation in *Banking, Money and Credit in Eastern Europe*, NATO Economics Directorate, Brussels.

Carr, E.H. (1958) *Socialism in One Country, 1924–1926, Vol. 1*, Macmillan, London.

Carr, E.H. and Davies, R.W. (1969)*Foundations of a Planned Economy, 1926–1929, Vol. 1*, Macmillan, London.

Cave, M. and Have, P. (1981) *Alternative Approaches to Economic Planning*. Macmillan, London.

Chapman, J. (1963) *Real Wages in Soviet Russia since 1928*, Harvard University Press, Cambridge.

Charemza, W. (1981) *Ekonometryczne Modele Nierownowagi*, University of Gdansk Press, Sopot.

Charemza, W. and Gierusz, B. (1978) The estimation and application of a supply/demand quarterly model of retail sales in Poland, paper presented at the conference on Problems of Building and Estimation of Large Econometric Models, Blazejewko.

Charemza, W. and Gronicki, M. (1988) *Plans and Disequilibria in Centrally Planned Economies*, North-Holland, Amsterdam.

Charemza, W. and Kiraly, J. (1988) Plans and exogeneity: the genetic – teleological dispute revisited, Discussion paper no. 71, Department of Economics, University of Leicester, Leicester.

Clower, R.W. (1965) The Keynesian counterrevolution: A theoretical appraisal in F.P.R. Brechling, and F.H. Hahn, (eds) *The Theory of Interest Rates*, Macmillan, London.

Collier, I.L. (1985) *Connections, Effective Purchasing Power and Real Product in the German Democratic Republic*, Berichte des Osteuropa-Institut an der Freien Universitat Berlin, Reiche Wirtschaft und Recht, Heft 135, West Berlin.

Collier, I.L. (1986) Effective purchasing power in a quantity constrained economy: An estimate for the German Democratic Republic, *The Review of Economics and Statistics*, **LXVIII** (1) 24–32.

Cuddington, J.T., Johansson, P.-O. and Loefgren, K.-G. (1984) *Disequilibrium Macroeconomics in Open Economies*, Basil Blackwell, Oxford.

Culbertson, W.P. and Amacher, R.C. (1972) Inflation in the planned economies: Some estimates for Eastern Europe, *Southern Economic Journal* **45**, (2) 380–93.

Daianu, D. (1987) On aggregate (dis)equilibrium and performance. Supply- and demand-constrained economies, *Revue Roumaine des Sciences Sociales. Series des Sciences Economiques*, **31**, (1) (January–June).

Davies, R.W. (1980) *The Socialist Offensive: The Collectivization of Soviet Agriculture, 1929–1930*, Macmillan, London.

Davies, R.W. (1988) *The Soviet Economy in Turmoil, 1929–1930*, Macmillan, London.

Davis, C. (1983) Economic problems of the Soviet health service: 1917–1930, *Soviet Studies*, **XXXV**, 343–61.

Davis, C. (1987) Developments in the health sector of the Soviet economy, 1970–90 in *Gorbachev's Economic Plans*, US Congress, Joint Economic Committee, USGPO, Washington DC.

Davis, C. (1988a) The second economy in disequilibrium and shortage models of centrally planned economies, *Berkeley-Duke Occasional Papers on the Second Economy in the USSR* No. 12, July.

Davis, C. (1988b) The Soviet medical system during 1928–32: Development strategy, resource constraints and health plans, in S. Gross Solomon and J.F. Hutchinson (eds) *Social Medicine in Revolutionary Russia*, Indiana University Press, Bloomington.

Dlouhý, V. (1981) Nerovnovazne modely a odhad parametru poptavkove funkce, *Ekonomicko-Matematicky Obzor*, **17**, 74–85.

Dobb, M. (1948) *Soviet Economic Development since 1917*, Routledge and Kegan Paul, London.

Dobb, M. (1970) *Welfare Economics and the Economics of Socialism*, Cambridge University Press, Cambridge.

Drazen, A. (1980) Recent developments in macroeconomic disequilibrium theory, *Econometrica*, **48**, 283–306.

Dreze, J.H. (1975) Existence of an exchange equilibrium under price rigidities, *International Economic Review*, **16**, 301–20.

Duchene, G. (1984) Marches paralléles et deséquilibres dans les économies socialistes in X. Richet (ed.) *Crises à l'Est*, Presses Univ. de Lyon, Lyon.

Dyba, K. (1973) Odhad parametru poptavkove funcke na nerovnovaznem trhu, *Politicka Ekonomie*, **21**, 1123–32.

Ellman, M. (1973) *Planning Problems in the USSR*, Cambridge University Press, Cambridge.

Ellman, M. (1979) *Socialist Planning*, Cambridge University Press, Cambridge.

Erlich, A. (1960) *The Soviet Industrialization Debate, 1924–1928*, Harvard University Press, Cambridge.

Fair, R.C. and Jaffee, D.M. (1972) Methods of estimation for markets in disequilibrium, *Econometrica*, **40**, 497–514.

Fallenbuch, Z.M. (1986) The economic crisis in Poland and prospects for recovery in *East European Economies: Slow Growth in the 1980's*, US Congress, Joint Economic Committee, USGPO, Washington DC.

Feshbach, M. and Rapawy, S. (1973) Labor constraints in the Five-Year Plan in *Soviet Economic Prospects for the Seventies*, US Congress, Joint Economic Committee, USGPO Washington DC.

Gaertner, W. and Wenig, A. (eds) (1985) *The Economics of the Shadow Economy*, Springer Verlag, Berlin/New York.

Goldfeld, S.M. and Quandt, R.E. (1975) Estimation in a disequilibrium model and the value of information, *Journal of Econometrics*, **3**, 325–48.

Gomulka, S. (1985) Kornai's soft budget constraint and the shortage phenomenon: A criticism and restatement, *Economics of Planning*, **19**, (1), 1–11.

Gourieroux, C., Laffont, J.J. and Monfort, A. (1980) Disequilibrium econometrics in simultaneous equation systems, *Econometrica*, **48**, 75–96.

Grandmont, J.M. (1977) The logic of the fix-point method, *Scandinavian Journal of Economics*, **79**, 169–86.

Granick, D. (1954) *Management of the Industrial Firm in the USSR*, Columbia University Press, New York.

Green, D.W. and Higgins, C.I. (1977) *SOVMOD I: A Macroeconometric Model of the Soviet Union*, Academic Press, New York.

Gregory, P.R. and Stuart, R.C. (1981) *Soviet Economic Structure and Performance*, Harper & Row, New York.

Grossman, G. (1953) Scarce capital and Soviet doctrine, *Quarterly Journal of Economics*, August.

Grossman, G. (1977) The 'second economy' of the USSR, *Problems of Communism*, **26**, (5) 25–40.

Grossman, G. (1979) Notes on the illegal private economy and corruption in *The Soviet Economy in a Time of Change*, US Congress, Joint Economic Committee, USGPO Washington DC.

Grossman, G. (ed.) (unpublished) *Studies in the Second Economy of Communist Countries*, University of California Press, Berkeley.

Harrison, M. (1985) *Soviet Planning in Peace and War 1938–1945*, Cambridge University Press, Cambridge.

Hewett, E.A. (1981) Lessons of the 1970's and prospects for the 1980's in *East European Economic Assessment*, US Congress, Joint Economic Committee, USGPO, Washington DC.

Holzman, F. (1955) *Soviet Taxation: The Fiscal and Monetary Problems of a Planned Economy*, Harvard University Press, Cambridge.

Holzman, F. (1960) Soviet inflationary pressures, 1928–1957: Causes and cures, *Quarterly Journal of Economics*, **74** (2) 167–87.

Howard, D. (1976a) A note on hidden inflation in the Soviet Union, *Soviet Studies*, **XXVIII**, no. 4.

Howard, D. (1976b) The disequilibrium model in a controlled economy: An empirical test of the Barro-Grossman model, *The American Economic Review*, **66**, no. 5.

Howard, D. (1979) *The Disequilibrium Model in a Controlled Economy*, Lexington Books, Lexington.

Hulyak, K. (1985) An econometric disequilibrium macromodel for Hungary, Paper presented at the World Congress of the Econometric Society Cambridge, Mass.

Hunter, H. (1961) Optimum tautness in developmental planning, *Economic Development and Cultural Change*, **9**, no. 4.

Ito, T. (1980) Methods of estimation for multi-market disequilibrium models, *Econometrica*, **48**, 97–126.

Jasny, N. (1961) *Soviet Industrialization 1928–1952*, University of Chicago Press, Chicago.

Katsenelinboigen, A. (1975) Disguised inflation in the Soviet Union in *Economic Aspects of Life in the USSR*, NATO Economics Directorate, Brussels.

Katz, B.G. (1979) The disequilibrium model in a controlled economy: Comment, *American Economic Review*, **69**, (4), 721–725.

Kazmer, D. (1986) The adjustment of the Polish economy to scarcities in the 1970's in *East European Economies: Slow Growth in the 1980's*, US Congress, Joint Economic Committee, USGPO, Washington DC.

Kemme, D.M. and Winiecki, J. (1985) Disequilibrium in centrally planned economies, University of North Carolina, Center for Applied Research, Working Papers Series no. E85 0501, Greensboro, North Carolina.

Klacek, J. and Klaus, V. (1970) Inflationary gap on the consumer goods market, *Czechoslovak Economic Papers*, **12**, 19–31.

Kolupa, M. (1961) Eliminacja obciazen wspolczynnikow elastycznosci w przypadku organiczen po stronie podazy, *Przeglad Statystyczny*, Vol 8.
Kornai, J. (1959) *Overcentralisation of Economic Administration*, Oxford University Press, London.
Kornai, J. (1971) *Anti-Equilibrium*, North-Holland, Amsterdam.
Kornai, J. (1980) *Economics of Shortage*, North-Holland, Amsterdam.
Kornai, J. (1986a) The soft budget constraint, *Kyklos*, **39**, (1) 3–30.
Kornai, J. (1986b) The Hungarian reform process: Visions, hopes and reality, *Journal of Economic Literature*, **XXIV**, December, 1687–1737.
Laffont, J.-J. (1985) Fix-price models: A survey of recent empirical models, in K.J. Arrow and S. Houkapohja (eds) *Frontiers of Economics*, Basil Blackwell, Oxford.
Leijonhufvud, A. (1968) *On Keynesian Economics and the Economics of Keynes*, Oxford University Press, London.
Levine, H.S. (1959) The centralized planning of supply in Soviet industry in *Comparisons of the United States and Soviet Economies*, US Congress, Joint Economic Committee, USGPO, Washington DC.
Levine, H.S. (1969) Pressure and planning in the Soviet economy in H. Rosovusky (ed.) *Industrialization in Two Systems: Essays in Honor of Alexander Gerschenkron*, John Wiley & Sons, Inc., New York.
Levine, H.S. (1983a) Book review of *Economics of Shortage*, *Journal of Economic Literature*, **XXI** (March).
Levine, H.S. (1983b) Possible causes of the deterioration of Soviet productivity growth in the period 1976–80 in *Soviet Economy in the 1980's: Problems and Prospects*, US Congress, Joint Economic Committee, USGPO, Washington DC.
Libura, U. (1979) *Konsumpcja a wykorzystanie mozliwosci produkcyjnych w gospodarce socjalistycznej*, PWN, Warsaw.
Maciejewski, W. (1976) *Zastosowania ekonometrycznych modeli rozwoju gospodarki narodowej*, PWE, Warsaw.
Maddala, G.S. (1986) Disequilibrium, self-selection and switching models, in Z. Griliches and M.D. Intriligator (eds) *Handbook of Econometrics*, Vol. 3, Part 7.
Maddala, G.S. and Nelson, F.D. (1975) Maximum likelihood methods for models of markets in disequilibrium, *Econometrica*, **42**, 1013–30.
Malinvaud, E. (1977) *The Theory of Unemployment Reconsidered*, Basil Blackwell, Oxford.
Malle, S. (1985) *The Economic Organisation of War Communism 1918–21*, Cambridge University Press, Cambridge.
Mandel, E. (1968) *Marxist Economic Theory (Vols. I and II)*, Monthly Review Press, London.
Marrese, M. and Mitchell, J.L. (1984) Kornai's resource-constrained economy: A survey and appraisal, *Journal of Comparative Economics*, **8**, 74–84.
McIntyre, J.F. (1987) 'The USSR's hard currency trade and payments position' in *Gorbachev's Economic Plans*, US Congress, Joint Economic Committee, USGPO, Washington DC.
Montias, J. (1959) Planning with material balances in Soviet-type economies, *American Economic Review*, **XLIX**, no. 5.
Nissanke, M.K. (1979) The disequilibrium model in a controlled economy: Comment, *American Economic Review*, **69**, (4), 726–32.
Nove, A. (1969) *An Economic History of the USSR*, The Penguin Press, London.
Nuti, D.M. (1982) Open and repressed inflation in Poland 1975–1981, Cambridge, Paper presented at the Annual Conference of the National Association of Soviet and East European Studies.
Pawlowski, Z. (1960) Problem nieobciazonosci parametrow funkcji poptyu a przypadki przejsciowej niedosstateoznej podazy dobr na rynku, *Przeglad Statystyczny*, **7**, 41–50.

Pickersgill, J. (1977) Soviet inflation: causes and consequences, *Soviet Union/Union Sovietique*, **4**, 297–313.

Pindak, F. (1983) Inflation under central planning, *Jahrbuch für Wirtschaft Osteuropas*, **10/2.**

Podkaminer, L. (1985) Investment cycles in the centrally planned economies: An explanation invoking consumer market disequilibrium and labor shortage, *Acta Oeconomica*, **35**, nos. 1–2.

Podkaminer, L. (1986a) Measurement of macroeconomic disequilibrium in centrally planned economies: A critical review of Professor Portes' results, Paper presented at the IIASA Meeting 'Applied General Equilibrium Modelling,' Lexanburg.

Podkaminer, L. (1986b) Persistent disequilibrium in Poland's consumer markets: Some hypothetical explanations, *Comparative Economic Studies*, **XXVIII**, no. 3.

Polterovich, V.M. (1980) Optimal'noe raspredelenie blag pri neravnovesnykh tsenakh, *Ekonomika i Matematicheskie Metody*, **XVI**, (4) 746–59.

Polterovich, V.M. (1982) Ratsionirovanie resursov i vtorichoe ravnovesie, *Ekonomika i Matematicheskie Metody*, **18**, (4) 684–98.

Polterovich, V.M. (1986) Ekonomiccheskoe ravnovesie pri gibkoi skheme ratsionirovaniya, *Kibernetika*, **1**, 109–15.

Portes, R. (1978) Macroeconometric modelling of centrally planned economies: Thoughts on SOVMOD I, Harvard Institute of Economic Research, Discussion Paper no. 621, Cambridge

Portes, R. and Winter, D. (1980) Disequilibrium estimates for consumption goods in centrally planned economies, *Review of Economic Studies*, **47**, 137–59.

Portes, R., Quandt, R.E., Winter, D. and Yeo, S. (1987) Macroeconomic planning and disequilibrium: Estimates for Poland, 1955–1980, *Econometrica*, **55**, 19–41.

Quandt, R.E. (1978) Tests of the equilibrium vs. disequilibrium hypothesis, *International Economic Review*, **19**, 435–52.

Quandt, R.E. (1982) Econometric disequilibrium models, *Econometric Reviews*, **1**, 1–63.

Romanski, J. and Welfe, W. (1986) Simulation minimodel W–6 of the Polish economy under disequilibrium, Szczyrk, Paper presented at the conference 'Problems of Building and Estimation of Large Econometric Models'.

Schaffer, M. (1986) Budget softness and innovation, LSE Discussion Paper, London.

Schroeder, G.E. (1975) Consumer goods availability and repressed inflation in the Soviet Union in *Economic Aspects of Life in the USSR*, NATO Economics Directorate, Brussels.

Schroeder, G.E. (1979) The Soviet economy on a treadmill of 'reforms' in *Soviet Economy in a Time of Change*, US Congress, Joint Economic Committee, USGPO, Washington DC.

Shapiro, H.T. and Halabuk, L. (1976) Macroeconometric model building in socialist and non-socialist countries: A comparative study, *International Economic Review*, **17**, (3) 529–65.

Silvestre, J. (1982) Fix-price analysis in exchange economies, *Journal of Economic Theory*, **26**, 28–58.

Simon, A. (1980) A model of Hungarian investment cycles, Polanica Zdroj, Paper presented at the Seventh Conference on Problems of Building and Estimating Large Econometric Models.

Soos, K.A. (1976) Causes of investment fluctuations in the Hungarian economy, *Eastern European Economics*, **XIV**, no. 2.

Soos, K.A. (1984) Apropos the explanation of shortage phenomena: Volume of demand and structural elasticity, *Acta Oeconomica*, **33** (3–4) 305–20.

Stalin, I.V. (1949) *Sochineniya (Vol. 12)*, Gosudarstvennoe Izdatel'stvo Politicheskoi Literatury, Moscow.

Starzec, K. (1983) L'economie Polonaise vue à travers des circuits paralleles. Discussion Paper, Paris.

Stupnicki, K. (1985) The disequilibrium theory: R. Portes vs. J. Kornai – A critical comparison, Working Paper no. 2, ICiZ, SGPiS, Warsaw.

Sutela, P. (1984) *Socialism, Planning and Optimality: A Study in Economic Thought*, The Finnish Council of Sciences and Letters, Helsinki.

Toms, M. (1983) Disequilibrium, price system and value, *Czechoslovak Economic Papers* **21**, 31–54.

Treml, V.G. and Hardt, J.P. (eds) (1972) *Soviet Economic Statistics*, Duke University Press, Durham.

Treml, V.G. Kostinsky, B.L. and Gallik, D.M. (1973) Interindustry structure of the Soviet economy: 1959 and 1966 in *Soviet Economic Prospects for the Seventies*, US Congress, Joint Economic Committee, USGPO, Washington DC.

Vanous, J. (1985) Macroeconomic adjustment in Eastern Europe in 1981–83 *East European Economies: Slow Growth in the 1980's*, US Congress, Joint Economic Committee, USGPO, Washington DC.

Wagener, H.-J. (1982) The economics of socialism as economics of shortage: A critical review, *De Economist*, **132**, no. 4.

Welfe, W. (1973) A medium-term econometric model of the Polish economy, Prace IEiS UL, no. 2, Series D, Lodz.

Welfe, W. (1978) Disequilibria in econometric macromodels of centrally planned economies, Paper presented at the conference 'Colloques sur Structures Economiques et Econometrie', Lyons.

Welfe, W. (1985) Econometric macromodels of unbalanced growth, Prace IEiS UL No. 29, Series D, Lodz.

Winiecki, J. (1982) Investment cycles and an excess demand inflation in planned economies: Sources and processes, *Acta oeconomica*, **28**, 147–52.

Winiecki, J. (1985) Portes ante Portas: A critique of the revisionist interpretation of inflation under central planning, *Comparative Economic Studies*, **27**, 25–52.

Zaleski, E. (1971) *Planning for Economic Growth in the Soviet Union, 1918–1932*, University of North Carolina Press, Chapel Hill.

Zaleski, E. (1980) *Stalinist Planning for Economic Growth, 1933–1952*, University of North Carolina Press, Chapel Hill.

2

The theory and measurement of macroeconomic disequilibrium in centrally planned economies

RICHARD PORTES

2.1 INTRODUCTION

The motivation for my research on CPE macroeconomics has been to elucidate the general problem of macroeconomic equilibrium and, in a specific application, to evaluate the evidence for repressed inflation in the consumption goods markets of the 'classical' Soviet-type economies of Eastern Europe. This empirical testing required a more elaborate and general macroeconomics than was then available for the CPEs. A prerequisite was to clarify the appropriate concept of 'equilibrium' (see Muellbauer and Portes, 1978, for market economies, and Dlouhy, 1984, in the CPE context). An equilibrium with quantity rationing is a state in which markets do not clear, but the economy is at rest, and we can in principle measure the magnitude of 'disequilibrium' in terms of the distance from a market-clearing allocation or price vector. Similarly, Kornai (1980) discusses a self-reproducing equilibrium state of an economy characterized by pervasive shortages which he subsequently seeks to measure (1982).

A corresponding theoretical framework is that of recent non-Walrasian general equilibrium macroeconomics, which has stressed the need for consistent microeconomic foundations. To call this the 'Clower (1965) – Barro–Grossman (1971, 1974) school' is perhaps too selective and omits some key elements which appear in the sequence Hansen (1951) – Benassy (1975) – Grandmont–Laroque (1976) – Malinvaud (1977) – Muellbauer–Portes (1978) – Neary–Stiglitz (1983),

International Studies in Economic Modelling Series.
Models of Disequilibrium and Shortage in Centrally Planned Economies.
Edited by C. Davis and W. Charemza.
Published in 1989 by Chapman and Hall Ltd, 11 New Fetter Lane, London EC4P 4EE.

such as the aggregation problem and expectations of quantity constraints.

Any application to socialist CPEs of contemporary neoclassical theory, developed in and for capitalist market economies, must be informed by a thorough knowledge of East European economies and their institutions. Without that background, one cannot sensibly choose the simplifying assumptions required to get any useful theorizing off the ground. Even with it, one may take a set of assumptions which are not universally regarded as 'realistic'. But reasonable and well-informed economists can differ (indeed, agree to differ) on the important and justifiable simplifications. These will in any case depend on the objectives of the particular model. One modeller's choice of assumptions that seem prima facie inappropriate to another may not reflect ignorance or disregard of the relevant institutions. Serious criticism requires demonstrating rigorously either the inconsistency of the assumptions or (with evidence) that they lead to inappropriate conclusions.

The macroeconomics of CPEs has attracted more than its share of unsupported assertions, purely a priori reasoning, 'stylized facts' not based on the data, and confident, elaborate verbal arguments founded on the most casual empiricism. For a theorist to use 'obviously' in argument or to dismiss empirical results as 'absurd' is only one level below offering various general qualitative considerations to support an essentially quantitative proposition. This is no longer acceptable. We have for some time had reasonably long and fairly reliable macroeconomic time series for several CPEs. Empirical assertions about the behaviour of CPE households, planners, enterprises or markets should not be taken seriously unless they are supported by the data. Disagreements over empirical issues can be resolved only by testing well-defined hypotheses.

All this does not deny but rather supports the validity of alternative approaches. One is a variety of institutionalism, seeking distinctive general lessons in the specificities of socialist planning, finding those generalities in a formalism which is not primarily directed towards proving theorems or generating testable hypotheses. The symbols used in this formalism are often difficult to translate into measurable variables, and the research strategy comes close to rejecting the standard model–hypotheses–tests procedure (Kornai, 1980; less so in 1982). It yields insights nevertheless, as do a range of models and empirical work within the more standard neoclassical canons (see below).

The following sections consider issues raised by my own research programme on macroeconomic equilibrium in CPEs and comparisons with alternative views. I defend the explicit aggregative, macroeconomic approach in theory, institutional relationships and measurement. It has offered a fresh, coherent framework for the analysis of many CPE phenomena, opened up a range of possibilities for empirical investigation, and generated several important spin-offs: work on planners' behaviour; insights into CPE policy problems of the 1970s and early 1980s, which centred on macroeconomic equilibrium and threats to it; and some developments in market economy macro theory and econometrics. Throughout, applications are to household behaviour, consumption goods markets and the period from the mid-1950s to the beginning of the 1980s.

2.2 THEORY

I take as common ground for this survey that CPEs exhibit behavioural as well as technological regularities; stable economic behaviour of CPE households, planners and enterprises is part of our maintained hypothesis. I limit the discussion here to the consumer goods and labour markets, while recognizing that relations between enterprises and planners and disequilibria within the state productive sector have been the subject of much interesting theory and empirical work (e.g. on investment cycles). I accept the consensus from this work that CPEs normally exhibit generalized excess real demand (qualified by the nonmarket character of transactions) in the state productive sector, for intermediate and investment goods. It is not relevant here, however, to assess the relative importance of overly taut initial plans (including systematic optimism about unknown data and exogenous shocks) and systemic features like an output-based incentive system.

Both demand for and supply of consumption goods and labour will depend on events in the state productive sector. Total demand for consumption goods will be affected indirectly by events in the labour market and directly by the demand from state enterprises for some consumption goods. The latter is probably of secondary importance (Kornai, 1980, p. 497), but the interplay between planners, enterprises and banks which determines wage plan (over)fulfilment is a key element of the story (Holzman, 1956, 1960; Portes, 1983). We discuss it and the supply side further below. Whether they do jointly yield excess demand is an empirical question. To answer it does require specifying demand and supply functions and hence the transmission of behaviour in the state productive sector through those functions, but not the internal processes which generate that behaviour.

Although the labour market is equally important, both theory and data are weaker, so assessing the nature and extent of disequilibrium there is much less advanced. This is particularly regrettable since the spillovers from the consumption goods market on to labour supply could in principle reveal significant information about the former, although evaluating it might be difficult too (Portes, 1974; Wiles and Rostowski, 1979). Moreover, one of the chief reasons why the planners themselves as well as external observers should be interested in consumption goods market disequilibrium is this spillover on the incentive to work (Hansen, 1951; Galbraith, 1952). Excess demand will also typically result in an inefficient allocation of goods and affect the distribution of real income; both will concern the planners. But the effects on economic welfare of excess demand or supply do not concern us here.

2.2.1 Definitions

What is macroeconomic disequilibrium and how do we recognize it? When prices are free to vary, even with some stickiness, we expect movements of the price level to provide evidence of excess demand or supply, even if such price flexibility may not eliminate the disequilibrium (as it does not in market economy hyperinflations). When prices are endogenously sticky or controlled, we may observe either hidden or repressed inflation (I consider here only positive excess demand). Some carelessly conflate the two (Winiecki, 1985), but this unnecessarily gives up a useful conceptual and practical distinction. A price index may be falsified, may cover an unrepresentative basket of goods, and may not reflect quality change; any of these phenomena may hide underlying inflation, i.e. a rise in an accurately calculated, appropriately weighted, quality-corrected price index. Repressed inflation, on the other hand, is simply

> a situation in which, at existing wages and prices, the aggregate demands for current output and labour services exceed the corresponding aggregate supplies... [so] purchases of goods and labour services are rationed (Barro and Grossman, 1974, p. 87).

This excess real demand (a flow) is the inflationary gap whose consequences for prices are suppressed. But the 'gap' terminology can be misleading, and it is at best unhelpful to say that rationing hides inflation (rather than suppresses it), since this suggests that the price level, appropriately defined as above, is actually increasing. That usage also obscures an important point, that repressed inflation may cause hidden inflation.

Rationing is one consequence of repressed inflation. At the aggregate level, its mirror image is forced saving (note that as a global monetary magnitude, this is inherently an aggregate concept). This is a flow of saving rather than a stock (the so-called monetary overhang). Flows are determined by desired and actual stocks and by many other factors. When we investigate equilibrium on the consumption goods market, we look at the balance of flow demand and supply, of which the stocks of household liquid assets and sellers' inventories are only partial determinants.

Simply put, forced saving is that part of their flow of disposable income which households would like to spend but cannot. Indeed, putting it so simply immediately prompts the question why it should ever occur – forcing people to save sounds odd, rather like pushing on a string. There are only two reasons: households voluntarily (maximizing subject to available supplies of goods) substitute (expected) future consumption for unavailable current consumption; or they involuntarily supply more labour than they would like, i.e. they are constrained from reducing their labour supply and hence income as much as the reduced (by rationing) marginal utility of income would otherwise induce. Note, therefore, that if there actually has been chronic, sustained repressed inflation since the beginning of central planning and households rationally expect it to continue, the voluntary component of forced saving will be negligible, and its only cause must be inhibitions on cutting labour supply.

Households can of course reduce their labour supply qualitatively (effort) as well as quantitatively, or quantitatively in ways that will not be recorded in the data. This in no way affects the essential argument, nor theoretical modelling of it: excess demand for consumption goods should have a spillover effect in reducing effective (quality-corrected) labour supply, and hence output; this is the basis of the supply multiplier.

How important is this effect? Wiles and Rostowski (1979) query it on mainly a priori grounds, except in so far as they claim that wage-elasticities of labour supply are observed to be low or negative. A simple calculation from equation (4) of Portes (1981) shows, however, that the response of labour supply to (rationed) consumption will depend on the difference between the wage-elasticities of unconstrained and constrained labour supply functions. These are likely to vary together, so the Wiles–Rostowski claim shows nothing. Kornai recognizes that 'the issue is basically an empirical one', but then without examining evidence asserts 'that the intensity of shortage is not one of the variables explaining labour supply either positively or negatively' (1980, p. 479). My prejudice goes the opposite way, along with Hansen and Galbraith; but we should remain agnostic until the evidence is in. Howard's (1976) results suggest a significant effect, but I cannot accept them (nor his general argument) for the reasons suggested in Katz (1979), Nissanke (1979), and Portes and Winter (1980).

In theory, then, how should we measure macroeconomic disequilibrium in the consumption goods market? The definitions above suggest trying to measure the difference between aggregate demand and aggregate supply, or alternatively, forced saving (again, the difference between actual desired holdings of financial assets is not what we want, although it is one determinant of excess demand). Three major questions immediately arise, of which we mention two and examine the third at length. First, should we use notional or effective demand and supply? The quantity-rationing macro literature suggests the latter (Muellbauer and Portes, 1978); but there are counterarguments, and there are several alternative specifications of effective demands (Martin and Portes, 1987). Second, the formal or informal rationing schemes used under excess demand may be manipulable, in the sense that a household may influence the amount it receives, in particular by presenting a demand exceeding what it actually wants, knowing that the probability of success or the fraction of revealed demand which it will achieve is less than unity. This would exaggerate measured excess demand. Weinrich (1984) considers the theoretically appropriate measure of excess demand in these circumstances for the stochastic rationing case, but we have as yet no practical procedure for making the correction (but see Burkett, 1988, who deals empirically with the converse, the discouraged consumer effect).

2.2.2 The aggregation problem

Much the most fundamental and important issue is the aggregation problem. Tartarin (1982) goes so far as to say, 'Only micro disequilibria are real, macro disequilibrium is an artifact, which exists only in the domain of the national accounts'. This suggests not only that our measurement problem is in principle insoluble – even the most ambitious assistant professor might decline a research grant for measuring statistical artifacts – but also that neither monetary theory nor macroeconomic policy are of any interest in considering disequilibria, at least in CPEs. Having written on both (Portes, 1980, 1983), I clearly do believe there are interesting theoretical propositions and policy interpretations in the aggregative economics of CPEs which involve macroeconomic disequilibrium in essential ways.

Then how can we distinguish between distorted relative prices and macroeconomic-monetary causes of the phenomena we identify with disequilibrium in CPEs? We have suggested thought experiments to clarify the theoretical distinction (Portes and Winter, 1980), and in policy terms we might consider the separate responsibilities of the Price Office and the Ministry of Finance (monobank) and Ministry of Labour. Macroeconomic policy cannot suffice to deal with problems created by the structure of relative prices, the system of incentives, etc.

Operationally, however, we must come to terms with the problem of identifying aggregate excess demand when, at the micro level, there are both positive and negative excess demands, shortages and slacks. Kornai insists that they must remain disaggregated, as vectors, and that it would be a 'grave mistake... to "net out" shortage and slack...' (1980, p. 45). He would accept that aggregate excess demand might be well defined in an economy whose normal state was not pervasive shortage. In a chronic shortage economy, however,

> 'Buyers have adjusted their attitudes to chronic shortages... where massive forced substitution, forced spending and queueing are to be found... aggregate excess demand is not an operational category...' (Kornai, 1980, p. 477).

The key point in this argument is the change in buyers' behaviour, which we interpret to mean that their quantity (constraint) expectations are different. These expectations are arguments of their demand functions, but the preferences underlying those functions are not altered by chronic shortage. Consider the case of long-run mass unemployment. Workers will adjust, in the sense of changing quantity (constraint) expectations. The discouraged worker effect is one manifestation; some people will leave the labour market. Others will substitute (accept) inferior jobs for those preferred but unavailable at existing relative wages. Liquidity preference and consumption behaviour are also affected. We do not normally suppose, however, that individual labour supply functions have altered. Should we say that aggregate excess supply of labour is not in these circumstances an operational category? If so, how do we characterize macroeconomic relationships in the 1930s? Moreover, would not the vector of disaggregated indicators with which Kornai characterizes shortage also be affected by quantity expectations? We would agree that aggregate excess demand cannot be defined independently of the normal state of the economy, on the understanding that this includes the quantity constraint expectations, if any, held by agents in the economy.

The existence of forced substitution with excess demand is not a major theoretical problem. After all, at the aggregate level, both forced saving and the response of labour supply to a constraint in the goods market are examples of forced substitution. Spillover effects of this kind have been modelled extensively, and all quantity-rationing macro deals with transactions taking place at false (disequilibrium) prices. In practice, forced substitution is a cause for worry only if elasticities of substitution in demand are low and it is nevertheless extensive. Although it is always welfare-reducing, the importance of that effect will depend on the elasticities, and it should not in any case stop us from doing applied macroeconomics.

In seeking to measure aggregate excess demand, we cannot avoid netting out shortages and slacks. This does indeed present a problem in theory, as suggested by the Barro–Grossman definition of repressed inflation quoted above. In a two-market model, they require that both be in excess demand. This stipulation that the vector of excess demands be semi-positive (so no netting out at false prices is necessary) goes back at least as far as Hansen (1951), and it is reflected in the consistent usage of the quantity-rationing macro literature: excess demand for goods with excess supply of labour is 'classical unemployment', a quite different regime (the converse is 'underconsumption'). In practice, however, the early period of central planning, when most goods were in excess demand and labour in excess supply, is better characterized as 'repressed inflation on the goods market' (Kornai, 1980, section 11.7; 1982, pp. 102–8).

We all need to use macroeconomic aggregates, even Kornai (1980, sections 16.2–16.5, 17.6, 18.5, 21.1–21.2, and chapter 19; and the explicit macro modelling of 1982, *passim*). Yet he would take the aggregation problem as sufficient reason to reject the central concepts of quantity-rationing macroeconomics: the short-side rule (that all agents on the short side of the market achieve their desired transactions), the minimum condition (that the aggregate quantity transacted is the minimum of supply and demand), and 'discrete switching' between excess supply and excess demand regimes. Even for market economies not subject to chronic shortages, however, it is quite impossible to do applied macroeconomics with a purist approach to the aggregation problem (see Deaton and Muellbauer, 1980, chapter 6 and section 12.1). Instead, one must simply do macroeconomics carefully, always remaining open to empirical evidence (not *a priori* assertion) that its results are contradicted when one allows explicitly for micro disequilibria. In fact, however, Burkett's (1988) study using Kornai's concepts gives results perfectly consistent with our own; see section 2.3.4 below.

One internally consistent way of approaching the problem in our context is 'smoothing by aggregation', whose lengthy pedigree is sketched by Muellbauer (1978). If we aggregate over many micro labour markets, for example, some in excess supply and some in excess demand, we find that total demand is the sum of employment and vacancies, total supply is the sum of employment and unemployment, so total excess demand equals vacancies minus unemployment; as the real wage rises, aggregate excess demand falls in a continuous manner, with a smooth unemployment–vacancies trade-off. Aggregate employment will always lie to the left of the wedge formed by the aggregate demand and supply curves.

A diagram in Portes and Winter (1980) applies this to the aggregate consumption goods market in CPEs. Note that the larger the elasticities of substitution, and hence the stronger is forced substitution, the closer will observed aggregate behaviour approximate the underlying aggregate demand and supply curves which would be observed in full Walrasian equilibrium, and consequently the more suitable is the discrete-switching, min-condition empirical model. The smoothing by aggregation approach has been developed further by Malinvaud (1981), Kooiman (1984), Gourieroux and Laroque (1985) and Martin (1986). It appears especially promising for applications where there are data indicating the proportions of micro markets in excess demand and in excess supply.

2.2.3 Macroeconomic models

I do not have space to discuss in detail the alternative complete models within which one might consider the theory and measurement of macroeconomic disequilibrium in CPEs. I have set out my own in Portes (1979, 1981); Wolf (1985) discusses it and others. It has only two aggregate markets, for consumption goods and labour; only households hold money; there is no parallel (flex-price) market; there are only two categories of agents, households and planners; inventories and foreign trade are incorporated, but only to a limited extent; households' quantity expectations are not treated explicitly; and there is no attempt to specify a complete macroeconometric model based on this theoretical framework. It nevertheless provides a coherent theoretical representation of macroeconomic disequilibrium and an adequate basis for measuring it in the consumption goods market, as we have done. Adding the specification of the plan construction process developed further in Portes *et al.* (1984a, 1987), Dlouhy (1984) suggests three major similarities between this model and that of Kornai (1982): both are nonWalrasian general equilibrium models; both deal with the 'real' and the 'control' spheres; and both have endogenous mechanisms for adjustment towards an equilibrium. I would add that Kornai (1980, sections 19.4 – 19.6) gives a similar representation of plan construction, especially in assuming that the planners are rational and do try *ex ante* to avoid excess demand, *ceteris paribus*. Stupnicki (1985) carefully compares my model with that of Kornai.

Several alternative models have been proposed with different assumptions. Pickersgill (1980a), Duchene (1984), Henin (1984) and Nuti (1985) explore the implications of a parallel market. Hare (1982) incorporates some of Kornai's ideas into the framework of Malinvaud (1977), which is similar to our own. Brada (1982) assumes that enterprises also hold money. Hulyak (1983) specifies a larger macroeconometric model whose consumption block resembles our treatment. There remains much work to be done with quantity constraint expectations, inventories, and foreign trade. For our purposes, however, I see little advantage in breaking up the state productive sector into planners and enterprises and modelling the interaction between them. This is clearly essential in modelling investment behaviour (see Chapter 14), but not for the consumption goods market. I follow Campbell (1970) in believing that

> the state production establishment [SPE] can best be considered as a single gigantic firm under the unified direction of the planners ... intra-SPE

> transactions ... lacking the market characteristics of those between households and the SPE, are of less interest from the point of view of pricing and macroeconomic equilibrium (Campbell, 1970, pp. 255, 258).

There are no general, a priori guidelines for the essential elements of a CPE macro model. All simplifying assumptions are wrong, so one chooses a minimal coherent set sufficient for one's objectives. Mine were to formalize some propositions and to provide a basis for econometric hypothesis testing. I would claim that the institutional framework is just about rich enough for the purpose and not seriously flawed. Some have queried the basic strategy of applying to CPEs the neoclassical macro theory, in its quantity-rationing version, as well as disequilibrium econometrics. I believe, however, that they and a simple monetarism are often better suited to CPEs than to comparable market economies.

2.2.4 Behaviour of households and planners

Others take a similar line in this context. Wiles and Rostowski (1979) describe the institutional characteristics of CPEs that 'led him [Portes] correctly to suppose that Western micro- and macroeconomics ... would be fully applicable to households in CPEs (p. 62)'. Pickersgill (1983) states,

> The work I have done so far increases my confidence in the similarity of behavioural relationships and therefore of the appropriateness of using Western methodologies in the study of Soviet household behaviour (Pickersgill, 1983).

Asselain (1981) takes the same view of saving behaviour in STEs.

Modelling the planners is more difficult, and there are no obvious analogies in neoclassical theory. We extend our basic plan-construction process with an explicit planners' loss function (Portes *et al.*, 1987), which incorporates a steady growth objective like that in Kornai (1980, p. 490; 1982). This appears to be successful empirically.

On the other hand, Kornai specifically rejects the analogy between repressed inflation in a fix-price capitalist economy and the suction system of a socialist economy, 'for the latter has different institutional conditions and its behavioral regularities are therefore different, too' (Kornai, 1980, p. 558). Of the differences he cites, however, all but one relate to aspects of firms' behaviour not relevant to the consumption goods market, while the last concerns the adjustment of buyers' behaviour to shortage, discussed above. I therefore maintain that this analogy is indeed instructive, and our intuition is not misleading.

Although Kornai does accept the symmetry between Keynesian unemployment and repressed inflation in a capitalist economy, he rejects any symmetry between the former in a capitalist economy and the latter in a socialist economy (p. 478). He maintains that suction in a socialist economy is *sui generis* because its

> mode of operation, its management system, and the behavior of the agents within it have adjusted to chronic shortage, which is continuously maintained by the institutional relationships (Kornai, 1980, p. 478).

Again, we suggest above that similar adaptation occurs in long-run mass unemployment, and one might think that Marxists would go even further to say that the institutional relationships of a capitalist economy are such as to maintain unemployment substantially above the natural rate continuously.

Moreover, just as the rate of unemployment is useful though crude as an index of excess supply in the labour markets of capitalist countries and is usable in cross-country comparisons despite data comparability problems, so we can construct a similar index for excess demand on the consumption goods markets of CPEs. We take the demand and supply functions obtained by disequilibrium estimation (see below), calculate fitted demand and supply for each year, and take their difference as a percentage of observed consumption (Portes and Winter, 1980 – an alternative version, based on stochastic simulation of the fitted functions, is calculated in Portes *et al.*, 1987). This measure can also be used in intertemporal and international comparisons.

Finally, the relations between CPE institutions and the theory of macroeconomic equilibrium appear in my analysis of how excess demand is generated, reproduced and controlled (Portes, 1977a, 1983). The planners' objectives and the enterprise incentive system generate taut plans in the state productive sector and pressure to overfulfil them, and these forces are exacerbated by errors and exogenous shocks. This may affect the demand for and supply of consumer goods, but less than is commonly thought. Although I accept much of Kornai's analysis of plan construction for consumption goods (1980, sections 16.1–16.4), my model explicitly admits the possibility of an *ex ante* 'unbalanced' plan. This gives some insights but has no effect on the empirical work, nor does the precise source of consumption goods market disequilibrium.

The view that consumption plans are chronically unrealistic (unbalanced, optimistic) and hence underfulfilled (Winiecki, 1985) is decisively rejected by aggregate time-series data for actual and planned consumption in three CPEs. Some argue a priori that supply (CS) will always fall short of the plan (C^*) and demand (CD) will always exceed it, so $CS < C^* < CD$. Then $C = \min(CD, CS)$ would suggest that actual $C < C^*$ always. In Poland, however, we find $C > C^*$ for 10 of the 19 years 1957–75, until the Gierek policies ensured $C < C^*$ throughout the second half of the 1970s. For Czechoslovakia, $C > C^*$ in 15 of the 24 years 1957–80; and in Hungary, $C > C^*$ in 11 of the 19 years 1957–75, then once more in the following quinquennium. Similarly, Kornai's general view that wage plans are fulfilled very closely (1980, section 16.1) is borne out by Farrell's (1975) study of Poland. He shows that the average amount by which the plan was exceeded in 1957–70 was 2.6%, although Poland is generally thought to have been rather lax in monetary control.

I cannot accept Kornai's views, however, on the roles of relative prices and of price rigidity in creating and maintaining shortages. He cites, for example, groups of products for which the relationship between price and average cost goes in the opposite sense of the intensity of shortage as 'a new empirical argument against the view that "shortage exists when relative prices are too low"' (1980, p. 500). That argument ignores the demand side and is hence a glaring *non sequitur*. Equally, as Gomulka (1985) points out, Kornai's assertion that 'there is no behavioral regularity ensuring that a rise in the price level of consumption goods eliminates the shortage' (p. 497) simply does not follow from the a priori arguments given (no evidence is cited); and it is inconsistent with his view that the household faces a hard budget constraint. Yet these are key elements in his downgrading of the role of prices and money because of the systemic features and institutional regularities in a shortage economy. I regard them as unsupported assertions.

2.3 MEASUREMENT

Only empirical testing can ultimately resolve the basic issues to which this analysis is directed. Assertion and casual empiricism will not do. To arrive (ideally) at an agreed interpretation of the results of such tests, however, the theoretical framework within which they are carried out must be broadly agreed as well. The discussion above is intended to extend the area of agreement. The discussion of empirical work below refers only to the consumption goods market and household behaviour.

2.3.1 Econometric results

Our own examinations of the data, tests, results and conclusions are set out in a series of papers (Portes, 1974, 1977a; Portes and Winter (P–W), 1977, 1978, 1980; Portes and Santorum, 1987; Portes *et al.*, 1987, 1984a, 1984b, 1988). I shall not attempt to summarize them, although I shall reply below to the relatively few specific criticisms which they have attracted. There seems to be an irresistible temptation, however, for those who reject our methods or our conclusions to misrepresent both by setting up straw men which can easily be tumbled without taking the effort to address specific issues of theory, econometrics or data. It may therefore be useful simply to quote for the record a few summary statements from these papers:

> 1. ... there was significant inflation, both repressed and sometimes open, in the CPEs until that time [mid-1950s]. Moreover, there may have been brief episodes in particular countries during particular periods since then ... [but] there is no evidence of sustained repressed inflation in any of the Soviet-type CPEs since the mid-1950s (Portes, 1974).
>
> 2. We have specified and estimated [in an equilibrium framework] household demand for money and savings functions in four CPEs. The successful estimation ... and the plausibility of the estimates support our

> underlying hypothesis that the behaviour of the household sector in CPEs can be analysed using the same methods that are used for this sector in Western countries ... [and] testify to the quality of the data. ... The residuals can tell us little about the overall degree of market equilibrium during the period, however ... [and] we can draw no further conclusions about repressed inflation until we treat both sides of the market in a disequilibrium framework (P–W, 1978, p. 17).
>
> 3. ... Begin with the estimated probabilities [that an observation was generated by an excess demand regime]. ... Simply allocating each year ... on the basis of [the probability exceeding or falling short of] 0.5 ... [indicates] excess supply was the dominant regime in three out of the four countries. This is a very crude description of the results, however. More precise efforts at hypothesis testing are made [below, from which] ... we cannot draw strong, unambiguous empirical conclusions ... [but] we believe that the evidence ... clearly justifies rejecting the hypothesis of sustained repressed inflation in the market for consumption goods and services since the mid-1950s [to the mid-1970s] in our four CPEs (P–W, 1980, pp. 149, 155–6).

Note that we have not stated any such view for the period since the mid-1970s, especially in respect of the USSR (to which we cannot extend our econometric analysis because of data problems); and other work suggests rising pressure of demand on the Soviet consumption goods market in recent years (see below).

Contrast our actual conclusions with Wiles and Rostowski's interpretation of them: 'In Portes' view there is no demand inflationary pressure in STEs because of successful planning ...' (p. 61). Tartarin (1982): 'P–W emphatically conclude [that there was] general excess supply' (p. 25). Kornai (1982):

> I shall quote one of their important findings... "On this basis excess supply was the dominant regime in three out of the four countries." In my *opinion* [emphasis added], the conclusion is absurd. All four countries should be considered chronic-shortage economies... (Kornai, 1982, p. 35).

Such distortion of serious empirical work, with the (widely quoted) dismissal of it as 'absurd', is not worthy of reply as such.

2.3.2 Validity of macroeconometrics for CPEs

Let us now turn to substance. Kornai rejects from the outset all macroeconometrics for CPE markets in which he believes shortages to be chronic and widespread, for in these circumstances no econometric technique can work: 'the inference from y [actual trade] to d [initial demand] becomes inadmissible' (Kornai, 1980, p. 98); it is a 'vain effort' (Kornai, 1982, p. 98). Consider the contrapositive of this proposition, however: if we and others do obtain results which satisfy *a priori* restrictions drawn from theory as well as conventional econometric criteria, then shortages must be either not widespread or not chronic. Yet this does seem to be the case. The large

CPE macroeconometric models of Welfe, Szakolczai, Sujan, Wolfling and others are well known and not easily dismissible. For careful estimation and hypothesis testing on a smaller scale which seems to yield coherent results on consumption goods markets, we cite (only by way of example) Charemza and Gronicki (1983), Hulyak (1982), Klaus and Rudlovcek (1982), Podkaminer (1982), and Timmer (1982); this work covers Hungary, Poland and Czechoslovakia.

How do we and others obtain well-behaved functions, significant and plausible parameter estimates, reasonable out-of-sample performance, etc.? Are all these results merely statistical artifacts? If one does admit they might have some validity, however, even some of the most cherished 'stylized facts' must be queried. Podkaminer (1982), for example, using a model totally different from our own, gets results which suggest aggregate excess supply of consumption goods in Poland up to the early 1970s (revised estimates in Podkaminer (1984) still allow equality of aggregate demand and supply in the late 1960s). The conclusion may look rather less 'absurd' if reached by various routes with different data sets. Klaus and Rudlovcek (1982) summarize their results unambiguously:

> We strongly believe that global equilibrium at the consumption goods market was the prevalent characteristic of the Czechoslovak economy in the last quarter of the century (Klaus and Rudlovcek, 1982).

Others are simply not willing to accept what Kornai regards as obvious:

> ...we do not find useful (and even realistic) to accept the assumption of permanent (or 'normal') shortage on these markets [consumption goods and labour]... (Dlouhý, 1984).

2.3.3 Hidden inflation and forced savings

We now take a few specific topics before passing to the contributions of disequilibrium macroeconometrics on the one hand and indicators of shortage on the other. First, hidden inflation: Portes (1977a) cites estimates by others of slightly over 1% p.a. for the USSR over two decades. More recent work by Alton *et al.* (1981) shows their alternative consumer price index growing faster than the official implicit consumer price index by 2.4% p.a. for Bulgaria, 1.6% p.a. for Czechoslovakia, 1.3% p.a. for the GDR, 1.1% p.a. for Hungary, and 1.6% p.a for Poland. These are unlikely to be underestimates. They should be contrasted with unsupported remarks like that in Nove (1979) putting hidden inflation in Poland in 1975 at 8% (which, with official estimates of nominal incomes and the CPI, would suggest that real incomes actually fell in that year).

Savings behaviour of CPE households has been investigated extensively with conventional methods. Some authors (e.g. Winiecki, 1985) continue the confusions arising from theoretically unjustifiable measures like the ratio of savings (flow) to change in income, or wealth to retail sales (consumption flow), long after Portes (1974, 1977a) and Asselain (1981) have clearly demonstrated that they give meaningless answers. Thus Asselain shows impeccably that a constant savings rate implies a long period of considerable increase in the wealth–income ratio, if initial stocks are low; and that comparing two economies with the same savings rate and rate of growth of real income but different inflation rates, the economy which is inflating faster will exhibit a lower wealth–income ratio. In the view of Wiles and Rostowski,

> He [Portes] rightly ridicules the Soviet-type habit of quoting savings/(change in income) as if it were the Keynesian marginal propensity. . . . Portes may even be on secure ground in setting up his judgment against Soviet-type bankers, who are alarmed by the ratio of the stock of liquid savings to the flow of retail trade (Wiles and Rostowski, 1979, pp. 64–5).

Serious work on household assets and savings in CPEs seems remarkably unanimous in not showing much evidence of forced saving. Pickersgill (1976) is clear on this for the USSR, but her later time-series results (1980) indicate that there may have been a substantial upward shift in the MPS in the late 1960s and early 1970s which could show some forced saving. Her cross-section work with Ofer (Ofer and Pickersgill, 1980) 'does not support the proposition that Soviet households save inordinately large amounts of their income for lack of consumer goods and services to purchase', and she expresses similar views in her later survey of both time series and budget study evidence (1983). Asselain's (1981) careful theoretical and empirical analysis leads him to conclude for several CPEs that

> The full set of characteristics of a typical situation of forced saving . . . are thus never found together, and one should not be surprised that econometric work seeks first to test the hypothesis of a savings–income relation of the same type as in Western market economies (Asselain, 1981, p. 43).

Klaus and Rudlovcek (1982) get good results for a standard savings function for Czechoslovakia, 1965–81; the forecasting performance was good, and the saving ratio was not high by international comparison and did not seem to depend on supply-side events. Peebles (1981), who constructs estimates of household cash holdings in the USSR, finds an overall trend of decline in the currency to deposit ratio (as elsewhere in Eastern Europe), and that 'Soviet liquidity is not excessive' (p. 75) by comparison with other CPEs. There is no agreement among authors, however, on whether we should expect savings rates in CPEs to be lower, higher, or about the same as in their market-economy counterparts. It should be stressed that none of these studies pretends to test for excess demand, nor should they be so interpreted. They and the work cited in section 2.3.2 are simply consistent with taking equilibrium as a working hypothesis. Again, it should be stressed that this work too extends only to the beginning of the 1980s.

Kornai (1980) asserts that the 'Clower–Barro–Grossman train of thought' provides 'theoretical background for the argument' that the

> time path of savings in the household sector [can indicate] whether the consumer goods market on the macro-level is in a state of aggregate excess demand or aggregate excess supply (Kornai, 1980, pp. 477–8).

This is puzzling. He is presumably referring to our work, yet the precise object of our analysis in Portes and Winter (1978, 1980) is to use the quantity-rationing macro model to show why and how in disequilibrium, this sort of argument may give quite the wrong answer. Even more inexplicable, therefore, is his own use of departures from trend of household savings as one of his indicators of shortage (Kornai, 1982, p. 139) – an aggregate indicator, at that!

2.3.4 Disequilibrium econometrics

Now we turn to disequilibrium econometrics (see Quandt, 1982, for a survey). Part of the maintained hypothesis is that one observes only the quantity transacted, and that supply and demand are latent variables to be estimated, along with the functions which generate them. Kornai, too, believes that it is impossible to separate consumption supply and demand in the observed data, that they are latent variables, and that they combine to determine actual consumption. Yet he concludes that there is only one consumption function, with both demand-side and supply-side variables on the right-hand side. It is difficult to justify such a quasi-reduced form if one really admits the possibility of disequilibrium. We believe it is preferable to estimate the latent variables directly – they are in principle measurable, as Kornai must agree, since he regards shortage as a latent variable and seeks himself to measure it.

Proper hypothesis testing in this area requires a clear statement of the maintained hypothesis, the null, and the alternative. The maintained hypothesis must explicitly admit an alternative to excess demand – otherwise, it is logically impossible to test for its presence. This fundamental point leads us to reject Howard's (1976) results and to find unhelpful the model of Brada (1982), which in this respect begs the basic question. Our own maintained hypothesis does allow excess demand, equilibrium (on a set of measure zero), and excess supply in the consumption goods market.

In specifying our consumption demand equation, however, we have chosen a Keynesian form which implicitly assumes that households are constrained in the labour market. This was not an empirical judgement. We chose the Houthakker–Taylor function because we needed a specification which was easy to estimate, did not exhaust too many degrees of freedom or require great attention to the error structure, and gave a few clear prior restrictions on the coefficients, violation of which would tell us that something was seriously wrong. This standard function permitted ready international comparisons and avoided difficult data problems. If one were to reject it on the grounds that the estimates will be biased unless there actually was excess supply of labour, one would also have to reject almost all consumption function estimates on Western economies which might ever have experienced 'full employment'.

As remarked above, our estimates have not been challenged in detail. We ourselves recognize important extensions to the empirical model which should be a high priority for future research: to incorporate foreign trade and inventories. To dispel any misconceptions, we have never used inventory data (which raise great problems – e.g. what is 'saleable'?) in our econometric work.

Ours are not the only versions of disequilibrium models for the consumption goods markets of CPEs. Charemza and Quandt (1982) suggest several alternatives; and Charemza and Gronicki (1983) implement a quasi-disequilibrium model for Poland. Within the models we use, however, we have performed a wide variety of tests, and both our theoretical framework and our results seem quite robust. For example, it is possible to test the Kornai hypothesis of continuous excess demand against our more general alternative. Our results clearly reject the restriction (Portes, Quandt and Yeo, 1988). Kornai would doubtless reject the test, because he believes that the aggregation problem invalidates any discrete-switching, min-condition model. As suggested above, taken too seriously, the aggregation problem invalidates all empirical macroeconomics. On his hypothesis, however, we could never obtain information about the latent consumption demand variable. But it is then inexplicable how we (and others) continue to obtain well-identified, well-behaved demand functions, if they do not exist or are in principle unobservable.

Burkett (1985) has made Kornai operational by ingeniously developing methods for dealing with the discouraged consumer effect and with the coexistence of shortage and slack. His specification directly estimates potential demand (allowing for the discouraged consumer effect), total shortage, and total slack, all separately, allowing for spillovers like forced substitution. His results, covering five East European countries, are entirely consistent with those we obtained using the category of aggregate excess demand. He finds that total slack exceeds shortage throughout 1957–78 in the GDR and Hungary and for 1956–80 (except 1972) in Poland, while shortage exceeds slack for 13 of the 24 years 1955–78 in Czechoslovakia. His direct estimates of shortage, net of slack, correlate remarkably well with our estimates of aggregate excess demand. Even allowing for potential (but discouraged) demand does not change this picture.

Thus Burkett concludes,

> One cannot reject the discrete switching hypothesis. Shortages exist but except in the case of Czechoslovakia are usually overshadowed by slack. Potential consumer demand, purged of the discouraged consumer effect, is substantially greater than actual demand but seldom as great as supply. Thus application of techniques suggested by Kornai's theory . . . can yield conclusions consistent with Portes' evidence of prevalent aggregate excess supply (Burkett, 1985).

Nuti (1985) claims that 'the spurious precision of advanced econometrics . . . almost tautologically defines imbalance out of existence'. This cavalier approach to rigorous empirical testing is unsubstantiated by any direct criticism of the aggregate discrete switching model, and it is *a fortiori* rejected by Burkett's results from a model which includes imbalances at the micro as well as the macro level.

Finally, we turn to Kornai's indicators of shortage. He says we need many of them, at a disaggregated level, in order properly and comprehensively to characterize shortage. Yet he expects the individual indicators to be highly correlated (Kornai, 1982, p. 15). Moreover he finds his macro (synthetic) index of shortage for Hungary close to our own series of estimated excess demands (Portes and Winter, 1980). The simple correlation is in fact $r = 0.604$, and the turning points are quite similar. He regards ours as invalid, because they are derived from a discrete-switching macro model. We had argued previously, however, on smoothing-by-aggregation grounds, that one could expect a well-defined correspondence of just this kind (P–W, 1980), so we were not surprised. Burkett's (1988) results are further, perhaps definitive confirmation that our approach using aggregate excess demand produces results which, so far from being invalid or absurd, are in fact consistent with and complementary to more disaggregated methods.

We finish on this reassuring point of agreement. The theoretical and empirical problems of measuring macroeconomic disequilibrium in CPEs may now be somewhat clearer, the areas of agreement and disagreement better defined. The main practical question, with policy implications, is still to distinguish between the effects of excess aggregate demand and distorted relative prices (or other problems like the inadequacies of the distributive network – see Pryor, 1977, Skurski, 1972, and Turcan, 1977). We may be getting closer.

2.4 CONCLUSION

The line of research surveyed here was originally motivated by scepticism towards the conventional wisdom that the household sector of CPEs has been subject to sustained repressed inflation since central planning began. This conventional wisdom has since been elaborated and defended by Kornai, indeed elevated with considerable flourish into a general characterization of CPEs as shortage economies. There is much which is stimulating in this analysis, which has gained currency from the worsening macroeconomic disequilibria in some CPEs (especially Poland) since the early 1970s. But it distorts the facts over the whole period since the early 1950s and thereby reinforces confused, incorrect theorizing, inaccurate forecasts and unjustified policy conclusions. As with most conventional wisdom, the underlying truths are belied by exaggeration and become immune to challenge from the data. This is buttressed by the evident political appeal in such a critique of central planning.

These are not just academic disputes. It is important to get the macroeconomics of these countries right. Macroeconomic developments heavily conditioned the evolution of the Hungarian microeconomic reforms. Proper understanding of internal and external balance and the macroeconomic bases of trade and foreign borrowing was essential in foreseeing and interpreting CPE external debt problems and the Polish crisis (Portes, 1977b; Marer, 1985). Those who took the conventional view of consumption as a buffer would have misunderstood CPE macroeconomic adjustment of the early 1980s. The proposition that investment cycles are the primary cause of macro disequilibria in the household sector would have been equally misleading in interpreting the past several years. And as Stupnicki (1985) concludes in his Kornai–Portes comparison, 'only Portes presents clues how to improve the situation'.

I have set out the theoretical innovations in Portes (1981, 1983). There are, however, equally significant practical lessons to be learned from eschewing the simple certainties of the 'shortage economy' in favour of a more complex, more nuanced and better balanced view of the realities of CPE macroeconomics.

ACKNOWLEDGEMENTS

The author is Director of the Centre for Economic Policy Research, London; Professor of Economics, Birkbeck College, University of London; and Directeur d'Etudes Associé, Ecole des Hautes Etudes en Sciences Sociales, Paris. My research on this topic has been supported by the Economic and Social Research Council, most recently under grant B00230048. I have benefitted greatly from collaborative work on the problems discussed here with Richard Quandt, David Winter, and Stephen Yeo, none of whom bears any responsibility for this paper. Nor do those who have commented on the paper, including Joe Brada, Wojtek Charemza, Paul Hare, Ed Hewett, Alec Nove, and Tom Wolf.

REFERENCES AND FURTHER READING

Asselain, J. C (1981) Mythe ou réalité de l'épargne forcée dans les pays socialistes, in *Travail et Monnaie en Systeme Socialiste* (ed. M. Lavigne), Economica, Paris.

Alton, T. *et al.* (1981) Official and alternative consumer price indices in Eastern Europe, Research project on National Income in East Central Europe, OP-68.

Barro, R. and Grossman, H. (1971) A general disequilibrium model of income and employment. *American Economic Review*, **61**, 62–83.

Barro, R. and Grossman, H. (1974) Suppressed inflation and the supply multiplier. *Review of Economic Studies*, **41**, 87–104.

Benassy, J. P. (1975) Neo-Keynesian disequilibrium theory in a monetary economy. *Review of Economic Studies*, **42**, 503–24.

Brada, J. (1982) Real and monetary approaches to foreign trade adjustment mechanisms in centrally planned economies. *European Economic Review*, **19**, 229–44.

Burkett, J. (1988) Slack, shortage and discouraged consumers in Eastern Europe. *Review of Economic Studies*, **55**, 493–506.

Campbell, R. (1970) Macroeconomic models and central price-setting in the Soviet economy, in *Essays in Economic Analysis and Policy* (eds F. Gehrels *et al.*), Indiana University Press, Bloomington, pp. 253–70.

Charemza, W. and Quandt, R. (1982) Models and estimation of disequilibrium for centrally planned economies. *Review of Economic Studies*, **49**, 109–16.

Charemza, W. and Gronicki, M. (1983) Rational expectations, wage illusion and consumption excess demand: an empirical investigation for Poland, Discussion Paper in Economics No. 143, Birkbeck College, London.

Clower, R. (1965) The Keynesian counter-revolution: a theoretical appraisal, in *The Theory of Interest Rates* (eds F. Hahn and F. Brechling), Macmillan, London.

Deaton, A. and Muellbauer, J. (1980) *Economics and Consumer Behaviour*, Cambridge University Press, Cambridge.

Dlouhý, V. (1984) On the problem of macroeconomic equilibrium in centrally planned economies, paper presented to Econometric Society European Meeting, Madrid.

Duchene, G. (1984) Marchés paralléles et déséquilibre dans les économies socialistes, in *Crises a l'Est* (ed. X. Richet), Presses Universitaires de Lyon, Lyon.

Farrell, J. (1975) Bank control of the wage fund in Poland: 1950–70. *Soviet Studies*, **27**, 265–87.

Galbraith, J. K. (1952) *A Theory of Price Control*, Harvard University Press, Cambridge, Mass.

Gardner, R. and Strauss, J. (1981) Repressed inflation in the Soviet Union. *European Economic Review*, **16**, 387–404.

Gomulka, S. (1985) Kornai's soft budget constraint and the shortage phenomenon: a criticism and restatement. *Economics of Planning*, **19**, 1–11.

Gourieroux, C. and Laroque, G. (1985) The aggregation of commodities in quantity rationing models. *International Economic Review*, **26**, 681–700.

Grandmont, J. M. and Laroque, G. (1976) On temporary Keynesian equilibria. *Review of Economic Studies*, **43**, 53–67.

Hansen, B. (1951) *A Study in the Theory of Inflation*, Allen and Unwin, London.

Hare, P. (1982) Review article: Economics of shortage and non-price control. *Journal of Comparative Economics*, **6**, 406–25.

Henin, P.-Y. (1984) Equilibres avec rationnement d'une économie à planification centralisée et secteur parallele: une analyse macroéconomique, Document No. 90, UER Analyse et Politique Economique, Université de Paris I.

Holzman, F. (1956) Financing Soviet development, in *Capital Formation and Economic Growth*, Princeton University Press, Princeton, for the National Bureau of Economic Research.

Holzman, F. (1960) Soviet inflationary pressures. 1928–1957. *Quarterly Journal of Economics*, **74**, 167–88.

Howard, D. (1976) The disequilibrium model in a controlled economy: an empirical test of the Barro–Grossman model. *American Economic Review*, **66**, 871–9.

Hulyak, K. (1982) Disequilibrium models for consumption goods in Hungary, paper presented to International Conference of Applied Econometrics, Budapest.

Hulyak, K. (1983) An experimental disequilibrium macromodel for Hungary (*mimeo*).

Katz, B. (1979) The disequilibrium model in a controlled economy: comment, *American Economic Review*, **69**, 721–5.

Klaus, V. and Rudlovcek, V. (1982) Savings function as an inverse problem of disequilibrium consumption modelling: a case study of Czechoslovakia, paper presented to International Conference of Applied Econometrics, Budapest.

Kooiman, P. (1984) Smoothing the aggregate fix-price model and the use of business survey data. *Economic Journal*, **94**, 899–913.

Kornai, J. (1980) *Economics of Shortage*, North-Holland, Amsterdam.

Kornai, J. (1982) *Growth, Shortage and Efficiency*, Basil Blackwell, Oxford.

Malinvaud, E. (1977) *The Theory of Unemployment Reconsidered*, Basil Blackwell, Oxford.

Malinvaud, E. (1981) Econometric implications of disequilibrium theory, INSEE Working Paper No. 8114.

Marer, P. (1985) Les relations commerciales Est–Ouest. *Revue d'etudes comparatives Est–Ouest*, **16**, No. 3, 5–16.

Martin, C. (1986) Disequilibrium models with dispersed trading, PhD thesis, Birkbeck College, University of London.

Martin, C. and Portes, R. (1987) Effective demands and spillovers, Discussion Papers in Economics 87/16, Birkbeck College, London.

Muellbauer, J. (1978) Macro theory vs. macroeconometrics: the treatment of disequilibrium in macroeconomic models, Birkbeck College Discussion Paper in *Economics* No. 59.

Muellbauer, J. and Portes, R. (1978) Macroeconomic models with quantity rationing. *Economic Journal*, **88**, 788–821.

Neary, J. P. and Stiglitz, J. (1983) Towards a reconstruction of Keynesian economics: expectations and constrained equilibria. *Quarterly Journal of Economics*, **98**, 199–228.

Nove, A. (1979) Inflation in communist countries, in *Political Economy and Soviet Socialism*, Allen and Unwin, London, pp. 178–91.

Nissanke, M. K. (1979) The disequilibrium model in a controlled economy: comment. *American Economic Review*, **69**, 726–32.

Nuti, D. M. (1985) Hidden and repressed inflation in Soviet-type economies: definitions, measurement and stabilization (*mimeo*).

Ofer, G. and Pickersgill, J. (1980) Soviet household saving: a cross-section study of Soviet emigrant families. *Quarterly Journal of Economics*, **94**, 121–44.

Peebles, G. (1981) Money incomes and expenditures of the population of the Soviet Union: an East European comparison. *Hong Kong Economic Papers*, **14**, 53–78.

Pickersgill, J. (1976) Soviet household saving behaviour. *Review of Economics and Statistics*, **18**, 139–47.

Pickersgill, J. (1980a) Repressed inflation and price controls in the Soviet household sector, Conference Paper, Kennan Institute.

Pickersgill, J. (1980b) Recent evidence on Soviet household saving behaviour. *Review of Economics and Statistics*, **62**, 628–33.

Pickersgill, J. (1983) Household saving in the USSR, in *The Determinants of National Saving and Wealth* (eds F. Modigliani and R. Hemming), Macmillan, London.

Podkaminer, L. (1982) Estimates of the disequilibria in Poland's consumer markets 1965–1978. *Review of Economics and Statistics*, **64**, 423–31.

Podkaminer, L. (1984) The political economy of permanent disequilibrium in Poland's consumer markets (*mimeo*).

Portes, R. (1974) Macroeconomic equilibrium under central planning, Seminar Paper No. 40, Institute for International Economic Studies, University of Stockholm.

Portes, R. (1977a) The control of inflation: lessons from East European experience. *Economica*, **44**, 109–30.

Portes, R. (1977b) East Europe's debt to the West. *Foreign Affairs*, **55**, 751–82.

Portes, R. (1979) Internal and external balance in a centrally planned economy. *Journal of Comparative Economics*, **3**, 325–45.

Portes, R. (1980) Effects of the world economic crisis on the East European economies. *The World Economy*, **3**, 13–52.

Portes, R. (1981) Macroeconomic equilibrium and disequilibrium in centrally planned economies. *Economic Inquiry*, **19**, 559–78.

Portes, R. (1983) Central planning and monetarism: fellow travellers? in *Marxism, Central Planning and the Soviet Economy* (ed. P. Desai), MIT Press, Cambridge, Mass., pp. 149–65.

Portes, R. and Santorum A. (1987) Money and the consumption goods market in China. *Journal of Comparative Economics*, **11**, 354–71.

Portes, R. and Winter, D. (1977) The supply of consumption goods in centrally planned economies. *Journal of Comparative Economics*, **1**, 351–65.

Portes, R. and Winter, D. (1978) The demand for money and for consumption goods in centrally planned economies. *Review of Economics and Statistics*, **60**, 8–18.

Portes, R. and Winter, D. (1980) Disequilibrium estimates for consumption goods markets in centrally planned economies. *Review of Economic Studies*, **47**, 137–159.

Portes, R., Quandt, R., Winter, D. and Yeo, S. (1984a) Planning the consumption goods market: preliminary disequilibrium estimates for Poland 1955–80, in *Contemporary Macroeconomic Modelling* (eds P. Malgrange and P.-A. Muet), Blackwell, Oxford, pp. 254–71.

Portes, R., Quandt, R., Winter, D. and Yeo, S. (1984b) Estimating the size of plan errors, Research Memo No. 48. Princeton University Financial Research Center; appeared (in French) in *Annales de l'INSEE*, **55/56**, 245–55.

Portes, R., Quandt, R., Winter, D. and Yeo, S. (1987) Macroeconomic planning and disequilibrium: estimates for Poland, 1955–1980. *Econometrica*, **55**, 19–41.

Portes, R., Quandt, R. and Yeo, S. (1988) Tests of the chronic shortage hypothesis: the case of Poland. *Review of Economics and Statistics* (in press).

Pryor, F. (1977) Some costs and benefits of markets: an empirical study. *Quarterly Journal of Economics*, **91**, 81–102.

Quandt, R. (1978) Tests of the equilibrium vs. disequilibrium hypotheses. *International Economic Review*, **19**, 435–52.

Quandt, R. (1982) Econometric disequilibrium models. *Econometric Reviews*, **1**, 1–63.

Skurski, R. (1972) The buyers' market and Soviet consumer goods distribution. *Slavic Review*, **31**, 817–30.

Stupnicki, K. (1985) The disequilibrium theory: R. Portes v. J. Kornai, a critical comparison, Working Paper No. 2, Institut Cybernetyki i Zarzadania, Szkola Glowna Planowania i Statystyki, Warsaw.

Tartarin, R. (1982) La théorie du déséquilibre appliquée aux économies du type socialiste (*mimeo*).

Timmer, J. (1982) Disequilibrium and buffer stocks, Erasmus University.

Turcan, J. (1977) Some observations on retail distribution in Poland. *Soviet Studies*, **29**, 128–36.

Weinrich, G. (1984) On the size of disequilibrium in an equilibrium with quantity rationing, CORE Discussion Paper No. 8418.

Wiles, P. (1982) Inflation in USSR 1982, paper presented to National Association of Soviet and East European Studies Conference.

Wiles, P. and Rostowski, J. (1979) Soviet type inflation: an open debate. *Est-Ouest*, **10** (2/3), 59–75.

Winiecki, J. (1985) Portes *ante portas*: a critique of the revisionist interpretation of inflation under central planning. *Comparative Economic Studies*, **27**, 25–52.

Wolf, T. (1985) Economic stabilization in planned economies. *International Monetary Fund Staff Papers*, **32**, 78–131.

3

The economics of shortage in the centrally planned economies

PAUL HARE

3.1 INTRODUCTION

The purpose of this chapter is to introduce, and critically review, an approach to the analysis and interpretation of the Eastern European economies largely based on the work of János Kornai and his colleagues at the Institute of Economics, Budapest. Kornai's ideas have been maturing since his early research on the behaviour of the centrally planned, socialist economy, based on Hungarian experience in the mid-1950s (Kornai, 1959). The general aim of his research programme since the 1950s has been to develop a conceptual framework for understanding the functioning of the traditional socialist economy. The main finding, a conception which now forms the central maintained hypothesis of Kornai's school of thought, is that the socialist economy is characterized by endemic and persistent shortage; moreover that this shortage is maintained over time by a variety of mechanisms all grounded in rational behaviour by enterprises, central planners and other agents given their information and expectations, the constraints they experience, and the organizational structures which tie the system together. Some of Kornai's early work is discussed in section 3.2 of the present chapter, to provide the reader with background information on the development of Kornai's thinking. Although Kornai's approach is constantly being developed and refined, both theoretically and through empirical work (some of which is reported later in the book), most of the recent contributions are based on Kornai's own book, *Economics of Shortage* (Kornai, 1980). Accordingly, section 3.3 of the present chapter is devoted to an account of the theoretical model set out in that book.

International Studies in Economic Modelling Series.
Models of Disequilibrium and Shortage in Centrally Planned Economies.
Edited by C. Davis and W. Charemza.
Published in 1989 by Chapman and Hall Ltd, 11 New Fetter Lane, London EC4P 4EE.

The original discussion of shortage-type models was mainly conducted in the context of a static economy, at least in the sense that the growth process was not explicitly modelled. Hence it was easy for critics of Kornai's approach to agree with his analysis and findings as a short-run explanation of certain behaviour and phenomena, while casting doubt on the asserted tendency for a centrally planned economy not only to generate shortage in the first place, but also to reproduce and maintain it over time. It was therefore important to develop more dynamic models in which shortage could occur along an equilibrium path. In a limited way, this was done in some of the contributions to *Non-price Control* (Kornai and Martos, 1981), but a fuller model was provided in *Growth, Shortage and Efficiency* (Kornai, 1982). Both these works are reviewed in section 3.4.

Much of Kornai's analysis is based on a view of the socialist economy quite different in several important respects from that derived from recent Western work on disequilibrium macroeconomics (for examples of the latter, both theoretical and applied, see elsewhere in the present volume). Not surprisingly, therefore, the two approaches have clashed on occasion, and section 3.5 attempts to set out the key differences between them. The remaining sections provide a critical review of Kornai's shortage model (section 3.6), offer an outline of an alernative model of shortage (section 3.7) and finally, in section 3.8, set out some brief conclusions.

3.2 BACKGROUND TO THE ECONOMICS OF SHORTAGE

Kornai has long been a critic of the economic functioning of centrally planned economies. His first detailed study (Kornai, 1959) investigated Hungarian light industry in the 1950s and concluded that the then prevalent extremes of centralization resulted in highly inefficient production, with inflexibility of output, a tendency to hoard input stocks to guard against erratic deliveries, reluctance to innovate, shortages of some goods accompanied by surpluses of others, and so on. Moreover, the internal logic of the centralized system appeared to mean that partial reforms (e.g. reducing the number of plan indicators, simplifying the economic control system, improving incentives at enterprise level) would at best only succeed temporarily, for they would soon be reversed in response to unforeseen difficulties. Such difficulties, however, should more properly be regarded as inherent in the centralized system, rather than as accidental consequences of plan errors. Despite his keen observation of the economic behaviour of Hungarian enterprises, based on very detailed knowledge, Kornai (1959) cannot be said to have developed a complete theory of the centralized socialist economy, though he has some extremely interesting analysis of enterprise/planning hierarchy relationships.

However, in a later work, Kornai (1970) did make some important steps towards a new theory. The first part of that book is a vigorous and trenchant attack on neoclassical general equilibrium theory which, as Hahn (1973b) among others has pointed out, is not wholly compelling because much of the fire-power is directed at the straw man of a highly simplified textbook general equilibrium model. The second part is more interesting for present purposes, since it develops an analysis of markets and the behaviour of firms in different types of market situation.

For Kornai, markets are rarely, if ever, in equilibrium in the Walrasian sense. Instead, he distinguishes between states of *suction* and *pressure* in the market. The former, which in later work is described as resource constrained, refers to a market situation where on average buyers are looking for sellers, while the latter rarely experience any difficulty in selling what they can produce (and hence have no need to incur substantial marketing costs). This situation can arise even when additional production would be profitable at prevailing prices, if the sellers are also facing supply constraints on their own input markets. Such an allocation, in which agents face constraints in at least two interlinked markets, can form an equilibrium in the sense that no individual agent is able to improve his position given others' behaviour, even though the allocation is clearly inefficient from a social point of view. Thus in this analysis there is no necessary connection between a suitable notion of equilibrium, and efficient resource allocation. An aspect of this in the resource constrained case is that, since firms have no need to seek and attract customers, they are likely to be relatively uninterested in innovation, either in the form of new products (because it is easy enough to sell the established ones) or new processes (because the pressure to cut costs is not especially strong). Hence considered dynamically, an economy experiencing persistent suction in a wide range of important markets can be expected to be a poor innovator, producing mainly traditional goods with outdated technology. Production will often be inefficient and the quality of output low. According to Kornai, this would not be an unreasonable characterization of much of Eastern Europe.

The position is entirely different when markets are subject to pressure, described in later work as demand constrained. These terms refer to markets in which sellers are seeking out and trying to attract buyers, while buyers can typically find what they want without difficulty and may even have a choice between a number of alternative suppliers. At prevailing prices, sellers may be perfectly willing to produce more, and could certainly obtain additional inputs. What prevents them from doing so is quite simply the demand constraint, the inability to sell more. Moreover, suppliers of inputs (labour, intermediate goods) also face demand constraints of their own because of the original constraint. Thus just as for the resource constrained economy, demand constraints tend to occur in a set of interlinked markets, rather than only in a single market. Again, an economy facing such restrictions can settle down into an equilibrium position in which no agent wishes to change his decision given others' decisions, though as before the resulting allocation of resources is not efficient. The allocation that results in this case, associated with unemployment, excess capacity and general underutilization of resources is more familiar when referred to as a Keynesian equilibrium.

While an economy may be obviously inefficient in the short run, Kornai is more optimistic about the long term, dynamic efficiency of an economy experiencing persistent pressure across a range of markets. For, as he emphasizes, when firms have difficulty in selling their goods they have an incentive to engage in activities that make them more attractive to customers (free delivery and installation, easy maintenance, and generally good consumer service); this also includes innovation to improve product characteristics (appearance, durability, special features, etc.), introduce new products or lower production costs. Thus over a period, a demand constrained economy will provide a relatively varied and up-to-date output mix, and will satisfy customers at least in the fundamental sense that queues, waiting lists and other indicators of shortage conditions will be relatively infrequent. Most of the time the majority of customers will be able to satisfy their requirements in the market, and their economic behaviour will be guided by that expectation.

Many of these ideas have been refined, developed and incorporated into a more consistent and complete model of the economy since 1970, especially in the resource constrained case. Elaborating such a model was the principal task undertaken by Kornai in his *Economics of Shortage*, as we shall see in the next section.

3.3 ECONOMICS OF SHORTAGE

This section outlines the principal arguments of *Economics of Shortage*, generally in a descriptive way, since most critical analysis is deferred to sections 3.5 and 3.6 below. To summarize a massive, two-volume work in a few pages is no easy task, but I shall try to pick out what I regard as the key ideas rather than attempt to précis everything. Given Kornai's experience of living in a centrally planned economy with a great deal of bureaucratic control over production, including strict price controls, it is not surprising that volume A on non-price adjustment should be most interesting for our purposes. However, volume B on adjustment in the presence of prices includes important material on budget constraints and macroeconomic relationships which must also be referred to. Most of the analysis is intended to apply to Hungary prior to that country's 1968 economic reforms, though Kornai argues that shortage conditions (albeit somewhat less severe and general) applied thereafter as well. *Mutatis mutandis* it would apply equally well to the other, still unreformed, economies of Eastern Europe and the Soviet Union.

Concerning the rigour and generality of its analysis, first, it is obvious that *Economics of Shortage* is not a revised version of Debreu (1959) and nor does it pretend to be. In Kornai's words it is a 'descriptive-explanatory theory', rooted in concrete observations of the functioning of socialist economies over many years. It makes no normative claims, and much of the analysis is still partial and incomplete, though in the more mathematical literature a few more general results are available (e.g. Kornai and Martos, 1981). Secondly, throughout the book Kornai seeks to formulate empirically testable hypotheses and is at pains to show how various aspects of shortage-related phenomena could be measured in practice. But he doesn't actually test hypotheses or collect the data needed to measure shortage, so the book is really setting out a major research programme, some initial elements of which have been undertaken since 1980 (see other chapters in the present book).

Firms and households are the principal economic factors in *Economics of*

Shortage, though attention is also paid to the state bureaucracy above enterprise level, especially in connection with investment behaviour and macroeconomic policy. Surprisingly, in view of Hungary's economic circumstances, only passing mention is made of international transactions; in most of the analysis, Hungary might as well be a closed economy.

Since prices are fixed most of the time, agents must respond and adjust to other types of economic signal. Such signals, collectively referred to as quantity signals, include information and observations about stocks, orders, queue lengths, waiting times, the availability of substitutes. Agents also respond to instructions and commands, though usually imperfectly since the combination of real constraints on transactions, and the incentive system prevents exact fulfilment.

In the explanation of shortage, production and the behaviour of firms (including their interaction with other parts of the state bureaucracy) play the leading role, with investment behaviour being the most influential factor. This is not to say that households are irrelevant to the economics of shortage. However, rather than being a fundamental cause of the shortage syndrome, households principally suffer its consequences in terms of queues, forced substitution, unavailable goods and services, and the demeaning relationships between sellers and buyers that prevail in a shortage economy. In short, they experience the substantial welfare losses associated with persistent shortage.

In theorizing about the socialist firm, three kinds of constraint are recognized:

Table 3.1 Key differences between capitalist and socialist firms in respect of the constraints they face

Type of constraint	*Classical capitalist firm*	*Traditional socialist firm*
Resource constraint	Rarely effective	Nearly always effective, more restrictive than demand
Demand constraint	Nearly always effective, more restrictive than resource constraint	Rarely effective
Budget constraint		
Hard	Production plan autonomous: the firm lays it down at the level of demand constraints; within resource constraints	
Soft		Production plan directive prescribed by superior authority at the level of resource constraints; within demand constraints

Source: Kornai (1980, table 2.1).

resource constraints (labour, capital, intermediate inputs), demand constraints (applying to the various outputs), and the budget constraint (finance). Whereas under capitalist conditions it is demand constraints that normally limit production, under socialism it is the resource constraints, hence the terminology of a resource constrained economy. Moreover, whereas a capitalist firm nearly always experiences a hard budget constraint, the corresponding constraint for a socialist firm tends to be very soft. This is why appeals for financial discipline are so widespread throughout Eastern Europe; equally, however, the persistent and institutionalized softness explains why such appeals fall on deaf ears. Table 3.1 shows the key differences between capitalist and socialist firms in respect of all these constraints. Under conditions of shortage (never mind, for the moment, how such conditions came about in the first place), the behaviour of firms and households has a number of important characteristics, which it is simplest to list, with brief explanations.

3.3.1 Shortage and slack normally occur together

In the case of a firm, the typical situation is one where a firm may have adequate supplies of all inputs except one which is in short supply. Since in short run production functions inputs are usually highly complementary, the consequence is a shortage of that particular input and a (hopefully temporary) surplus of others, these surpluses being an example of slack. In an interesting paper, Manove (1973) studied the implications of this kind of complementarity for the economy as a whole using an input–output framework of analysis. He found that the imbalances (shortages and surpluses) showed no tendency to increase without limit, and instead remained within bounds that depended on the pattern of final demand and the technical coefficients: so although resource allocation was inefficient, the economy was certainly viable. For Kornai's analysis, however, the essential point is that in a shortage economy one would expect to observe the simultaneous occurrence of shortage and slack.

3.3.2 Both firms and households engage in forced substitution

Firms use whatever inputs they can obtain to produce output. This output may be of lower quality than desired, or simply involve a different mix of output than had been planned. In either case, households failing to purchase their most preferred combination of goods and services will buy more of what is actually produced. Finally, some of the inputs firms buy may differ from what they originally intended to buy. All these cases are instances of forced substitution.

3.3.3 Characteristic composition of stocks

Whereas capitalist firms tend to hold relatively large and varied output stocks in order to be able to satisfy customers and respond quickly to competition in the market, and hold the lowest possible input stocks in order to keep their costs down, the opposite is the case with firms experiencing shortage. For such firms can sell virtually anything as soon as it is produced (there is, of course, some unsaleable output, which Kornai refers to as an unproductive slack) and so hold very little output stock. On the other hand, given firms' uncertainties about availability of inputs and their experience of shortages, they tend to demand more inputs than they really need and hoard those that they succeed in getting. The resulting high levels of input stock give firms some protection against poor supplies of other inputs in future periods and also, unofficially, give them something to trade with other firms to get their inputs into the proper proportions. Given suitable statistics on input and output stocks, it would be possible to study the prevalence of shortage by investigating this characteristic stock-holding behaviour. However, shortage is fundamentally a problem of the control sphere (information, regulation etc.) rather than the real sphere (production, transactions) of the economy, since it is always a difference between an intention and a realization. Hence data on stocks can only tell part of the story.

3.3.4 Adjustment without prices

As noted above in a fix-price economy experiencing shortage, the economically relevant signals perceived by firms, households and planners are basically quantity signals: stocks, queue length, orders received, waiting time, and so on. Production does adjust to these signals, albeit with some delay: production of inputs in short supply is raised, resources are directed towards the output of goods for which there are long queues and away from other products. This kind of adjustment occurs both in the allocation of current inputs (e.g. material balances, to formulate the annual plan) and in investment decision-making. One might expect that such adjustments should gradually eliminate an initial shortage situation, but there are several reasons why this does not happen, as indicated in the next three points.

3.3.5 Quantity drive

The central planning systems established throughout Eastern Europe in the late 1940s were modelled on that of the Soviet Union, the main features of which were firmly settled in the 1930s. One such feature was a tendency towards very taut planning, based on the priority allocation of resources to a limited number of key sectors; this feature was certainly transmitted to Eastern Europe. It used to be argued that shortage in Eastern Europe was either the result of accidental errors in planning or a consequence of excessively taut planning on the part of the central authorities. While there may have been some force in the latter view in the early 1950s, Kornai is surely right to stress the need to look beyond such a superficial perspective in order to understand the persistence of shortage in economies where central planning is no longer so taut. Stemming from their experience in the 1950s, the basic motivating force for enterprises is a strong quantity drive, an urge to increase output as far as possible given available resources, and with little regard to cost (see section 3.3.6 below).

The drive stimulates enterprises to demand both current and investment resources, to an extent described by Kornai as almost insatiable demand. To some extent, the demand is tempered by tactical considerations: each enterprise has some idea what level of demand would be considered reasonable by higher bodies and knows that high demands for inputs will only be accepted if a correspondingly higher output target can be agreed on. In addition, if the enterprise claims too much and promises too much in the current period, the demands placed on it in the future will be all the greater, in accordance with the usual bureaucratic principle of planning from the achieved level (the ratchet principle). Despite these reasons for caution on the part of the enterprise, Kornai sees the quantity drive operating at enterprise level as the key driving force behind persistent shortage. In its favour is the fact that both managerial rewards and an enterprise's strength in the usual plan-bargaining process are likely to be enhanced by the achievement of larger size: quite literally, 'big is beautiful' in the socialist world.

3.3.6 Soft budget constraint

What makes the quantity drive possible for enterprises is the softness of the budget constraints under which they operate. Four conditions contribute to the softness of the constraint for a state enterprise, which may hold to different extents in different countries or at different times: these are (Kornai, 1980, pp. 306–9):

1. price-making, in the sense that sooner or later enterprises are able to impose cost increases on their customers
2. soft tax system, with enterprises able to negotiate special rates or exemptions, or influence the formulation of tax rules
3. free state grants available to enterprises for a variety of purposes
4. soft credit system, with loans only loosely related to future sales revenue and with only mild repayment conditions and/or weak penalties for non-repayment

Obviously, under these conditions, the survival of a firm is hardly at all contingent on its ability to cover all its costs out of its sales proceeds since grants, subsidies, tax favours etc. can be negotiated to fill the gap. Likewise, growth of the firm does not depend only on internally generated funds supplemented by hard loans, but can be supported from a variety of softer financial means. State enterprises in this kind of environment are also not obliged to react to price changes as one would expect a capitalist firm to do. They may react by changing input combinations, or production, (real sphere), but they may just as well negotiate a special tax regime or exemption from a credit repayment obligation with a higher authority (control sphere).

Furthermore, enterprises with soft budget constraints face a special kind of uncertainty. Although survival is virtually assured (no bankruptcy), an enterprise can never be sure that it will be allowed to keep and use any additional funds that it manages to accumulate. There is an ever-present risk that surplus cash may be siphoned off to help out other enterprises in a less fortunate situation.

Consequently, whenever enterprises have financial resources, or see the possibility of getting them, they will go all out to spend and commit the funds as quickly as they can: hence the almost-insatiable demand referred to above. To sum up the soft budget:

> The soft budget constraint – as opposed to the hard one – is unable to act as an effective behavioural constraint, but exists only as an accounting relationship (Kornai, 1980, p. 309).

3.3.7 Control by norms

In the light of these observations, we are now ready to sketch the adjustment mechanism of a shortage economy, to explain how shortage may be reproduced, rather than be eliminated by the adjustments described in section 3.3.4. The essential notion is that of a norm. One aspect of a norm is simply an intertemporal average of some economic variable. However, not all such averages can be regarded as norms. A norm has the additional feature that any departure of the economy from the normal value of the variable(s) in question sets up a behavioural response – on the part of households, firms, central agencies – tending to restore the normal state. This mechanism is called control by norms. Similar mechanisms can be envisaged, and are discussed by Kornai, in which the control process is governed by upper and lower acceptance limits of some economic variable, but the basic idea is much the same.

If shortage becomes a regular, practically institutionalized feature of an economy, then a normal degree of shortage can become established, in the above sense. Then if for some reason shortage becomes more intense, central authorities are likely to receive more complaints about poor quality and unavailable goods. They can react by diverting goods from exports, by restricting investment, by concentrating investment on factories producing those goods in most severe shortage, by raising prices (though this will only be effective for households, and is politically hazardous), and by cutting back on input deliveries to sectors/enterprises not producing shortage goods. Sooner or later, a combination of central measures of this kind will cut back the shortage intensity to an acceptable level and the economy can return to the more routine regulation and resource allocation of the centralized model.

This is a sketch of a possible centralized mechanism whereby a given and established normal intensity of shortage can be reproduced over time through control by norms: it does not, of course, explain how that shortage intensity came to be the normal one. But in some ways more interesting is Kornai's demonstration that a variety of decentralized mechanisms can achieve the same result. Here we merely set out one such mechanism, based on stock and order signals. In this case the model is a multisector Leontief model, and the norms concern input stocks, output stocks and the backlog of unfilled orders. Four equations define the model, two of which represent resource balances, two of which are behavioural. The first two are balance equations for input and output stocks respectively:

$$\dot{v}_{ij} = Y_{ij} - a_{ij}X_j \qquad i,j = 1,2,\ldots m \tag{3.1}$$

where Y_{ij} is purchase of good i by branch j, a_{ij} is the usual input coefficient and X_j is gross output (here we are using the normal notation employed in input–output analysis); and and v_{ij} is the input stock of good i held by sector j (as usual, a dot denotes differentiation).

$$\dot{U}_i = X_i - \sum_{j=1}^{m} Y_{ij} - f_i \qquad i = 1,2\ldots m \tag{3.2}$$

where U_i is the output stock of good i held by producers, f_i is sales to households (Kornai, 1980, p. 139).

The control sphere of the model is represented by two equations; one describes firms' purchases of inputs:

$$\dot{Y}_{ij} = a_{ij}\dot{X}_j - 2\theta\lambda\dot{v}_{ij} + \lambda^2(v_{ij}^{\circ} - v_{ij}) \qquad i,j = 1,2\ldots m \tag{3.3}$$

where θ, λ are control parameters. In words, this states that firms purchase more inputs if output has increased, input stocks have fallen, or actual input stocks are below their desired levels, v_{ij}°. The second control equation concerns the adjustment of production,

$$\dot{X}_i = \sum_{j=1}^{m} \dot{Y}_{ij} + \dot{f}_i - 2\theta\lambda\dot{U}_i + \lambda^2(U_i^* - U_i) \quad i = 1,2\ldots m \tag{3.4}$$

where U_i^* is the norm for output stocks. Thus output is increased when sales to other firms and to households have increased, when output stocks have declined, and when actual output stocks are below their normal level (Kornai, 1980, pp. 140, 141).

The system defined by Equations 3.1–3.4 has a unique normal state in which household demands are satisfied and $v_{ij} = v_{ij}^{\circ}$, $U_i = U_i^*$. Moreover, if the system is disturbed it is stable: the control mechanism drives it back to the normal state. Finally, a similar model with order backlogs instead of stock signals has the same properties and Kornai conjectures that a mixed system (with both types of signal) would be similarly well behaved. To sum up, control by norms is both viable and stable, and the normal state may well be quite distant from the more familiar Walrasian equilibrium.

The above control mechanism is decentralized in the sense that it uses entirely decentralized information which each firm/sector can observe for itself, and each sector makes its own decisions independently of what others have decided. In a model with order signals there is also some horizontal communication of information. But in both cases, this kind of control is described as vegetative control. Thus what Kornai has shown here is that a decentralized economy can function adequately in the absence of price signals, through the operation of vegetative control, and that this control mechanism permits a state which may be characterized by a high degree of shortage continually reproduced. Kornai suggests that in a more complex control system with strong vertical information flows (e.g. plan instructions), the vertical flows are superimposed upon an underlying vegetative control mechanism; also that in post-reform Hungary, where plan instructions are greatly diminished but the market remains weak, the role of vegetative control is much increased.

3.3.8 Friction

All economies contain elements of friction, phenomena that impede or introduce errors and uncertainty into processes of economic adjustment. In a very general way, it makes sense to think of a tripartite relationship between intensity of shortage, η, productive slack, λ and some measure of friction, ε:

$$\eta = \phi(\lambda, \varepsilon) \tag{3.5}$$

(Kornai, 1980, p. 161 and elsewhere).

Such friction functions, as Kornai calls them, can arise in a number of different ways: for example, imperfect information on the part of buyers or sellers, volatility of demand, delays and rigidity in sellers' adjustment. In most cases, the notion of friction seems to arise in situations where agents have some difficulty (have to incur some costs) in coming together to effect a transaction either because of uncertainty, or because of poor and/or costly information. Thus the idea is analogous to the concept of transactions costs which has been developed in a number of partial models in Western economies, though not yet properly integrated into Walrasian general equilibrium theory (Williamson, 1986; Ulph and Ulph, 1975). The other aspect of friction has to do with change, and resistance to it on the part of one or more agents, usually firms. This has to do partly with the simple point that no-one likes to change what they are used to doing, and a firm operating with a relatively soft budget constraint is not under the same pressure to change that a capitalist firm

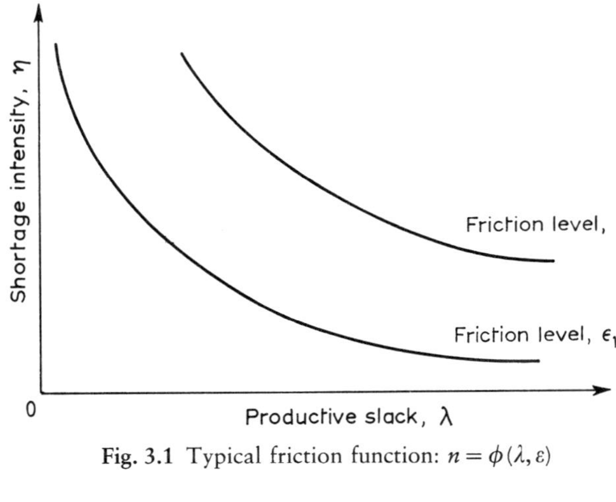

Fig. 3.1 Typical friction function: $n = \phi(\lambda, \varepsilon)$

would be. But also, it reflects the fact that change frequently involves real resource costs and may therefore be resisted on that account too.

In all the examples presented by Kornai, the friction function takes the following simple form. From Fig. 3.1 it is clear that for any given level of friction in the economy, shortage intensity and productive slack are inversely related. And as friction increases from ε_1 to ε_2, the whole curve shifts outwards, away from the origin.

All three variables involved in the friction functions, η, λ and ε, should be measured by means of vectors of suitably chosen indicators, so

$$\eta = f_1(\mathbf{Z}), \quad \lambda = f_2(\mathbf{q}) \quad \text{and} \quad \varepsilon = f_3(\mathbf{W}) \tag{3.6}$$

where **Z** and **q** are vectors of shortage and slack indicators respectively, **W** is a vector of indicators of friction. In practice, as suggested previously, the aggregation implied by Equation 3.6 may involve losing too much information, and functions of this kind may have to be specified separately for a number of partial markets.

3.3.9 Investment and innovation

Within a centrally planned economy, the institutional framework through which investment is determined involves the central planning office, branch ministries, banks and the ministry of finance, other central agencies (e.g. ministry of foreign trade if imported machinery is required), and the enterprises themselves. At national level, both in five year and annual plans, the real and financial resources available for investment are identified or estimated and either parcelled out according to a preassigned sectoral division, allocated to specific major projects, or left unallocated initially but made available to support enterprises' and their superiors' own investment proposals.

In principle, approved projects (whether state or enterprise initiated) have to satisfy appropriate investment criteria which are similar, at least in recent years, to those recommended in Western countries (on investment in Hungary, see Hare (1981); on the Soviet Union, Dyker (1983)). Nevertheless, despite the clear structure for determining the level and pattern of investment, and the need for individual projects to satisfy economically rational criteria, centrally planned economies persistently suffer from shortage which extends into the sphere of investment. Moreover, as emphasized earlier, the problem is at most only partly explained by taut planning imposed from on high since, to a large extent, plan tautness is nowadays a consequence of the prevailing normal intensity of shortage rather than an independently chosen variable.

The same quantity drive that we discussed above, becomes, in the investment context, an expansion drive. Virtually all socialist enterprises wish to initiate and undertake investment projects to expand their scales of operation and because of their soft budget constraints they are not normally restrained by fears of loss or failure. Although there is some self-restraint for tactical reasons, the demand for investment is almost insatiable and is highly insensitive to price-type instruments such as the rate of interest. Even after the Hungarian economic reforms, budget constraints remained soft and so the same mechanism continued to reproduce investment tension.

Shortage has a number of specific manifestations in the investment sphere. First, project appraisal using approved criteria becomes almost a meaningless exercise because investors do not bear any serious responsibility for losses and evaluators lack reliable information: hence the widespread use of formal criteria tells us almost nothing about real investment efficiency. Also, the existence of shortage is often used as a justification for investing in some field, while its absence can be a reason for postponement. Second, since investment resources can be spread more or less thinly, it is easy for higher authorities to approve too many projects without (initially) realizing the probable effects: longer than planned completion times, cost overruns. Third, enterprises wish to expand, but preferably wihout disrupting current production, so scrapping rates are inefficiently low. Fourth, the structure of investment becomes extremely rigid, partly because in any given year a very high proportion of the available resources is devoted to continuing work on already committed projects, and partly because competition for resources among the central agencies tends to repeat resource shares established in earlier periods. Lastly, a consequence of these points is that enterprises are commonly unwilling to undertake innovation of a particularly fundamental kind since they can easily convince themselves, within the established framework of constraints and opportunities, that a more conservative investment policy would be quite adequate.

Once a normal degree of investment tension has become established, the centrally planned economy's adjustment mechanisms will generate a cyclical path of development around the norm. For detailed empirical studies of this type of process, see the monumental work of Bauer (1981), as well as Soós (1975, 1983).

Overall, therefore, this approach to the analysis of a centrally planned economy under conditions of shortage accounts for many of the features of economic behaviour in such economies which have commonly been attributed to irrationalities or accidental errors on the part of planners or lower organs of regulation. Quite the contrary, shortage and its associated phenomena must be regarded as a perfectly logical outcome of rational economic behaviour under suitable conditions. Let us now, in the following section, investigate some of these phenomena in a more dynamic context.

3.4 SHORTAGE IN A GROWING ECONOMY

In *Non-price Control* (Kornai and Martos, 1981), Kornai and Simonovits prove some results about the normal path in a von Neumann model of the economy. Such a path, with control by stock signals, is shown to be stable and viable under suitable conditions, the model being essentially that of section 3.3.7 above. Similar results are obtained for a model with order signals. But neither of these models is truly dynamic in the sense of incorporating growth in the supplies of primary factors of production, notably labour and capital. However, one might suspect that a model which also dealt with the factor markets properly would no longer exhibit persistent shortage. Initial shortage should be alleviated and eventually eliminated altogether by increases in factor inputs. The objective of Kornai (1982) is to demonstrate that this view is incorrect, and that shortage can persist even in a growing economy in which the supplies of capital, labour, the level of output, incomes and living standards are all regularly increasing.

The model is a closed, linear macroeconomic model without money, with firms, households and a labour market. It is set out in a detailed version, with 26 mainly linear or linearized equations (11 describing the real sphere, 15 describing the control sphere), but can also be compressed into a set of six equations in six fundamental variables. The compressed version of the model is useful for the mathematical analysis of growth paths, but is not very convenient from the point of view of economic interpretation. Accordingly we focus here on the detailed model but only cover certain aspects of it, because of space limitations. Throughout, the price level is assumed to be given and fixed, though it would be interesting to rework the analysis with a varying price level.

The first key equation explains actual shortage, as measured by a suitable macroindex, as:

$$\begin{aligned} Z(t) = Z^*(t) &+ a_K[K(t) - K^*(t)] - a_u[U(t) - U^*(t)] \\ &- a_V[V(t) - V^*(t)] + a_Z[Z(t-1) - Z^*(t-1)] \end{aligned} \tag{3.7}$$

where

$K(t)$ is the economy's investment commitment at time t
$U(t)$ is the input stock held by firms
$V(t)$ is the corresponding output stock

and an asterisk denotes a normal value (Kornai, 1982, p. 24). Thus Equation 3.7 indicates that shortage is greater if investment commitments are above normal, or if stocks are below normal, of if shortage was relatively intense in the previous period.

The prevailing shortage then affects the demands of households and firms. Thus for households:

$$H(t) = H^*(t) - b_Z[Z(t) - Z^*(t)] \tag{3.8}$$

and for firms:

$$Y(t) = Y^*(t) - c_V[V(t) - V^*(t)] - c_Z[Z(t) - Z^*(t)] \tag{3.9}$$

where

$H(t)$ is household purchases of goods and services (aggregate commodity)
$Y(t)$ is the firms' purchases

and other notation is as above (Kornai, 1982, pp. 27, 30). Then production,

$$X(t) = X^*(t) - d_u[U(t) - U^*(t)] + d_Z[Z(t) - Z^*(t)] \tag{3.10}$$

Normal production,

$$X^*(t) = p(t)N(t) \tag{3.11}$$

where

$p(t)$ is standard productivity at t,
$N(t)$ is actual employment

(this equation is one of only two remaining nonlinearities in the model) (Kornai, 1982, p. 31).

To simplify the treatment of investment it is assumed that project gestation periods are fixed and constant at G years, that the economic life of all projects is T years after construction and that spending on any project takes place according to the fractions $\beta_i (i = 1, 2 \ldots G)$ in year i of the construction period ($\sum \beta_i = 1$, of course). With these assumptions, the investment commitment at t (i.e. the resources required to complete all projects approved before date t),

$$K(t) = \sum_{\theta=0}^{G-1} \sum_{t=\theta+1}^{G} \beta_t M(t-\theta) \tag{3.12}$$

where

$M(t)$ is the volume of the tth investment vintage, i.e. the spending estimated to be necessary to complete all projects started at date t.

Then the control equation of the investment process takes the form:

$$\begin{aligned} M(t) = M^*(t) + e_{\mathrm{H}}[H(t-1) - H^*_{\mathrm{p}}(t-1)] - e_{\mathrm{K}}[K(t) - K^*(t)] \\ - e_{\mathrm{Z}}[Z(t) - Z^*(t)] \end{aligned} \tag{3.13}$$

where the only notation not yet established is $H^*_{\mathrm{p}}(t)$, which represents the normal, or planned value of consumption from the standpoint of the central agencies; it may differ from $H^*(t)$ referred to in Equation 3.8, which is the normal value of consumption from the households' point of view (Kornai, 1982, p. 44).

What Equation 3.13 says is that the volume of investment will be raised if consumption is higher than normal, if investment commitments are below normal, or if shortage is below its normal value. Thus three separate non-price indicators influence the fluctuations and growth in $M(t)$ about its normal value $M^*(t)$. Notice that, in contrast to the theory of investment in a capitalist economy, none of these is closely related to the expected future demand: socialist firms expect to be able to sell whatever they can produce, so investment is rarely constrained from the side of demand. In addition, the financial state of enterprises (their current or prospective profits) has almost no effect on investment under socialism, so such factors are also omitted from Equation 3.13 (see also section 3.6.2 below).

Each vintage is associated with a demand for labour,

$$J(t) = fg^t M(t) \quad 0 < g < 1 \tag{3.14}$$

where f is a constant and $(1 - g)$ measures the rate of fall of the labour requirements per unit of investment. (Kornai, 1982, p. 42). Then total labour demand,

$$L_D(t) = \sum_{\theta=G}^{T+G-1} J(t-\theta) \tag{3.15}$$

and in the *extensive* growth version of the model, this in turn equals actual employment, $N(t)$ (Kornai, 1982, pp. 54, 55). In the *intensive* growth version of the model, the labour market also experiences shortage and so the labour market has to be modelled somewhat differently. However, that need not detain us here. The important point to note is that Equations 3.14 and 3.15, together with 3.16 and 3.11 establish a connection between investment and subsequent output, via employment creation.

The equations presented so far fall well short of a complete statement of this dynamic model, but they are sufficient to illustrate its general structure, and show some of the main relationships. What is given above has to be supplemented with resource balances, a wage equation, a productivity equation and a variety of other items in order to close the model.

If the various normal values in the model grow at compatible (in most cases, identical) rates over time, then it can be shown that this model:

1. has a normal path along which consumption, investment, output, etc., grow at constant rates, maintaining a constant (in relative terms) intensity of shortage;
2. follows a growth path which fluctuates around the normal path if the model is not already established in its normal state; for small deviations from the normal path, the normal path is asymptotically stable.

This means, therefore, that economic growth, with capital accumulation and improving living standards, is perfectly compatible with persistent shortage. There is no necessary connection between living standards and the intensity of shortage, though it is obviously the case that a high intensity will impose substantial welfare losses on consumers – queueing, delays, forced substitution, search – as they seek to achieve the established standard of consumption. Such costs are not directly allowed for in the model, though their effects are picked up through the impact of shortage on other economic variables.

3.5 CONTRASTS BETWEEN ECONOMICS OF SHORTAGE AND THE DISEQUILIBRIUM APPROACH

In *Economics of Shortage* as well as in his earlier work, Kornai has been at pains to distance himself from Walrasian general equilibrium theory and the more recently developed constrained equilibrium models derived from this theory through the work of Clower, Barro, Grossman, Malinvaud and others, and in its application to Eastern Europe, the work of Portes and his associates (see elsewhere in the present book for more details about this approach). It is all too easy, though in my view it would be a serious mistake, to dismiss this as mere product differentation for, as

Table 3.2 Points of difference between constrained equilibrium and shortage models

Feature	*Constrained equilibrium*	*Shortage*
(1) Excess demand	Uses the concept	Rejects the concept or uses with extreme caution
(2) Behaviour of firms	Profit maximizing (hard budget)	Quantity drive (soft budget)
(3) Stability of equilibrium	Dynamic behaviour not usually discussed	Control by norms; interested in existence, reproduction and stability of an equilibrium with shortages.
(4) Explanation of shortage	Incorrect relative prices	Non-price behaviour of firms and planners
(5) Micro and macro interactions	Only through aggregation; no behavioural link	Macro states affect agents' expectations, hence micro behaviour

should be clear from the last two sections, Kornai's analysis departs from the Walrasian framework in several important ways and is not just a minor variant on the same theme. For the moment, let us simply summarize the principal differences between constrained equilibrium models and what we may call (Kornai's) shortage models.

For convenience and as a guide to our subsequent discussion, these differences can be grouped into five broad categories: excess demand; the behaviour of firms; the stability of equilibrium; the explanation of shortage; interactions between micro and macro outcomes. Table 3.2 summarizes the key points that we need to discuss, in relation to these categories.

For economists brought up in the Walrasian tradition, the notion of excess demand is familiar, straightforward and reasonably unproblematic; but for Kornai it is misleading and confusing, because of the way in which it aggregates economic information illegitimately. Consider the following simple example, concerning a market for a single product. Suppose firm i offers supply y_i on to the market ($i = 1, 2, \ldots, I$) and household h demands a quantity $X_h (h = 1, 2, \ldots, H)$. Let total supply be y, and total demand X. Then according to the conventional definition, excess demand,

$$Z = X - y \tag{3.16}$$

A more detailed story would be along the following lines. Household h may find that it is only able to purchase an amount $\bar{X}_h$, which may be strictly less than X_h; it would then be proper to say that household h experiences (partial) shortage if excess demand,

$$Z_h = (X_h - \bar{X}_h) > 0 \tag{3.17}$$

Combining these gives us a vector of shortage indicators, $\mathbf{Z} = (Z_1 \ldots, Z_H)$. Similarly, firm i may find that it can only sell $\bar{y}_i$ which may be less than y_i. If the firm has actually decided to produce y_i and has the resources to do so, then it may accumulate stocks, a form of slack. The amount of such slack at firm i, is

$$q_i = (y_i - \bar{y}_i) \tag{3.18}$$

and again we can form a vector of these slack indicators, $\mathbf{q} = (q_1, \ldots, q_I)$.

Now, of course, if x_h and y_i represent actual transactions, as we have suggested, then

$$\sum_h \bar{x}_h = \sum_i \bar{y}_i \tag{3.19}$$

and

$$Z = \sum_h Z_h - \sum_i q_i \tag{3.20}$$

where Z is the notion of excess demand defined in Equation 3.16. Kornai's point, in this context, is that it is frequently illegitimate to carry out the aggregation entailed in Equation 3.20. This is partly because the behaviour of any given market is likely to depend in important ways on both shortage and slack indicators, partly because not all market participants will have the same experience (the short-side rule may not hold), and partly because the usual aggregation required to obtain the Walrasian excess demand imposes unduly stringent requirements on market information.

On the behaviour of firms, Kornai argues that neoclassical theory is inappropriate for a socialist economy. In the neoclassical theory, firms maximize profits at given prices subject to remaining within their respective production possibility sets. The only difference that occurs in constrained equilibrium models is that firms have to solve this problem in the presence of additional constraints resulting from rationing or other restrictions on the set of feasible transactions. But to the extent permitted by the constraints, they still seek to produce any given output vector at minimum cost, and to produce a profit maximizing output vector too.

In Eastern Europe, however, the bulk of production has been nationalized and subject to highly centralized state planning since 1950, with an emphasis on increasing output as rapidly as possible and with little reliance on price-type market signals to guide resource allocation. It would be wrong to blame all the mistakes of Eastern European planning on this expansionary pressure from above, since the resulting quantity drive (to use Kornai's illuminating terminology) has now been internalized by enterprises themselves, and persists even in post-reform Hungary. This is so despite the fact that in most of the region, some measure of profit has been one of, and in some cases, the leading indicator of enterprise performance since reforms were introduced in the 1960s.

The implication of this is that the approach to modelling an enterprise functioning in a socialist economic environment must be substantially different from the familiar profit maximizing story of a capitalist firm.

The third point in Table 3.2 concerned the stability of equilibrium. In Walrasian theory, as well as in the newer constrained equilibrium models, the focus is on equilibrium: the characterization of such states and then demonstrating that a given model possesses at least one equilibrium. Relatively little attention is paid to out-of-equilibrium behaviour, not least because it is extremely difficult and quickly runs into apparently insurmountable problems, as argued for instance in Arrow (1959) and Hahn (1973a). Moreover, even the stability analysis that has been done (see Arrow and Hahn, 1971; Fisher, 1984) mainly concentrates on the so-called tâtonnement case in which no transactions take place until the equilibrium has been found. Only a few results are available in non-tâtonnement stability analysis, for although it is clearly more realistic to allow transactions to occur while the economy remains out of equilibrium, models allowing this have not proved especially tractable.

Contrary to what he would usually concede, Kornai himself is principally interested in the characterization of equilibrium positions, in his case equilibrium with shortage. As always, equilibrium is a state at which effective demands and supplies are equalized in every market, and where no agent can achieve a better allocation for himself through his own actions, in view of the prevailing real and perceived constraints. An equilibrium with shortage, therefore, is one in which on average agents perceive supply constraints in the markets that concern them, the constraints occurring both in output markets and in the markets for factors and intermediate inputs.

Where Kornai departs from equilibrium analysis is when he seeks to explain and describe the mechanisms according to which a state of shortage can be maintained and reproduced over time, not through errors of planning but through the rational behaviour of economic agents interacting together under specific conditions. This requires him to consider what happens when an economy with shortage is disturbed: will it gravitate towards the (or a) Walrasian equilibrium, or is it possible to identify economic forces and feedback loops through which the economy will be restored to an equilibrium with shortage? Under the socialist conditions taken for granted throughout his analysis, Kornai argues that there will indeed be forces tending to reproduce a shortage equilibrium, to explain which he invokes such concepts as normal shortage, normal slack and control by norms, as we saw in sections 3.3 and 3.4. The approach is at once powerful and problematic: powerful, in that it yields the results Kornai wants; but problematic, in that, as I pointed out in an earlier paper (Hare, 1982), it seems to imply that equilibrium is what agents expect it to be. In common with other bootstrap approaches to equilibrium (e.g. Hahn, 1982), it opens up a serious possibility of multiple equilibrium; we discuss this further in the next section.

On the fourth point in Table 3.2, constrained equilibrium and shortage models offer very different explanations of shortage. In the former, shortage is a consequence of certain configurations of relative prices (and hence can be remedied merely by bringing about appropriate adjustments in these prices), while in the latter prices are at most a supplementary factor. Other factors, as we have seen, include the institutional framework within which production decisions are taken, the softness of budget constraints and the persistent quantity drive.

A great deal of the recent work on constrained equilibrium models has had the objective of providing a coherent microfoundation to Keynesian macroeconomics, e.g. the survey in Weintraub (1979), Muellbauer and Portes (1978), Barro and Grossman (1976) among many others. The approach is synthetic, in that it explains macroeconomic magnitudes entirely in terms of an aggregation of individual behaviour, but there is usually little or no feedback from macroeconomic states to the behaviour of individual firms and households. Moreover, vigorous application of the short side rule and the usual assumption of identical preferences means that one might as well (with very little loss) restrict attention to models with a single household and a single firm, as is sometimes done. At this point, however, it is surely pertinent to ask what has happened to the microfoundation, though one could still discuss this in terms of optimal behavioural rules.

In contrast, Kornai's notion of a normal state recognizes explicitly that the behaviour of individual agents (firms or households) may be conditioned by their appreciation of the macroeconomic environment in which they operate. To use the terminology of Hahn (1973a), some expectations about shortage form part of every agent's theory of the economy in a centrally planned economy; more important, these expectations are likely to affect behaviour even if particular agents are not currently experiencing shortage on the markets in which they currently wish to trade. Thus, while Kornai can hardly reject the straightforward use of aggregation to form macroeconomic quantities (like consumption, output, employment) from individual agents' behaviour, he would also insist on a causal link in the opposite direction which has not often been taken into account in constrained equilibrium models.

Aside from their usual failure to allow for the simultaneous occurrence of shortage and slack, either in individual markets or in aggregate, Walrasian models also typically neglect frictions in the economy. Using the notation introduced in section 3.3.8, Kornai therefore characterizes a Walrasian equilibrium as that state of the economy in which

$$Z=0, \quad q=0 \quad \text{and} \quad W=0 \tag{3.21}$$

i.e. no shortage, no slack and no friction. While formally correct, it is not hard to see this to some degree as a straw man set up by Kornai only to knock down. For there is certainly no real economy in which all conditions are satisfied simultaneously (3.21), but even the most ardent neoclassical theorist is well aware of the fact and would readily concede that some resources are absorbed by frictions of various kinds. However, in neoclassical models such frictions have no effect on the fundamental behaviour of economic agents, while Kornai's important point here is that frictions do affect behaviour.

In the terminology of Hahn (1973a), for a Walrasian equilibrium, agents must hold a theory of the economy according to which their own decisions will not influence market prices (price-taking) and in which they do not expect to encounter shortages, or experience the inability to sell output or factor supplies, in the normal course of business (supply = demand); this does not exclude occasional and temporary shortages and surpluses since these will be insufficient to oblige agents to modify their theories, nor does it exclude the absorption of some resources to deal with friction in the economy. Consequently, Kornai's assertion of Equation 3.21 is an overstatement of the requirements for Walrasian equilibrium. However, in the context of his own theory this is in any case a side issue, for he is principally concerned with economies in which agents expect to, and actually do encounter constraints on their transactions, so any agents foolish enough to adopt the Walrasian conjecture would quickly have it falsified by experience. This is one of the important factors distinguishing Kornai's theory from the Walrasian model of general equilibrium, though the treatment of friction is also a significant difference as we have seen.

However, part of the reason Kornai's emphasis on the differences between his approach and Walrasian theory can be found in another direction. This is that *Economics of Shortage* is devoted to explaining not only the existence and features of normal states with shortage, but also as we saw in section 3.3.7 above, the dynamic behaviour of an economy when disturbed from a normal state, where Walrasian models almost invariably focus on equilibrium states alone. Such dynamic behaviour involves changes in economic decisions by both firms and households, in response to whatever signals are included in a given analysis, and hence the account of friction given earlier arises quite naturally. At the normal state itself, agents no longer seek to change their decisions and everyone has found partners to carry out those transactions which are feasible: hence once such a state has been established for some time, logically, friction should drop to zero. But in an economy that is constantly changing this situation is hardly likely to occur and friction will remain an essential and significant feature of economic adjustments, as Kornai quite properly emphasizes.

Moreover, according to a less rigid interpretation of the normal state, perhaps closer in spirit to Kornai (and possibly also to Hahn, 1973, interestingly), even at the normal state micro-level transactions will be changing as shortages affect varying inputs at different firms; again, it would be wrong to neglect friction in this situation.

Philosophically, the approach outlined here is not far removed from that of the Austrian school (see Kirtzner, 1973, for example; and in the context of debates about planning, Lavoie, 1985) which treats the market as a process and price signals as stimuli to economic change. Although we are not so interested in price signals in this chapter, the Austrian emphasis on dynamic aspects of markets is clearly analogous to the adjustment processes going on in centrally planned economies experiencing shortage, as described and modelled by Kornai. The processes described by the latter for a planned economy are not, however, so benign as the market processes described by the Austrians are supposed to be: in particular, socialist enterprises are less interested in innovation.

These points also have major implications for the conduct of empirical work. As is made clear elsewhere in this book, empirical studies based on Barro–Grossman models incorporate 'min' conditions and admit several types of equilibrium within any given model. Hence estimating the model simultaneously identifies the equilibrium regime or regimes existing at various times. In particular, as Portes and Winter (1978) sought to show, one can conclude that shortage conditions (equivalent to repressed inflation in their approach, though not for Kornai) existed in some years though not in others. Kornai has expressed a good deal of scepticism about this kind of conclusion. Empirical work based on his approach can either endeavour to measure shortage (an aspect discussed later on) or it can estimate all or parts of the economic control mechanisms that reproduce shortage, as in some of the work on investment of Lackó (1986). In either case, there is a working assumption that shortage conditions are present, and only if the estimation failed very badly would one question the validity of that assumption.

3.6 CRITICAL REVIEW OF THE SHORTAGE MODEL

In this section, I select for detailed discussion a few issues related to Kornai's models of shortage. The issues are chosen either because they are helpful in locating Kornai's theories in relation to other work, including the constrained equilibrium models examined elsewhere in this book (and in section 3.5 above); or because they point to problematic aspects of his approach; or finally, because they are important from an empirical standpoint. The selected topics are: the nature of equilibrium; the soft budget constraint; and the measurement of shortage under various conditions.

3.6.1 The nature of equilibrium

From the Barro–Grossman analysis of fix-price macroeconomic models it is well known that particular constellations of prices are only compatible with particular types of equilibrium. For example, Keynesian equilibrium can only occur while prices lie in a certain region of the price space and repressed inflation equilibrium can only occur in another, normally disjoint region. If the household sector contains more than one household and income effects are significant, multiple equilibria are possible and it is even quite easy to construct cases for which different types of equilibrium are compatible with the same price vector, e.g. Hare (1982). Similarly, if goods and/or labour markets are disaggregated, multiple equilibria also become more likely. But in all these situations, there is a well-defined correspondence from price vectors to sets of equilibria, the correspondence depending entirely on the shape of the underlying utility and production functions.

For the shortage equilibria that arise in Kornai's analysis, the position is completely different. On the one hand, prices seem not to matter at all and play no active role in most of the discussion, while on the other hand the equilibrium itself seems to be determined entirely by what agents expect it to be, as reflected in the norms of the system (i.e. normal shortage, normal input and output stocks, etc.). This is most clear in the formal models of sections 3.3.7 and 3.4 above, in which normal shortage, Z^*, is an exogenous variable. If agents (notably firms) expect shortage to be very intense (high Z^*) then in equilibrium it certainly will be and everyone is proved correct. Equally, with an identical structure but a general expectation of a low intensity of shortage, that is how things turn out. Hence in these models we have multiple equilibria with a vengeance, for equilibrium can be practically anything we choose it to be. What the models actually tell us is twofold: first, that Kornai's story of a shortage economy is consistent in the sense that formal models with shortage equilibria can be constructed; and second, that the equilibria in such models are stable as a result of the feedback control mechanisms built into them.[1] Thus once established, shortage can be reproduced over time, as Kornai argued. However, the formal analysis does not indicate how shortage emerges in the first place, nor why the normal intensity of shortage comes to have a particular value.

This implies, therefore, that the mathematical models do not explain shortage, but only its reproduction. Yet Kornai is insistent that shortage is indeed the state that occurs in practice, so something must be missing from the models. The missing factor must have something to do with the quantity drive or expansion drive of socialist enterprises which, in conjunction with the ever-present soft budget constraint, are what generate shortage conditions, according to his arguments. Yet in the model of section 3.3.7 there is nothing that can be attributed to a quantity-drive *per se*; the absence of prices and financial constraints from the equations is surely not enough. Furthermore, the situation is no better in the dynamic model of section 3.4, since this also fails to model carefully the particular factors said to generate shortage. However, it is important to emphasize that these deficiencies are only shortcomings in the modelling. In his more descriptive analysis of shortage economics Kornai certainly pays attention to the historical/institutional factors which generated shortage in the first place. Moreover, in view of the limited powers of the planners and the strength of autonomous processes in the economy, even the role of expectations may be less dramatic than the formal models would imply. It may not be easy to make firms believe that normal shortage has fallen.

The distinction between the equilibria arising in *Economics of Shortage* and those occurring in the Barro–Grossman analysis can be brought out more forcibly if we consider their respective policy implications. In the latter case, policy recommendations would tend to focus on the price structure: thus if the economy is stuck in a repressed inflation equilibrium, one would normally advise measures to raise prices and money wages, while reducing the real wage. Such measures would be effective in the consumer goods sector in a socialist economy, since consumers have hard budget constraints, but they would have scarcely any effect on the sphere

[1]It is perhaps worth adding here that most models of the Barro–Grossman type do not investigate the second property very fully. An interesting micro-model which does investigate the dynamics of fix-price equilibria is given in Picard (1983).

of production; since any resources released in consumption would be absorbed by producers in such a way as to maintain the established degree of shortage. This at least appears to be the logic of Kornai's position, though the apparent implication that shortage cannot be overcome is hard to accept.

The principal positive policy implication from *Economics of Shortage* is that if we wish to abolish shortage, then agents (principally enterprises) have to be convinced that shortage will not occur. Then the formal models will immediately, and with no real change in the underlying economic conditions, generate and reproduce an equilibrium without shortage. The difficulty in practice, of course, would be to convince agents that anything at all had changed given their long previous experience of shortage. In any case, this is surely not a very plausible story as it stands. So we must seek factors more fundamental than agents' expectations as the basis for policy action. Suitable factors were already alluded to above, namely quantity drive and soft budgets, and these we shall discuss further below. The essential point, however, is that a real change in firms' expectations may only be achievable as a result of a quite fundamental economic reform involving both policy instruments and the institutional structure of the economy. Thus a careful consideration of Kornai's approach makes clear that its implications are far more radical than those of the constrained equilibrium school.

3.6.2 The soft budget constraint

In the previous subsection it was argued that the notions of quantity drive and soft budget constraint were not modelled satisfactorily in *Economics of Shortage* and hence that, after all, we still lack a convincing formalization of the theory of shortage. Section 3.3.6 reviewed Kornai's original arguments about soft budget constraints, which have since been refined and developed, though without fundamental change, in Kornai (1986). The arguments have also been discussed, with some critical comment, in Soos (1984), Gomulka (1985) and Szabó (1985). The object of this section is to review these contributions and to assess the present state of the debate about budget constraints of the firm.

The idea of a soft budget constraint explains both too little and too much: too little because, as Gomulka points out, it is not easy to find a firm with a budget constraint that is not at least partly soft, and yet one does not observe shortage everywhere; and too much because the implied quantity drive is neither as universal nor as strong as Kornai has claimed. The first point suggests that it may only be certain aspects or features of soft budget constraints which are essential in the generation of shortage, while the second indicates that these constraints may be an oversimplification of a much more complex institutional environment for the traditional socialist firm.

As compared with a typical capitalist firm, the key features of socialist enterprises are the lack of entry and exit from the market (hardly any firms allowed to go bankrupt, hardly any new firms established in competition with existing ones), and their insertion into a network of predominantly vertical relationships with upper levels of the planning and control hierarchy. Far from always generating shortage, Gomulka claims that this aspect of the soft budget constraint results in a loss of efficiency (X-inefficiency) which may be substantial, but will be kept within limits by the authorities because of their natural insistence that the economy remain productive. Moreover, shortage may be reduced or eliminated altogether using the price mechanism, provided that enterprises expect a delay between an increase in their costs and an offsetting relaxation in their budget constraint (in Gomulka's terminology, budget flexibility < price flexibility). Kornai is far less hopeful about the efficacy of the price mechanism, of course, except in the sphere of personal consumption. However, logically, Gomulka's argument is surely only correct if the authorities are prepared to entertain the possibility of bankruptcy: for otherwise enterprises would always be able to rely on higher organs to act as lender of last resort, hence undermining the authority of price signals.

Soós is concerned about the generality of the soft-budget constraint concept, as shown in the conditions that were listed in section 3.3.6, for it is not clear that the presence or absence of each of these conditions has the same economic effects, nor which combinations are specially decisive for shortage. To give an example: before 1968, Hungarian enterprises operated in an environment in which credits were relatively easily available, and there was no bankruptcy; since then, at least for some of the period, credit and other financial conditions have appeared to be tighter while bankruptcy was still almost unheard of. Does this mean that enterprise budget constraints became softer? Has the economy's tendency to generate shortage diminished? The evidence is actually quite mixed in this respect. The point, however, is that the different conditions listed in section 3.3.6 are not merely variations on a theme. They are different in the sense that they are brought about by different institutional arrangements within the system, one or more could be changed without changing others and they are likely to have different effects. It is therefore important to analyse these conditions separately, especially from the point of view of policy-makers seeking advice on how they might eliminate shortage.

A further puzzling issue in the analysis of shortage, which cuts across the budget constraint problem, is the following. If firms experience an increase in demand, it is accepted that they will normally, perhaps with a lag, adjust by seeking to increase supply. Hence in order to explain persistent shortage, it has to be shown that these increases in supply will always be less than the initial increase in demand; in other words, that supply is relatively inflexible. Otherwise, supply adjustments would eventually catch up with demand and eliminate the shortage. Soós argues that inflexible prices and the institutionalized inelasticity of supply are the key explanatory variables here. Firms have an interest in preserving shortage because it makes life easy, so there is a lack of motivation to adjust, especially in the presence of a soft budget constraint.

Although I suspect that Kornai would accept the above point, his own analysis leads to a formal model of shortage in which friction is apparently zero (section 3.3.7). Indeed friction is not even discussed in his book until after the main analysis of shortage, though it does play an important part in the later analysis. Yet as Szabó quite rightly emphasizes, if there is no friction it is hard to see how the required inflexibility of supply can be explained and therefore unclear how Kornai is able to justify his view of demand constantly outstripping supply. While the soft budget constraint allows socialist firms to behave in ways that differ from their profit-maximizing capitalist counterparts, it is certainly not enough to explain shortage. We shall have more to say about friction in section 3.7.

Aside from these theoretical points, what is the evidence for soft budget constraints? In Kornai (1986) some results are presented on the relationships between enterprise profits, fiscal redistribution of profits (taxes and subsidies) and enterprise investment in Hungarian state-owned enterprises (of which there were 1755 in 1982); these results are based on Kornai and Matits (1983) and Matits (1984). The evidence shows that there is a strong tendency to redistribute profits from 'the strong to the weak', and the correlation between original profits and final profits (after taxes and subsidies) was practically zero in 1982. Of course, even the data for original profits has to be regarded as suspect from an economic point of view, because of distortions in the price system. Nevertheless, the negligible reported correlation is a striking result. Equally striking is the additional finding that investment activity by firms is hardly correlated with their profitablity in earlier years, again suggesting that the state will help out firms which manage to get permission to undertake investment, irrespective of their profitability.

These observations are said to demonstrate that enterprise budget constraints, even in post-reform Hungary, are indeed soft: but is this a correct interpretation of the results? In my view, the situation is a good deal more complex than Kornai suggests.

First, the observed relationships, with the obviously highly differentiated tax and subsidy system, show one way in which enterprises in Hungary continue to be part of an administrative hierarchy. There are vertical channels other than this financial one, and in this respect Hungary remains close to the traditional centrally planned economy. Secondly, if firms really have soft budget constraints in Kornai's sense, it is not obvious that any particular significance can be attached to profits before or after tax; all we are entitled to conclude is that firms experience considerable financial intervention. Thirdly, being *ex post* magnitudes the reported figures tell us nothing directly about quantity drive: however, it is curious to say the least that the redistribution works as it does (giving additional resources to weaker firms, taking from the more profitable) if quantity drive is the fundamental urge of all enterprises, unless some other more rational allocation principle is also operating in the background. Fourthly, and in support of the last point, even if each individual enterprise perceives that it has a soft budget constraint, for the sphere of production as a whole this is clearly not correct, because an overall real resource constraint comes into effect. Hence, to the extent that enterprise behaviour is influenced by their expectations about the prevailing aggregate and individual constraints, both must be taken into account in any modelling exercise. Furthermore, one way of regarding the reallocation that occurs is as a means of ensuring that enterprises possess the resources allowing them to develop in a reasonably balanced way. True, the tax/subsidy system that achieves this is absurdly complex, but in a system where prices don't matter very much and may be quite irrational and where there is an emphasis on planning in quantitative terms, it is not an implausible story. Hence rather than supporting the notion of a soft budget constraint, Kornai's observations may merely reflect the prevailing irrationality of the price system.

What remains of the soft budget constraint concept is its origin in the institutional structure within which enterprises operate: a vertically oriented hierarchy (in which a system of sectoral priorities plays an important role in resource allocation), with no bankruptcy for the individual firm, and almost no entry of new firms into an established branch. These are the key features of Eastern European economies which give rise to the behaviour discussed by Kornai. In the next section of this chapter we discuss an alternative approach to modelling these factors.

3.6.3 Measurement of shortage

This topic arose earlier, including in section 3.5, where it was emphasized that shortage should be regarded as a vector quantity, $\mathbf{Z} = (Z_1, \dots Z_N)$, that it should not be confused with Walrasian excess demand, and that slack indicators, $q = (q_1, \dots q_m)$ should be measured separately, rather than being merged into the measure of shortage by means of aggregation. Finally, section 3.3.8 reviewed the fundamental inverse relationship between shortage intensity $\eta = f(Z)$, slack $\lambda = g(q)$ and some measure of fraction, ε, in particular markets (Fig. 3.1).

In practice, the injunction against aggregation has not prevented Kornai himself from engaging in it when convenient and appropriate, as for instance in the growth model of section 3.4. However, he has not tended to combine Z and q. On the other hand, in view of the relationship between η and λ just mentioned, it is hard to see why one should not aggregate, unless there is some reason to expect the degree of friction in the economy to change significantly over the period being studied. For any relationship involving both η and λ obviously reduces to one in η alone with an appropriate change of functional form. Hence in the present discussion, we lose very little by concentrating on shortage.

To measure shortage in a particular market we have to identify a suitable list of partial indicators, Z, concerning the actual market situation; to study the behaviour of the market, it is also important to measure normal shortage. The latter can either be done in terms of the same list of indicators or, more likely, in terms of the shortage intensity, $\eta = f(Z)$. In this case, the correct way of combining components of Z into a single measure is itself a matter for empirical estimation. η has to be a function of the partial indicators, but it must also have behavioural significance in that firms (and possibly households, and government agencies) react to the deviation $(\eta - \eta^*)$, where η^* is the required normal shortage. It is perhaps pertinent to point out here that Kornai himself provides little guidance on the measurement of normal shortage.

Most partial indicators that are readily measureable lack an unambiguous interpretation. For instance, for a product purchased by households we should try to measure the quantities (same notation as earlier):

$$Z_h = \max [X_h - \bar{X}_h, 0] \tag{3.22}$$

However, when we do so it is unavoidably necessary to aggregate over households to some degree and the X_h that is observable is unlikely to be the unconstrained, Walrasian demand, but is almost certainly influenced by forced substitution into or away from the market being studied. Hence depending on the situation in related markets, X_h may under- or overestimate the true demand in the given market, and the need to aggregate also loses some information. Both factors mean that Z_h as measured in Equation 3.22 can change when the underlying shortage conditions are unchanged; equally, Z_h can remain constant while shortage really is changing. Consequently without a well-specified model which includes an explanation of the process generating the observations, Z_h, and hence imposes a well-defined, 'correct' interpretation, it is impossible to measure shortage satisfactorily.

A further point on measurement, one recognized by Kornai to some extent, is that even for a market that one would not wish to characterize as a market with shortage, measured shortage (Z or η) would normally not be zero. In terms of Fig. 3.1, such an outcome would be represented by a point towards the bottom right of the diagram (low η, high λ). Presumably, however, the normal shortage, η^* should be zero (or very close to zero) in such cases, so that agents would behave as if they did not expect to encounter shortage in the normal course of events (in line with Hahn's (1973a) view of agents forming theories about the economy in which they operate). What is unknown, without detailed empirical study, is how often and with what intensity agents can experience shortage before changing their theory of the economy.

3.7 AN ALTERNATIVE MODEL

Some of the key points in *Economics of Shortage* can be accommodated in the following rather simple model, which shows that under the conditions held by Kornai to characterize the socialist economies, shortage is more likely to occur than it would in competitive capitalism. In general terms, a set of input–output balances for an economy can be written in the form:

$$\mathrm{X} = A\mathrm{X} + \mathrm{y} + \mathrm{f} \tag{3.23}$$

where

$\mathbf{X}$ = vector of gross output
y = vector of final demand (consumption plus investment plus government spending plus net exports)
A = matrix of technical input coefficients

} All this is the standard terminology and notation of input–output analysis

f = vector of resources absorbed by adjustment frictions in the economy

In Equation 3.23, although it is a static model, it is perhaps best to think of **f** as a measure of the costs of getting from last year's allocation of resources to the current year's. This is, of course, an extremely crude way of representing friction, but it does at least recognize that the operation of any resource allocation mechanism (including a supposedly perfect market) absorbs real resources. It contrasts with standard accounts of perfect markets in which the costs of transmitting, receiving and processing economically relevant signals are wholly ignored.

An equation such as 3.23 holds for any economy. From the arguments in *Economics of Shortage* we can conclude that in a socialist economy, the lack of competition (no bankruptcy, hardly any new entry, import controls) allows current costs to be higher than they would otherwise be, while adjustment frictions are likely to be higher too, partly for lack of proper incentives, partly due to distorted prices and partly due to the complexity of the administrative structure through which adjustments are implemented. These differences can be modelled by assuming

$$A_s > A_c \quad \text{and} \quad f_s > f_c \tag{3.24}$$

where the suffix, s, refers to a socialist economy, and c refers to a capitalist one.

Suppose the capitalist and socialist economies being compared both possess the same capacities in each industry; let the common capacity vector be **k**. In addition, suppose both economies aim to produce the same final demand vector, **y**. Then for the capitalist economy,

$$X_c = \min\,(A_c X_c + \mathbf{y} + f_c, \mathbf{k}) \tag{3.25}$$

and for the socialist economy,

$$X_s = \min\,(A_s X_s + \mathbf{y} + f_s, \mathbf{k}) \tag{3.26}$$

both equations taking account of capacity constraints. If some capacity constraints are binding, then the actually produced final output will be less than **y** in some components:

$$y_c = (I - A_c)X_c - f_c \leqq \mathbf{y} \tag{3.27}$$

and

$$y_s = (I - A_s)X_s - f_s \leqq \mathbf{y} \tag{3.28}$$

If **y** is set at a level which is just feasible for the capitalist economy ($y_c = \mathbf{y}$ in 3.27), it follows from Equations 3.24 and 3.28 that $y_s < y_c$, so that the corresponding socialist economy will experience general shortage. Hence, given higher production costs and higher adjustment frictions, the socialist economy is substantially more prone to shortage, just as Kornai has argued. However, this argument took it for granted that the socialist economy would seek to achieve the same output vector as its capitalist equivalent, making no allowance at the planning stage for higher input costs and losses due to friction. To this extent, one could equally well attribute the shortage that results under socialism to over-ambitious planning. Or alternatively, this approach reinstates Kornai's view by requiring some form of quantity drive to make the model work.

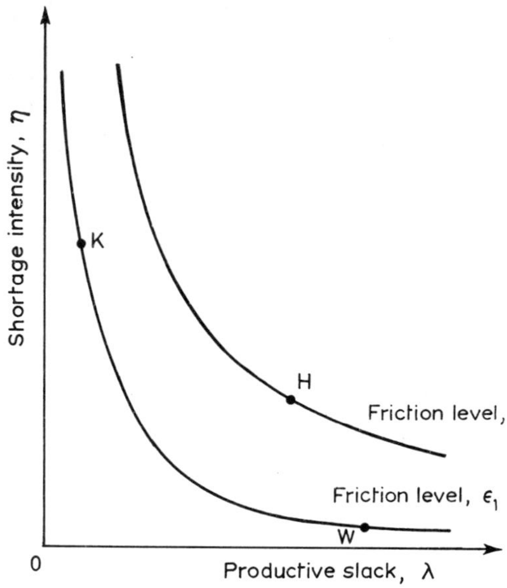

Fig. 3.2 Alternative states of the economy: friction, slack and shortage. W = Walrasian equilibrium, K = Kornai's shortage economy, H = Hare's economy with adjustment friction.

At this point it is convenient to return to Fig. 3.1 which is reproduced as Fig. 3.2, together with some additional information. Three points are shown in the diagram. W is essentially a Walrasian equilibrium, with negligible or small intensity of shortage and relatively large productive slack (strictly, O is the Walrasian equilibrium, but as pointed out earlier such an outcome is impossible). K represents Kornai's view of an economy with shortage; η is large, λ is small so that for a given level of friction in the economy we simply move along the friction function. Finally, H is the present author's view of an economy with shortage: as compared with the competitive economy it is characterized by a higher level of adjustment friction.

We now seem to have a story of shortage in which lack of competition and higher adjustment friction – and quantity drive – are jointly sufficient to explain shortage in a socialist economy. Moreover, the factors emphasized here are not only consistent with much of Kornai's argument, but also have the virtue of setting out before us a clear agenda for further work: both in modelling their effects on the economy, and in identifying the kinds of economic reform that might allow shortage to be overcome.

3.8 CONCLUDING OBSERVATIONS

This chapter has reviewed Kornai's important contribution to the study of Eastern European economies, noting his rigorous analytical approach and his willingness to overturn traditional dogmas. However, this review has encountered a number of lacunae in his theory. This concluding section briefly summarizes our main findings. For simplicity, I merely list the key points.

1. Despite the attention quite properly given to the notions of expansion drive, quantity drive and soft budget constraint, these are not formally modelled in *Economics of Shortage* or in other work by Kornai and his research group;
2. The concept of control by norms enables models to be developed in which the reproduction of shortage is explained through a feedback mechanism based on agents' expectations. However, the formal analysis does not explain why or how a particular intensity of shortage comes to be established as the norm under given conditions, though Kornai's descriptive theorizing does draw attention to the relevant historical and institutional factors.
3. Kornai lays considerable emphasis on adjustment frictions in the economy, including the fundamental relationships of the form, $\eta = \phi(\lambda, \varepsilon)$, yet his models of control by norms are entirely frictionless, as is much of the discussion of shortage. Again, however, his descriptive analysis gives friction its proper place in the story;
4. The measurement of shortage in the presence of adjustment frictions implies that it will never be zero. But not all economies develop feedback control mechanisms based on a deviation between actual and normal shortage. It is not apparent from Kornai's analysis under what conditions such a mechanism can emerge and persist (this is analogous to the occurrence of a particular equilibrium regime in the Barro–Grossman story). Consequently, it is not clear that we are entitled to conclude that centrally planned economies will always experience shortage in this sense;
5. In seeking to develop a theory of shortage in terms of rational behaviour by individual agents (principally households and firms), Kornai quite correctly wishes to avoid explanations based on planners' errors or overambition. However, the cost of this approach is a neglect of the role played by the vertical planning hierarchy in guiding the economy (and hence, among other things, in generating shortage), except where it impinges directly on the enterprise. Although horizontal information flows, and the vegetative control they make possible, suffice to explain the reproduction of shortage (see point 2 above, as well as section 3.3.7), they do not explain its normal intensity, as Kornai does indeed recognize;
6. To explain the latter, one would have to appeal to a variety of historical and institutional factors going back to the origins of the centrally planned economic systems in Eastern Europe;
7. Finally, to eliminate, or significantly reduce the degree of shortage experienced in Eastern Europe, requires major reforms of the economic management structure; it would not be sufficient merely to correct distortions in the structure of relative prices.

ACKNOWLEDGEMENTS

This chapter has benefited greatly from comments from participants at the Birmingham Conference (see Foreword), as well as from the book's editors. In addition, I received very useful comments from seminar discussions at the Universities of Bristol, Oxford and Rome, and from the European University Institute (EUI). I should particularly like to thank Mario Nuti (EUI), and Philip Welham (Heriot-Watt University). Remaining errors and omissions are entirely my own responsibility.

REFERENCES AND FURTHER READING

Arrow, K. J. (1959) Toward a theory of price adjustment, in *The Allocation of Economic Resources* (eds M. Abramovitz *et al.*), Stanford University Press, Stanford.

Arrow, K. J. and Hahn, F. (1971) *General Competitive Analysis*, Holden Day, San Francisco.

Barro, R. and Grossman, H. (1976) *Money, Employment and Inflation*, Cambridge University Press, Cambridge.

Bauer, T. (1981) *Tervgazdaság Beruházás, Ciklusok*, Közgazdasági és Jogi Könyvkiadó, Budapest.

Debreu, G. (1959) *Theory of Value*, Cowles Foundation Monograph 17, John Wiley, New York.

Dyker, D. A. (1983) *The Process of Investment in the Soviet Union*, Cambridge University Press, Cambridge.

Fisher, F. (1984) *Disequilibrium Foundations of Equilibrium Economics*, Cambridge University Press, Cambridge.

Gomulka, S. (1985) Kornai's soft budget constraint and the shortage phenomenon: a criticism and restatement. *Economics of Planning*, **19**(1), 1–11.

Hahn, F. (1973a) *On the Notion of Equilibrium in Economics*, Cambridge University Press, Cambridge.

Hahn, F. (1973b) The winter of our discontent. *Economica*, **40**, 322–30.

Hahn, F. (1982) *Money and Inflation*, Basil Blackwell, Oxford.

Hare, P. G. (1981) The investment system in Hungary, in *Hungary: a Decade of Economic Reform*, (eds P. G. Hare, H. K. Radice and N. Swain), Allen and Unwin, London.

Hare, P.G. (1982) Economics of shortage and non-price control (review article). *Journal of Comparative Economics*, **6**, 406–25.

Hey, J. D. (1982) Review of *Economics of Shortage*. *Economic Journal*, **92**, 712–3.

Kirtzner, I. (1973) *Competition and Enterpreneurship*, University of Chicago Press, Chicago.

Kornai, J. (1959) *Overcentralisation in Economic Administration*, Oxford University Press, Oxford.

Kornai, J. (1970) *Anti-equilibrium*, North-Holland, Amsterdam.

Kornai, J. (1980) *Economics of Shortage*, North-Holland, Amsterdam.

Kornai, J. (1982) *Growth, Shortage and Efficiency*, Basil Blackwell, Oxford.

Kornai, J. (1986) The soft budget constraint. *Kyklos*, **39**(1), 3–30.

Kornai, J. and Martos, B., (1981) *Non-price Control*, Akadémiai Kiadó, Budapest.

Kornai, J. and Matits, A. (1983) A költségvetési korlát puhaságáról vállalati adatok alapján, *Gazdaság*, **16**, 3–30.

Lackó, M. (1986) *Feszültségek és normák szerepe a beruházások szabályozásában Magyarországon (1961–1980)*, MTA Közgazdaságtudományi Intezete, (*mimeo*) (draft Candidate's dissertation, to be submitted by end 1986).

Lavoie, D. (1985) *Rivalry and Central Planning*, Cambridge University Press, Cambridge.

Levine, H. S. (1983) Review of *Economics of Shortage*. *Journal of Economic Literature*, **21** (1), 95–8.

Malinvaud, E. (1984) *Mass Unemployment*, Basil Blackwell, Oxford.

Manove, M. (1973) Non-price rationing of intermediate goods in centrally planned economies. *Econometrica*, **41** (5), 829–52.

Matits, A. (1984) *A redisztribúció szerepe az állami vállalatok jövedelemezóségének alakulásában: második beszámoló*, (*mimeo*). Ministry of Industry, Budapest.

Muellbauer, J. and Portes, R. (1978) Macroeconomic models with quantity rationing. *Economic Journal*, **88**, 788–821.

Picard, P. (1983) Inflation and growth in a disequilibrium macroeconomic model. *Journal of Economic Theory*, **30**, 266–95.

Portes, R. and Winter, D. (1978) The demand for money and for consumption goods in centrally planned economies, *Review of Economics and Statistics*. **60**, 8–18.

Soós, K.A. (1975) Causes of investment fluctuations in the Hungarian economy. *Eastern European Economics*, 1975/6, No. 2.

Soós, K. A. (1983) The problem of time lags in the short-term control of macroeconomic processes, *Acta Oeconomica*, **30** (3–4), 369–80.

Soós, K. A. (1984) A propos the explanation of shortage phenomena: volume of demand and structural inelasticity, *Acta Oeconomica*, **33**(3–4), 305–20.

Szabó, J. (1985) Inelastic supply, suction, slack and shortage. *Közgazdasági Szemle*, No. 7–8, pp. 951–60.

Ulph, A. M. and Ulph, D. T. (1975) Transaction costs in general equilibrium theory. *Economica*, **42**, 355–72.

Weintraub, E. R. (1979) *Microfoundations*, Cambridge University Press, Cambridge.

Williamson, O. (1986) *Economic Organization*, Wheatsheaf Books, Brighton.

4

The chronic excess demand hypothesis

DAVID M. KEMME

4.1 INTRODUCTION

In modelling the macroeconomic phenomenon of centrally planned economies (CPEs)[1] economists are sharply divided on whether or not chronic shortage is a pervasive and persistent phenomenon at the macroeconomic level. One approach, the so-called chronic shortage or chronic excess demand model, essentially assumes that the demand for goods and services continually outstrips supply, not only for certain individual goods but also in the aggregate. Chronic excess demand is a maintained hypothesis in the majority of these models. The second approach, that of the disequilibrium macroeconometric model, generally makes no assumption concerning the existence of disequilibrium but attempts to detect and measure it if found.[2] As a result of the different underlying assumptions and methodologies, there is confusion and disagreement over the existence and significance of shortages in CPEs. Further, the issue of macroeconomic disequilibrium in CPEs (excess aggregate demand coupled with forced savings, in particular), its existence and

[1] In this chapter CPE refers to the socialist centrally planned economies of Eastern Europe (Poland, GDR, Czechoslovakia, Hungary, Bulgaria, and Romania) and the USSR.

[2] Macroeconometric analysis refers to the use of traditional macroeconomic data in both equilibrium and disequilibrium models. Here I am not referring to disequilibrium and discrete switching models when applied to single markets in a microeconomic framework. Several examples of the later appear in chapters of this book which follow. Portes, in Chapter 2, discusses the macroeconometric modelling approach in detail. The dichotomy presented may be too simplistic, however. A. Welfe, in Chapter 11, bridges the gap by estimating a disaggregated econometric model with excess demand as the maintained hypothesis.

International Studies in Economic Modelling Series.
Models of Disequilibrium and Shortage in Centrally Planned Economies.
Edited by C. Davis and W. Charemza.
Published in 1989 by Chapman and Hall Ltd, 11 New Fetter Lane, London EC4P 4EE.

significance, has often inappropriately been viewed as equivalent to the notion of chronic shortage.[3]

This chapter examines primarily those models developed on the basis of the chronic excess demand hypothesis highlighting the important methodological differences and conclusions reached *vis-à-vis* the macroeconometric modelling approach. The discussion below considers macroeconomic models of the CPE, including Kornai (for essential details see the previous chapter), Winiecki (1982, 1985), Bauer (1978), Beksiak (1986), and Kemme and Winiecki (1984), *inter alia*, as well as several microeconomic or sectoral models which tend to confirm the existence of shortages in CPEs (many of which follow in later chapters of this book). These proponents of the chronic excess demand hypothesis argue that there are system-specific elements which generate persistent excess demand, initially in the production sphere then spilling over into the household sphere. It is the system-specific features which the macroeconometric models, to date, fail to incorporate and this failure, therefore, leads to results contrary to CPE reality.[4]

Below, the fundamental concepts of the macroeconomic models which implicitly assume excess demand are presented, noting the system-specific features that generate the excess demand. Sectoral models which are based on the notion of chronic shortages or simply confirm the existence of shortages are then presented. The fourth section discusses the implications of the chronic excess demand approach with regard to macroeconometric modelling, questioning the methodological validity of the macroeconometric approach and the interpretation of macroeconometric results.

4.2 MACROECONOMIC MODELS WITH CHRONIC EXCESS DEMAND

The chronic shortage model, developed most fully by Kornai and discussed in the previous chapter, outlines a situation in which both producers and consumers are faced with shortages.[5] Similar to Kornai, Winiecki (1982, 1985) and Kemme and

[3]In Kemme (1987) I argue that chronic shortage is primarily a microeconomic phenomenon that defies measurement via traditional macroeconomic tools. Portes and Winter (1980) also do not question the existence of 'numerous chronic and substantial *micro*-level disequilibria . . .' or '[t]hat *relative* prices are distorted . . .' (p. 137, their emphasis). I argue that it is the existence of these chronic and substantial microeconomic disequilibria that makes the aggregation problem difficult, if not impossible, to solve in the CPE.

[4]It is this perceived specification error and the aggregation problems discussed below which proponents of chronic excess demand models question most.

[5]The models discussed below are generally not formal, mathematical models as developed by Kornai, but descriptive and heuristic. There is also a group of formal macroeconometric models, such as Charemza and Dlouhy (1986) which assume excess demand for consumption goods. These are not discussed in this section since I consider them a subset of the macroeconometric modelling approach and they are subject to the same criticisms (aggregation problems). Along these same line see Lackó, Chapter 10, section 10.2, for a list of 'verbal', 'mathematical – theoretical', and econometric models by Hungarian researchers. Again, A. Welfe's chapter is a unique hybrid which takes a first step toward resolving the conflict between the two approaches.

Winiecki (1984) also argue that these shortages are primarily the result of systemic features which must be considered explicitly in the modelling of CPEs. In part these works were a response to the literature following Portes' (1974) challenge to the conventional wisdom that aggregate excess demand or repressed inflation was the norm for CPEs.[6] Portes' challenge advanced the application of traditional macroeconometric modelling where others were reluctant to tread. The macroeconomic approach, however, ignores much of the microeconomic behaviour considered so important by proponents of the chronic excess demand hypothesis.

The system of managerial incentives, which tends to emphasize output fulfilment criteria rather than cost control or profitability, is of primary importance. Managers, in an attempt to minimize the effects of late material deliveries and uncertainty as to future input availabilities, gather excessive labour, spare parts, and stocks of materials. This hoarding is possible due to the soft budget constraint, as developed by Kornai (1980, *inter alia*). Balicki (1983, pp. 25–7) also emphasizes the nonsecurity of the plan and the malallocation of resources in the plan that managers must protect themselves against.

The ability to stockpile factor inputs, along with the plan formation process, generates excess demand for investment. Bauer (1978) and Winiecki (1982) note the cyclical nature of investment expenditures and the rationale for investment cycles which represent varying levels of disequilibrium. Podkaminer (1985) argues that it is the effects of consumer goods shortages on the supply of labour which prevents the investment plan from being fulfilled and initiates a downturn in investment expenditures. Further, innovative activity which would normally increase output levels without additional investment expenditures, or at least without a different composition of expenditures, is rare due to the risk aversion of managers and the fear of ratcheting after successful innovation.

These systemic features introduce distortions which, over time, exert disequilibrating pressures not common to market-type economies. It is the

> existence of an institutional and policy setting that by its very nature permanently reproduces persistent disequilibrium of varied intensity, quite independently of 'taut,' 'mobilizing' or otherwise over-optimistic plans... (Winiecki, 1985, p. 30)

which is at the source of the chronic excess demand in CPEs. These systemic features do not play a prominent role in the macroeconometric modelling approach which focuses on traditional macroeconomic aggregates, ignoring the underlying microeconomic phenomena.

The approach of those assuming chronic excess aggregate demand also differs from the macroeconometric modelling approach in its rejection of what Winiecki (1985) calls 'the undivided trinity of planners' beliefs' implicit in the macroeconometric models of Portes (1977, 1979), Portes and Winter (1980) and some later

[6] A significant body of literature existed to advance this notion, including, for example, Bush (1973), Holzman (1960), and Schroeder (1975). Extensions and updates of this line of research are numerous.

works.[7] First of this trinity is that excess demand for consumer goods appears only as a result of planners' errors. It was argued above, however, that excess demand in the production sphere of the economy is generated by systemic features. This does not preclude planners' errors or the potentially severe effects of errors, but planners' errors are not necessary for disequilibrium to arise.

The second item of the trinity is the separability of the producer goods market from the consumer goods market.[8] While it is argued that the nonmonetization of the production sphere – or the use of enterprise deposits rather than currency – prevents the spillover of excess demand from the production sphere to the consumption goods market, proponents of the chronic excess demand position reject these claims. Winiecki (1985) argues that there is a strong link through the labour supply process as wages are bid up by enterprise managers seeking workers in short supply. Further, producers of investment goods and producers of consumer goods compete for the same limited amount of factor inputs. Given the preferences of planners, investment goods producers are more readily subsidized and, therefore, able to bid away inputs from enterprises producing consumer goods. The level of output of consumer goods is more likely to fall short of planned, or market clearing, levels as a result. Complicating the process is the phenomenon of siphoning introduced by Kornai (1979). Essentially, many of the goods sold on the consumer goods market are also purchased by enterprises (gasoline, spare parts for automobiles, etc.). As a result, households are competing with enterprises for certain consumption goods. Households are at a severe disadvantage, however, since they do not enjoy the soft budget constraint that enterprises do. When all is said and done, it is impossible to prevent the excess demand for intermediate products and labour in the production sphere from spilling over to the household sphere.

Now, if excess demand does spill into the household sphere, either in the aggregate or for individual goods, the third item of the trinity comes into play. This is the belief that planners control prices and, therefore, provide a final barrier to inflation in consumer goods markets. But, again, proponents of the chronic excess aggregate demand model argue simply that planners can control changes in the official price indices but inflation manifests itself in other ways. These ways are generally acknowledged and consist of changes in assortment and quality of goods, the introduction of new goods (that differ little if at all from the old) at higher prices, and secondary market activity, legal and otherwise. These result in hidden inflation, not revealed by changes in the official price indices. In addition, there may be repressed inflation, an excess of liquidity in the household sector. The existence

[7]This doesn't mean, of course, that macroeconometric modellers are bound to these assumptions. Recent attempts are abandoning them and often in earlier works there is discussion of possible extensions and solutions to the problems implicit in the models.

[8]For example. Portes (1981, p. 560) recognizes 'that there is undoubtedly substantial excess demand (planners' tension) *within* (his emphasis) the state productive sector (i.e. in interenterprise transactions) . . .'. But the spillover from the production sector to the consumption sector is not accepted without question.

or nonexistence of open inflation is of little consequence. As long as the enterprise budget constraint is soft, enterprises can finance expenditures in excess of revenues and the later two forms of inflation will reflect changes in aggregate shortage if the official price index does not.[9]

The excess demand in consumer goods markets has a secondary effect on production to the extent that households react to the shortages by reducing their supply of labour, in terms either of quantity or quality. Libura (1979) describes the effects of frustrated consumption expectations on the implementation of plan targets and the effects on output. If there is a significant divergence between planners' preferences and household expectations there may be a reduction in work effort. The reduction in work effort, or reduction in the actual supply of labour in the extreme case, shifts the production possibility frontier inwards. The net result is that neither the planned output mix nor the output mix desired by households can be produced.[10] The disequilibrium worsens and becomes self-perpetuating.

In addition to Winiecki's trinity, Podkaminer (1986b) adds that central planners are ultimately responsible for the disequilibria which arise. Relative price distortions – especially between luxuries and necessities – are of primary importance in the case of consumer goods markets in Poland. He argues that while it is unclear whether price distortions arise from the intentional practice of central planners or from their incompetence, it is none the less central planners who create the price distortions. They do not act to eliminate them but may in fact tolerate them and benefit by them. They benefit by them because they allow planners or party officials to allocate luxuries in short supply outside normal market channels. This behaviour of planners is certainly contrary to that of planners as specfied in most macroeconometric models where planners are seen as a means of balancing the plan and market forces.

The evidence of chronic excess demand at the aggregate level presented by adherents of the model is limited and indirect. Most argue that as a result of the disequilibrium important economic aggregates are not entirly satisfactory measures of economic activity.[11] Contrary to the Portes (1981) and Portes and Winter (1980) contention that the aggregation problem is not qualitatively more difficult in CPEs than in market economies, Kornai (1979, 1980) argues that it is essentially

[9]Nuti in Chapter 5 offers a clarification of the alternative means by which inflation manifests itself in a CPE.

[10]See Winiecki (1985, pp. 37–9) for a more detailed explanation of the impact of shortages of consumer goods on the labour supply and the inward shift of the production possibility frontier. Also see Podkaminer's (1985) discussion of the shortage of consumer goods as the initiator of investment cycles.

[11]It is important to maintain the distinction between microeconomic effects, or shortages on individual markets, and macroeconomic effects, or excess aggregate demand. Bornstein (1976, pp. 51–2) recognizes this distinction in his discussion of repressed inflation in the USSR. Evidence of what he calls limited repressed inflation, a situation in which there are shortages of some goods at the same time as there are surpluses of other goods, is evidenced by widespread shortages, formal rationing schemes, distribution by queues, sales personnel demanding and receiving payments greater than official prices indicate, the existence of black markets, and rising prices on collective farm markets. Many of these are identical to the indicators of chronic shortage discussed by Kornai.

impossible to aggregate in CPEs.[12] As an alternative to traditional aggregates Kornai (1980) discussed queueing and search time as potential microeconomic indicators and Kornai (1982a) proposed a macro-index of shortage intensity.[13] Winiecki (1982) and Jansen (1982) developed an indicator based on the ratio of saving to retail sales at current prices. Winiecki (1985) uses this indicator to argue that for six CPEs – Bulgaria, Czechoslovakia, GDR, Hungary, Poland, and the USSR – during the 1960–80 period there is evidence of growing repressed inflation. He does not argue that the evidence is firm and incontrovertible, only that these partial and indirect indicators highlight interesting regularities.[14]

Winiecki's results are not without critics, however.[15] Peebles (1987) notes that since Winiecki did not include household currency holdings his ratio may overstate the rate of growth of household liquidity. Peebles examines measures of household liquidity, labour productivity, state employment, and open inflation. He determines that the inverse relationship that is found to exist between household liquidity growth and the constancy of official price indices is 'exactly what you would expect if you believe in the existence of sustained excess demand' (Peebles, 1987, p. 88). Peebles, like Winiecki, qualifies his results:

> '[t]he simple statistics presented here could be accused of oversimplification, incompleteness, arbitrariness . . . and inaccuracy . . . [h]owever, . . . they are sufficient to illustrate Winiecki's "interesting regularity"' (Peebles, 1987, p. 90).

In addition to the above, evidence of aggregate excess demand is provided by such works as Holzman's (1960) classic study of the USSR, Schroeder (1975) and numerous earlier studies. None the less evidence of chronic excess aggregate demand remains at the level of interesting regularities and may be labelled unsystematic. Empirical difficulties arise because shortage is primarily a microeconomic or market by market phenomenon. If there are markets only in a state of shortage, or only in a state of surplus, with others in equilibrium then it may be possible to aggregate.[16] The Barro–Grossman model requires all excess demands to be semi-positive. Sneessens (1981, 1983) also requires market by market

[12]Portes and Winter (1980) argue that Kornai's measures of microlevel shortages should 'correspond' to macroeconomic diseqiuilibrium. But if the microeconomic indicators imply the existence of shortage the aggregation to the macrolevel becomes problematic and the measures of macro-disequilibrium are questionable.

[13]See Kapitany, Kornai, and Szabó (1984) and Chapter 15 for an application of these methods.

[14]Winiecki presents this evidence as part of his critique of Portes' approach. For further direct methodological criticisms see those sections of Winiecki (1985) entitled Resurrected beliefs, Misapplied theory, and Misunderstood data. For a more lively criticism of the macroeconometric approach see Podkaminer (1986a).

[15]Other aspects of Winiecki's argument – extreme inelastic demand, for example – may also be criticized as in Portes' Chapter 2. But these aspects of his (and Kornai's) general arguments are simply matters of degree and their conclusions are not inextricably linked to them.

[16]It is possible if the disequilibrium is temporary and the economy is moving towards a state of equilibrium – not unlikely for a market economy, less likely for a CPE. If this condition is not met then very important assumptions about the rationing scheme and elasticities of substitution must be made.

efficiency – only one side of the market is rationed.[17] If there are goods in excess demand and other goods in excess supply, spillover effects are directly linked to the degree of substitutability between commodities for individual agents. And, if the spillover effects vary across agents, the rationing scheme utilized to allocate the shortage goods among individuals and across markets must also be specified.[18] These relationships should be explicitly specified in macroeconomic models, but in CPE models to date, they have not.

Rather than specifying explicit rationing schemes and elasticities of substitution, CPE macroeconometric models are typically models which estimate effective supply and demand (those which are functions of both prices and quantities) rather than notional supply and demand (those which are functions of prices only). A temporary equilibrium with rationing or a nonWalrasian equilibrium is an equilibrium in effective demand and suggests involuntary underconsumption in the CPE case, analogous to the involuntary unemployment found in the market-type economy. When individuals face constraints on the consumer goods market, their excess demands persist because they cannot be translated into an effective supply of goods.[19] Assuming that CPE macroeconometric models are properly specified and that there are no data problems, findings of effective equilibrium are consistent with involuntary underconsumption in the aggregate. Problems of incorrect relative prices and the simultaneous existence of excess demand for some goods and excess supply of other goods is another matter.

Kornai (1982b) emphasizes that the existence of a shortage of a particular factor input which creates a constraint on production necessarily implies the existence of a surplus of another factor. Similar conditions exist in the market for consumption goods. The existence of simultaneous shortage and slack in both consumer goods markets and factor input markets is impossible to reconcile when aggregating data without very specific assumptions concerning the elasticities of substitution and the

[17] It is the market by market efficiency which leads to the use of the familiar min condition. Sneessens (1983) reminds us that not all the underlying assumptions of this type of analysis are well suited for macroeconomic analysis and Sneessens (1981) shows that these assumptions lead to ill-defined trade offer curves.

[18] See Gourieroux and Laroque (1985) for an example. They explicitly formalize the degree of substitutability between commodities and the spillover effects across markets. They argue that the aggregate traded quantity is not the minimum of the aggregate demand and supply. To my knowledge such a model has not been applied to CPEs.

[19] The CPE analogue of the Dreze market economy model depends crucially on the rationing scheme. How shortage goods are allocated is critical in determining the existence of an effective equilibrium and failure to consider actual methods of rationing is a drawback to most of the macroeconometric models. Also in the Drezeian analogue agents express demands satisfying perceived constraints and this does not generate an exchange of information concerning the magnitude of rationing agents are experiencing. In the Benassy-type model, which most CPE models follow, the rationing scheme must satisfy certain conditions: actual trades balance, trade is voluntary, and the short side of the market realizes its transactions. The specification of the actual and perceived rationing mechanism is critical because individuals can express effective demands in excess of their constraints. Problems of manipulable rationing schemes and the ability of individuals to break their constraints in market economy models are discussed by Drazen (1980), section 2. This is specially important in the CPE production sphere, and generally has not been addressed in the macroeconometric literature.

rationing scheme.[20] One must wonder, then, what CPE data on aggregate consumption, for example, really mean if there is persistent disequilibrium in individual markets.[21] Although long time series of macroeconomic aggregates are available for most CPEs their meaning must be questioned. It simply means that planners have utilized *a* price vector to aggregate. Aggregate consumption expenditures may absorb income net of savings and desired money balances but if there is not market by market efficiency nothing can be said about the existence of market shortages, or shortage in the aggregate from this information. The discovery of well-behaved econometric relationships among the aggregates is of no significance whatsoever. Gorman (1968) notes when aggregating outside equilibrium that there are many equally acceptable versions of the aggregate relationships. Thus, many traditional macroeconomic aggregates lose their meaning and we must seek evidence of chronic shortage or persistent excess demand elsewhere.

An additional problem in interpreting officially reported macroeconomic aggregates arises if second economy or black market activity is considered. Since official data on consumption, for example, generally do not include black-market activity there may be a significant measurement bias depending on the degree of black market activity in the country in question. Without explicitly modelling black-market activity it may be that equilibrium or excess supply is all that may be observed with official data if black-market activity absorbs all excess demands. Nuti points this out in the next chapter and Charemza addresses this issue in section 11.4 of Chapter 11.[22]

4.3 MICROECONOMIC AND SECTORAL MODELS SUPPORTING THE CHRONIC EXCESS DEMAND HYPOTHESIS

Above, the meaningfulness of traditional macroeconomic aggregates was questioned and it was argued that alternative types of evidence for or against the chronic excess demand hypothesis must be presented. While the evidence of aggregate excess demand presented above tends to confirm interesting regularities in the macroeconomic behaviour it was argued that chronic excess demand is primarily a microeconomic problem. Is there evidence of chronic excess demand on individual markets or in particular sectors in CPEs? Yes, abundant evidence. Kemme (1987) provides a brief survey of several studies which confirm the existence of chronic shortage. Here we also briefly consider some of this evidence. Similar studies are

[20] The recent work on market-type economies of Sneessens (1981, 1983, 1985) and Gourieroux and Laroque (1985) emphasizes this. Strict assumptions about the elasticities of substitution are implicit in Burkett's (1985) CPE model.

[21] One condition for aggregating outside equilibrium is that the economy be approaching (instantaneously) a competitive equilibrium. This is not the case if shortages in individual markets are chronic. As a result Walras' law is not applicable to the CPE in this situation, not even to compute gross shortages (ignoring excess supplies), much less including excess supplies.

[22] Charemza, though, is considering aggregate consumption.

found in Chapters 15 (by Kapitany, Kornai and Szabo), 11 (by Charemza) and 17 (by Davis).

As an alternative to macroeconomic aggregates, Kornai (1980), *inter alia*, has suggested measurements of search time, queueing, etc., as indicators of shortage.[23] There is ample evidence in the form of press accounts and personal observation, most often with reference to the consumption goods market. It is difficult to evaluate these measures if they remain in a purely qualitative form, however. This does not mean this evidence is not valid, certainly no less so than evidence from macroeconometric models based on questionable aggregates, only that it is difficult to interpret and make comparisons across countires and over time. Kapitany, Kornai and Szabo (1984, and updated in Chapter 15) provide quantitative measures of the lengths of queues and queueing times, and survey measures of forced substitution for the automobile market in Hungary. They find, unequivocally, evidence of persistent chronic shortage in this market.

In addition to the automobile market, another market in Hungary for which there is indirect evidence of shortage is the housing market. Daniel (1985, p. 391), while not interested in measuring shortage *per se*, in an examination of the allocation of housing in Hungary reports that there 'is a chronic urban housing shortage'. For state-owned housing[24] rents are centrally administered at a noneffective very low level, generating queues with an average waiting time of eight to ten years in the 1970s. The market for private dwellings absorbs some of the excess demand for housing and the imputed monthly rent on a privately owned dwelling may be eight to ten times higher than the rent of a state owned dwelling of comparable quality.[25] If the market were frictionless it would clear at these higher prices but this market is poorly organized and information is costly.

Thus, there is evidence of chronic shortage for two major budget items, automobiles and housing, throughout the 1970s and later for Hungary. Similar evidence exists for Poland, the GDR, Czechoslovakia and the USSR.[26] Charemza, Gronicki and Quandt (1988) and Chapter 16, focus on the automobile market in Poland. While their methodology is quite different from Kapitany, Kornai and Szabo they also find chronic but varying excess demand for automobiles from the mid-1970s to late 1982.

Rather than focusing on a single market Podkaminer examines several consumption goods categories, including housing and food, in Poland. Podkaminer (1982) finds that relative prices for consumer goods are so seriously distorted during the 1965—78 period that observed shortages in food are attributable to demand spillovers from other categories of consumption, primarily housing. Forced

[23]See Kornai (1980, pp. 464–6) for additional indicators which may be utilized. Kornai has also considered a macro-indicator based on many of these.

[24]State-owned housing accounted for 25.5% of the housing stock in 1980 for the entire country and nearly 60% for Budapest, (Daniel, 1985, p. 392).

[25]See Daniel (1985, p. 392) for further details.

[26]I discuss the GDR and Czechoslovakia shortly and Davis discusses medical services in the Soviet Union in Chapter 17.

substitution was taking place and shortages of food appeared even though the price of food was estimated to be at least 15% too high. Podkaminer argues that aggregate equilibrium could have been obtained even with lower food prices if prices or supplies of housing and other services were increased.

Updating the analysis to consider the 1965–86 period, Podkaminer (1986b) examines relative price distortions using three alternative methodologies: transplanted extended linear expenditure demand system estimates, intercountry consumption comparisons, and information on black market activity. His conclusions remain unchanged – shortages of food are the result of inter-market demand spillovers. He extends the results by computing an index of aggregate consumption based on estimated equilibrium notional demands, and a price index based on the notional equilibrium prices. His results imply the existence of aggregate excess demand for household consumption goods and repressed inflation which was rather low for 1965, increasing to a moderate level by 1970, and rather stable from 1975 to 1985, peaking in 1981.

The Economic Commission for Europe has also examined cross-country consumption patterns for several countries. Askanas and Laski (1985) report the results of a comparison between Poland and Austria.[27] For both countries they calculate price ratios and consumption ratios (both Paasche and Laspeyres indices) for nine individual groups of goods and services and overall private consumption. Their results indicate that the implicit average annual rate of inflation in Poland during the 1964–78 period is significantly higher than officially reported: 5.2% versus 2.9%. As a result the rate of growth of real per capita private consumption in Poland was overestimated by official sources: 5.4% versus 3.2% found by them.[28] They conclude that

> official data grossly understate price inflation and overstate real growth of consumption in Poland. The causes of this discrepancy are mainly seen in the systemic difficulty of properly measuring price changes in Poland's 'shortage economy' (Askanas and Laski, 1985, p. 164).

These difficulties include the usual problems of deteriorating quality of goods with fixed prices, exaggerated price increases for improvements in quality, and, importantly, the accessibility of the representative goods and forced substitution on a seller's market. Again, systemic factors are critical.

While there have been numerous studies of Poland, all indicating chronic shortage, these problems have only recently been examined in the GDR and only indirectly. With a methodology similar to Podkaminer, Collier (1986) estimates the level of disequilibrium for households in the GDR when they face both budget and quantity constraints. He defines the gap between effective purchasing power and notional purchasing power as the amount that the total family budget could be

[27] Also see Economic Commission for Europe (1977) for further details.

[28] The possibility that Austrian official data overestimate the rate of inflation and overestimate the rate of growth of consumption in Austria is discounted since similar comparisons between Western market-type economies yield much smaller discrepancies between official and implicit rates of growth.

reduced without lowering its utility given a movement from the budget-and-quantity constrained environment to the budget-constrained environment relative to the original total expenditures. The typical GDR family in 1977 would have been willing to give up 13% of its total expenditure in order to attain its notional demand at official prices. Or, the average family would have been willing to pay 157 of its 1244 Mark total monthly expenditure – roughly what it spent on electric durable goods, motor vehicles and parts, gasoline and housing together – just to avoid quantity constraints (Collier, 1986, p. 29). This implies there was significant forced substitution taking place at this time in the GDR.

The possibility of chronic shortage in the market for producer goods has also been examined, but to a lesser extent than consumer goods markets. Major (1982) focuses on the transport sector of the Hungarian economy. Even though the transport sector in Hungary is not as well developed as those in West European countries Major argues that production and consumption in Hungary are transport intensive. There exist plan tensions – an underplanning of transport services – in the medium term plan which flow downward to become market tensions, i.e. excess demand. Throughout the reform period in Hungary plan directives were not replaced by a real market for transport services and prices did not clear the market. Both planners and enterprises regarded transport constraints as soft and higher transport tariffs faced by enterprises were passed on to their customers and ultimately to the central budget with little difficulty.[29]

A final piece of evidence of excess demand is official government statistics collected for various markets with the intent of ascertaining the market situation. Pindak (1983) provides an overview of inflation in CPEs and evidence of market difficulties in Czechoslovakia. He reports official data for various markets (both for the consumption goods market and the producer goods market) indicating the share of goods being provided without supply difficulties. During the 1972–78 period from one-third to two-thirds of all foodstuffs and from one-half to over three-quarters of industrial products experienced supply difficulties.[30]

The microeconomic studies cited above and those that follow in this book are limited in number and in scope. The evidence provided is clear, however. All imply excess demand for the market or groups of commodities examined.[31] The chronic excess demand hypothesis has been confirmed for several commodities, countries and time periods. Persistent and chronic shortages at the microlevel provide prima facie evidence that elasticities of substitution are such that aggregation is not appropriate. Aggregate demand, aggregate supply and aggregate consumption, therefore, are macroeconomic categories which may be constructed on the basis

[29] See Major (1982, p. 99) for details.

[30] Pindak (1983, p. 117). The methodology used in the original study was not discussed by Pindak but the trend in the data and the fact that such a large portion of both markets experienced some difficulty are notable.

[31] There are numerous macroeconomic studies, including Charemza and Gronicki (1988), Nuti (1985), Brada and King (1986), and several chapters below which also reveal excess demand for consumer goods. I do not include those works here since I am concerned with microeconomic studies or those dealing with more disaggregated data.

of prices which are not market clearing and do not represent scarcity values. They are not derived in the context of a notional equilibrium. It is not clear that they are even based on an effective equilibrium because the constraints generating the effective supply and demand functions are not temporary but changing over time, and not necessarily in a fashion which moves the economy towards an effective (or notional) equilibrium. Therefore, even if macroeconometric regularities exist, their meaning is difficult to evaluate, given the evidence of severely distorted relative prices in some cases and chronic shortage of major budget items such as housing and automobiles in others.

4.4 IMPLICATIONS AND CONCLUSIONS

While neither the chronic shortage model nor macroeconometric analysis are novelties much is yet to be done to reconcile the two approaches. It was argued above that chronic shortage is a microeconomic phenomenon generated primarily by systemic features; the planning and managerial system that leads to soft budget constraints, taut plans, and rigid and incorrect relative prices. There are numerous indicators of shortage and the bulk of empirical evidence suggests that shortages are persistent with varying degrees of severity, at least during certain long time periods for those countries studied.

Associated with shortage and forced substitution is a welfare loss arising in each market. If aggregate consumption is constrained, a similar result – welfare losses – can be derived for the aggregate economy in the Barro–Grossman context.[32] While the theoretical result can be derived the empirical estimation of aggregate welfare losses is elusive due to aggregation at false prices. Alternative means of measuring welfare losses for the economy as a whole must be devised using data as disaggregated as possible, or synthetic indicators, as Kornai suggests, must be constructed.

All this indicates the need for more study, primarily at the microeconomic level. Kornai argues that as a result of chronic shortage there is a change in the behavioural relationships among economic agents in the economy which deserves careful study. The new behavioural relationships among economic agents must be explored further. The existence of secondary markets, legal or otherwise, has only recently been taken into account. Empirical estimates of the ability of secondary market activity to absorb excess demand in the state sector must be provided in a more comprehensive fashion. Is this activity prevalent in all CPEs, or do secondary markets arise only in certain economies at certain times, or is it a permanent and

[32]The possibility of aggregating out of equilibrium also exists for the market-type economy but it is argued that in the market economy there are forces which generally tend to move the economy towards an equilibrium. Disequilibrium is less severe, less persistent and therefore equilibrium and disequilibrium analysis (as in Sneessens, 1983 and Gourieroux and Laroque, 1985) is less problematic. Nuti, in the next chapter, expresses a similar view when comparing the intensity of disequilibrium in market and planned economies.

essential feature of the CPE? Theoretical and empirical research in this area is so limited that it is impossible at this time to reach firm conclusions concerning the significance of secondary markets as an equilibrating mechanism.

While little is said about the potential welfare losses even less is said about the opposite side of the coin: are there benefits to chronic shortage that are not being considered? For example, has economic growth really been greater than otherwise possible and are consumers actually better off despite the existence of shortages of particular goods? Much more development of the microeconomic foundations of macroeconomics in CPEs must be undertaken.

Given the limited power of macroeconometric analysis in the CPE disequilibrium setting, much more research must be done on the problem of aggregation outside equilibrium in CPEs.[33] I believe lack of attention to this problem represents the death-knell of the current line of disequilibrium macroeconometric research. However, believing in the advancement of science and life after death I look forward to its resurrection in later chapters of this book.

ACKNOWLEDGEMENTS

This chapter is a companion to the paper 'The chronic shortage model' presented at the Conference on Modelling of Disequilibrium and Shortage in Centrally Planned Economies, Birmingham, England, April 6–8, 1987. I would like to thank Wojciech Charemza, Christopher Davis and Karoly A. Soós for their comments. I would also like to thank the organizers of the conference and the many organizations providing financial support for this research project.

REFERENCES AND FURTHER READING

Askanas, B. and Laski, K. (1985) Consumer prices and private consumption in Poland and Austria. *Journal of Comparative Economics*, **9** (2), 164–77.

Bajt, A. (1971) Investment cycles in European socialist economies: a review article. *Journal of Economic Literature*, **9** (1).

Balicki, W. (1983) Theory of disequilibrium in centrally planned economies. *Jahrbuch der Wirtschaft Osteuropas*, **10** (2).

Barro, R. J. and Grossman, H. I. (1976) *Money, Employment and Inflation*, Cambridge University Press, Cambridge.

Bauer, T. (1978) Investment cycles in planned economies. *Acta Oeconomica*, **21** (3).

Beksiak, J. (1986) Inflacja w gospodarce socjalistycznej, *Economista*, pp. 91–107.

Beksiak, J. and Libura, U. (1974) *Rownowaga Gospodarcza w Socjalizmie*, PWN, Warsaw.

Bliss, C. J. (1975) *Capital Theory and the Distribution of Income*, North Holland, Amsterdam.

Boot, P. (1984) Industrial cycles in the German Democratic Republic and Professor Wiles' thesis. *Association for Comparative Economic Studies Bulletin*, **26**, 1–26.

[33]In particular, explicit specification of rationing schemes, empirical estimation of elasticities of substitution among commodities or commodity groups, and modelling and testing of spillover effects.

Bornstein, M. (1976) Soviet price policy in the 1970s. *Soviet Economy in a New Perspective*, US GPO, Washington, DC, pp. 17–66.

Brada, J. and King, A. (1986) Macroeconomic Disequilibrium and Soviet Agricultural Imports (*mimeo*).

Brada, J. and King, A. (1986) Taut plans, repressed inflation and the supply of effort in centrally planned economies. *Economics of Planning* (to be published).

Burkett, J. P. (1985) Slack and shortage in socialist economies, Paper prepared for the 1985 American Economics Association meetings.

Bush, K. (1973) Soviet inflation, in *Banking, Money, and Credit in the USSR*, (ed. Y. Laulin) NATO, Brussels, pp. 97–106.

Campbell, R. (1970) Macroeconomic models and central price setting in the soviet economy, in *Essays in Economic Analysis and Policy* (F. Gehrels *et al.*), Indiana University Press, Bloomington, pp. 253–70.

Champernowne, D. G. (1953–54) The production function and the theory of capital: comment. *Review of Economic Studies*, **21**, 112–35.

Charemza, W. and Dlouhý, V. (1986) Two alternative disequilibrium models for a planned economy, paper presented at the IIASA meetings, revised December 1986.

Charemza, W. and Gronicki, M. (1988), *Plans and Disequilibrium in Centrally Planned Economies*, North Holland, Amsterdam.

Charemza, W. Gronicki, M. and Quandt, R. E. (1988) Modelling parallel markets in centrally planned economies: the case of the automobile market in Poland. *European Economic Review* (to be published).

Charemza, W. and Quandt, R. E. (1982) Models and estimation of disequilibrium for centrally planned economies. *Review of Economic Studies*, **49**, 109–16.

Collier, I. L. Jr (1986) Effective purchasing power in a quantity constrained economy: an estimate for the German Democratic Republic. *Review of Economics and Statistics*, **68** (1), 24–32.

Daniel, Z. (1985) The effect of housing allocation on social inequality in Hungary. *Journal of Comparative Economics*, **9** (4), 391–409.

Dirksen, E. (1981) The control of inflation? Errors in the interpretation of CPE data, *Economica*, **48**, 305–8.

Divisia, F. (1925) L'indice monetaire et la theorie de la monnaie. *Revue d'Economie Politique*, **39** (4, 5, 6), 842–61, 980–1008, 1121–51.

Drazen, A. (1980) Recent developments in macroeconomic disequilibrium theory. *Econometrica*, **48** (2), 283–306.

Economic Commission for Europe (1977) *Comparison of Consumer Prices and Levels of Consumption between Austria and Poland, 1973*. ECE, Vienna-Warsaw.

Gacs, J. and Lackó, M. (1973) A study of planning behavior on the national economy level *Economics of Planning*, **13**, 1–2.

Gardner, R. and Strauss, J. (1981) Repressed inflation in the Soviet Union: a temporary equilibrium approach. *European Economic Review*, **16**, 387–404.

Goldman, J. (1969) Fluctuation in the growth rate in a Socialist economy and the inventory cycle, in *Is the Business Cycle Obsolete*? (ed. M. Bronfenbrenner), John Wiley, New York.

Gomulka, S. (1985) Kornai's soft budget constraint and the shortage phenomenon: a criticism and restatement. *Economics of Planning*, **19** (1).

Gorman, W. M. (1968) Measuring the quantity of fixed factors, in *Value, Capital and Growth* (ed. J. N. Wolfe), Edinburgh University Press, Edinburgh.

Grosfeld, I. (1986) Endogenous planners and the investment cycle in the centrally planned economies, *Comparative Economic Studies*, **28** (1), 42–53.

Gourieroux, C. and Laroque, G. (1985) The aggregation of commodities in quantity rationing models. *International Economic Review*, **26** (3), 681–99.

Grossman, G. (1966) Gold and the sword: money in the Soviet command economy, in *Industralization in Two Systems* (ed. H. Rosovsky), John Wiley, New York, pp. 204–36.

Hare, P. G. (1982) Economics of shortage and non-price control. *Journal of Comparative Economics*, **6**, 400–25.

Holzman, F. D. (1960) Soviet inflationary pressures, 1928–1957: causes and cures. *Quarterly Journal of Economics*, **74** (2), 167–87.

Hunter, H. (1961) Optimum tautness in developmental planning. *Economic Development and Cultural Change*, **9** (4).

Ickes, B. W. (1984) The Constrained Production Function and Economic Activity: Foundations of the Shortage Multiplier (*mimeo*).

Ickes, B. W. (1985) Cyclical Fluctuations in Centrally Planned Economies: Some Cross Country Comparisons (*mimeo*).

Ickes, B. W. (1986) Cyclical fluctuations in centrally planned economies: a critique of the literature, *Soviet Studies*, **38** (1) 36–52.

Ickes, B. W. (1986) A Macroeconomic Model for Centrally Planned Economies (*mimeo*).

Ickes, B. W. (1987) On the economics of taut plans. *Journal of Comparative Economics*, **10** (4), 388–99.

Jansen, P. (1982) *Das Inflations problem in der Zontralvrvaltungswirtschaft*, G. Fisher, Stuttgart.

Jutta-Pietsch, A. (1982) Shortage as a fundamental problem of centrally planned economies and the Hungarian reform: interview with Prof. Janos Korani, *Economics of Planning*, **18** (3) 103–13.

Kapitany, Z., Kornai, J. and Szabó, J. (1984) Reproduction of shortage in the Hungarian Car Market, *Soviet Studies*, **36**, 236–56.

Kemme, D. M. (1980) A disequilibrium macroeconomic model of the modified centrally planned economy: Poland, unpublished PhD dissertation, University Microfilms International, Ann Arbor.

Kemme, D. M. (1982) Microeconomic efficiency in the Polish enterprise during the NEM period. *Economics of Planning*, **18** (1) 11–21.

Kemme, D. (1987) The chronic shortage model, Paper presented at the Conference on Modelling of Disequilibrium and Shortage in Centrally Planned Economies, Birmingham, England, April 6–8, 1987.

Kemme, D. M. and Winiecki, J. (1989) Disequilibrium in centrally planned economies. *Jahrbuch der Wirtschaft Osteuropas*, **12** (2) 65–96.

Keren, M. (1972) On the tautness of plans. *Review of Economic Studies*, **39** (4), 469–86. *Society*, MIT Press, Cambridge, MA.

Kornai, J. (1972) *Rush Versus Harmonic Growth*, North Holland, Amsterdam.

Kornai, J. (1979) Resource-constrained versus demand-constrained systems. *Econometrica*, **47**, 801–19.

Kornai, J. (1980) *Economics of Shortage* (Volumes A and B), North Holland, Amsterdam.

Kornai, J. (1982a) *Growth, Shortage and Efficiency*, Basil Blackwell, Oxford.

Kornai, J. (1982b) Hard and Soft Budget Constraint. *Acta Oeconomica*, **25**, 231–45.

Kornai, J. (1985) Gomulka on the soft budget constraint: a reply. *Economics of Planning*, **19** (2).

Kornai, J. (1986a) The soft budget constraint. *Kyklos*, **39** (1).

Kornai, J. (1986b) *Contradictions and Dilemmas: Studies on the Socialist Economy and*

Kornai, J. and Weilbull, J. W. (1978) The normal state of the market in a shortage economy: a queue model. *Scandanavian Journal of Economics*, **80**, 375–98.

Lacko, M. (1980) Cumulating and easing of tensions: a simple model of the cyclical development of investments in Hungary *Acta Oeconomica*, **24**, 2–4.

Leontief, W. (1947) Introduction to a theory of the internal structure of functional relationships. *Econometrica*, **15**, 361–73.

Levine, H. S. (1966) Pressure and planning in the Soviet economy, in *Industrialization in Two Systems: Essays in Honor of Alexander Gerschenkron* (ed. Henry Rostowski), John Wiley, New York, 266–85.

Libura, U. (1979) Konsumpcja a wykorzystanie mozliwosci produkcyjnych gospodarce Socjalistycznej, PWN, Warsaw.

Major, I. (1982) The reproduction of tensions in transport in Hungary. *Economics of Planning*, **18** (2), 85–100.

Marrese, M. and Mitchell, J. L. (1984) Kornai's resource-constrained economy. A survey and appraisal. *Journal of Comparative Economics*, **8**, 74–84.

Martin, C. (1986) General linear rationing models, Discussion Paper No. 184, Birkbeck College, London.

Nataf, A (1948) Sur la possibilité de la construction de certains macromodels. *Econometrica*, **16**, 232–44.

Nove, A. (1979) Inflation in Communist countries, in *Perspectives on Inflation* (ed. David Heathfield), Longman Press, London, pp. 217–30.

Nove, A. (1985) A note on errors and their causes. *Soviet Studies*, **37**, (2), 276–80.

Nuti, D. M. (1985) Hidden and repressed inflation in Soviet-type economies: definitions, measurement and stabilization, Paper presented at the 3rd World Congress for Soviet and East European Studies, Washington DC, 30 October – 4 November 1985.

Peebles, G. (1987) On the importance of establishing the inverse relationship between open inflation and household liquidity growth under socialism: A critique of Jan Winiecki's savings deposit data. *Comparative Economic Studies*, **28**, 85–91.

Pindak, F. (1983) Inflation under central planing. *Jahrbuch der Wirtschaft Osteuropas*, **10**(2).

Podkaminer, L. (1982) Estimates of the Disequilibria in Poland's consumer markets, 1965–1978. *Review of Economics and Statistics*, **64**, 423–31.

Podkaminer, L. (1983) Estimates of disequilibria in Poland's consumer markets, in *Modelling Growing Economies in Equilibrium and Disequilibrium* (eds A. Kelley *et al.), Duke University Press, Durham, NC.*

Podkaminer, L. (1985) Investment cycles in centrally planned economies: an explanation invoking consumer market disequilibrium and labour shortage. *Acta Oeonomica*, **35** (1–2), 133–44.

Podkaminer, L. (1986a) Measurement of macroeconomic disequilibrium in centrally planned economies: a critical review of Professor Portes' results, Paper presented at the Conference on Applied General Equilibrium Models, Laxenburg.

Podkaminer, L. (1986b) Pesistent disequilibrium in Poland's consumer markets: some hypothetical explanations, *Comparative Economic Studies*, **28**, (3), 1–16.

Podkaminer, L. Finke, R. and Theil, H. (1984) Cross-country demand functions and centrally planned economies. *Economic Letters*, **16**, 269–71.

Portes, R. (1974) Macroeconomic equilibrium under central planning, Seminar paper no. 40, Institute for International Economic Studies, University of Stockholm.

Portes, R. (1977) The control of inflation: lessons from East Europe experience. *Economica*, **44**, 109–30.

Portes, R. (1979a) Internal and external balance in a centrally planned economy. *Journal of Comparative Economics*, **3**, 325–46.

Portes, R. (1979b) Review article: *Sovmod: A Macroeconometric Model of the Soviet Union* by Donald W. Green and Christopher Higgins. *The Economic Journal*, **89**, 668–72.

Portes, R. (1981) Macroeconomic equilibrium and disequilibrium in centrally planned economies, *Economic Inquiry*, **19**, 559–78.

Portes, R. (1983) Central planning and monetarism: fellow travellers? in *Marxism, Central Planning and the Soviet Economy* (ed. P. Desai), MIT Press, Cambridge, MA, pp. 149–65.

Portes, R. (1986) The theory and measurement of macroeconomic disequilibrium in centrally planned economies, Discussion Paper No. 91, Centre for Economic Policy Research. London.

Portes, R. and Winter, D. (1980) Disequilibrium estimates for consumption goods markets in centrally planned economics. *Review of Economic Studies*, **47**, 137–59.

Portes, R., Quandt, R., Winter, D. and Yeo, S. (1983) Macroeconomic planning and disequilibrium: estimates for Poland, 1955–80. *Econometrica*, **55** (1), 19–41.

Portes, R., Quandt, R., Winter, D., and Yeo, S. (1984) Planning the consumption goods market: preliminary disequilibrium estimates for Poland 1955–1980, in *Contemporary Macroeconomic Modelling* (ed. P.Malgrange and P.-A. Muet) Basil Blackwell, Oxford, pp. 254–71.

Portes, R. and Santorum, A. (1986) Money and the consumption goods market in China, Discussion Paper No. 193, Birkbeck College, London.

Robinson, J. V. (1953–1954) The production function and the theory of capital. *Review of Economic Studies*, **21**, 81–106.

Samuelson, P. (1960) An extension of Le Chatelier's principle. *Econometrica*, **28**, 368–79.

Schroeder, G. (1975) Consumer goods availability and repressed inflation in the Soviet Union, in *Economic Aspects of Life in the USSR*, NATO, Brussels pp. 30–47.

Sneessens, H. R. (1981) *Theory and Estimation of Macroeconomic Rationing Models*, Springer-Verlag, Berlin.

Sneessens, H. R., (1983) A macroeconomic rationing model of the Belgian economy. *European Economic Review*, **20**, 193–215.

Sneessens, H. R. (1985) Two alternative stochastic specification and estimation methods for quantity rationing model – A Monte Carlo Comparison. *European Economic Review*, **29**, 111–36.

Soos, K. A. (1975) Causes of investment fluctuations in the Hungarian economy. *Eastern European Economies*, **14** (2).

Stupnicki, K. (1985) The disequilibrium theory: R. Portes vs. J. Kornai – critical comparison, Central School of Planning and Statistics, Warsaw, Poland.

Welfe, A. (1986) Intensity of disequilibrium and changes in inventories, Collaborative Papers, International Institute for Applied Systems Analysis, Laxenburg, Austria.

Welfe, A. (1985) The system of demand equations in a state of disequilibrium, Collaborative Papers, International Institute for Applied Systems Analysis, Laxenburg, Austria.

Wiles, P. (1982) Are there any communist economic cycles? *Association for Comparative Economic Studies Bulletin*, **24**, 1–20.

Wiles, P. J. D. and Rostowski,J. A. (1979) Soviet-type inflation: an open debate, *Est–Oust*, **10** (2–3), 59–75.

Winiecki, J. (1982) Investment cycles and an excess demand inflation in planned economies: sources and processes. *Acta Oeconomica*, **28**, 147–60.

Winiecki, J. (1985) Portes ante Portes: a critique of the revisionist interpretation of inflation under central planning. *Comparative Economic Studies*, **27** (2), 25–52.

Wolf, T. A. (1985) Economic stabilization in planned economies – toward an analytical framework. *International Monetary Fund Staff Papers*, **32**, (1), 78–131.

5

Hidden and repressed inflation in Soviet-type economies: definitions, measurements and stabilization

DOMENICO MARIO NUTI

5.1 INTRODUCTION

Official price trends in the USSR and in the Soviet-type economies have changed markedly through time but have followed a roughly uniform general pattern: hyperinflation at times of war, systemic transition and reconstruction; inflation at times of accelerated industrialization; stabilization through currency reform and fiscal measures, followed by modest deflation and a record of substantial price stability, recently broken by the spreading – with fewer and fewer exceptions – of renewed open inflation. This general pattern is due to fairly uniform trends in both policy stances and objective conditions: in theory central planning of both physical and financial flows should enable Soviet-type economies to achieve price stability; in practice the persistence of downward rigidity of money wages, the ambitious growth and accumulation targets, as well as adverse exogenous and systemic factors, have frequently necessitated planned or unplanned price increases.

What is particularly striking, however, is not the frequent departure of Soviet-type economies from price stability, since by and large their price record has been comparatively creditable and often even impressive, but the persistence of unsatisfied demand at the official prices (especially but not only for consumer goods) regardless of whether these prices were being kept stable, or falling, or rising at slow, fast or very fast rates.

Market clearing at official prices has been a rare and short-lived phenomenon in the economic life of socialist countries, where shortage of both production and

International Studies in Economic Modelling Series.
Models of Disequilibrium and Shortage in Centrally Planned Economies.
Edited by C. Davis and W. Charemza.
Published in 1989 by Chapman and Hall Ltd, 11 New Fetter Lane, London EC4P 4EE.

consumption goods has been the norm (Kornai, 1980), while access to goods and services at the official prices has been the privilege of priority sectors and élite groups. Shortages have taken one or more of the following forms: long searches; queues and waiting lists (including sometimes queues to join waiting lists and waiting lists to join queues); transactions at prices higher than official norms even in state shops (sometimes disguised by nominal changes in product or labels) but more often in alternative black, grey and multicoloured markets; forced substitution of available for unavailable goods; inventories below norm held by distributors and above norm by consumers; extensive holdings and use of foreign exchange; excess liquidity; rationing; barter (which through effective demonetisation suggests the existence of hyperinflationary excess demand). The difference between countries and between periods has been one of degree, ranging from market near-clearing at stable and even falling prices in the early 1950s almost throughout the area after the tough measures of post-war stabilization, to the simultaneous and dramatic occurrence of shortages in all the guises listed above together with high official open inflation in post-1980 Poland.

These manifestations of shortage, i.e. excess demand at official prices, are often discussed in Western literature under the heading of 'hidden' and 'repressed' inflation. In Eastern Europe they are rarely discussed, mostly in Hungarian literature as 'shortage' (Kornai, 1980, 1982) and in Polish literature (e.g. Herer and Sadowski, 1981), including official documents, as 'inflationary gap' (*luka*) between the value of available goods and intended purchases out of income, and inflationary 'overhang' (*nawis*, or the cumulated gaps over time). This terminology requires two important qualifications. First, current usage in the literature has been inconsistent but it seems useful and appropriate to define in general 'unofficial inflation' or more precisely 'inflation other than open official', as including both 'hidden' open inflation, i.e. higher price indices than officially recorded and 'repressed inflation', i.e. rising excess liquid balances in the hands of the population with respect to what they would wish to hold if markets cleared at official prices (see below, section 5.4 on definitions). Second, concepts of hidden inflation require the presence not just of shortage but of increasing shortage over time (i.e. increasing excess demand); if excess demand was constant a once-for-all increase in official price would be sufficient to eliminate it, and we would not call 'inflation' a once-and-for-all price rise (Portes, 1977; though there is inconsistency in usage, even by the same author in the same year, see Portes–Winter, 1977, p. 352, who define repressed inflation as 'the existence of excess demand in the consumption goods market'). Thus endemic and large shortages at official prices are consistent with falling rates of hidden and repressed inflation (if shortages are increasing at a decreasing rate) or even negative rates (if shortages, though large, are falling).

This chapter deals primarily with hidden and repressed inflation in the consumption goods market; it will consider inflationary pressure in general (whether or not it takes the form of open official inflation) with special reference to system-specific sources (section 5.2); discuss briefly the Soviet and Eastern

European record of open official inflation (section 5.3); propose a set of definitions for alternative forms of unofficial inflation (section 5.4); review critically a number of existing estimates of 'hidden' (section 5.5) and 'repressed' inflation (section 5.6); sketch an alternative model for analysing repressed inflation (section 5.7) and draw policy implications relevant for the possible paths towards stabilization (section 5.8), desirable in view of the costs of unofficial inflation (section 5.9). The main propositions of the chapter are summarized in section 5.10.

5.2 INFLATIONARY PRESSURE IN THE SOCIALIST ECONOMY

The Soviet-type socialist economy is a monetarist paradise where the quantity theory of money applies in its strictest traditional form, but where there is no scope for monetary policy because the quantity of money is automatically adjusted by monetary authorities to planned physical flows (given planned prices) and to the degree of their actual implementation (Lavlan, 1973; Garvy, 1977; Zwass, 1979; Portes, 1981a). There are two monetary circuits, separate in principle and to a broad extent also in practice. The first consists of cash and cash-convertible accounts for the payment of incomes to households (wages, salaries, scholarships, pensions, state purchases from the private sector; liquid assets of the population) who use them for their purchases primarily of consumption goods and services (and also of some physical and financial assets) from the state sector, as well as for all transactions within the private sector. The second monetary circuit consists of bank-money which enterprises can use for transactions within the state sector but not for the payment of personal incomes. The separation of the two circuits, and the planners' a priori attempts at balancing incomes and expenditures of the population through wages policy and control (Adam, 1979) is the foundation of the traditional claim that inflation can be eliminated under socialism. The instruments for the a priori balancing of incomes and expenditures, however, are fiscal instruments and direct control of wages and of consumption supply; if institutional or physical constraints in the use of these instruments leave an extant inflationary gap the planners have only one instrument of domestic monetary policy, namely the net sale of financial assets to the population,[1] which is however blunted by the traditional ideological resistance to the payment of attractive enough interest rates for fear of creating a new rentier class. Hence the justification for the neglect of monetary policy (as distinct from the behaviour of some monetary aggregates such as money incomes and cash balances of the population) in the analysis of inflationary pressure in the Soviet-type economy (on Soviet-type banking and credit see also: Podolski, 1972; Lavlan, 1973).

Since the last war, in Soviet and East European countries price stability has been

[1]Net of government debt redemptions and increases in bank lending to the public; private loans however are modest and do not give banks tangible opportunities of coping with inflationary pressure by means of a credit squeeze.

such a major policy commitment that traditionally victory over inflation is regarded as a systemic feature of socialism and one of its claims to superiority over the capitalist system. For instance Kronrod (1960, p. 364) claims that 'inflation has been eliminated in a socialist society'; Dudinsky (1976) lists lack of inflation as one of the features of socialism (together with full employment, steady economic growth, confidence in the future and social justice) which are most attractive to workers in capitalist countries. Glushkov (1983, p. 24.) claims that

> in the developed socialist society . . . prices are planned in accordance to the demands of objective economic laws. Planned production and unplanned prices in the market would be incompatible with one another. Stability of the system of prices is assured by the unified and centralised plan for economic development.

Similar claims abound. There have been times, however, when open inflation has been the planned instrument for expropriating the private sector or financing investment and defence, or the unplanned result of below-plan economic performance (see next section).

As in the case of capitalist countries, the source of inflationary pressure in Soviet-type economies is to be found in cost factors or in the shortfall of supply with respect to demand, except that these factors

1. exercise a direct influence on market imbalance and neither affect nor are affected by monetary policy; and
2. are influenced by the specific systemic features of the Soviet-type economy.

On the supply side, inflationary pressure can arise from rising costs or from quantity shortfalls. Rising costs may be due to the exhaustion of accessible natural resources, wage growth outstripping labour productivity growth, or imported inflation. Quantity shortfalls arise from:

1. the systematic underplanning (except for the present decade) and underfulfilling of the production of consumption goods and services, built into the traditional principles of Soviet planning, of priority to heavy industry (or also to the so-called A-sector producing means of production, as opposed to the B-sector of consumption goods);
2. the failure to recognize the existence of downward trends in both labour and capital productivity growth (including occasional negative growth) which have been recorded in Soviet-type systems since the mid-1960s, and which are associated partly with the rising costs caused by exhaustion of labour and natural resources reserves, partly with the failure to adjust the centralized allocation system to these changed circumstances;
3. slow growth of agricultural production due to system-specific neglect of investment in agriculture and the adoption of organizational forms (large state

and collective farms) which are ill suited to a spatially-diffused and effort-responsive economic activity such as agriculture.

These adverse systemic supply factors are sometimes made worse by random supply shocks such as adverse terms of trade (e.g. those experienced by East European countries over the last decade, with the exception of the oil-exporting USSR), world recession depressing export below plans; natural conditions (drought or flood, heat or frost) affecting especially agriculture, building and transport.

On the demand side, inflationary systemic factors include:

1. Pressure on wage rates (under the guise of job classification, career structure, and other discretionary decisions of enterprises causing wage drift) because of full employment of labour, through high labour turnover and informal plant bargaining.
2. Over-ambitious investment plans exceeding the absorption capacity of the economy (i.e. its ability to finance, install, operate all the new plant and to import necessary inputs, sometimes investment plans at the lower operational levels adding up to more than the overall centrally planned investment target, and also being systematically overfulfilled).
3. The increasing demands of all other items of expenditures (which like investment might crowd out private consumption), such as public consumption and defence.

Cost-push pressures can be and frequently have been absorbed in Soviet-type economies by means of subsidies from the state budget, either boosting the revenue side of enterprise accounts with subsidies on the goods produced or distributed (e.g. foodstuffs) or relieving the expenditure side of enterprise accounts with grants for investments or other costs. A typical instrument of price stability in this respect is the 'price equalization' subsidy (or tax, in the case of deflationary trends; see Wolf, 1980) for imported commodities, absorbing into the state budget the effects of imported inflation; Wiles (1973, 1974) advocated its introduction as an anti-inflationary measure in Western countries. All these subsidies, however, do not eliminate inflationary pressure but simply transform a cost-push into a demand-pull; to avoid inflation the subsidy component of the state budget will raise the government borrowing requirement or crowd out alternative items of expenditure. Thus subsidies can be a convenient instrument of inflationary control when cost-push pressures are of a cyclical nature, but are ill suited to deal with permanent shifts (for instance in terms of trade) and even less with sustained trends (such as rising costs of natural resources, especially in the USSR).

The appearance of a shortfall of supply with respect to demand at constant prices – whether due to demand or supply factors, or the transformation of cost-push into demand-pull through state subsidies; whether planned or unplanned – leaves the planners with a very narrow range of choices: additional external finance, and/or inventory falls below normal levels and/or open inflation, recorded or unrecorded, in state sales or in secondary markets and/or additional cash holdings in the hands of the population over and above what would be held if markets cleared at official prices.

5.3 OPEN OFFICIAL INFLATION IN THE USSR AND EASTERN EUROPE

Open official inflation in the USSR and the Eastern European countries has occurred at hyperinflationary rates at times of war and reconstruction, especially during the transition to the new system, and at lower but still high rates at times of accelerated industrialization. For instance in the USSR the official price index, which had risen from 100 in June 1914 to 630 in November 1917, had risen to 6200 in July 1918, 60 500 in July 1919, 129 000 in July 1920 and 1 290 000 in January 1921 (Pindak, 1983). This was partly due to the destruction of productive capacity and transport, partly the result of a deliberate monetary policy – theorized among others by Preobrazhensky – aimed (especially during war communism 1918–21) at gaining state access to real resources, protecting the workers from inflation (whose full blast was felt by the bourgeoisie and the kulaks) through higher wages which fuelled further inflation, and abolishing money through demonetization of the economy. The New Economic Policy (1921–28) restored a measure of price stability

Table 5.1 Prices and nominal wages in the USSR, 1928–47

	Average percentage annual increments				
	Retail prices in state and co-operative shops			*Wholesale prices of basic industrial goods*	*Average nominal wages*
	Official	*Chapman*	*Holzman*		
Period	*(1)*	*(2)*	*(3)*	*(4)*	*(5)*
1928–47	17.2	16.2–17.4[a]	21.3	5.6	12.8
of which:					
1928–37	20.5	22.5–26.3	24.0	12.2	17.6
1938–40	5.9	8.0– 9.7	13.0	6.5	10.3
1941–47	18.1	11.5–12.3[a]	21.4	0.8	8.0

[a]Chapman's data refer to 1948 instead of 1947; they do not comprise, therefore, the peak price level of 1947.

Sources: Average percentage annual increments were computed from the following data:
Col. (1): Chapman (1963, p. 157) for 1928–40; Malafeyev (1964, p. 258) for 1940–47.
Col. (2): Chapman (1963, pp. 81 and 87) for 'all commodities'; her two sets of indices in col. (2) reflect different weights, 1928 and 1937 respectively.
Col. (3), (4), (5): Holzman (1960, pp. 168 and 169); for 1940 his January retail price index – and not that of July – was employed.
From: Pindak (1983, pp. 5–6).

through conventional fiscal and monetary policies, but a further inflationary wave was brought about by accelerated industrialization and the need to finance it in spite of downward rigidity of money wages, as well as by collectivization of agriculture (through its adverse impact on agricultural output) and full employment of labour (which caused high labour turnover and wage drift). Table 5.1 indicates retail price indices from official sources and Western reconstructions for the period 1928–47, at a yearly rate of the order of 17–21% *vis-à-vis* a parallel increase of 12.8% in average nominal wages (the implied trend in real wages grossly underestimating average standard of living growth due to the parallel growth of wage labour employment during the period at a rate faster than population growth), wholesale prices of basic industrial goods remaining much more stable thanks to state cross subsidies from consumption to production goods, to supplement the revenue of enterprises producing production goods. 1947 was an anomalous year, with peak prices and a monetary reform aimed at confiscating much of the purchasing power of the population through a change of the currency that converted prices, wages, savings and cash at rates diversified so as to favour wages and saving deposits with respect to cash.[2]

The East European countries that joined the socialist bloc after the last war experienced a similar pattern: inflation at a fast and occasionally hyperinflationary rate during the period of transition to the new system and reconstruction and the early stages of Soviet-type ambitious investment policy (for instance, retail prices in Poland increased by 62.4% between 1946 and 1949, and again by 78.5% in 1950–53; in Czechoslovakia by about 64% between 1948 and 1953; the GDR experienced early hyperinflation), followed by monetary stabilization mainly through Soviet-type currency conversion at diversified and progressive rates. In Poland in 1950 all prices, wages and saving deposits were converted at a rate of 3 new to 100 old zlotys, while cash was converted at a 1 to 100 rate, thus confiscating two thirds of the cash in the hands of the population (see Montias, 1962, pp. 69–70). In 1952 in Romania and Bulgaria cash holdings were reduced to one fifth and one fourth respectively of their previous value; conversion rates were progressively unfavourable with respect to the amount converted;[3] a similar reform took place in Czechoslovakia in 1953 (see ECE-UN, 1953, pp. 32).

There followed, throughout the area, three distinct periods (see Table 5.2), a period of falling prices from the time of stabilization roughly to the late 1950s; a period of remarkable price stability until after the first oil shock and often well beyond it into the late 1970s (thanks to the delayed impact of oil price rises within

[2]In the Soviet Union in 1947 all roubles in circulation were withdrawn and replaced with new roubles at a 10 to 1 ratio, while for bank savings and state bonds the conversion ratio was more favourable. On the day after the reform workers were paid their two-weekly wage in new roubles, at a one to one ratio.

[3]In 1952 in Romania prices, wages and state funds were converted at the rate 100 to 5; cast at a progressive rate ranging between 100 to 1 and 100 to 0.25, savings at progressive rates ranging between 100 to 2 and 100 to 0.50. In Bulgaria in the same year the conversion rate was 100 to 4 for prices, wages and state funds, 100 to 1 for cash, while savings were converted at progressive rates ranging between 100 to 3 and 100 to 1.

Table 5.2 Retail prices and nominal wages in USSR and Eastern Europe, 1948–80

	Average percentage annual rates + increase/ − decrease						
	USSR	*GDR*	*Czech*	*Poland*	*Hungary*	*Bulgaria*	*Romania*
(1) Deflationary period:	1948 –1957	1951 –1960	1954 –1961	1954 –1961	1954 –1956		
retail prices	−8.2	−6.0	−3.5	−3.9	−1.0	–	–
nominal wages[a]	+3.3[b]	+2.9	+3.1	+7.2	+6.7[c]		
(2) Period for retail price stability	1958 –1976	1961 –1978	1962 –1973	1957 –1972	1957 –1968	1958 –1972	1961 –1972
retail prices	0.0	−0.2	1.0	1.3	0.6	0.5	0.4
nominal wages[a]	3.6	3.2	4.0	5.2	2.7[d.f.]	5.4	–
(3) Period of open inflation:	1977 –1979	1979	1974 –1980	1973 –1979	1969 –1980	1973 –1979	1973 –1979
retail prices	0.7	0.2	1.7	5.3	4.1	1.1	1.0
nominal wages[a]	2.6	3.0	2.9	10.7[e]	4.8[d]	4.5[d]	5.0[d]
of which: retail prices in 1979	1.1	0.2	3.8	7.1	8.9	4.6	2.0
in 1980	–	–	3.4	–	10.9	–	–
nominal wages[a] in 1979	2.2	3.0	2.5	8.8[d]	5.8[d]	5.5	4.8[d]
in 1980	–	–	2.4	13.5[d]	2.6[d]	–	–

[a]Gross nominal wages in the socialist sector, unless otherwise stated.
[b]Instead of the non-available average nominal wage in 1947, A. Nove's estimate – 550 roubles – was used.
[c]Average rate of 1955 and 1956.
[d]Wages in non-agricultural sector.
[e]Net wages.
[f]Average rate for 1958–68.
Sources: Computed from official data of national statistical yearbooks, various issues; International labour Office, *Yearbook of Labour Statistics*, various issues, and International Labour Office, *Supplement 1981*, April 1981.
From: Pindak (1983, p. 10).

CMEA trade, the initial absorption of imported inflation through state budget subsidies, and the rise of external borrowing); and a period of open inflation, fuelled by the gradual diffusion of oil price rises, the mounting cost of price subsidies, the slowdown and occasional fall of external finance and the rising pressure of wages. This tendency has become much clearer in the last couple of years, but (see Table 5.3) has not reached two digit rates with the exception of Romania in 1982 and Poland in 1980 to date (primarily due to the accelerated growth of money wages in conditions of falling labour productivity, the external credit squeeze and its negative cascade effects on supplies through the curtailment of essential imports, the excessive weight of over-ambitious investment plans ineffectively executed, as well as exogenous factors; see Nuti, 1981, 1982). By 1983, when major price increases have been recorded even in Bulgaria, though their

Table 5.3 Retail prices in the USSR and Eastern Europe 1974–83 (*average annual percentage growth rates*)

Country	*1971–75*	*1976–80*	*1981*	*1982*	*1983*
All items[a]					
Bulgaria	0.2	4.0	0.5	0.2	n.a.
Czechoslovakia	0.1	2.1	0.8	5.7	0.9
East Germany	−0.3	0.1	0.2	0.0	0.0
Hungary	3.0	7.1	5.0	6.6	7.2
Poland	2.5	6.8	21.2	101.0	25.0
Romania	0.5	1.4	2.0	16.0	5–7
Soviet Union	−0.1	0.7	1.4	3.4	n.a.
Food[b]					
Bulgaria	0.6	6.2	0.3	0.2	n.a.
Czechoslovakia	−0.1	1.2	0.0	10.6	0.3
East Germany	0.2	0.0	0.0	0.0	0.0
Hungary	2.2	7.3	3.1	8.2	5.6
Poland	2.9	7.5	30.3	125.0	26.0
Romania	1.1	1.0	1.7	n.a.	n.a.
Soviet Union	0.3	0.4	1.9	3.8	n.a.
Non-food items[c]					
Bulgaria	−0.1	2.4	0.7	0.1	n.a.
Czechoslovakia	0.2	2.8	1.5	1.5	1.5
East Germany	−1.0	0.2	0.5	0.0	0.0
Hungary	3.7	6.9	6.6	5.5	8.8
Poland[d]	2.5	6.5	13.1	85.0	19.0
Romania	−0.3	0.6	2.3	n.a.	n.a.
Soviet Union	−0.3	0.8	1.6	2.3	n.a.

[a]In most cases including services.
[b]In most cases including prepared food in restaurants and enterprise cafeterias. Alcohol is included in this category (even in the case of Poland).
[c]In most cases, this item does not include service. It covers primarily prices of industrial consumer goods.
[d]Excluding both services and non-consumption goods.

Sources: 1980–82: official East European and Soviet statistics.
1983: estimates prepared by the Centrally Planned Economies Service of Wharton Econometrics on the basis of preliminary official statistics published in 1983 plan fulfilment reports and other official East European statistical sources.

From: Vanous (1984).

impact is still not officially assessed, the GDR is the only East European country that can claim an uninterrupted quarter century spell of price stability.

5.4 DEFINITIONS: UNOFFICIAL, HIDDEN AND REPRESSED INFLATION

Inflationary pressure which does not take the form of open official inflation (including official open inflation in the collective farm market) can express itself through the market in the form of open non-official (unrecorded) and therefore hidden inflation and/or be repressed and remain at the purely potential state of unused liquid assets in the hands of the population over and above their desired level. Both types of inflation are 'hidden' until they are reconstructed from the inconsistencies of official records, fragmentary direct observations and indirect evidence.

Unrecorded – and therefore unofficial – open but hidden inflation can occur for three different reasons. First, because of price rises above the official level, either in state shops or in private markets. Official price lists may lag behind actual prices charged by retailing organizations within their discretionary powers (e.g. for new products, including pseudo-novelties). Quality may deteriorate at constant prices. State goods may be sold at prices higher than state prices by dishonest retailers charging a premium to customers or by buyers reselling in a private market; private producers may supply additional consumption goods and services also at a price higher than the official price (of course profits from private production or redistribution should then be included on the income side of household accounts). When these price premia appear or rise, actual open inflation is higher than officially recorded.

Second, unrecorded open inflation can arise from quantity weights in official price indices understating the relative weights of goods whose actual market prices (whether or not equal to the official price) rise relatively faster. In principle quantity weights different from those appropriate to actual prices could lead to a price index bias either way, but systematic official understating is the likely result of cosmetic data manipulation by the authorities, information lags and higher than official price rises. Third, unrecorded open inflation can arise from progressive forced substitution by consumers of goods which are available at their official price for goods which are not available; usually higher priced substitutes replace cheaper (often subsidised) goods. The use of actual quantity weights instead of dubious and conveniently changing official baskets captures both the possible understating of inflation with respect to equilibrium weights and the inflationary effect of forced substitution over time (though not the loss of consumer surplus involved in the limited access to shortage goods, implicit in forced substitution). As Nove puts it, 'In *any* system when price controls exist and the product mix can be varied, the price index will *always* understate price rises...' (1981a, p. 145).

If, directly or indirectly, a price index is constructed reflecting actual prices and actual (Laspeyres or Paasche) quantity weights, the actual open inflation rate p can be decomposed into its recorded and unrecorded components, respectively p_r and p_u, given the relationship $(1+p)=(1+p_r)(1+p_u)$. Rates of unrecorded open inflation calculated for each period putting the previous period's index equal to 100 are cumulative over time, but unrecorded open inflation is not 'repressed' in any conceivable sense, as it is an effective outlet for inflationary pressure, and therefore is not cumulative in the sense of contributing to inflationary pressure in the following period through accumulated pent-up demand, as in the case of excess liquid assets of the population (inflationary gap and overhang).

Repressed inflation is rising excess demand which does not take the form of higher prices but remains in the hands of the population as excess liquid assets over their desired level. Thus its definition and measurement require the prior definition and measurement of 'desired' liquid assets. In a trivial sense all liquid assets could be considered as desired because asset holders can always convert any surplus holdings into any good in the free (legal or black) market or into non-shortage goods in the official market. Therefore in the strictest sense there could be 'excess' asset holdings only for inflexibly law-abiding citizens with zero storage facilities, or for people deterred from transacting in the markets to the full extent of their intended purchases because of particularly high transaction costs (in the state market because of abortive searches, queues, waiting, in the free market because of legal and moral restraints, its thinness and inconvenience). Otherwise liquid asset holders will have expressed effective demand up to the point where the marginal utility of current purchases of higher-priced goods in the free market or of forced substitution goods in the state market is equal to the discounted marginal utility of future prospective purchases of goods at official prices (weighed by the probability of their future availability) or at expected free prices, or of additional leisure at the going wage (at least within the period obtainable through voluntary unemployment in between jobs). By and large, any intended purchasing power which cannot satisfy itself in the state market will push up prices in the free (legal and black) market where goods purchased from the state market are retraded and newly produced goods from the private sectors may be sold, up to the point where demand for liquid assets for transaction and speculation purposes is raised to the level of existing liquid assets. Barring extreme hyperinflationary situations where the economy is effectively demonetized, monetary equilibrium in the economy as a whole will be reached or (within the limits set by high transaction costs) approached. Nevertheless, a significant part, if not all, of liquid assets held by the population in a situation of imbalance in the state market at official prices will be excessive with respect to the assets which the population would wish to hold if it could convert them into goods in that market a those prices.

Government internal and external borrowing can ease inflationary pressure but the size (and growth) of debt cannot be properly included in excess demand (and repressed inflation), except for the component of debt (and its growth) on which the government may have defaulted.

Once excess demand in the state sector at official prices rises above a critical threshold, repressed inflation steps in; open inflation in free markets can lead to overall monetary equilibrium in the economy as a whole but this is consistent with persistent disequilibrium in the state sector. In fact the entire excess of liquid assets over the level which would be desired at the given state price level if the state market had been clearing for a long time, plus the liquid assets which the population – having experienced market imbalance instead – would wish to convert speculatively into goods if supply suddenly became infinitely elastic at state prices, at any moment of time is pressing on the state market backing intended purchases with the necessary spending power. Moreover, any unsatisfied demand (overhang) existing at the beginning of a period plus any inflationary gap (intended purchases out of disposable income not matched by goods at official prices) will be carried over into the next period; a sequence of small gaps can therefore cumulate over time to a serious overall level of excess demand. While it is correct to define as repressed inflation, strictly speaking, only the growth rate of this excess demand, past repressed inflation is cumulatively pressing in current markets and would have to be added to current open inflation in its cumulative entirety if it were to be transformed into open inflation (its additional contribution to open inflation of course depending on the length of the period over which it was made open). A zero or negative current rate of repressed inflation is therefore consistent with the possibility of massive current imbalance in the state market, which would require an injection of goods, or a tax on liquid assets or a higher rate of open inflation than would otherwise be the case, at rates inversely proportional to the period of time over which the stabilization is attempted.

There is a connection between unofficial/unrecorded inflation, plus officially recorded inflation in nonstate markets (such as kolkhozian markets), and repressed inflation. Namely, if the sum of the first two types of inflation is zero (or small), it can be conjectured that the monetary magnitude of excess demand has not grown (much) relatively to money income; however excess demand can still be large, and – following this line of argument – it will have grown approximately as fast as money income.

5.5 ESTIMATES OF HIDDEN INFLATION

Three basic methods have been used to estimate inflation rates over and above official levels (including official inflation in recorded legal markets): the direct computation of price indices from available information about quantities and prices of individual categories of goods; the calculation of deflators implicit in indices of monetary and real magnitudes; the purchasing power parity approach to black market exchange rates.

Western economists started recomputing their own price series in the early 1960s: Michal (1960) for Czechoslovakia, Nutter (1962) and Chapman (1963) for the Soviet Union. Michal, for instance, found that retail prices rose in Czechoslovakia by about 64% between 1948 and mid-1953, instead of the officially reported stability depicted in Table 5.2; Chapman's estimates for Soviet retail prices in 1928–47 diverge less from official indices but are also higher than them (except for 1941–47 due to noncomparability of periods; see Table 5.1). The long duration of the price stability achieved in the mid-1950s somewhat lessened interest in independent computations of this kind. When the reliability of official statistics began to be suspect again, after incidents such as Polish under-recording of the 1970 price increases and similar prima facie cases for challenging official data, convenience and lack of data led Western economists to use indirect methods.

It is well known that dividing an index of expenditure at current prices by a

Table 5.4 Howard's estimated and official indices for prices and sales, USSR 1955–72

	State retail price index			*General consumer price index*			*Real sales index on state and co-operative retail market*		
	Official	*Minimum estimate*	*Maximum estimate*	*Official*	*Minimum estimate*	*Maximum estimate*	*Official*	*Minimum estimate*	*Maximum estimate*
1955	100.0	100.0	100.0	100.0	100.0	100.0	100.0	100.0	100.0
1956	100.0	101.9	102.5	98.8	100.9	101.5	108.7	106.4	107.0
1957	100.0	104.8	108.7	98.4	103.3	106.9	124.0	114.6	118.8
1958	102.2	106.7	107.8	101.0	105.4	106.4	131.7	125.2	126.4
1959	101.4	106.1	107.3	100.1	104.7	105.8	142.3	133.5	135.1
1960	100.7	106.0	108.0	99.6	104.7	106.5	156.7	144.9	147.6
1961	100.0	104.5	105.7	99.7	103.7	104.8	162.5	152.8	154.6
1962	101.4	107.2	108.0	101.7	106.8	107.5	172.6	161.0	162.2
1963	102.2	108.7	108.9	102.9	108.6	108.8	180.8	167.7	168.0
1964	102.2	108.3	108.9	103.3	108.5	109.0	190.4	176.3	177.3
1965	101.4	109.6	111.6	101.6	108.9	110.9	209.1	187.0	190.4
1966	100.7	108.7	110.4	100.4	107.7	109.3	227.4	203.9	207.1
1967	100.7	109.5	112.2	100.4	108.4	110.9	248.6	219.4	224.8
1968	100.7	110.1	113.5	100.5	109.1	112.2	269.2	235.5	242.8
1969	100.7	111.3	115.3	101.4	110.9	114.5	288.9	249.5	258.5
1970	100.7	112.3	117.3	102.0	112.2	116.7	310.6	263.6	275.3
1971	100.7	114.1	120.5	101.4	113.4	119.3	331.7	273.8	289.1
1972	100.7	115.1	122.6	102.1	114.8	121.7	354.3	286.7	305.4

From: Howard (1976a, p. 607).

Laspeyres quantity index one obtains an implicit Paasche price index. Howard (1976) used real consumption estimates independently constructed by Bronson and Severin (1973) and current consumption expenditure indices to calculate lower and upper limits for Soviet state retail and general consumer true or actual price indices. The lower limit is calculated by assuming that there was no hidden inflation in the food sector; the higher limit by assuming that the hidden inflation in food is equal to that in durables and soft goods (see Table 5.4). Howard found true price indices substantially but not devastatingly higher than the official indices, true inflation still being contained within 0.8–1.2% per annum in 1955–72; Howard also found that state and co-operative retail sales have not increased as much as suggested by official indices over the period.

Howard's approach was severely criticized by Rosefielde (1980) and further debated by the two authors (Howard, 1980; Rosefielde, 1981a). The main point of the debate is the legitimacy of using, for the calculation of an implicit deflator, a quantity index which does not embody the same weights used to construct the monetary expenditure index, and therefore the reliability of Howard's estimates.

Table 5.5 East European alternative consumer prices indices

	Price indices 1970 = 100		*Average annual change* %			
	1975	*1978*	*1971–75*	*1976–78*	*1978*	*1979*
Bulgaria:						
Official: state retail trade[a]	100.9	102.8	0.2	0.5	1.2	b
Official: co-operative market[a]	123.6[c]	137.6	4.3[d]	5.5	b	b
Alternative	116.0	129.1	3.0	3.2	2.5	b
Czechoslovakia:						
Official: consumer retail prices[a]	100.7	103.6	0.1	0.9	0.9	2.5
Alternative	111.4	116.2	2.2	1.6	1.8	b
German Democratic Republic:						
Official: consumer retail prices[a]	98.4	98.2	−0.3	0.0	0.0	b
Alternative	103.3	108.6	0.7	1.3	0.6	b
Hungary:						
Official: consumer prices[a]	114.5	130.8	2.9	4.3	4.6	9.0
Alternative	122.4	139.8	4.1	4.4	4.5	b
Poland:						
Official: consumer retail prices[a]	113.3	134.1	2.5	5.9	8.1	b
Alternative	131.8	168.5	5.7	8.0	8.9	b
Romania:						
Official: commodity prices and service tariffs	102.6	105.4	0.5	0.7	1.5	2.5
Alternative	b	b	b	b	b	b

[a]The official price indices are described in detail in Alton (see source).
[b]Not available.
[c]1977.
[d]1976–77.
Source: Alton *et al.* (1979).

Admittedly the true price index calculated by Howard is, as Rosefielde contends, a quantity/price hybrid; it remains to be shown, however, that the use of more accurate weights would yield drastically different results. As a 'rough estimate', in Howard's words, the true price index provides an interesting check to the hypothesis of a close-to-true Soviet official index. (Similar rates were computed by Schroeder and Severin, 1976, at the yearly rate of 1.3% in 1955–75.)

The same line of reasoning followed by Howard is implicit in Portes (1977), who compared directly average increases in money wages (which nobody ever suggested are under-recorded) and real consumption (which he takes from official data, regarded as reliable) in order to obtain indirect confirmation of the thesis that officially recorded inflation was by and large correct in the 1955–75 period. Alton *et al* (1979) used their own independent estimates of real consumption growth for the CMEA Six to establish the true inflation rates for 1970–78, which are substantially higher (though still modest by Western standards and in the one-digit range) than official rates except for Hungary (see Table 5.5; for a discussion of Alton's results, see also Kohn, 1980). Presumably Rosefielde's objection of Howard applies with the same strength to Alton's estimates.

If the purpose of the true indices was that of assessing the truthfulness and efficiency of statistical offices in recording prices and relative quantities of all goods leaving the state shops and all those exchanged in the official non-state sector, then the use of own recalculations or implicit deflators could be good enough. But for discussing hidden and repressed inflationary processes, however, these indices have a major defect. A true price index in fact ought to record ideally the price actually paid by ultimate users, both for goods originally purchased in state shops and for goods which private producers might supply. At the same time, true income indices ought to include the total increment that the value of state goods acquire in the (legal or illegal) process of redistribution, and the total value added in private production, so as to make price and income indices mutually consistent.[4]

The existence of a two-tier market, consisting of controlled and (legal and illegal) free segments (see Katsenelinboigen, 1977; Morton, 1980; Bauer, 1983), is fully recognized and modelled in the literature on the second economy (e.g. G. Grossman, 1977, 1985; O'Hearn, 1980) but is not fully modelled and measured in the literature on unofficial inflation, where the dominant model is that (exemplified by Barro and H. Grossman, 1974) of the rationed quantity-constrained consumer (and firm) and the measurement of market phenomena excludes both the private retrading of state goods and non-state production outside the collective farm sector (which is played down anyway in view of its small relative weight in total consumption expenditure, e.g. by Portes, 1979, p. 329). Yet consumers in Soviet-type economies are not at all technically rationed in their overall budget

[4]Nove's guess-estimate of Polish hidden inflation at 8% in 1975 (Nove, 1979) is rejected by Portes because, given official estimates of nominal incomes and consumption price index, would imply 'that real incomes actually fell that year!' (Portes, 1984). Yet if, say, half the hidden inflation was due to transactions in private and secondary markets, nominal incomes would also be higher, by 4%, and Nove's assertion – though unsupported – would at any rate be perfectly plausible.

constraint; they are only rationed (whether through coupons or at random) in their purchase from state shops at official prices. Their position is that of a consumer facing a two-part tariff with a high marginal price, and with an option on purchases at the lower price of the first part of his tariff – an option which however is not being honoured and which he relentlessly seeks to realize ready to switch all his higher part tariff expenditure and buy more if only he could at any time. Such a consumer would not be regarded as quantity rationed, nor should the consumer in Soviet-type economies. The thinness of the free (legal and illegal) market where state goods are retraded and non-state goods are sold – if that market is thin – makes its quantitative impact on hidden inflation small, but the very fact that consumers can, if they wish, convert liquid assets into goods and services in free markets has a major implication also for the definition and measuring of repressed inflation, because the opportunity of transacting in secondary and primary non-state markets naturally makes all liquid assets desired.

Culberston and Amacher (1972) apply the purchasing power parity theory of exchange rates to black markets for foreign exchange in CMEA countries and estimate actual inflation rates by comparing changes in the black rate of the US dollar – regarded as an equilibrium rate – with trends in the US consumer price index. Any depreciation of the national currency *vis-à-vis* the US dollar is attributed to differential inflation with respect to the US. Their estimates are summarized in Table 5.6; like the other estimates reviewed here, actual inflation is estimated at a substantially higher but still one-digit level. If it worked, this method of indirect estimate of actual inflation rates would reflect trends in the marginal cost to consumers of acquiring goods in free markets, i.e. the inflation rate in those markets rather than in the average cost of living. This method, however, is subject not only to the customary reservations applicable on the purchasing power parity approach to market economies (the neglect of capital flows, the choice of weights, the use of average instead of marginal concepts, etc.) but also to system-specific reservations:

Table 5.6 Culberston and Amacher's estimated average annual rates of inflation, 1963–70 (purchasing power parity method)

Country	*Officially reported open inflation (%)*	*Culbertson and Amacher's estimates (%)*
Bulgaria	1.1	3.4
Czechoslovakia	1.2	5.0
East Germany	0.0	3.4
Hungary	0.9	4.1
Poland	1.2	5.5
Romania	0.4	6.3
USSR	0.1	8.6

Source: Culberston and Amacher (1972). Estimated actual inflation is equal to US rate of inflation plus domestic currency depreciation with respect to the US dollar in the black foreign exchange market.

1. No black-market price in a two-tier black/white market can be regarded as an equilibrium price; it is usually higher than the equilibrium price but not necessarily by the same proportion at different times;
2. Black-market rates are affected by a number of variable demand and supply factors, which have nothing to do with underlying trends in hidden inflation, such as remittances from abroad, degree of foreign tourist surveillance, toughness of penalties, foreign travel opportunities, official attitude towards emigration, etc.;
3. Black market rates have a floor below which the dollar will not fall, given by the ratio between the domestic price of goods (like vodka and cigarettes) widely consumed and with semi-monetary qualities in personal exchange, and their dollar price in special state shops (Beriozka, Pewex, Tuzek, etc.) for sales in hard currencies.

Thus if the dollar price of a bottle of vodka in a Pewex shop is $1 and its zloty price in ordinary shops is zl. 700, the black market rate for the US dollar will not fall substantially below zl. 700. Thus trends in black market exchange rates reflect the vodka purchasing power parity in domestic ordinary and special state shops, or the equivalent rate for other goods supplied in both kinds of shop at more advantageous rates if supplied and demanded in sufficiently high quantities to make them the marginal rate, rather than any comparison of generalized purchasing power parity. Vanous (1984) suggests the same approach with reference to the Deutschmark, but the same reservations apply; since arbitrage in black markets for foreign currency is far from perfect, cross rates and therefore inflation estimates will differ with the hard currency actually used in the comparisons.

A more interesting and imaginative use of purchasing power parities, which does not lend itself to the criticisms raised here, is made by Askanas and Laski (1985) who, starting from a bilateral comparison of consumption levels in Poland and Austria in 1964, 1973 and 1978 (estimated by the Central Statistical Offices of the two countries in the currencies of either country) calculate the implicit price indices for both countries. They find that implicit indices differ from official statistics and infer that over the period considered either Austrian actual inflation has been over-recorded and has taken place at 3.0% per year instead of the official 5.3%, or Polish actual inflation has been at 5.2% instead of the official under-recorded 2.9%. Since this kind of problem does not arise in similar comparisons among Western countries, and the authors have no reason to believe that Austrian inflation – lowest among market economies in Europe – has been overvalued, they conclude that Polish inflation has been 5.2% instead of 2.9% over the period.

In general, there is an 'informal consensus of experts' that 'official inflation statistics are most reliable for Hungary, followed by Czechoslovakia, East Germany, Bulgaria, Poland and Romania' (Vanous, 1984). When inflation is officially understated, if this is due to actual prices in state markets being higher than official price lists no policy measure is required except official admission that prices are higher than previously claimed; a once-for-all uplift in official inflation (which can be substantial due to the cumulative effects of even small discrepancies over time) will suffice without altering the population's purchasing power. If official understatement of inflation is due to free (black and legal) markets being under-recorded, the case for closing the gap between the two price levels is made stronger. Even the total coincidence of true and official inflation rates does not rule out the existence of a possibly large inflationary overhang, already in existence at the beginning of the period for which the comparison is made and growing through time at the same pace as income and expenditure.

The existence and measurement of hidden inflation outside the sphere of consumption goods has also been widely debated (see for instance Becker, 1974; Nove, 1981a,b; Cohn, 1981; Rosenfielde, 1981b, 1983; Steiner, 1982, 1983). The question is interesting for the overall assessment of economic performance in Soviet-type economies, and in particular of the actual size and growth of investment and defence outlays. Inflationary pressures outside the consumption goods sector are bound to spill over into it through direct competition between consumers and firms for some goods and through substitution processes; non-open inflation for non-consumption is likely to take more the form of repressed than hidden inflation, in view of greater restrictions and easier monitoring of financial transactions between firms with respect to redistributive transactions between consumers.

5.6 ESTIMATES OF REPRESSED INFLATION

Visible signs of frustrated spending intentions, such as queues, are not by themselves conclusive evidence of repressed inflation. They are not necessary, because spending intentions may be so totally frustrated that queueing is not worth one's time.[5] They are not sufficient, because queues may be due to the inefficiency[6] or the underprovision of distributive services (Turcan, 1977), or to relative prices (instead of the overall price index) differing from market-clearing prices (Delhaes, 1978; Podkaminer,[7] 1983), and not necessarily to the presence of excess demand at official prices. Even if there are visible signs of frustrated spending intentions, they

[5] 'Do you often have queues as long as these?' 'No, not very often. Only when there are goods in the shops' (Polish humour).

[6] Often simply moving few people from the queue to behind the counter would eliminate queues and save time all round. Turcan (1977) suggests that in Poland in the mid-1970s queues were due to 'organisational and human factors' more than to shortages (p. 136). Underprovision of distributive services is of course by definition excess demand but only for the services, not the goods; a small excess demand for distributive services may appear as large excess demand for goods, because these goods must be jointly purchased together with the distributive service associated with their sale.

[7] Comparing Polish with Italian and Irish data Podkaminer estimates excess demand for commodity groups and the price changes necessary to eliminate them in Poland in the period 1965–78. He claims that food demand would have been consistently lower than supply had it not been for spillover demand (forced substitution) due to excessively low prices in the rent and rest commodities. To eliminate imbalance food prices should have been lowered by 15%, provided that other low-price goods were made correspondingly dearer (Podkaminer, 1983, p. 179).

can be misleading for the measurement of excess demand and above all of its rate of change (which is, strictly speaking, repressed inflation: declared demands may be greater than intended purchases, because they discount a scaling down process (just as in an expectedly oversubscribed share issue in a market economy, a household may be queueing in more than one queue for a single intended purchase); queues may get longer because of an improvement rather than a worsening of the state of supplies.

By itself the increase (even if substantial) of liquid balances in the hands of the population, relative to income or sales, is also neither necessary not sufficient as an indication of repressed inflation. Liquid balances may be accumulated because of systemic lack of access to less liquid forms of holding wealth (land, plant, non-state bonds, shares, etc.). Open (official or unrecorded) inflation might require the accumulation of liquid assets to raise transaction balances whose real value has been lowered by inflation. Cash requirements relative to purchases may increase because of the increasing share of transactions taking place in the second economy (Grossman, 1977). A more erratic pattern of supply, without any alteration in the overall balance between supply and demand, may raise required cash holdings, desired for speculative purposes; Green (1978) likens this phenomenon to 'the demand for cash balances among New York drug addicts' – or, one might add less dramatically, among antique dealers everywhere.

Nevertheless, the coexistence of visible officially recognized signs of frustrated spending intentions and the parallel accumulation of increasing substantial liquid assets relative to incomes and sales has been widely taken as a prima facie case for presuming excess demand and repressed inflation, or at any rate for considering it as a hypothesis worthy of testing and measurement.

Pindak (1983) is typical of a simple and direct approach to repressed inflation (another instance is Cassel, 1984). First indirect evidence of excess demand is considered, such as the falling share of goods officially recognized as not experiencing 'supply difficulties' (for Czechoslovakia in the 1970s, for instance, see

Table 5.7 Share of goods without supply difficulties in Czechoslovakia, 1972–78

Period	*Foodstuffs* (%)	*Industrial products* (%)
1972	58.5	52.0
1973	72.5	43.2
1974	75.0	36.6
1975	66.7	28.2
1976	47.1	28.7
1977	27.9	20.7
1978	35.9	19.0

Source: *Politicka ekonomie*, Prague 1979, 27, p. 458.
From: Pindak (1983, table 4).

Table 5.7); then the ratios of cash and non-cash financial assets to other variables are regarded as indicators which 'express the processes of repressed inflation'... 'Even if the exact level of forced savings remain unclear...' (p. 33). These indicators are the average saving propensity, the ratio between cash increments and total income, and especially the income velocity of financial assets (which is reproduced from Pindak in Table 5.8). However, as we have just noted, there is no systematic necessary relation between these indicators and repressed inflation; moreover Pindak's conclusion that 'the GDR emerges as the country with the most virulent repressed inflation' is contrary to casual observation and other evidence (Flaetgen, 1985; see below). Birman (1980a) claims that 'an overwhelming part of

Table 5.8 Pindak's income velocity of household financial assets, 1955–79 (for Czechoslovakia, the GDR, Hungary, Poland and the USSR)

	Ratio between income and household cash				*Ratio between income and household noncash financial assets*				
Year	*Czech*	*GDR*	*Hungary*	*Poland*	*Czech*	*GDR*	*Hungary*	*Poland*	*USSR*
1955	7.4	4.7	—	10.7	15.0	8.5	—	83.1	10.1
1956	6.1	4.1	9.9	8.1	11.1	7.1	119.8	58.2	9.2
1957	5.5	3.7	10.3	7.9	8.6	5.2	60.5	32.0	7.8
1958	4.7	3.9	9.3	7.3	7.0	4.5	36.6	26.3	7.7
1959	4.3	3.0	7.6	6.9	6.1	3.9	23.3	19.0	7.2
1960	4.2	2.6	6.8	5.5	5.7	3.3	17.0	13.6	7.1
1961	3.9	2.4	6.4	5.5	5.2	2.9	14.6	13.6	7.2
1962	3.6	2.3	5.9	5.0	4.8	2.7	11.7	11.0	7.1
1963	3.5	2.1	5.1	4.6	4.5	2.5	9.0	8.7	6.9
1964	3.2	2.0	4.5	4.2	4.2	2.3	7.1	7.6	6.7
1965	3.1	1.8	4.1	4.0	3.9	2.1	6.1	6.9	6.0
1966	2.9	1.7	3.9	3.8	3.7	1.9	5.8	6.2	5.4
1967	2.7	1.6	3.9	3.5	3.5	1.8	5.8	5.5	4.9
1968	2.7	1.5	3.6	3.3	3.5	1.7	5.3	5.2	4.5
1969	2.8	1.4	3.3	3.1	3.6	1.6	4.8	4.7	4.0
1970	2.5	1.3	3.0	3.0	3.2	1.5	4.9	4.4	3.6
1971	2.3	1.3	3.0	2.9	2.9	1.5	4.3	4.2	3.3
1972	2.1	1.3	2.9	2.8	2.6	1.4	4.1	3.9	3.1
1973	2.0	1.2	2.7	2.5	2.4	1.4	3.9	3.6	2.9
1974	1.9	1.2	2.6	2.3	2.3	1.4	3.8	3.2	2.7
1975	1.8	1.2	2.5	2.3	2.2	1.3	3.6	3.2	2.4
1976	1.8	1.2	2.4	2.2	2.2	1.3	3.4	3.2	—
1977	1.7	1.1	2.3	2.3	2.1	1.3	3.2	3.3	—
1978	1.7	1.1	2.2	2.2	2.1	1.2	3.0	3.2	—
1979	1.7	1.1	—	2.2	2.1	1.2	—	3.2	—

Sources: Computed from the following data:

(a) For Czechoslovakia, the GDR, Hungary and Poland for 1956–75, Rudcenko (1979). The corresponding data for 1976–79 were taken from national statistical yearbooks, various issues.

(b) For the USSR, the data on savings deposits of the households from *Narodnoe khoziaistvo SSSR*, various issues; estimates of household incomes from Schroeder and Severin (1976) (years 1956–59 and 1961–64 were extrapolated).

From: Pindak (1983, table 5).

Table 5.9 Fundamental magnitudes illustrating the formation of market equilibrium in Poland, 1970–80 (current prices, zl. bn)

	1970	*1971*	*1972*	*1973*	*1974*	*1975*	*1976*	*1977*	*1978*	*1979*	*1980*	*Second half year 1980*	*First quarter 1981*
1	2	3	4	5	6	7	8	9	10	11	12	13	14
Sales of goods to the population	395	429	482	544	625	713	813	920	1000	1102	1207	630	309
Sales of services	70	75	82	92	104	119	134	150	162	175	192	98	48
Total goods and services	465	504	564	636	729	832	947	1070	1162	1277	1399	728	357
Dynamics (previous year = 100)	—	108.4	111.9	112.8	114.6	114.1	113.8	113.0	108.6	110.0	109.5	108.0	111.2
Total net incomes	489	543	626	717	824	933	1040	1169	1271	1397	1535	797	447
Dynamics (previous year = 100)	—	111.0	115.3	114.5	114.9	113.2	111.5	112.4	108.7	109.9	109.8	109.6	119.4
Monetary balances of the population													
Total (zl. bn)	171	198	239	300	370	435	486	538	600	675	766	766	843
of which – savings	117	137	170	213	264	307	339	376	415	464	500	500	542
– cash	54	61	69	87	106	128	147	162	185	211	266	266	301

Source: Raport (1981, p. 127).

all savings is forced' (p. 88), talks of 'the colossal surplus of money' and the 'crisis in the circulation of cash' (p. 90; see also 1980b).

Official recognition of excess monetary balances of the population in Soviet-type economies first came in the 1981 official report by the Polish government on the state of the economy (from which Table 5.9 is taken; Raport, 1981), echoing the approach developed by Polish economists (for instance, Herer-Sadowski, 1981) who carefully distinguish inflationary gap (*luka*), i.e. the share of current personal incomes not covered by consumption goods and services at official prices, and inflationary overhang (*nawis*, or current demand out of liquid assets built out of past inflationary gaps and carried over from previous periods). The sum of the two adds up to more conventional excess demand; the distinction between the two components is notional and arbitrary, but it is useful in discussing time processes and trends. Without reference to additional evidence, however, or without the support of models and estimates of demand for financial assets, both notions of gap and overhang are also arbitrary and so are their measurements. The same applies to calculations embodying an allowance for an arbitrarily estimated growth of desired financial holdings with respect to a period of market balance and tranquillity (Gomulka, 1979).

The first attempt at building and index of repressed inflation is by Holzman (1960) who interprets the ratio between free market price (in the kolkhozian market) and state price for foodstuffs, weighted by the share of output sold in the free market, as an indicator of overall excess demand (which he identifies with repressed inflation); Holzman's series have been updated by Garvy (1977) and Dirksen (1981; see Table 5.10, where the value of the index in 1955 is made equal to 100; see also Portes, 1977, 1981b). The index suggests a fall in excess demand (i.e. negative repressed inflation) between 1955 and the early 1970s in the USSR, followed by a modest positive rise in the rest of the decade. The notion that, if there is excess demand at state prices and some of it is repressed, the amount of repressed excess demand will be directly related to the impact of nonrepressed excess demand on free prices is ingenious and appealing. If this is the underlying rationale of Holzman's original approach, an even better indicator would be provided by the

Table 5.10 Holzman's indicator of repressed inflation for the USSR, 1955–79 (1955 = 100)

Year	*Indicator*	*Year*	*Indicator*	*Year*	*Indicator*
1956	59	1964	39	1972	23
1957	40	1965	28	1973	25
1958	41	1966	24	1974	23
1959	32	1967	24	1975	27
1960	30	1968	22	1976	30
1961	38	1969	25	1977	30
1962	37	1970	22	1978	34
1963	36	1971	22	1979	34

Source: Holzman (1960); Garvy (1977); Dirkens (1981).

ratio between true price indices such as those discussed in the previous section, and official price indices in state shops (i.e. excluding the kolkhozian market from the official price index) because in that way the varying intensity of excess demand in nonfood as well as food intended expenditure is taken into consideration. If the reasoning is accepted, the constancy of the index from one year to the next indicates the constancy of relative excess demand with respect to current sales, i.e. the growth of absolute excess demand by the same rate as sales – which is convenient for preserving the possibility of multiplying indices of true open inflation and repressed inflation to obtain total inflation.[8] Zero repressed inflation, however, simply means the preservation of the relative excess demand existing in the base period, and if this is large it is hardly a consolation for the planners facing the task of markets stabilization. Holzman's index, unfortunately, does not indicate the size of that excess demand, only its time trend.

Kornai claims that in a situation of chronic shortages '... aggregate excess demand is not an operational category ...' (1980, p. 477) and proposes a composite 'shortage index' (Kornai, 1982). Partial indicators of shortages are identified: the orders refused by the building industry, unfilled orders for cars, building materials shortage, deviation of household savings from the trend; a macroindex of shortage is then constructed on the basis of time series for the partial indicators (see Table 5.11 for Hungary 1965–78). Like Holzman's index, Kornai's macro-index

Table 5.11 Kornai's macro-index of shortage (based on Hungarian data, 1965–78)

Year	*Index based on 14-year time-series*	*Index based on 9-year time-series*
1965	0.412	—
1966	0.430	—
1967	0.429	—
1968	0.544	—
1969	0.600	—
1970	0.548	0.548
1971	0.498	0.486
1972	0.400	0.404
1973	0.401	0.401
1974	0.456	0.447
1975	0.508	0.496
1976	0.505	0.503
1977	0.553	0.531
1978	0.496	0.479

From: Kornai (1982).

[8] As the analogous relationship between the recorded and unrecorded components of open inflation, in section 5.4 above. Dirksen contends that the validity of Holzman's indicator rests on state prices having a zero rate of open inflation (1981, p. 307, footnote 1); this is unjustified and would violate the multiplicative relation required by consistent accounting.

indicates the dynamics of the phenomenon under investigation, not its absolute size at any time; also, it combines consumption and production goods; these appear as advantages from Kornai's theoretical stance, since they avoid the notion of aggregate excess demand and the problem of netting out positive and negative excess demands in different sectors, and allow for the generalized spreading of shortage, when it occurs, across sectors; they seem, however, drawbacks with respect to the observer's and the planner's task of assessing both performance and the difficulty of tasks ahead. Moreover, the particular version of macroindex constructed by Kornai – which shows shortage to be a cyclical phenomenon rather than a trend – may reflect the choice of savings deviation from trend as a partial indicator of shortage, which already presumes consumer shortage to be cyclical without a long-run trend.

Attempts at measuring excess demand and the implicit repressed inflation, as well as testing the hypothesis of its existence at the same time, have been developed in econometric studies, and have recently grown considerably with the application of disequilibrium econometrics. Soviet household savings were found to conform to patterns well explored for market economies in early attempts by Volkonskii *et al.* (1976) using a Houthakker–Taylor savings function and by Pickersgill (1976) using the permanent income hypothesis (see also Pickersgill, 1970, 1977, 1980). Portes and Winter (1978) provided an estimate of household demand for money and saving functions well tried for market economies (log-linear function for desired balances with a stock-adjustment process; and savings functions based on the absolute income hypothesis and on the Houthakker–Taylor function); they reported successful estimation and found '. . . no evidence of abnormal behaviour that might suggest sustained disequilibrium . . .' (p. 17). Green (1978) replicating a slightly modified version of the same analysis for the GDR reached similar conclusions but found that the attainment of equilibrium appeared to depend upon the availability of food sold at official retail prices, since households accumulate fewer financial assets when food supplies are tight in the GDR (which therefore indicates greater consumption disequilibrium in spite of lower savings when that occurs).

In a subsequent paper Portes and Winter (1980) apply techniques of disequilibrium econometrics to the same problems, applying a discrete-switching disequilibrium model with a household demand equation for consumption goods, a planners' supply equation and 'min' condition stating that the actual quantity transacted is the lesser of the quantity demanded and supplied; they reject 'the hypothesis of sustained repressed inflation in the market for consumption goods and services since the mid-1950s in our four CPEs' (1980, pp. 155–6); indeed, excess supply is regarded as the most probable dominant regime in three out of the four countries (1980, p. 149). A further development of this model was based on the approach of Charemza and Quandt (1982) who postulated the adjustment of planned quantities instead of actual prices; Portes *et al.* (1983) used new data about plan fulfilment to model an elaborate (plausible though arbitrary) objective function for socialist planners which include plan response to plan fulfilment, and

Table 5.12 Average simulated excess demands (%), Poland 1957–80 (Portes *et al.*)

	Model I	*Model IIa*	*Model III*	*Model Vb*
1957	3.0	3.7	0.7	−3.6
1958	−4.1	−1.8	−2.5	−3.9
1959	0.8	0.7	−0.8	1.0
1960	−6.8	−5.5	−5.2	−4.4
1961	5.1	2.2	1.7	1.4
1962	−4.1	−3.1	−2.8	−2.0
1963	−0.3	−1.5	−0.8	−0.6
1964	−4.2	−2.5	−3.1	−1.6
1965	−1.1	−0.8	−1.0	1.8
1966	−3.5	−2.0	−3.0	1.0
1967	−3.0	−1.1	−2.1	−0.7
1968	0.4	1.8	0.5	2.5
1969	−0.9	1.6	−0.8	0.2
1970	−0.2	0.9	0.1	−1.0
1971	4.5	3.8	2.2	2.5
1972	0.7	5.0	0.9	7.0
1973	−3.0	1.8	−2.3	2.7
1974	−0.9	3.6	−0.3	1.2
1975	1.0	2.2	−0.4	2.8
1976	−1.0	−0.8	−2.4	−2.9
1977	−1.5	−1.3	−2.6	−0.1
1978	−10.7	−7.1	−7.2	−8.7
1979	1.5	0.4	0.4	1.2
1980	1.2	−0.1	0.5	1.2

From: Portes *et al.* (1983).

applied the model to Poland in 1955–80. Alternative models are generated, according to slightly different ways of defining and computing excess demand terms; average simulated excess demands for Poland in 1957–80 using the four best performing variants of the model are reproduced in Table 5.12. The startling result is the persistence of low estimates of excess demand (and of the average rates of repressed inflation which are implicit in them over the period), even for the country and the years where indirect evidence suggests that excess demand and repressed inflation were at their peak. Thus in 1980 Poland simulated excess demand for the best four models ranges from a maximum of 1.2% in two of the models to −0.1% (i.e. excess supply, though at a negligible rate).

Charemza and Gronicki (1982, 1984) and Charemza (1984) apply the 'rational' expectations hypothesis (pioneered by Muth, 1961) to a disequilibrium model à la Barro and Grossman (1976), embodying households demand for the consumption and money and supply of labour; wages adjust to workers' expected excess demand and primary plans drawn on enterprises' expectations of labour shortages and aggregated and adjusted to labour supply by central planners. Estimated excess demands for consumption and labour, absolute and relative to the quantities

Table 5.13 Charemza and Gronicki's estimates for absolute and percentage excess demand for consumption and labour

	Absolute estimates		*in percentage of quantity transacted*	
Year	*Consumption in bn zl.*	*Labour in ml. of employees*	*Consumption*	*Labour*
1960	15.1	0.373	6.20	3.01
1961	16.6	0.351	6.29	2.79
1962	16.1	0.532	5.74	4.15
1963	13.2	0.234	4.43	1.79
1964	15.2	0.309	4.82	2.35
1965	17.2	0.140	5.04	1.03
1966	20.3	0.089	5.53	0.64
1967	20.2	0.365	5.07	2.54
1968	19.6	0.265	4.51	1.80
1969	16.2	0.428	3.54	2.84
1970	13.9	0.269	2.84	1.77
1971	30.8	0.039	5.81	0.25
1972	48.4	0.261	8.08	1.65
1973	62.8	0.971	9.30	6.01
1974	79.7	0.919	10.34	5.58
1975	81.4	2.273	9.27	7.68
1976	86.0	1.313	8.48	7.93
1977	87.6	0.826	7.72	5.00
1978	80.4	1.169	6.51	7.03
1979	92.1	1.985	6.81	12.00
1980	117.7	2.133	7.74	12.93

From: Charemza and Gronicki (1984, table 2, p. 82).

transacted, are included in Table 5.13; rising trends can be observed for all variables, with a U-shaped pattern for percentage excess demands (with turning points in 1970 for consumption and 1971 for labour).[9] Charemza and Gronicki also define balancing price p_B as the price level which would have equated the value of actual purchases to the value of intended purchases at actual prices. The difference between the p_B index and the actual price index p_{60} obtained by putting 1960 = 100

[9]From the data reproduced in Table 5.13 Charemza and Gronicki calculate an index of degree of repressed inflation (RID) defined as the difference between the excess demand index evaluated in constant 1960 prices and the price index for particular years (Charemza and Gronicki, 1984, p. 88). The RID index has the drawbacks of comparing indices (of excess demand and of prices) not weighted by their respective volumes of transactions, of not possessing the multiplicative properties of price indices (differences between indices being very rough approximations to the trend of their ratios for the large relative changes recorded) and of measuring changes in excess demand with respect to the base year and not its degree. The Charemza and Gronicki index for the degree of repressed inflation (1984, table 4) is therefore meaningless.

Table 5.14 Charemza and Gronicki's balancing and observed price indices for consumption goods, Poland 1960–80

Year	P_B	P_{60}	$P_B - P_{60}$
	1960 = 100.0		
1960	106.2	100.0	6.2
1961	107.6	100.9	6.7
1962	109.4	103.5	5.9
1963	108.9	104.3	4.6
1964	111.1	106.0	5.1
1965	113.0	107.6	5.4
1966	114.6	108.6	6.0
1967	116.0	110.4	5.6
1968	117.4	112.3	5.1
1969	118.0	114.0	4.0
1970	118.8	115.5	3.3
1971	123.4	116.6	6.8
1972	126.2	116.8	9.4
1973	131.7	120.5	11.2
1974	141.6	128.3	13.3
1975	144.2	132.0	12.4
1976	151.3	139.5	11.8
1977	158.2	146.9	11.3
1978	169.0	158.7	10.3
1979	180.8	169.3	11.5
1980	200.6	186.2	14.4

From: Charemza and Gronicki (1984, table 5, p. 91).

is regarded as a measure of excess demand (Table 5.14), better than that indicated in Table 5.13 in that it conveniently peaks in 1980 as direct observation would have suggested.[10]

There are a number of criticisms and reservations which could be raised against the econometric estimates of excess demand and its rates of change reported here, to try and reconcile the generally low estimates with the more widespread impressions of large scale and growing imbalance. Thus, for instance, the use by Portes *et al.* (1983) of deviations from trend in households' financial assets could be thought to rule out from the outset the possibility of a trend of increasing excess demand and explain the closeness of their results to Kornai (1982) who does the same. Labour supply responding negatively to market imbalance (Charemza and

[10] As in the RID index, mentioned in the previous footnote, differences between indices only approximate trends in the ratios between them; even accepting the proposed approach the appropriate measure of the proportional excess of P_B over P_{60} is $(P_B/P_{60} - 1)$ and not $(P_B - P_{60})$, which here systematically overestimates excess demand. When properly calculated exess demand still peaks in 1974 at 10.4%, later falls to 6.5% in 1978 and rises again to 7.7% in 1980. Cumulative rates of repressed inflation can be calculated as the growth rates of these excess demand indices.

Gronicki, 1984) in spite of constraints in workers' choice of working time may have overestimated excess demand in the labour market and underestimated it for consumption goods. Rational expectations – which many would regard as inappropriate to model market economies, let alone centrally planned ones – leaving hardly any room for the unexpected might have been crucial to the results reported by Charemza and Gronicki (who specifically use them) and Portes *et al.* (who approximate them in two of their model variants).

Even without this kind of criticism, however, it is perfectly possible to reconcile estimates of low excess demand and repressed inflation with the widespread concern for market imbalance expressed by many observers and officials alike: consumers may be quantity rationed in their purchases from the state sector at state prices, but are not rationed in their purchases in the nonstate sector where nonstate goods may be supplied and state goods may be retraded at higher prices. Excess demand at state prices is channelled into the non-state sector driving prices there up to the point where liquid balances for transactions plus those speculatively held are voluntarily demanded by the population. This point is made strongly and persuasively by Gregory Grossman:

> . . . it would seem that the very presence of a large secondary economy, and in particular of a black market, in a sense does away with repressed inflation, despite fairly rigid control of official retail prices. In the second economy, prices tend to be high enough to eliminate any overall 'monetary overhang' (that is, excess of purchasing power over the total supply of goods and services at effective prices) and to forestall a repressed inflationary situation in relation to the controlled and noncontrolled sectors taken together (Grossman, 1977, in Tanzi, 1982, p. 259).

Hartwig (1983) also stresses that involuntary real money balances are removed already in the short run for their overwhelming part, 'partly by increasing prices in the second economy, partly by a renewed rise in the demand for liquid assets'; the rise in transaction costs (queueing, bribes, information costs, etc.), growing informal transactions and 'the desire to possess liquid assets in order to take advantage of unexpected opportunities cause economic agents to convert involuntary into voluntary money balances' (Hartwig, 1983, p. 104). Brus and Laski (1983), though concurring with this general view, regard the speculative element in money holdings as dependent on the maintenance of lower state prices and therefore properly classified under excess demand; this is an important condition for the stability of the free market which, however, does not make speculative cash holdings any less voluntarily held. By and large, except for temporary excess balances due to time lags in adjusting to monetary shocks, 'enormous amounts of involuntary liquid assets in the private sector seem very implausible'[11] (Hartwig, 1983, p. 105).

[11] With the existence of an extended second economy, postulated generally for the CPEs of Eastern Europe, growing cash and savings bank deposits do not indicate forced saving, but only show that private households are unwilling to pay the higher prices on the different types of markets outside the

If there is no hidden or open hyperinflation, and no differential transaction costs in the free and the controlled market, a properly constructed and specified econometric model should give zero excess demand all the time and therefore zero repressed inflation. Since we know from direct evidence that there was no hidden or open hyperinflation in the period and countries considered, estimates of positive excess demand must be due to factors such as the use of price indices different from the true indices, consumer restraint from transacting fully on free markets (out of Party loyalty, fear of penalties or lack of storage space) and more generally differential transaction costs. However, information costs and any other inconvenience in the free market cannot be much greater than the cost of searching and queueing in the controlled market. Party-loyal citizens are likely to have privileged access to goods and large storage space, while less enlightened citizens are likely to be less deterred by penalties and to be willing to go to great lengths to realize their purchasing power (within the limits set by expected future prices, probability of supply and time preference, as indicated in the previous section): excess demand in the economy as a whole is likely to be small.

This is the good news. The bad news is that most and possibly all of the liquid assets in the hands of the population are pressing at any time on to the lower priced and quantity constrained state sector.[12] Intended purchases at state prices will be equal to the quantity which would have been demanded at those prices if no market imbalance had been experienced recently, plus speculative demand[13] expressed on the reasonable expectation that recent market imbalance at those state prices might reoccur in the near future (within the constraints set by storage opportunities and by expectations about likely appreciation at future free prices relative to interest foregone plus physical depreciation and other storage costs). Excess demand in the state sector is equal to these intended purchases minus actual supplies, and this can be large and increasing fast even if excess demand in the economy as a whole is

---

public sector, to pay bribes, or to incur other forms of transaction costs. They prefer liquid assets. Likewise queueing does not indicate that people cannot buy what they wish and therefore cannot spend their money, as is often argued. Queueing only shows that private households evaluate the disutilities of queueing less than the disutilities arising from higher prices in the second economy or from paying bribes. Every economic unit can choose to spend its money outside the public sector and obtain goods more rapidly, but at higher costs, or to wait for a favourable occasion. The respective behaviour is determined by various expectations. However, the decision to accumulate money and near-money under these circumstances is voluntary' (Hartwig, 1983, p. 104).

[12]'. . . repressed inflation (excess demand in our terminology) *in the state goods market* persists even when the collective farm market clears' (Gardner and Strauss, 1981, pp. 392–3). These authors, however, infer from this the existence of excess (involuntary) money balances in the economy as a whole, which as we have seen is not necessarily the case. See also Wimberley (1981) on the involuntary nature of most transactions in the second economy.

[13]In the summer of 1985 water authorities facing drought in Florence did not ration water in spite of water reserves gradually falling to zero because they expected the introduction of rationing would raise current demand over the normal level through consumers building up their own reserves, filling up their bath tubs and therefore demanding more water than usual at the same price. In the same way, once supply of any good falls substantially below equilibrium at the going price, consumers will demand at that price more than the normal equilibrium level, lowering their liquid asset holdings below the normal equilibrium level.

constantly held at zero by inflation in the nonstate sector whether or not officially recognized. In the presence of this phenomenon, which has very considerable economic costs to the economy (not to count political costs to the regime), claims about the absence or low incidence of excess demand sound somewhat frivolous and the rough and ready estimates of experts in the field (whose main fault is that of not distinguishing between the state sector and the whole economy) appear more credible, as an indication of the existence and size of market imbalance and the gravity of the problem confronting central planners, than the spurious precision of advanced econometrics, which almost tautologically defines imbalance out of existence.

5.7 AN ALTERNATIVE MODEL

Let us consider a Soviet-type economy in a state of persistent equilibrium: consumption markets clear and have cleared without inflation for the last *h* years of relevant consumer memory (*h* being of the order of magnitude of, say five years), at a uniform price level in state and non-state sectors. Money (i.e. liquid financial assets) is demanded both to finance current transactions and as a component of consumers' portfolio, i.e.

$$M_d = M_T + M_K = M(Y, P, K) \tag{5.1}$$

$$M_S = 0 \tag{5.2}$$

where

M_d = consumers' demand for money
M_T = consumers' demand for transaction purposes
M_K = consumers' demand for money as an asset
Y = money incomes
P = the price level
K = the money value of consumers' total assets
M_S = consumers' speculative demand for money

Differences between money incomes and current expenditure add to (or diminish) the stock of money held by consumers adjusting it to the desired level over time. In the position considered here equilibrium has prevailed for *h* periods during which

$$P = P^* \tag{5.3}$$

$$M = M^*(Y, P^*, K) \tag{5.4}$$

where M is the money in existence and the asterisk indicates equilibrium magnitudes in conditions of persistent equilibrium. These are the conditions which prevailed, or were approximated, in the Soviet Union and Eastern Europe roughly in the second half of the 1950s, after the experience of price stability and mild deflation.

Next we can imagine the appearance at time t of creeping repressed inflation, i.e. excess demand at prices P^*_{t-1} prevailing previously. The price level is held at P^*_{t-1} but $P^*_t > P^*_{t-1}$. An excess demand appears and may even grow over time, but the following conditions are fulfilled:

1. excess demand appears and remains at below a critical fraction of current consumption supplies (or, which is the same, the value of excess demand is below a critical fraction of M_T); thus excess demand grows no faster than consumption;
2. no open inflation appears (whether officially recorded or unrecorded);
3. shortages are sporadic, not persistently concentrated; in the same sectors; they are accompanied by excess supply in other sectors and shortages of individual commodities can fall as well as increase.

Welfare is reduced because of the absolute shortfall and because of forced substitution, but the persistent expectation of availability of goods in the future at constant prices and at a small but positive real interest rate leads consumers to hold speculatively any surplus money over and above desired transaction and asset demand. Within the bounds of the three conditions listed here, there is infinitely elastic demand for money on the part of consumers for speculative purposes, a kind of liquidity trap which, however, in the circumstances of the Soviet type is not at all a trap (catching the economy on its way to full employment) but rather a shelter protecting it from open inflation.

$$M_S = M - (M_T + M_K) \text{ subject to conditions (1)–(3)} \qquad (5.5)$$

These conditions prevailed, or were approximated, in Eastern Europe in the 1960s, and well into the 1970s for the Soviet Union and especially the GDR (thanks to its special relation with the FRG which transforms structural imbalances into a less intense overall imbalance, preventing spillover effects of concentrated shortages).

The acceleration of the relative weight of shortages, their concentration, or generalized upward trend, lead to open inflation, whether hidden or officially recorded. When inflation first appears, or when it accelerates, if the price increase is initially calibrated to respond to the scale of the shortfall of consumption with respect to current income there will remain residual inflationary pressure because at least some of the former M_S will be switched towards current purchases. This is why higher prices, when decreed after a spell of creeping repressed inflation, are either pitched at too low a level with respect to equilibrium in current markets or, if prices are pitched at the equilibrium level, they will have to be higher than the long-term equilibrium level consistent with the given flow of real consumption and of monetary income, in which case they are perceived as an unwarrantedly drastic cut in the standard of living and strongly resisted.

Open inflation may and often does lead to the establishment or rapid growth of a two-tier market: the state sector where price P_1 prevails, and price P_2 in the second economy where a fraction of the goods originally supplied by the state sector are retraded and additional goods are supplied. Thus (for each commodity, though here we assume for simplicity that there is a homogeneous single consumption good) we have:

$$P = P_1 C_{S1} + P_2(C_{S2} + C_K) \tag{5.6}$$

where C_K is the consumption supplied by the non-state (kolkhozian and other) sector, C_{S1} and C_{S2} are the parts of state-supplied consumption C_S respectively consumed directly and retraded in the second economy. Retrading and nonstate output raise income by the same amount of the value added by these activities, i.e.

$$Y = Y_S + (P_2 - P_1)C_{S2} + P_2 C_K \tag{5.7}$$

Therefore the second economy does not eliminate excess demand by its direct contribution (if any) to the reduction of supply shortfalls. It may contribute to the absorption of excess demand through its higher transaction requirements of cash per value of turnover; this is likely to be the case but is not necessary to the equilibrating role of the second economy. Ultimately, excess demand is eliminated by P_2 rising high enough to raise the demand for money for transaction purposes and (given time preferences, expected free prices, and expected state prices times the likelihood of their availability at those prices) speculative demand for money. An equilibrium is reached at prices P_2^{**} such that:

$$M = M_T + M_S = \tilde{M}(Y, P_2^{**}, \bar{P}_1) = M^{**} \tag{5.8}$$

The process through which equilibrium is reached is similar in the capitalist and the Soviet-type economy. In a capitalist economy with a full-fledged monetary and financial system there cannot be too much money in the sense of excess liquid assets in the hands of the public with respect to their desired level. Any excess bonds can be converted into cash; any excess cash can be spent on goods driving up their price (which also indirectly reduces the excess by raising cash requirements for transactions); any remaining excess cash can be used to reduce outstanding debt or raise net credits *vis-à-vis* the banking system, any excess cash within the banking system can be deposited with the Central Bank; the interest rate and its structure will be influenced by all these moves, but in this way excess liquid assets will ultimately be eliminated and can only appear as a temporary phenomenon due to the length and structure of lags involved in the adjustment process. In the Soviet-type economy there are constraints affecting this process: price controls lead to price rises only outside the formal sector, the interest is kept at low levels to prevent the rise of rentiers and the existence of two monetary circuits slows down both the transmission of shocks and the adjustment process; the two-tier formal and informal price system causes queues and other transaction costs but otherwise the adjustment process is the same.

It follows from the temporary and unlikely character of overall excess demand, and from the existence of a second economy where state goods are retraded and possibly additional goods supplied, that the consumer-rationing model à la Barro and H. Grossman is not at all suitable for Soviet-type economies outside the extreme case of distribution exclusively through actual ration coupons without retrading, i.e. war communism. While Barro and H. Grossman specifically stress their assumption that there is no secondary trading in a two-tier market, and do not claim their results to apply outside their model's assumptions, Howard's use of that model in spite of his recognition of 'the uncontrolled or free consumer goods market' (which, moreover, is restricted to the kolkhozian market) is illegitimate. The same consideration applies to Portes' theoretical (1979, which considers not only current consumption but also expectation of future consumption as quantity-rationed) and empirical work (Portes *et al.*, 1977, 1978, 1980, 1983, 1984), Charemza and Gronicki (1982, 1984) and others. Consumers are indeed quantity-rationed in the state sector but they are not subject individually to overall quantity constraints since they can always spend their money in the secondary market. It follows that the supply-multiplier, i.e. the rounds of reduction in labour supply (and therefore consumer goods supply) which are alleged as a consequence of quantity constraints, do not necessarily occur and can only be expected to be present in the same circumstances in which labour supply would respond negatively to open inflation. Moreover, demand functions in the presence of persistent and large quantity-rationing must differ from the case of sporadic, occasional and small rationing.

Under-recording of the actual price level faced by consumers (whose actual average income is inclusive of second economy value added) is due to four possible causes:

1. P_1 being higher than its recorded level;
2. P_2 being higher than its recorded level;
3. the relative weights of C_K with respect to C_S being under-recorded;
4. the neglect of C_{S2} being ultimately purchased by consumers at P_2 instead of P_1 (though, for the purpose of calculating consumers' total purchasing power income is also under-recorded unless Equation 5.7 is used to restate it).

In the extensive literature on hidden inflation considered above (section 5.5) the first three sources of under-recording are discussed and attempts are made for their estimate; while demand for money and the income to which demand for money is related should depend on P as defined in Equation 5.6 and not on $P_1C_S + P_2C_K$ as in standard work (Howard, Alton, Askanas and Laski etc.). The same, more comprehensively defined, price level should be used in econometric work à la Portes. Once hidden inflation is properly accounted for, any residual excess demand which might be detected must be due to adjustment not being instantaneous, because of lags in stock adjustment behaviour or to storage capacity constraints (restricting effective demand) and similar factors unlikely to be large. Hence it is not surprising to find econometric studies estimating it to be small, as in the literature surveyed above (in the previous section). Money which, in view of the opportunities afforded by the second economy, is voluntarily held by consumers is, however, involuntarily held in so far as consumers would only too willingly unload on to the state sector at state prices their entire cash holdings except for $\tilde{M}(Y_S + P_2C_2, P_1)$. Hence the alarmed assessment and the admittedly rough estimates of observers who claim there is large imbalance is also correct, and indeed has greater merit if the question is 'do Soviet-type economies face a large-scale stabilization task?' instead of 'are there forced savings?' or 'are official authorities lying?'. Cumulative hidden inflation can be dealt with at a stroke simply by updating official price lists and/or recalculating official price indices with the appropriate weights; but the quantification of the stabilization task, the policy measures and the alternative paths towards the restoration of market clearing at a uniform price and of price stability is much harder both to theorize and to implement.

There are very few a priori propositions which we can make about the size of the task and the price trends which would restore equilibrium:

1. The price level which would restore equilibrium is not a single valued magnitude but an inverse function of the period over which equilibrium is to be restored;
2. The range of prices over which this re-equilibrating price would have to be pitched, as a function of the period of transition to equilibrium, presumably has a maximum level of P_2 because, once consumers have to pay prices higher than P_1 in the state sector, their real income is reduced and they could not afford to pay P_2 for the rest of their purchases. However, an awkward complication is created by the possibility of consumers, formerly willing to hold cash balances in the expectation of future purchases at given expected state and free prices, now revise upwards their price expectations and unload at least some of their balances into the present market, possibly driving the range of re-equilibrating prices above the current P_2.
3. The re-equilibrating price will be greater than P^*, i.e. the price which would preserve equilibrium if it had been in force for the last h years;
4. The re-equilibrating price being a function of the period of transition to equilibrium, it cannot be identified with current P, which is *a* weighted average of P_1 and P_2, but not necessarily the average corresponding to the selected period of transition.

These propositions may not provide much assistance for actually assessing the gravity of the problem faced by planners, but at the same time are strong reasons for rejecting measurements of hidden and repressed inflation put forward in the literature reviewed above. Estimates of hidden inflation can be rejected because they neglect the weight of secondary transactions involving state goods, and because the actual price index relative to the official one does not indicate the true extent of current excess demand, or the true price level which if prevailing in the state sector would restore equilibrium. Diverging estimates of repressed inflation – whether crude or based on advanced econometric methods – can be reconciled in so far as crude pessimistic assessments can be referred to policy propositions and optimistic econometric assessments can be referred to a much stricter (and policy-irrelevant) definition of the problem; both approaches, however, neglect the problem's time dimension (being instantaneous measurements) and complexity (i.e. the difference between the equilibrium level which would preserve equilibrium and the time path of prices which would restore it). Both features of the problem have strong policy implications.

5.8 IMPLICATIONS FOR MARKET-CLEARING AND STABILIZATION

In market and Soviet-type economies the availability of monetary instruments is radically different. In the market economy the measures available include: direct controls over bank lending, raising of reserve requirements, special (compulsory and unprofitable) deposits to be held with the Central Bank by commercial banks, funding of government debt, open market operations, raising of the bank rate, etc. In the Soviet-type economy some of these monetary instruments can be used if the economy has got out of planners' control to reinstate that control and restore planning: controls over bank lending, interest rate policy for firms. Once the economy is back in the grip of planners, however, none of these instruments can be used. In the household sector the interest rate cannot be raised and net issues of government bonds at low interest rates do not find buyers unless an element of formal or informal compulsion is introduced in which case they become just a form of taxation. In the firms sector, if financial flows have already been dovetailed to plans for output, any sales of bonds, additional interest rate payment, direct control of credit, would have to be matched and therefore neutralised by a government subsidy to firms to enable them to fulfil plans. Thus money can be an instrument of monitoring plan execution (the 'control by the rouble' of Soviet textbooks), not a policy instrument. The same applies to taxation in the firms sector. The burden of market adjustment will fall on firms under the guise of changes in output plans, in which case financial adjustments can be made.

In the Soviet-type economy, therefore, once money incomes in the state sector have been fixed, market stabilization, understood as the restoration of a single market-clearing price, will take one or a combination of three policies which do not involve monetary means: (1) price rises, (2) taxation of wealth (not income tax because, if after-tax income could have been lowered, money incomes would have been fixed at a lower level); (3) supply rises. Both in academic and official circles the discussion is of effective stabilization policies in a search for the equilibrium price level, or the once-and-for-all capital levy (mostly in the form of a currency conversion at diversified rates, see section 5.3 above), or the level of the once-and-for-all or recurring increase in supplies which must take place to stabilize the market. The problem, however, is more complex and has a time dimension.

It is more complex because the price level at which markets would clear if they had been clearing for the last h years of consumers' relevant memory will only preserve market equilibrium, if everything else remains unchanged, after equilibrium has been reached and maintained for h years. That price level, however, will be too low to restore equilibrium even given all the time in the world. The time dimension is involved not only under the guise of h (a parameter of the formation of consumers' expectations, which is influenced not by economic but by political factors, i.e. the credibility of governments and of their commitment to market clearing) but as the length of the period over which first market clearing is to be achieved, and then a transition to equilibrium ratios between assets, income and sales is to be implemented. In fact the price level necessary to absorb all excess demand will be inversely related to the length of time over which such absorption takes place. For instance, if the liquid assets which the population would like to convert into goods in the state sector are equal to six months of state sales at current prices, it will take a price level twice the current level to reach equilibrium in six months, six times the current level to reach it in one month, fifty per cent higher to reach it in one year and only 10% higher to reach it in three years, with disequilibrium persisting at gradually falling levels during the period of transition. Once market clearing at a uniform price is reached, consumers having exhaused their excess liquid reserves will not be able to maintain their real purchases at the going price; the price level will have to fall. After h periods consumers will be persuaded that market clearing will persist and they will begin to replenish their liquid assets and reduce their inventories of goods towards their long-run equilibrium levels. For markets to clear at a constant rate of supplies and constant money incomes prices will have to fall again for a period of time V' equal to the ratio of the difference between equilibrium and current money holdings ($M^* - M_{V+h}$) to the decrease in the market value of the constant supply of goods (per unit of time) caused by the price fall. After a time V' has lapsed, consumers will have reached their long-term equilibrium holdings of liquid assets M^* but further cyclical perturbations will follow unless the authorities promptly decree a price increase that brings prices to their equilibrium P^*. At that point, and not before, market clearing is associated with price stability. The process is illustrated in Fig. 5.1.

Planners do not know beforehand the value of h, M^{**} or M^*; they need to know these values in order to complete the process neither later nor sooner than they intend, but they do not have to know them to execute a stabilization programme. As long as the planners set a price level at which consumers cannot afford to buy up current supplies out of income alone, sooner or later they will complete the first

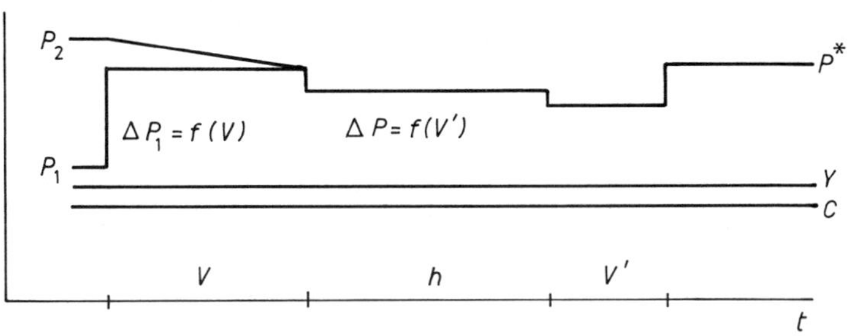

Fig. 5.1 Stabilization through price policy.

stage of market stabilization after a time V corresponding to the price increase and whatever $M - M^{**}$ is. If, as is likely, the strictest constraint on the stabilization programme is the rate at which the excess liquid assets can be depleted, i.e. the highest state price level the population is prepared to put up with, planners' ignorance of M^{**} does not matter at all. They will know that the length of time V has been completed when they experience a generalized rise in inventory levels. At that point they can let prices find their own level; they will know the second stage of stabilization is completed and the hth period has arrived after V when at those prices they will see a further generalized rise in inventories. Again they can let prices find their own level but after a while they will experience lower than normal inventories. The authorities will know from this sign that they have just overshot the end of the further V' periods. They will have to raise prices but if they let the market find its level a dampened fluctuation around the long-run equilibrium level of liquid assets and prices will follow; this might be preferable to attempts at guessing what the equilibrium level is, which might cause disturbances worse than dampened fluctuations. At the end of the V' periods after an initial mild price fall, which is desirable in itself to penalise speculative hoarding of goods and stabilize expectations, it may be preferable and more expedient to raise incomes instead of lowering prices further, in order to allow consumers to replenish their cash holdings. The trouble is that the lower prices cannot be maintained after the end of period $V + h + V'$ when consumers have replenished their cash holdings; for the higher income level to be maintained, prices would have to undertake a final hike, to the P^* equilibrium level. This policy variant mixing prices and incomes measures is illustrated in Fig. 5.2.

Reliance on a wealth tax or on increased supplies to restore equilibrium can now be easily illustrated by highlighting the differences with respect to the previous case during the three stages of stabilization, of lengths V, h and V'. A conversion of the currency confiscating the excess part of real value of liquid assets would shorten period V to zero. Beside the advantage of speeding up the stabilization process, the conversion has other enormous advantages: it can confiscate purchasing power at progressive rates; it confiscates totally those liquid assets which originated from illegal profits to the extent that they cannot be laundered; even if a constant (i.e.

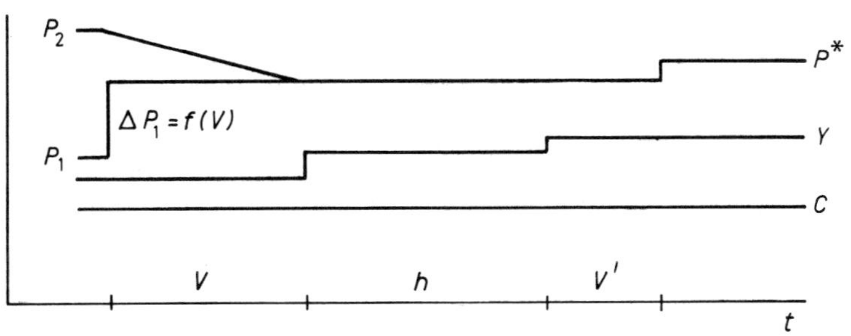

Fig. 5.2 Stabilization through prices and incomes policy.

non-progressive) part of liquid assets are confiscated, the measure is not regressive, as a generalized price increase which forces greater consumption cuts on people who do not hold liquid assets would be. The only disadvantage of a confiscatory currency conversion is that it takes longer to print the new currency and execute the manoeuvre than to decree higher prices; there is a greater chance of a leak, in which case all liquid assets would become instantaneously surplus in the whole economy (including the nonstate sector): serious market disruptions and social disturbances might be set in motion unless the manoeuvre is swiftly and efficiently executed. Also, there may be – without real justification – greater odium for the government in a single large-scale confiscatory move than in an equally confiscatory price rise diluted over a period of time V. If the conversion is pitched at the right rate, equivalent to M^{**}/M units of the new currency for one old, market-clearing will be instantaneously reached and from that point onwards the story will continue as in the previous case: prices will fall for h periods, fall further for another V' periods, after which equilibrium can be attained (see Fig. 5.3), or approached through dampened fluctuations (or incomes may undertake compensatory inverse changes, as in the previous case). The period of mild open deflation experienced by the Soviet Union and other East European countries after the currency conversions at the turn of the 1950s (see above, section 5.3 and Table 5.2) can well be attributed to the necessary overshooting implicit in achieving market-clearing, whether instantly or through a period of confiscatory inflation. The experience of those conversions suggest that the conversion had been pitched at rates which were either correct or slightly lower (i.e. more confiscatory) than necessary. If the conversion is not pitched at the appropriate rate, because planners do not know the value of M^*, other instruments (prices, incomes, income taxes or supply changes) will have to be used, rather than making further clumsy attempts at currency reform.

If stabilization were to be achieved through changes in supplies the corresponding path is the parallel of stabilization through price changes. Namely, incomes and prices remain unchanged while supplies must rise over and above what consumers can afford at current prices out of income, for a length of time V inversely related to the improvement in supplies, until liquid assets reach M^{**}. At that point, signalled by inventories rising above normal levels, supply could fall below the previous level for h periods, fall further for V' periods and then be pitched at the long-run

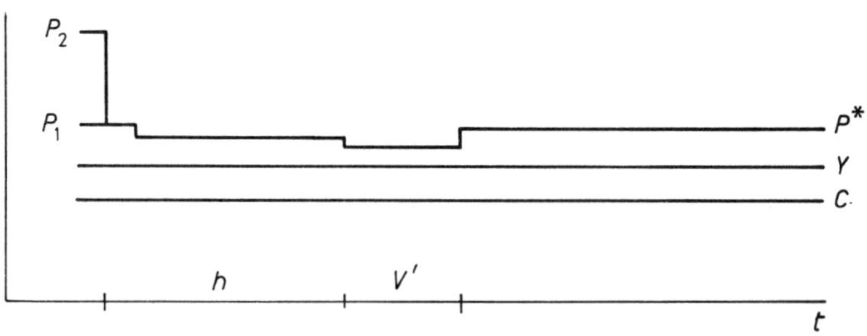

Fig. 5.3 Stabilization through currency reform.

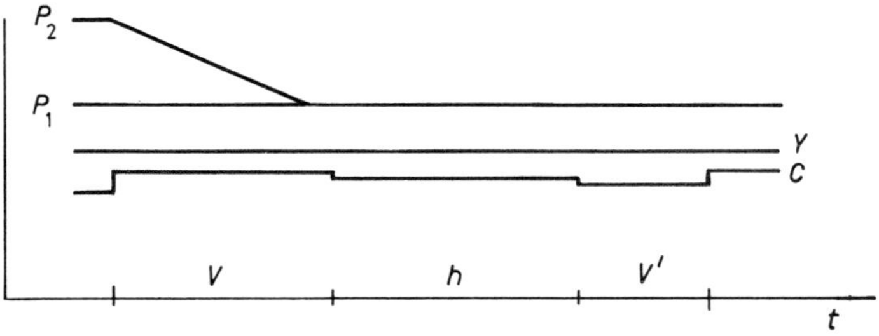

Fig. 5.4 Stabilization through supply policy.

equilibrium level exactly when consumers reach asset holdings M^*. The path is illustrated in Fig. 5.4.

An infinite number of alternative paths can be generated by mixing in different proportions the three policy alternatives. The best course would be the following:

1. Smooth the course of real consumption, switching resources from accumulation to consumption if this can be done without reducing growth – as it probably can if the structure of investment is more efficiently planned and executed – or at any rate without a growth cost greater than the value put on market stabilization.
2. Maintain the existing level of money incomes, or if necessary raise them at any time as little as possible consistently with the maintenance of social peace (or at least of law and order without extreme measures).
3. Execute a currency reform aimed at instantly obtaining market clearing at a uniform price level within and outside the state sector.
4. Aim at the maintenance of the price level during the h periods of the second stage of stabilization, lowering income tax rather than raising incomes or lowering prices (if social pressure imposes higher incomes than required by the manoeuvre, then maintain the increase as low as possible and allow prices to increase so as to preserve market-clearing).
5. Continue with this kind of policy during the V' periods of the third stage of stabilization.
6. Raise income tax and if necessary hike the price level once and for all at the end of the period to make current incomes and liquid assets consistent with the consumption flow.

Others might prefer other courses; the choice is political, but one thing is sure: the path to market-clearing and to stabilization is more complex than the simplistic pitching of an equilibrium price level as usually suggested in discussions about repressed inflation and in policy documents.

5.9 THE COST OF HIDDEN AND REPRESSED INFLATION

If overall excess demand in a two-tier Soviet-type economy is supposed to be small, in the limited sense that a sudden and extensive 'flight out of money' with hyperinflationary consequences is highly unlikely, the question might be asked whether hidden and repressed inflation have a cost for the economy which is high enough to warrant the effort of reaching and maintaining market clearing and stabilizing prices.

Hidden inflation has a distorting informational effect, if people actually believe official indices, but since consumer choice will be guided by actual prices the unreliability of official indices is likely to have little effect (except perhaps in wage negotiations). Excess demand at state prices, however, through queueing, informal exchanges and a two-tier price system, does have considerable costs. Queueing time is a direct and sizeable loss of leisure; it is also a transaction cost which, together with the other additional costs of the two-tier market (which often has three or many more tiers, neglected here for simplicity) reduce consumer welfare. There are illegal and immoral implications of two-tier markets and the nonstate production sector (pilfering to supply inputs in the non-state sector, etc.). Short of actually rationing supplies with coupons, distribution of real consumption acquires an element of chance which cannot be reduced by fiscal measures; it is as if money consisted of lottery tickets, a system which nobody has ever sponsored. Speculative stockpiling by consumers instead of shops can be wasteful. The persistence of a sellers' market depresses quality of goods and keeps producers out of touch with consumers' wishes, leading to unnecessary inefficiency; persistent excess demand is the main obstacle to the introduction and to the effectiveness of economic reforms, which is a considerable cost in itself. It is doubtful – or at any rate controversial – whether inflation has a real cost in a capitalist market-economy; apart from adverse distributional effects (which can be mitigated or eliminated by compensatory payments to the needier and weaker sections of society) the only tangible drawback being the restriction that inflation poses on the scope of division of labour and therefore welfare. These costs of open inflation are bound to be even smaller in a Soviet-type economy, where redistribution measures can be taken even more swiftly than in a market economy, and the degree of division of labour can be planned regardless of inflation. Open inflation, therefore, is undoubtedly a better alternative than persistent excess demand.

It may or may not be the case that an environment of generalized abundance will make it possible to reduce the scope of markets (of commodity production in Marxian parlance) without adverse effects on labour supply, moral hazards, etc., but it is absurd to pretend that abundance has already been achieved when it has not. As long as markets are used for the distribution of at least some consumption goods, there can be no doubt that the restoration of equilibrium in those markets at a single price is a desirable, indeed essential task for the socialist planner. Markets may be rightly distrusted in their role in investment allocation, and even in the allocation of current resources (for the neglect of externalities,inability to deal competitively with increasing returns, etc.), but remain unrivalled systems of distributing whatever amounts and kinds of consumption goods and services a socialist economy of the Soviet-type makes available to its citizens, apart from public goods (and possibly saturated needs). There is nothing socialist or even Marxist–Leninist about excess demand. There is nothing in the original design or in the philosophy of socialism that justifies the distribution of consumption goods and services to the population through the inefficient, unfair and degrading mechanisms of queueing, informal exchanges and barter. Marx incited his followers to change the world, not to pretend that they had changed it: if socialism, through a combination of bad luck and ill judgement, has failed to deliver price stability it is better to recognize it than to ignore this failure or treat it as a passing temporary self-regulating minor phenomenon. The persistence of excess demand, indeed the elevation of shortage to a systemic feature, leads to a prima facie case for suspecting that it is maintained primarily because it conceals the privileges of the élite through exclusive access to luxuries and necessities at abnormally low prices, while market-clearing prices would reveal and quantify privilege, as its maintenance would require drastically more unequal incomes and wealth.

5.10 SUMMARY AND CONCLUSIONS

In a centrally planned economy of the Soviet type, inflationary pressures may arise due to planning error, unfavourable exogenous factors and downward inflexibility of money wages (section 5.2). These pressures do not always take the form of open official inflation (section 5.3) in view of systemic commitment to the maintenance of price stability; in this case inflationary pressures take the form of hidden and/or repressed inflation.

Hidden inflation is a differential between official (state and nonstate) price indices and the actual cost of living to consumers; it arises from understatement of actual prices (in the state and nonstate sectors) and/or of the relative weight of the more inflationary goods or sectors. Repressed inflation is rising excess demand for goods and services at the actual prices in the state and nonstate sectors, i.e. rising excess liquid balances in the hands of the population (often the very existence of excess demand is defined as repressed inflation; see section 5.4).

Estimates of hidden inflation in the recent literature (reviewed in section 5.5) are based on direct recomputation of indices, the calculation of implicit deflators and international comparisons of purchasing power parities. These estimates neglect the impact on the price level of the retrading of goods originally supplied in the state sector and other production and distribution activities outside the state sector (except the collective farm market), i.e. they understate the price differential in the controlled and uncontrolled (legal and illegal) market. Also the actual price level, being an average of prices in a two-tier system, cannot be identified with an equilibrium level.

Excess demand and its rate of change (i.e. the rate of repressed inflation) are the object of wildly diverging estimates (section 5.6). These range from alarmed outcries of impending financial catastrophe, based on the visible growth of financial (cash and non-cash) assets with respect to income and expenditure, to cool dismissive counterclaims that excess demand is negligible, based on the observation of uncontrolled transactions in the second economy and on econometric testing and quantification. Econometric models based on quantity-rationing consumer behaviour models seem inappropriate in view of uncontrolled-price transactions where individual consumers and not rationed, and therefore the postulated adverse feedback on labour supply is not necessarily present. By and large excess money holdings in the economy as a whole (i.e. with reference to the two-tier market) can be deemed to be small regardless of reliance on econometric evidence, except for the presence of lags and other minor factors which might slow down the adjustment of money balances to the actual two-tier price level. However, the small size or even absence of excess money holdings (and therefore of repressed inflation) is consistent with large-scale imbalance in the state sector, since consumers who – given the two-tier market – voluntarily hold liquid assets, do not refrain voluntarily from converting most or all of them into goods in state shops where they *are* quantity-rationed.

A simple model has been presented illustrating the complexity and time profile of the problem under discussion, due to the difference between the price level which would preserve equilibrium if it had been prevailing for a sufficiently long period, and the price levels (depending on the speed of adjustment) and their time paths which can restore market clearing at single prices starting from a two-tier market in a state of imbalance (section 5.7). The model has been used to analyse states of persistent equilibrium, creeping repressed inflation (i.e. excess demand growth being tolerated because it remains diffused, fluctuating and not exceeding a critical average threshold), hidden inflation (as defined above in this section) and excess demand/repressed inflation which, when referred to the state sector, is large and endemic in Soviet-type economies.

Alternative stabilization paths and their combinations have been investigated, using as policy instruments the price level, a liquid wealth tax (or currency conversion) and supply improvements (section 5.8). All paths have three stages: transition to market clearing at a single price (instantly reached only with a currency conversion at the 'right' rate) which implies overshooting with respect to long-run equilibrium, restoration of market-clearing habits on the part of disillusioned consumers, and eventually the replenishing of liquid asset holdings by reassured consumers.

Finally, the costs of hidden inflation and above all of excess demand (and repressed inflation) in the state sector, and the two-tier market associated with excess demand at state prices, are discussed (transaction costs, misallocations and hindrance of economic reform, random distributive effects, etc., see section 5.9) and are found to be greater than the cost (if any) of open inflation. In sum, the task of stabilization in Soviet-type economies is seen as large scale, time consuming and complex, and its implementation imperative.

ACKNOWLEDGEMENTS

Presented at the Third World Congress for Soviet and East European Studies, Washington DC, 30 October – 4 November 1958. An earlier draft was presented at the AISSEC Conference (Turin, 25–26 October 1984), the EUI Working Group on Comparative Economic Systems (Florence, 22 November 1984), and the Department of Economics of Warsaw University (19 April 1985); acknowledgements are due to participants in these events for useful comments. This research is part of an EC funded project on 'East–West Trade and Financial Relations in the 1980s', conducted by the author at the EUI. Reprinted with permission from *Contributions to Political Economy*, 5, pp. 37–82 (© Academic Press Inc. [London] Ltd [1986]).

REFERENCES AND FURTHER READING

Adam, J. (1979) *Wage Control and Inflation in the Soviet Bloc Countries* Macmillan, London.

Alton, T. H. *et al.* (1979) Official and alternative consumer price indexes in Eastern Europe, selected years, 1960–78, Working Papers, LWIFR, New York, September.

Askanas, B. and Laski, B. (1985) Consumer prices and private consumption in Poland and Austria. *Journal of Comparative Economics*, **9** (2), June, 164–77.

Barro, R. and Grossman, H. (1974) Suppressed inflation and the supply multiplier. *Review of Economic Studies*, **41**, 87–104.

Barro, R. and Grossman, H. (1976) *Money, Employment and Infection.* Cambridge University Press, Cambridge.

Bauer, T. (1983) A note on money and the consumer in Eastern Europe. *Soviet Studies*, **35** (3), July, 376–84.

Becker, A. S. (1974) The price level of Soviet machinery in 1960. *Soviet Studies*, **26** (3), July, 363–73.

Birman, I. (1980a) The financial crisis in the USSR. *Soviet Studies*, **32** (1), January.

Birman, I. (1980b) A reply to Professor Pickersgill. *Soviet Studies*, **33** (4), October.

Bronson, D. and Severin, B. (1966) Recent trends in consumption and disposable money income in the USSR. *New Directions in the Soviet Economy*, US Congress, JEC, Washington, pp. 495–530.

Bronson, D. and Severin, B. (1973) Soviet consumer welfare: the Breznev era. In *Soviet Economic Prospects for the Seventies*, US Congress, JEC, Washington.

Brus, W. and Laski, K. (1983) Repressed inflation and the second economy under central planning, Conference on the Economics of the Shadow Economy, Bielefeld.

Cassel, D. (1984) Inflation in socialist planning economies – causes, effects and policy options of inflation in Soviet-type command economies, Working Paper No. 64, Fachbereich Wirtschaftswissenschaft Universität Duisberg-Gesamthochschule.

Chapman, J. G. (1963) *Real Wages in Soviet Russia since 1928*, Cambridge, Mass.

Charemza, W. (1984) The centrally planned economy in the international crisis: Poland 1974–80, CREES Annual Conference, Cumberland Lodge, Windsor, July.

Charemza, W. and Gronicki, M. (1982) Rational expectations in a disequilibrium model for centrally planned economies: the case of Poland, Discussion Paper on econometrics and statistics, No. 14, University of Gdansk, March.

Charemza, W. and Gronicki, M. (1984) Simultaneous consumption-labour-money analysis in disequilibrium framework: Poland 1960–1980, CREES Discussion Paper, University of Birmingham, March.

Charemza, W. and Quandt, R. E. (1982) Models and estimation of disequilibrium for centrally planned economies. *Review of Economic Studies*, **69**, 109–16.

Cohn, S. H. (1981) A comment on Alec Nove note on growth investment and price indexes. *Soviet Studies*, **33** (2), April, 300–1.

Culberston, W. P. and Amacher, R. C. (1972) Inflation in the planned economies: some estimates for Eastern Europe. *Sourthern Economic Journal*, **45** (2), 380–93.
Delhaes, K. V. (1978) Mis-allocations as cause of inflationary processes in centrally administered economic systems, analysis of a model. *Jahrbuch für Socialwissenschaft*, **29**, 282–6.
Dirksen, E. (1981) The control of inflation? Errors in the interpretation of CPE data. *Economica*, **48**, August, 305–8.
Drewnowski, J. (ed.) (1982) *The Polish Disease*, Croom Helm, London.
Dudinskiy, I. (1976). Ekonomicheskie zakonomiernosti mirovoi sistemy sotsialisma. *Voprosy ekonomiki*, No. 9, 23–33.
ECE-UN (1953) *Economic Survey of Europe since the War*, Geneva.
Flaetgen, S. (1985) Monetary stability in a centrally planned economy – the case of the GDR, PhD dissertation, European University Institute, Florence.
Gardner, R. and Strauss, J. (1981) Repressed inflation in the Soviet Union: a temporary equilibrium approach. *European Economic Review*, **16** (2–3), 387–404.
Garvy, G. (1977) *Money, Financial Flows and Credit in the Soviet Union*, Ballinger, Cambridge, Mass.
Glushkov, N. T. (1982) Tsenoobrazovanie i khoziaistvennyi mekhanizm. *Ekonomika i organizatssia promyshlennogo proizvodstva*, No. 9, 3–33; also in *Problems of Economics*, August, 1983, pp.24–56.
Gomulka, S. (1979) Poland's economic situation in the second half of the 1970s. *Osteuropa Wirtschaft*, **24**, 13–23.
Green, D. W. (1978) Household expenditures and the demand for money in the German Democratic Republic (*mimeo*).
Grossman, G. (1977) The second economy of the USSR. *Problems of Communism*, September–October, pp. 25–40; also in Tanzi (1982) pp. 245–69.
Grossman, G. (1985) The second economy in the USSR and Eastern Europe: a bibliography, Paper No. 1, Berkeley-Duke Occasional Papers on the Second Economy in the USSR.
Hartwig, K.-H. (1983). Involuntary liquid assets in Eastern Europe: some critical remarks, *Soviet Studies*, **35** (1), January.
Herer, W. and Sadowski, W. (1981) Nawis inflacyjny: I: Zjawisko i mechanizmy; II: Reforma cen. *Zycie Gospodarcze*, Nos 32 and 33.
Holzman, F. D. (1960) Soviet inflationary pressures, 1928–1957: causes and cures. *Quarterly Journal of Economics*, **74** (2), May, 167–88.
Houthakker, H. S. and Taylor, L. D. (1970) *Consumer Demand in the United States: Analyses and Projections*, Harvard University Press, Cambridge, Mass.
Howard, D. H. (1976a) A note on hidden inflation in the Soviet Union. *Soviet Studies*, **28** (4), October.
Howard, D. H. (1976b) The disequilibrium model in a controlled economy: an empirical test of the Barro–Grossman model. *The American Economic Review*, **66** (5), 871–9.
Howard, D. H. (1980) Hidden inflation in the Soviet Union: a reply to Professor Rosefielde. *Soviet Studies*, **32** (4), October.
Katsenelinboigen, A. (1977) Coloured markets in the Soviet Union. *Soviet Studies*, **29**, January, 62–85.
Kelly, A. C., Sanderson, W. C. and Williamson, J. G. (1983) *Modelling Growing Economies in Equilibrium and Disequilibrium*, Duke Press Policy Studies, Durham, NC.
Kohn, M. J. (1980) Consumer price developments in Eastern Europe, US Congress, JEC.
Kornai, J. (1980) *Economics of Shortage* (2 vols), North Holland, Amsterdam.
Kornai, J. (1982) *Growth, Shortage and Efficiency – a Macrodynamic Model of the Socialist Economy*, Blackwell, Oxford.
Kronrod, Ya. A. (1960) *Den'gi v sotsialisticheskom obschestve*, Gosfinizdat, Moscow.

Legget, R. E. (1981) Measuring inflation in the Soviet machine-building sector. *Journal of Comparative Economics*, 5.

Lavlan, Y. (ed.) (1973) *Banking, money and credit in Eastern Europe*, Brussels.

Malafeyev, A. (1964) *Istoriya tsenoobrazovaniya v SSSR*, Moscow.

Marer, P. and Montias, J. M. (1980) *East European Integration and East–West Trade*, Indiana University Press, Bloomington, Indiana.

Montias, J. M. (1962) *Central Planning in Poland*, Yale University Press New Haven, CT.

Morton, H. W. (1980) Who gets what, when and how? Housing in the Soviet Union. *Soviet Studies*, **32** (2), April, 235–59.

Muth, J. F. (1961) Rational expectations and the theory of price movements. *Econometrica*, **29**, 315–35.

Nove, A. (1979) Inflation in communist countries, in *Political Economy and Soviet Socialism*, Allen and Unwin, London, 178–91.

Nove, A. (1981a) A note on growth, investment and price-indexes. *Soviet Studies*, **33** (1), January, 142–45.

Nove, A. (1981b) A reply to Stanley Cohn. *Soviet Studies*, **33** (2), April, 300–1.

Nuti, D. M. (1981) The Polish crisis: economic factors and constraints. *The Socialist Register 1981*, Merlin Press, London; reprinted in Drewnowski (1982).

Nuti, D. M. (1982) Open and repressed inflation in Poland 1975–1981, NASEES Annual Conference, Cambridge, 27–29 March.

Nuti, D. M. (1984) Economic crisis in Eastern Europe: prospects and repercussions, Working Paper No. 26, European University Institute, Florence, Italy, July.

Nutter, G. (1962) *Growth of Industrial Production in the Soviet Union*, Princeton.

Ofer, G. and Pickersgill, J. (1980) Soviet household saving: a cross-section study of Soviet emigrant families. *Quarterly Journal of Economics*, **44** (4), November.

O'Hearn, D. (1980) The consumer second economy: size and effects. *Soviet Studies*, **32** (2), April, 218–34.

Pickersgill, J. (1976) Soviet household saving behavior. *The Review of Economics and Statistics*, **58** (2), May.

Pickersgill, J. (1977) Soviet inflation: causes and consequences. *Soviet Union/Union Soviètique*, **4**, 297–313.

Pickersgill, J. (1980a) A long-run demand function for money in the socialist economy. *Journal of Money, Credit and Banking*, No. 1, February.

Pickersgill, J. (1980b) The financial crisis in the USSR, a comment. *Soviet Studies*, **32** (4), October, 583–5.

Pindak, F. (1983) Inflation under central planning. Jahrbuch für Wirtschaft Osteuropas, **10/2**, reprinted in SUERF Reprint Series, No. 29, Tilburg, 194.

Podkaminer, L. (1983) Estimates of the disequilibria in Poland's consumer markets (1965–1978), in (A. C. Kelly *et al.*), *Modelling Growing Economies in Equilibrium and Disequilibrium*, Dube Press Policy Studies, Durham, NC.

Podolski, T. (1972) *Socialist Banking and Monetary Control*, Cambridge University Press, Cambridge.

Portes, R. (1977) The control of inflation: lessons from East European experience. *Economica*, **44**.

Portes, R. (1979) Internal and external balance in a centrally planned economy. *Journal of Comparative Economics*, **3** (4), December, 325–45.

Portes, R. (1981a) Central planning and monetarism: fellow travellers? Working Paper No. 782, NBER, Cambridge, Mass., October.

Portes, R. (1981b) Reply to E. Dirksen, 'The control of inflation? Errors in the interpretation of CPE data' *Economica*, **48**, August, 309–11.

Portes, R. (1984) The theory and measurement of macroeconomic disequilibrium in centrally planned economies (*mimeo*) CEPR, October.

Portes, R., Quandt, R., Winter, D. and Yeo, S. (1983) Macroeconomic planning and disequilibrium estimates for Poland, 1955–1980, Working Paper No. 1182, NBER, Cambridge, Mass., August.

Portes, R. and Winter, D. (1977) The supply of consumption goods in centrally planned economies. *Journal of Comparative Economics*, **4**, December, 351–65.

Portes, R. and Winter, D. (1978) The demand for money and for consumption goods in centrally planned economies. *Review of Economics and Statistics*, **60**.

Portes, R. and Winter, D. (1980) Disequilibrium estimates for consumption goods markets in centrally planned economies. *Review of Economic Studies*, **47**, 137–59.

Powell, R. P. (1972) Monetary statistics, in *Soviet Economic Statistics* (V. S. Treml and J. P. Hardt (eds)), (1972) Duke University Press, Durham, NC.

Rosefielde, S. (1980) A comment of D. Howard's estimate of hidden inflation in the Soviet Union retail sales sector. *Soviet Studies*, **32** (3), July.

Rosefielde, S. (1981a) Hidden inflation in the Soviet Union: a rejoinder to David Howard. *Soviet Studies*, **33** (4), October.

Rosefielde, S. (1981b) Are Soviet industrial-production statistics significantly distorted by hidden inflation? *Journal of Comparative Economics*, **5**.

Rosefielde, S. (1983) Disguised inflation in Soviet industrial sector: a reply to J. Steiner. *Journal of Comparative Economics*, **7**.

Rudcenko, S. (1979) Household money income, expenditure and monetary assets in Czechoslovakia, GDR, Hungary and Poland, 1956–75. *Jahrbuch der Wirtschaft Osteuropas*, **8**, 431–50.

Schroeder, G. E. (1975) Consumer goods availability and repressed inflation in the Soviet Union. *Economic Aspects of Daily Life in USSR*, NATO, Brussels.

Schroeder, G. E. and Severin, D. (1976) Soviet consumption and income policies in perspective, US Congress, JEC, *Soviet Economy in a New Perspective*, USGPO, Washington DC.

Severin, B. (1979) USSR: The all-Union and Moscow collective farm market price indexes. *ACES Bulletin*, **21**, 23–36.

Steiner, J. E. (1982) Disguised inflation in Soviet industry. *Journal of Comparative Economics*, **6**, September, 278–87.

Steiner, J. E. (1983) Disguised inflation in Soviet industry: a final reply to Steven Rosefielde. *Journal of Comparative Economics*, No. 4, December, 449–51.

Tanzi, V. (ed.) (1982) *The Underground Economy in the US and Abroad*, Lexington Books, Lexington, Mass.

Turcan, G. (1977) Some observations on retail distribution in Poland. *Soviet Studies*, **29** (1), January, 128–36.

Vanous, J. (1984) Macroeconomic adjustment in Eastern Europe in 1981–83. *Wharton Centrally Planned Economies Current Analysis*, **3** (28–29), 23 April.

Volkonskii, V. A., Solovyev, J. P. and Semenova, I. V. (1976) Voprosy modelirovaniya denezhnykh sberezhenii naseleniya. *Ekonomika i Mathematicheskiye Metody*, No. 3.

Wiles, P. J. D. (1973) Cost inflation and the state of economic theory. *Economic Journal*, **83**.

Wiles, P. J. D. (1974) The control of inflation in Hungary, January 1968–June 1973. *Economic Appliquée*, No. 1, Paris.

Wimberley, J. (1981) The Soviet financial crisis? A comment. *Soviet Studies*, **33** (3), July.

Wolf. T. (1980) On the adjustment of centrally planned economies to external economic disturbances, in (P. Marer and J. M. Montias), *East European Integration and East–West Trade*, Indiana University Press, Bloomington, Indiana.

Zwass, A. (1979) *Money, Banking and Credit in the Soviet Union and Eastern Europe*, London. *Report o stanie gospodraki* (cited as 'Raport') (1981). Nakladem Trybuny Ludu, Warsaw, July.

6

Disequilibrium econometrics for centrally planned economies

RICHARD E. QUANDT

6.1 INTRODUCTION

The equilibrium model has been the dominant economic paradigm and has been the foundation of the early Cowles Commission work on estimating the parameters of simultaneous equations models. According to this basic view, behavioural, technological and definitional relations among endogenous and exogenous variables exist and, if the model is correctly specified, describe the manner in which 'nature' generates the observations on the contemporaneous endogenous variables. Nature 'knows' the values of the parameters and the values of the predetermined (exogenous and lagged endogenous) variables; she generates in each time period some random shocks, and solves for the observable endogenous variables by solving a system of simultaneous (linear or nonlinear) equations. The models are called equilibrium models for two related but not wholly satisfactory reasons.

(1) The values of the endogenous variables observed by the econometricians satisfy the stochastic equations describing the economy, in the sense that price and quantity in an ordinary demand–supply model satisfy both the demand and supply equations and thus represent the intersection of the two – hence the price in such a model is market-clearing or equilibrating.
(2) To the extent that the behavioural relations in the model represent the consequences of optimization by economic agents, they succeed in solving their (constrained) optimization problems – hence no incentives are set up that force modification of their behaviour.

In contrast, disequilibrium models, according to the common terminology, are

International Studies in Economic Modelling Series.
Models of Disequilibrium and Shortage in Centrally Planned Economies.
Edited by C. Davis and W. Charemza.
Published in 1989 by Chapman and Hall Ltd, 11 New Fetter Lane, London EC4P 4EE.

those in which some agents are 'off their behavioural curves'. In the demand–supply model the price no longer corresponds to the intersection of the two curves. According to the common interpretation of the Walrasian auction, the presence of excess demand (supply) sets up forces in the flexible-price case to change prices; in the disequilibrium model the price is rigid and equilibrating does not occur. Hence agents presumably do not succeed in solving their constrained optimization problems which, *ceteris paribus*, ought to result in a revision of their plans.

In fact, this difference between equilibrium and disequilibrium models is more apparent than real. In the Walrasian case, consumers maximize a utility function $U(\mathbf{x})$ with respect to the elements of the commodity vector $\mathbf{x}$ subject to a budget constraint $p'\mathbf{x} \leqq y$, where y is their monetary endowment. If $U(\mathbf{x})$ has positive partial derivatives, at the optimum we shall have $p'\mathbf{x} = y$, and if we actually observe them buying a commodity bundle $\mathbf{x}^0$ such that $p'\mathbf{x}^0 < y$, we are inclined to say that they have not reached equilibrium (and should accordingly alter their behaviour). On the other hand, consider the possibility that we have forgotten to specify their budget constraint altogether, but have specified correctly the quantity constraint $\mathbf{x} \leqq \bar{\mathbf{x}}$. Under the same assumption about the utility function, the optimum occurs at $\mathbf{x} = \bar{\mathbf{x}}$; thus if in fact we observe that an element $\mathbf{x}_i < \bar{\mathbf{x}}_i$, we would deduce that they are in disequilibrium and ought to modify their behaviour. In both cases, the conclusion is due to incomplete specification; in the first case we have 'forgotten' the quantity constraints and in the second the budget constraint. A correct specification then is to maximize $U(\mathbf{x})$ subject to both $p'\mathbf{x} \leqq y$ and $\mathbf{x} \leqq \bar{\mathbf{x}}$. The principal theoretical difference between equilibrium and disequilibrium models is thus whether constraints are satisfied as equalities or strict inequalities and not whether forces are set up in the disequilibrium case to modify agents' behaviour. This view is broadly consistent with Kornai's (1979, 1980) perception that shortage is the normal state of the market in centrally planned economies: in such a case it is normal that the demands and/or supplies of agents continue to be rationed. A more practical difference, with profound econometric implications, is that not all endogenous variables are observable by the econometrician: considering these variables to be solution values from an equation system derived by imposing equality constraints, some agents are indeed 'off their curves'. This places the econometrics of disequilibrium models in the realm of latent variable models.

6.2 SPECIFICATION OF MODELS

We do not propose to discuss exhaustively all types of disequilibrium models, nor elucidate all the differences in applying them to free market versus centrally planned economies. We restrict ourselves to highlighting the principal issues and illustrating them with pertinent examples. All that follows should be viewed in the light of three general remarks on which we also elaborate later as necessary.

(1) The economic system (household, market, industry, national economy) that we

purport to describe may involve in our specification more or less aggregation. The volume of aggregation that is performed prior to estimation may be chosen for convenience or compelled by the nature of the data, but whatever the causes, aggregation of some amount will almost invariably be present. Unless some extremely strong conditions are imposed, aggregation will tend to bias the results and will do so whether we are dealing with free market or centrally planned economies. The extent of this bias is largely an empirical question and there is no a priori evidence that the bias is generally more or less serious in free market than centrally planned economies.

(2) The cause of disequilibrium in free market economies is frequently posited to be the stickiness in wage and price adjustment, and it is a defect of model specification if the sources of this stickiness are neglected. In CPEs the stickiness of wage and price adjustment may be more easily acceptable on institutional grounds and variables other than wage and price (such as the central plan) may play a larger role in equilibrating the market; but then it behoves us to inquire into the sources of the adjustment behaviour of these other variables. In both types of cases a formal specification of the adjustment mechanisms, if any, is required.

(3) It may be thought that CPE's are so different from free market economies, that the same econometric approaches cannot be applied to both. Kornai's (1979, 1980) theory of the soft budget constraint leads to the enterprise's input demand being (nearly) infinite. A concomitant of that is that the enterprise's supply function is not well defined. One would therefore deduce that an econometric apparatus built upon Barro–Grossman (1971) type foundations is likely to be unproductive or misleading. We reject this view here. If the firm's input demand is infinite and, consequently, its goods supply not well defined, it is difficult to comprehened how firms in CPEs can make determinate input and output decisions. What is a much more plausible interpretation is that the firm's supply function cannot be obtained from a mechanism as simple as profit maximization (or even sales maximization, or what have you). The appropriate theory of the firm or CPEs is perhaps yet to be specified (but see Goldfeld and Quandt (1988)), but *a* supply function is alive and well and there is no reason to eschew the conventional econometric disequilibrium approach. (For a detailed examination of Kornai's theory see also Chapter 3.)

6.2.1 Disequilibrium models in free market economies

(a) *The canonical model*

The simplest model is that of a single market in which purchase and selling intentions are represented by a demand and a supply function given by

$$D_t = \alpha_1 p_t + \beta_1' \mathbf{x}_{1t} + u_{1t} \tag{6.1}$$

$$S_t = \alpha_2 p_t + \beta_2' \mathbf{x}_{2t} + u_{2t} \tag{6.2}$$

where p_t denotes the price, $\mathbf{x}_{1t}$, $\mathbf{x}_{2t}$ are vectors of exogenous variables and u_{1t}, u_{2t} error terms (conventionally assumed to be distributed as $N(0, \Sigma)$).[1] In the equilibrium paradigm the observed quantity Q_t is given by $Q_t = D_t = S_t$ and Equations 6.1 and 6.2 represent a pair of simultaneous equations which determine the endogenous variables p_t, Q_t. Efficient and asymptotically normal estimates are obtained by maximizing the likelihood function (Chow, 1983)

$$L = \frac{|\alpha_1 - \alpha_2|^T}{(2\pi)^T |\Sigma|^{T/2}} \exp\{-\tfrac{1}{2}[\sigma^{11}(Q_t - \alpha_1 p_t - \beta_1' \mathbf{x}_{1t})^2 + \sigma^{22}(Q_t - \alpha_2 p_t - \beta_2' \mathbf{x}_{2t})^2$$
$$+ 2\sigma^{12}(Q_t - \alpha_1 p_t - \beta_1' \mathbf{x}_{1t})(Q_t - \alpha_2 p_t - \beta_2' \mathbf{x}_{2t})]\} \tag{6.3}$$

where σ^{ij} represents the elements of Σ^{-1}. Even if the likelihood function is condensed with respect to the elements of Σ^{-1}, nonlinear optimization techniques are required, if for no other reason than the presence of the Jacobian term $|\alpha_1 - \alpha_2|^T$.

If the price is completely rigid, i.e. exogenous, $D_t = S_t$ on a set of measure zero and the observed quantity Q_t is plausibly modelled as

$$Q_t = \min(D_t, S_t) \tag{6.4}$$

on the grounds that neither buyers nor sellers can be forced to transact larger quantities than they desire. Econometricians do not observe either demand or supply; moreover, in general, they do not know whether the Q_t they observe is lying on the demand curve or supply curve. The density function $h(Q_t)$ for the observable random variable Q_t is obtained from elementary probability considerations as follows.

Let $g(D_t, S_t)$ be the joint density of D_t, S_t, which can be obtained directly from Equations 6.1 and 6.2. Then we have

$$h(Q_t) = f(Q_t | D_t < S_t) \Pr\{D_t < S_t\} + f(Q_t | D_t \geqq S_t) \Pr\{D_t \geqq S_t\} \tag{6.5}$$

But the conditional density $f(Q_t | D_t < S_t)$ is given by

$$\int_{Q_t}^{\infty} g(Q_t, S_t)\, \mathrm{d}S_t / \Pr\{D_t < S_t\}$$

and similarly for $f(Q_t | D_t \geqq S_t)$; hence

$$h(Q_t) = \int_{Q_t}^{\infty} g(Q_t, S_t)\, \mathrm{d}S_t + \int_{Q_t}^{\infty} g(D_t, Q_t)\, \mathrm{d}D_t \tag{6.6}$$

In the case in which (u_1, u_2) are distributed as $N(0, \Sigma)$, (6.6) becomes

$$h(Q_t) = \theta_{1t}\theta_{2t} + \theta_{3t}\theta_{4t}$$

where

$$\theta_{1t} = \frac{1}{\sqrt{(2\pi)}\sigma_1} \exp\left[-\frac{(Q_t - \alpha_1 p_t - \beta'_1 \mathbf{x}_{1t})^2}{2\sigma_1^2}\right]$$

[1] At this point it is of no great importance to specify whether 6.1 and 6.2 refer to a single market or an aggregate of markets – some degree of aggregation already underlies these functions in any event.

$$\theta_{2t} = 1 - \Phi\left[\frac{Q_t - \alpha_2 p_t - \beta_2' \mathbf{x}_{2t} - \rho(\sigma_2/\sigma_1)(Q_t - \alpha_1 p_t - \beta_1' \mathbf{x}_{1t})}{\sigma_2(1-\rho^2)^{1/2}}\right]$$

$$\theta_{3t} = \frac{1}{\sqrt{(2\pi)}\sigma_2} \exp\left[-\frac{(Q_t - \alpha_2 p_t - \beta_2' \mathbf{x}_{2t})^2}{2\sigma_2^2}\right]$$

$$\theta_{4t} = -\Phi\left[\frac{Q_t - \alpha_1 p_t - \beta_1' \mathbf{x}_{1t} - \rho(\sigma_1/\sigma_2)(Q_t - \alpha_2 p_t - \beta_2' \mathbf{x}_{2t})}{\sigma_1(1-\rho^2)^{1/2}}\right]$$

where $\rho = \sigma_{12}/\sigma_1\sigma_2$ and $\Phi(\quad)$ is the cumulative standard normal integral. Hence the likelihood is $L = \prod_t h(Q_t)$.

(b) *A price-adjusting model*

In general, it may not be plausible to argue that prices are completely rigid, i.e. exogenous. A disequilibrium model with endogenous (but sticky) prices is obtained by adjoining to the model consisting of equations 6.1, 6.2 and 6.4 the equation

$$p_t = p_{t-1} + \gamma(D_t - S_t) + u_{3t} \tag{6.7}$$

The equation expresses the sensible enough observation that price tends to increase if excess demand is positive and declines otherwise (except for the effect of the error term u_{3t}, which might be suppressed altogether in some formulations). Models incorporating such a price adjustment equation have been estimated by Rosen and Quandt (1978). But the equation is somewhat peculiar, in that the agent who is responsible for changing the price from yesterday to today must, in order to do so, know today's excess demand, which depends on the price to be set for today. Although we are normally quite comfortable with simultaneity, the behavioural basis of the equation does not seem to rest on solid foundations.

A somewhat more solid foundation is obtained if the adjustment equation can be derived from some agent's optimizing behaviour. To be sure, the unscrupulous can play endless games here, since presumably any adjustment equation is derivable from optimizing *some* objective function – hence to provide better justification for the adjustment equation at hand, all we would have to do is to specify some suitable objective. But inspection of objective functions may in itself be revealing and may pinpoint absurdities in one's thinking; moreover, tests of the resulting adjustment equation then become tests about some agent's behaviour, which is an eminently sensible preoccupation for the econometrician.

The adjustment equation at hand can, as is well known, be derived by minimizing the loss function $(p_t - p_{t-1})^2 + \theta(D_t - S_t)^2$ with respect to p_t; it expresses that costs are incurred due both to price change and to the existence of disequilibrium. It is not our objective here to pursue further this issue with respect to free market economies. We now turn to the case of centrally planned economies.

6.2.2 Disequilibrium models for CPEs

(a) *Models with rigid prices*

The first model for CPEs to incorporate the technology of disequilibrium econometrics is that of Portes and Winter (1980). It attempts to explain the demand and supply of aggregate consumption in four CPEs (Poland, Hungary, Czechoslovakia and the GDR) on the basis of a simple disequilibrium model that incorporates, besides the demand and supply functions, only the min condition.[2] The central plan featured prominently in this model (although actual plan data were not yet available for its implementation). It was subsequently argued by Charemza and Quandt (1982) that central planners could reasonably be regarded as endogenous agents in the economy whose behaviour determines the level of the central plan. This central plan level is an endogenous variable that is affected by other economic variables and, in turn, affects others.

Let D_t, S_t denote, as before, the demand for and supply of some aggregate commodity, let Q_t be the level of actual transactions and let Q_t' be the planned level (i.e. the level of transactions envisaged by the central plan).[3] Then the demand and supply functions in Charemza and Quandt (1982) can be written as

$$D_t = \alpha_1 p_t + \beta_1' \mathbf{x}_{1t} + \psi_1 W_t + u_{1t} \tag{6.8}$$

$$S_t = \alpha_2 p_t + \beta_2' \mathbf{x}_{2t} + \psi_2 W_t + u_{2t} \tag{6.9}$$

where $W_t = Q_t^*$ if Q_t^* denotes the plan for period t and $W_t = Q_{t-1}^*$ if the convention is that Q_t^* is the plan announced in period t for the subsequent period. In Equations 6.8 and 6.9 we retain the explicit appearance of the price terms on the grounds that price does influence at least demand, even though the price may be truly exogenous here. W_t affects demand to the extent that it reveals something about the probability of rationing; this probability clearly does affect the effective demand.[4] W_t affects supply directly; in fact it may be argued that in the absence of surprises, the central plan *is* the supply. Hence both coefficients ψ_1 and ψ_2 are likely to be positive.

What determines Q_t^*? The simplest quadratic loss function attributed to planners by Charemza and Quandt has them suffer disutility from the existence of disequilibrium and from plan changes; thus the loss function is $V = (Q_t^* - Q_{t-1}^*)^2 + \mu(D_t - S_t)^2$.[5] Its minimization with respect to Q_t^* (in the event

[2]We return to a more detailed discussion of this later in this section.

[3]There is an obvious problem of timing here that must be paid attention in any practical implementation. Plans set in some period t are to be effective in some future period; moreover, plans are often revised and so a more careful notation would be $Q_{t|\tau}$ to represent the plan for period t as it is known or announced in period τ. We ignore this refinement here.

[4]This is the 'discouraged worker' syndrome in the labour market context: workers who expect to be unemployed may stop seeking jobs. See Eaton and Quandt (1983).

[5]This is only one of several possible approaches. A much more complete and econometrically ambitious approach is in Charemza and Dlouhý's (1987) quantity rationing model in which planners maximize a utility function defined over consumption and government expenditures, subject to an accounting identity, an export supply constraint and behaviour relations in consumption and labour markets.

that $W_t = Q_t^*$) yields

$$\frac{\partial V}{\partial Q_t^*} = 2(Q_t^* - Q_{t-1}^*) + 2\mu(D_t - S_t)(\psi_1 - \psi_2) = 0$$

which leads to the adjustment equation

$$Q_t^* = Q_{t-1}^* + \gamma(D_t - S_t).$$

For estimation purposes, an error term would ordinarily be added to this. Alternatively, planners may wish to minimize the discrepancy between plans and realizations (with due account being taken of excess demand as well). Letting Q_t^* denote the plan for period t as before, the objective function is $V = (Q_t^* - Q_t)^2 + \mu(D_t - S_t)^2$. Differentiating we obtain

$$\frac{\partial V}{\partial Q_t^*} = 2(Q_t^* - Q_t)\left(1 - \frac{\partial Q_t}{\partial Q_t^*}\right) + 2\mu(D_t - S_t)(\psi_1 - \psi_2) = 0 \qquad (6.10)$$

But $Q_t = D_t$ if $D_t < S_t$ and $Q_t = S_t$ otherwise; hence

$$\frac{\partial Q_t}{\partial Q_t^*} = \begin{cases} \psi_1 & \text{if } D_t < S_t \\ \psi_2 & \text{otherwise.} \end{cases}$$

The adjustment equation thus has two forms, depending on the sign of excess demand and yielding the type of adjustment function discussed, for example, by Laffont and Monfort (1976). In the present case it is

$$Q_t^* = Q_t + \left(\frac{\psi_2 - \psi_1}{1 - \psi_i}\right)\mu(D_t - S_t) \qquad (6.11)$$

where $i = 1$ if $D_t < S_t$ and $= 2$ otherwise. A priori, both ψ_1 and ψ_2 are positive and it is plausible to argue that the demand effect is smaller. Provided that $\psi_2 < 1$, the adjustment coefficient is positive in both regimes and is larger in the excess demand regime.

In models of this kind, two further aspects deserve exploration. We first rewrite the model in more compact notation as follows:

$$\begin{aligned} D_t &= m_{1t} + \psi_1 Q_t^* \\ S_t &= m_{2t} + \psi_2 Q_t^* \\ Q_t &= \min(D_t, S_t) \\ Q_t^* &= Q_t + \gamma_i(D_t - S_t) + m_{3t} \end{aligned} \qquad (6.12)$$

where $i = 1$ if $D_t < S_t$, and $= 2$ otherwise and where m_{1t}, m_{2t}, m_{3t} represent exogenous variables and error terms. The first possible problem is that the model may fail to be coherent (Gourieroux, Laffont and Monfort, 1980); i.e. if the mapping from the m's to D, S, Q^* is not one-to-one. This is equivalent to saying that there might not exist a well-defined reduced form. Let

$y_t' = (D_t S_t Q_t^*)$, and let $m_t' = (m_{1t} m_{2t} m_{3t})$. Then the regimes of the model are

$$A_1 y_t = m_t \quad \text{if} \quad D_t < S_t$$
$$A_2 y_t = m_t \quad \text{if} \quad D_t \geqq S_t$$

where

$$A_1 = \begin{bmatrix} 1 & 0 & -\psi_1 \\ 0 & 1 & -\psi_2 \\ -(1+\gamma_1) & \gamma_1 & 1 \end{bmatrix}, A_2 = \begin{bmatrix} 1 & 0 & -\psi_1 \\ 0 & 1 & -\psi_2 \\ -\gamma_2 & -(1-\gamma_2) & 1 \end{bmatrix}$$

The necessary and sufficient conditions for coherency, without which attempting estimation makes no sense, is that $\det(A_1)$ and $\det(A_2)$ have the same sign. But

$$\det(A_1) = 1 + \gamma_1(\psi_2 - \psi_1) - \psi_1$$
$$\det(A_2) = 1 + \gamma_2(\psi_2 - \psi_1) - \psi_2$$

The two determinants will have same sign if ψ_1, ψ_2 are near each other in value but may have opposite signs in the converse case.

The second issue pertains to the notion of equilibrium. Solving Equation 6.12 for D_t and S_t in the two regimes, we obtain

$$\begin{bmatrix} D_t \\ S_t \end{bmatrix} = \frac{1}{\psi_2 - \psi_1 + \dfrac{1}{\gamma_1}(1-\psi_1)} \begin{bmatrix} m_{1t}\left(\dfrac{1}{\gamma_1} + \psi_2\right) - m_{2t}\psi_1 + \dfrac{m_{3t}\psi_1}{\gamma_1} \\ m_{1t}\psi_2\left(\dfrac{1}{\gamma_1} + 1\right) + m_{2t}\left(-\psi_1 + \dfrac{1}{\gamma_1}(1-\psi_1)\right) + \dfrac{m_{3t}\psi_2}{\gamma_1} \end{bmatrix}$$

if $D_t < S_t$, and

$$\begin{bmatrix} D_t \\ S_t \end{bmatrix} = \frac{1}{\psi_2 - \psi_1 + \dfrac{1}{\gamma_2}(1-\psi_2)} \begin{bmatrix} m_{1t}\left(\dfrac{1}{\gamma_1}(1-\psi_2) + \psi_2\right) + m_{2t}\psi_1\left(\dfrac{1}{\gamma_2} - 1\right) + \dfrac{m_{3t}\psi_1}{\gamma_2} \\ m_{1t}\psi_2 + m_{2t}\left(\dfrac{1}{\gamma_2} - \psi_1\right) + \dfrac{m_{3t}\psi_2}{\gamma_2} \end{bmatrix}$$

if $D_t \geqq S_t$. If $\psi_2 \to 1$, i.e. if every dollar's worth of plan-change translates into a dollar's worth of supply-change, $\gamma_2 \to \infty$ and $D_t - S_t \to 0$; hence in the limit, the excess demand regime collapses to equilibrium.

It should be noted that the previous observations are reflections on properties of models and not the real world; the point is that model specification has to take the implied properties into account before sensible estimation may commence.

A more elaborate loss function was introduced by Portes *et al.* (1987) in their macromodel of consumption in Poland. Let D_t, S_t denote demand and supply as before, let C_t denote actual consumption $(=\min(D_t, S_t))$ and let C_t^* be the consumption plan for period t made and announced toward the end of period $t-1$. Then further define

$$x_1 = C_{t+1}^* - (1+g)[\rho C_t + (1-\rho)C_t^*]$$
$$x_2 = C_t - C_t^*$$
$$x_3 = D_{t+1} - S_{t+1}$$
$$x_4 = D_t - S_t$$

where $0 \leqq \rho \leqq 1$ and where g is a growth rate.

They posit that planners attempt to minimize the loss function

$$V = \tfrac{1}{2}a_1x_1^2 + a_2x_1x_2 + \tfrac{1}{2}a_3x_3^2 + a_4x_3x_4$$

where a_1, a_3 and a_4 are expected to be positive and a_2 is expected to be negative. The reasoning behind these expectations is as follows. (1) The argument x_1 represents the steady growth objective based on a convex combination of the current reality and the current period's plan incremented by a suitable growth rate; thus the plan for the next period should not depart too drastically from this growth objective. (2) If x_2 happens to be positive, i.e. if the current reality exceeds the current plan, an excess of next year's plan over the growth target is more easily accommodated (and conversely); this is expressed by the negative sign of a_2. (3) Planning should aim at not permitting disequilibrium in the period planned for to be excessive; this objective is represented by the third term. (4) Finally, persistent disequilibria in the same direction (e.g. excess demand in consecutive periods) are to be avoided, and this is expressed by the last term.

The plan adjustment equation is obtained by differentiating V:

$$\frac{\partial V}{\partial C_{t+1}^*} = a_1x_1 + a_2x_2 + a_3x_3\left(\frac{\partial D_{t+1}}{\partial C_{t+1}^*} - \frac{\partial S_{t+1}}{\partial C_{t+1}^*}\right) + a_4x_4\left(\frac{\partial D_{t+1}}{\partial C_{t+1}^*} - \frac{\partial S_{t+1}}{\partial C_{t+1}^*}\right) = 0.$$

The particular form assumed by this obviously depends on the specification of the demand and supply functions; in the case considered by Portes *et al.* (1987) it becomes

$$C_{t+1}^* = \delta_1 C_t^* + \delta_2 C_t + \delta_3(D_t - S_t) + \delta_4(D_{t+1} - S_{t+1})$$

where the δs depend on g, p and the as.[6]

As before, the adjustment equation is in effect a model of the planners. Other loss

[6]For estimation purposes this equation was further modified by including lagged consumption C_{t-1} (which turned out to be statistically insignificant) and by replacing the future value $D_{t+1} - S_{t+1}$ by a proxy.

functions will yield other adjustment equations. Some of these will be rejected by the data, and others will survive; the point is that the modelling effort concerned with CPEs should have as its target as much the discovery of the planners' behaviour as of the behaviour of other agents.

(b) *Models with flexible prices*

Models with endogenous plan setting of the type discussed above may not be numerous, but at least some have been explicitly formulated and estimated. Models in which there are endogenous price movements are even less well developed. We briefly review some of these.

To the extent that free markets are negligible in size in CPEs, it is plausible that prices represent conscious decisions by planners and their magnitudes may adjust over time as a result, again, of optimizing some suitable loss function. The empirical evidence on plan setting seems to suggest that planners are concerned with the magnitude of disequilibrium and as a first approximation one might posit that similar motivations drive price adjustment as well. The problem is that price changes may not occur as regularly as plan adjustments, which may make a formal description of the process difficult.

As a first approximation, Charemza and Quandt (1982) suggest grafting an ordinary price adjustment equation

$$p_t = p_{t-1} + \mu(D_t - S_t) + u_{4t} \tag{6.13}$$

on to a model of the general form given by Equation 6.12. Price then becomes endogenous and the likelihood function can ordinarily be derived, provided that the model can be shown to be coherent. As an alternative, they have suggested that pricing policy may be linked more directly to the planning process by an equation such as

$$p_t = p_{t-1} + \mu(Q_t^* - Q_t) + u_{4t} \tag{6.14}$$

Both Equations 6.13 and 6.14 imply that the timing of price changes coincides with the timing of plan changes. Since this may well be regarded as unrealistic, alternative formulations have used thresholds. As Charemza (1987) notes,

> Usually in a CPE aggregate price changes in time are imperceptible (but almost always positive) for a number of years and after that a general "price revolution" takes place, with a huge increase in the price level (Charemza, 1987).

One might consider, for example, the following formulation, similar to one suggested by Charemza and Quandt (1982):

$$\begin{aligned} p_t &= p_{t-1} + \lambda(D_t - S_t) + \beta_3' x_{3t} + u_{4t} \quad &\text{if} \quad D_t - S_t \geqq K \\ p_t &= p_{t-1} + \beta_3' x_{3t} + u_{4t} \quad &\text{if} \quad D_t - S_t < K \end{aligned} \tag{6.15}$$

where $K > 0$.

Accordingly, if excess demand is negative or positive and small ($< K$), price is set exogenously. If excess demand is positive and large, price is set in a manner that responds positively to the excess demand. Unfortunately, threshold models of this type are almost always incoherent. To show this, we respecify demand and supply as follows:

$$\begin{aligned} D_t &= \alpha_1 p_t + m_{1t} \\ S_t &= \alpha_2 p_t + m_{2t} \end{aligned} \tag{6.16}$$

where m_{1t}, m_{2t} are shorthand for exogenous variables and parameters, as well as the error terms. We omit the $\psi_1 Q_t^*$, $\psi_2 Q_t^*$ terms that appear in Equation 6.12 in order to simplify; in the present context these add nothing to our argument. Then the model consisting of Equations 6.15 and 6.16 has two regimes which we write in matrix form:

(1) If $D_t - S_t \geqq K$, we have

$$\begin{bmatrix} 1 & 0 & -\alpha_1 \\ 0 & 1 & -\alpha_2 \\ -\lambda & \lambda & 1 \end{bmatrix} \begin{bmatrix} D_t \\ S_t \\ p_t \end{bmatrix} = \begin{bmatrix} m_{1t} \\ m_{2t} \\ m_{3t} \end{bmatrix} \tag{6.17}$$

(2) If $D_t - S_t < K$, we have

$$\begin{bmatrix} 1 & 0 & -\alpha_1 \\ 0 & 1 & -\alpha_2 \\ 0 & 0 & 1 \end{bmatrix} \begin{bmatrix} D_t \\ S_t \\ p_t \end{bmatrix} = \begin{bmatrix} m_{1t} \\ m_{2t} \\ m_{3t} \end{bmatrix} \tag{6.18}$$

where we have abbreviated $p_{t-1} + \beta'_{3x3t} + u_{4t}$ by m_{3t}. Solving each of these systems, we can obtain the reduced form equations for D_t and S_t and hence we obtain expressions for $D_t - S_t$ as

$$D_t - S_t = \begin{cases} \dfrac{1}{\Delta}[m_{1t} - m_{2t} + (\alpha_1 - \alpha_2)m_{3t}] & \text{from (6.17)} \\ [m_{1t} - m_{2t} + (\alpha_1 - \alpha_2)m_{3t}] & \text{from (6.18)} \end{cases} \tag{6.19}$$

where $\Delta = 1 + \lambda(\alpha_2 - \alpha_1)$, which expresses excess demand entirely in terms of given exogenous variables, parameters and error terms. Equations 6.19 are in effect the algorithm that nature uses to solve for the excess demand. But the value of this excess demand is needed for knowing which formula in Equation 6.19 to employ! In order to ensure logical consistency, it must not be true that the first formula of Equation 6.19 gives $D_t - S_t \geqq K$ and the second gives $D_t - S_t < K$, for then both excess demand values would simultaneously be compatible with the exogenous variables, parameters, and error terms. Denoting $m_{1t} - m_{2t} + (\alpha_1 - \alpha_2)m_{3t}$ by A, it must therefore be true that either

$$A/\Delta \geqq K \quad \text{and} \quad A \geqq K$$

or

$$A/\Delta < K \quad \text{and} \quad A < K.$$

This is clearly not assured. Under normal circumstances $\Delta > 1$. Thus, $A/\Delta \geqq K$ implies $A \geqq K$, but $A \geqq K$ is compatible with $A/\Delta < K$. Coherency then would seem to depend on esoteric conditions on exogenous variables and error terms which are incompatible with the assumption of the normality of the errors. Moreover, essentially the same difficulty is encountered if one adopts Charemza's (1987) modification according to which (a) the criterion for regime choice is not $D - S$ but $D - Q$, (b) the endogenous and exogenous price setting modes are reversed in the sense that the Walrasian mechanism by which price adjustment depends on excess demand is in effect when $D - Q$ (or $D - S$) is small.

There is as yet no obvious solution to this quandary. Some advantage may be gained from specifying a different lag structure; for example, Equation 6.15 may be altered to read

$$\begin{aligned} p_t &= m_{3t} + \lambda(D_t - S_t) && \text{if} \quad D_{t-1} - S_{t-1} \geqq K \\ p_t &= m_{3t} && \text{if} \quad D_{t-1} - S_{t-1} < K \end{aligned}$$

for in this case, the coherency of the model is restored; however, it is at the cost of a substantially more complicated likelihood function. Formulation of a realistic and tractable model for introducing price adjustment with threshold effects is thus a matter of considerable priority in econometric research for CPEs.

6.3 ECONOMETRICS IN MODELS WITH CHRONIC EXCESS DEMAND

6.3.1 Introduction

It is often argued that two-sided disequilibrium models, i.e. those that permit both excess demand and excess supply to occur, make less sense in the context of CPE's than in the free market context. This is certainly the thrust of Kornai (1982) who criticizes the Portes–Winter (1980) model for its excess supply predictions.[7] We must therefore address the econometric problems if it is posited that excess demand (supply) was always observed. Superficially one might consider this to be an advantage, for a priori sorting of observations is generally helpful for estimation. Thus, consider the model given by Equations 6.1, 6.2 and 6.4. In general, as we showed in section 6.2.2 the density function $h(Q_t)$ of the observable random variable is given by Equation 6.6, and the likelihood is $L = \prod_t h(Q_t)$. However, if the econometrician has prior information that for $t \in T_1$, $S_t > D_t$ and for $t \in T_2$, $D_t > S_t$, the likelihood function is

$$L = \prod_{t \in T_1} \int_{Q_t}^{\infty} g(Q_t, S_t)\, \mathrm{d}S_t \prod_{t \in T_2} \int_{Q_t}^{\infty} g(D_t, Q_t)\, \mathrm{d}D_t. \tag{6.20}$$

[7]A separate, and here irrelevant, criticism is also the point that the supply function is not defined in socialist enterprises as a result of the soft budget constraint; then no model of the general Barro–Grossman type will be seen to have relevance.

This likelihood function has indisputable advantages over that based on Equation 6.6: the latter is known to be unbounded in parameter space, whereas Equation 6.20 is well behaved.

A particularly important case of Equation 6.20 arises if it is asserted that the market in question always exhibits excess demand or always exhibits excess supply. In that case, the index set T_1 is empty and Equation 6.20 becomes

$$L = \prod_{t=1}^{T} \int_{Q_t}^{\infty} g(D_t, Q_t)\, \mathrm{d}D_t \tag{6.21}$$

Cases of this type have been investigated by Hartley and Mennemeyer (1974) who concentrate the model with respect to the unobserved demand or supply and deduce the values of the structural parameters from full-information maximum likelihood method (FIML) estimates from the concentrated (partially reduced form) model and by Charemza, Gronicki and Quandt (1988; and Chapter 16).

In the present section we investigate with reference to the likelihood function under what circumstances the parameters of the model may remain estimable if there is prior information that excess demand or excess supply has always occurred. This approach highlights the reasons for the absence of estimability in certain cases. It has to be noted that the assumption that excess demand has always occurred represents knowledge about the sample of data. The assumption does not mean that we believe that excess supply can never occur even conceivably, for such an assumption would contradict all reasonable assumptions about the error terms.

6.3.2 The simple disequilibrium model

Let $u_t' = (u_{1t}, u_{2t})$ is distributed as $N(0, \Sigma)$, with diagonal Σ, which is the most common case. The likelihood function corresponding to the density Equation 6.6 becomes

$$L = \prod_{t=1}^{T} \frac{1}{\sqrt{(2\pi)}\sigma_2} \exp\left[-\frac{1}{2} \frac{(Q_t - \alpha_2 p_t - \beta_2' \mathbf{x}_{2t})^2}{\sigma_2^2} \right] \left[1 - \Phi\left(\frac{Q_t - \alpha_1 p_t - \beta_1' \mathbf{x}_{1t}}{\sigma_1} \right) \right] \tag{6.22}$$

where $\Phi(\ )$ is the cumulative standard normal distribution. From the form of Equation 6.22 we immediately derive

(a) *Proposition 1*

If the demand function contains a constant term, the parameters α_1, β_1, σ_1^2 are not estimable.

PROOF

The likelihood is the product over t of a density term and the $1 - \Phi(\ )$ term. These terms share no parameters. Hence L can be maximized by independently maximizing with respect to α_1, β_1, σ_1^2 on the one hand, and α_2, β_2, σ_2^2 on the other. But the value of $1-\Phi(\ )$ can be made to be arbitrarily close to its upper bound of 1.0 by assigning arbitrary values to the slope coefficients $\beta_{11}, \ldots, \beta_{1k}$ and making the constant term β_{10} arbitrarily large.

The same result holds if there is no constant term in the equation but there is a variable that does not change sign over the observations. The fact that makes estimation impossible under these circumstances is that we can increase the relevant coefficient value and thus make the long side of the market even 'longer', which does not violate the a priori restriction. For simplicity we posit henceforth, that all relevant equations contain a constant term.

The typical term of the likelihood function in this case and for many other types of disequilibrium models can always be written as

$$\Psi(\theta_1, \theta_2)\,[1-\Phi(\theta_2, \theta_3)] \tag{6.23}$$

where $\Psi(\ )$ is the density or density-like term, θ_1 is the subset of parameters occurring only in $\Psi(\ )$, θ_3 is the subset occurring only in $\Phi(\ )$, and θ_2 is the subset occurring in both.[8] A necessary condition for the estimability of the equation describing the long-side of the market is then given by:

(b) *Proposition 2*

A necessary condition for estimability is either that the set of coefficients θ_3 be empty or that they be the coefficients of variables that change sign over the sample period.

We conclude this section by relaxing the assumption that the covariance σ_{12} between u_{1t} and u_{2t} is zero. Equation 6.22 then becomes

$$L=\prod_{t=1}^{T}\frac{1}{\sqrt{(2\pi)}\sigma_2}\exp\left[-\frac{1}{2}\frac{(Q_t-\alpha_2 p_t-\beta_2'\mathbf{x}_{2t})^2}{\sigma_2^2}\right] \times\left[1-\Phi\left(\frac{Q_t-\alpha_{1t}-\beta_1'\mathbf{x}_{1t}-\rho(\sigma_1/\sigma_2)(Q_t-\alpha_2 p_t-\beta_2'\mathbf{x}_{2t}}{\sigma_1(1-\rho^2)^{1/2}}\right)\right] \tag{6.24}$$

where $\rho=\sigma_{12}/\sigma_1\sigma_2$, and it is clear that introducing a nonzero covariance is not sufficient for estimating the demand equation, although, as before, the supply equation is estimable.[9]

[8]Note that the parameterization into θ_1, θ_2, θ_3 need not be unique.

[9]It should be noted that sufficient conditions for estimability are in general harder to obtain. Consider the following (artificial) case in which we know a priori that $\alpha_1=\alpha_2$, $\beta_1=\beta_2$, the variance of u_{1t} is σ_1^2 and the variance of u_{2t} is $\sigma_1^2+\sigma_2^2$. Let $\hat{\alpha}_1$, $\hat{\beta}_1$ be the ordinary least squares (OLS) estimates of α_1, β_1 from the *supply* equation (having replaced S_t by Q_t) and imagine that it is the case that the residuals $Q_t-\alpha_1 p_t-\beta_1'\mathbf{x}_{1t}$ are all negative. Then it is clear from Equation 6.22 that the global maximum of the likelihood is obtained at $\hat{\alpha}_1$, $\hat{\beta}_1$ with σ_1 being set arbitrarily close to zero; hence σ_1 is not estimable, even though there are no parameters in $\Phi(\ )$ of Equation 6.23 that do not also appear in $\Psi(\ )$.

6.3.3 The model with price adjustment

As before, we now extend the model by adjoining a price adjustment equation; to simplify matters, we rewrite the model in compact notation as

$$D_t = \alpha_1 p_t + z_{1t} + u_{1t} \tag{6.25}$$

$$S_t = \alpha_2 p_2 + z_{2t} + u_{2t} \tag{6.26}$$

$$Q_t = \min(D_t, S_t) \tag{6.27}$$

$$p_t = \gamma_1 D_t + \gamma_2 S_t + z_{3t} + u_{2t} \tag{6.28}$$

where the zs represent linear functions of coefficients and of predetermined and exogenous variables. u_{3t} is assumed to be distributed as $N(0, \sigma_3^2)$ and independent of u_{1t}, u_{2t}. Given the joint pdf $g(D_t, S_t, p_t)$, the likelihood function under the restriction that excess supply has always occurred is

$$L = \prod_{t=1}^{T} \int_{Q_t}^{\infty} g(Q_t, S_t, p_t)\,\mathrm{d}S_t \tag{6.29}$$

and is

$$L = \prod_{t=1}^{T} \int_{Q_t}^{\infty} g(D_t, Q_t, p_t)\,\mathrm{d}D_t \tag{6.30}$$

for the case of perpetual excess demand. The typical term of Equations 6.29 and 6.30 is (for derivation, see Quandt, 1978)

$$h_i(Q_t, p_t) = \frac{|J_t|}{2\pi\sigma_i(\sigma_3^2 + \gamma_{3-i}^2\sigma_{3-i}^2)^{1/2}} \exp\left\{-\frac{1}{2}\left[A_{it} + \frac{C_{it} - B_{it}^2}{\sigma_{3-i}^2\sigma_3^2/(\sigma_3^2 + \gamma_{3-i}^2\sigma_{3-i}^2)}\right]\right\}$$
$$\times \left\{1 - \Phi\left[\frac{Q_t - B_{it}}{\sigma_{3-i}\sigma_3/(\sigma_3^2 + \gamma_{3-i}^2\sigma_{3-i}^2)^{1/2}}\right]\right\} \quad i = 1, 2 \tag{6.31}$$

where $h_1(h_2)$ denotes the density for the excess supply (demand) case, $|J_t| = |1 - \alpha_1\gamma_1 - \alpha_2\gamma_2|$ is the Jacobian, $A_{it} = (Q_t - \alpha_i p_t - z_{it})^2/\sigma_i^2$ and where

$$v_{1t} = \alpha_2 p_t + z_{2t} \qquad w_{1t} = \alpha_1 p_t + z_{1t}$$

$$v_{2t} = p_t - \gamma_1 Q_t - z_{3t} \qquad w_{2t} = p_t - \gamma_2 Q_t - z_{3t}$$

$$B_{1t} = \frac{\sigma_3^2 v_{1t} + \sigma_2^2\gamma_2 v_{2t}}{\sigma_3^2 + \gamma_2^2\sigma_2^2} \qquad B_{2t} = \frac{\sigma_3^2 w_{1t} + \sigma_1^2\gamma_1 w_{2t}}{\sigma_3^2 + \gamma_1^2\sigma_1^2}$$

$$C_{1t} = \frac{\sigma_3^2 v_{1t}^2 + \sigma_2^2 v_{2t}^2}{\sigma_3^2 + \gamma_2^2\sigma_2^2} \qquad C_{2t} = \frac{\sigma_3^2 w_{1t}^2 + \sigma_1^2 w_{2t}^2}{\sigma_2^2 + \gamma_1^2\sigma_1^2}$$

It is clear that the necessary condition for estimability of section 3.2 is met. In particular, it is met in the ordinary case when $\gamma_1 = -\gamma_2$. However, this does not ensure that all parameters can be estimated. Simplifying the $\exp\{\ \}$ term in Equation 6.31 for $i = 1$ gives $\exp\{-\frac{1}{2}[A_{1t} + (v_{2t} - \gamma_2 v_{1t})^2/(\sigma_3^2 + \gamma_2^2\sigma_2^2)]\}$. If $Q_t - B_{1t}$ is negative for all observations, as is possible if the values of the error terms are small but the excess supply is large, the likelihood may be monotone increasing as $\sigma_2^2 \to 0$ and this parameter cannot be estimated. Hence, even if the necessary condition is met, estimation may require an a priori assumption about the value of the variance associated with the variable that is never observed.

Now consider $\gamma_1 = 0$. In this event $h_1(Q_t, p_t)$ remains unchanged, but $h_2(Q_t, p_t)$ is

$$h_2(Q_t, p_t) = \frac{|J_t|}{2\pi\sigma_2\sigma_3} \exp\left\{ -\frac{1}{2}\left[A_{2t} + \frac{(p_t - \gamma_2 Q_t - z_{3t})^2}{\sigma_3^2} \right]\right\}\left[1 - \Phi\left(\frac{Q_t - \alpha_1 p_t - z_{1t}}{\sigma_1} \right)\right] \tag{6.32}$$

Thus the condition of Proposition 2 is violated for h_2 but not for h_1. Hence, if $\gamma_1 = 0$, and the 'all-excess-demand' restriction is imposed, the demand function cannot be estimated, whereas if the 'all-excess-supply' restriction is imposed, both functions are estimable. The converse holds if $\gamma_2 = 0$.

If $\gamma_1 = \gamma_2 = 0$, $h_2(Q_t, p_t)$ is the same as Equation 6.32 with $\gamma_2 = 0$ and

$$h_1(Q_t, p) = \frac{|J_t|}{2\pi\sigma_1\sigma_3} \exp\left\{\left[A_{2t} + \frac{(p_t - z_{3t})^2}{\sigma_3^2} \right]\right\}\left[1 - \Phi\left(\frac{Q_t - \alpha_2 p_t - z_{2t}}{\sigma_2} \right)\right]$$

and the equation representing the long side of the market cannot be estimated.

Finally, if $\gamma_1 = \gamma_2 = 0$, but $\sigma_{ij} \neq 0$ for $i \neq j$, the formulae for the density function become much more complicated (Quandt, 1978), but it can be verified that in the case of perpetual excess demand, the $\Phi(\ \ )$ in Equation 6.23 is a function of α_1 and z_{1t}, whereas all terms involving α_1 and z_{1t} cancel out in $\Psi(\ \ )$. The necessary condition for estimability thus fails again.

6.3.4 Estimation from a condensed model

In the case of perpetual excess demand, the model in Equations 6.25 to 6.28 may be condensed according to Hartley and Mennemeyer (1974) by replacing S_t by Q_t and substituting from 6.25 in 6.28, yielding

$$Q_t = \alpha_2 p_t + z_{2t} + u_{2t} \tag{6.33}$$

$$p_t = \gamma_1(\alpha_1 p_t + z_{1t} + u_{1t}) + \gamma_2 Q_t + z_{3t} + u_{3t} \tag{6.34}$$

The density function for the observable variables is

$$g_2(Q_t, p_t) = \frac{|1 - \alpha_1\gamma_1 - \alpha_2\gamma_2|}{2\pi\sigma_2(\sigma_3^2 + \gamma_1^2\sigma_1^2)^{1/2}} \exp\left\{ -\frac{1}{2}\left[\frac{(Q_t - \alpha_2 p_t - z_{2t})^2}{\sigma_2^2} + \frac{((1-\alpha_1\gamma_1)p_t - \gamma_2 Q_t - \gamma_1 z_{1t} - z_{3t})^2}{\sigma_3^2 + \gamma_1^2\sigma_1^2} \right]\right\} \tag{6.35}$$

By substituting in C_{2t} and B_{2t} in Equation 6.31, it is immediately evident $g_2(Q_t, p_t)$ is the same as $h_2(Q_t, p_t)$ without the $1 - \Phi(\ \ )$ term. Thus, basing estimation on this condensation is equivalent to the interpretation of the all-excess-demand hypothesis that $\Phi(\ \ )$ in (3–7) is equal to zero, this latter being the integral of the density of D_t, conditional on S_t and p_t, all evaluated at Q_t. It is clear from Equation 6.35 that the necessary condition for the estimability of demand is the same, i.e., γ_1 must be nonzero; it is further clear that σ_1 cannot be estimated in the condensed model.

The failure of estimability in a model subject to the restriction that there is always excess demand (supply) is due to inadequate feedbacks in the structural equations. Thus, for example, if in the price-adjusting model $\gamma_1 = 0$, there is no feedback in the model through D_t, although there is through S_t. The consequence is that arbitrary demand parameters are compatible with the restrictions and the data. Nor does it help to have a fully recursive model in order to encounter estimability failure. Therefore, it appears useful in empirical work, prior to estimation to check if at least the necessary condition for estimability is met in cases subject to the all-excess-demand restriction.

A case in point is the model of the automobile market in Poland due to Charemza, Gronicki and Quandt (1988). We simply restate the model here for the sake of illustration and refer the reader to the original for justification.

The essence of the model is that there are three interrrelated markets: (1) the market for new cars sold against zlotys; (2) the market for new cars sold against hard currencies; (3) the market for second-hand cars. The quantity of new zloty cars sold, q_t, and the prices of cars and dollars are observed, while the quantities of hard-currency and second-hand cars are not. The first market is assumed to exhibit excess demand, while the other two are assumed to clear. Letting y_{1t} denote the demand for zloty cars, y_{2t} the blackmarket price of dollars and y_{3t} the price of cars on the second-hand market, and letting z_{1t}, z_{2t}, z_{3t}, z_{4t} denote combinations of exogenous variables and parameters, the model is specified as

$$\begin{aligned} y_{1t} + \alpha_2 y_{2t} - \alpha_4 y_{3t} &= z_{1t} + u_{1t} \\ -\beta_1 y_{1t} + z_{4t} y_{2t} &= z_{2t} + u_{2t} \\ -\varepsilon_2 y_{2t} + y_{3t} &= z_{3t} + u_{3t} \\ y_{1t} &\geqq q_t \end{aligned}$$

Forming the joint density of y_{1t}, y_{2t}, y_{3t} from the first three equations and integrating out y_{1t} over the range defined by the fourth yields as the density function

$$h(y_{2t}, y_{3t}) = \frac{|z_{4t} - \beta_1(\varepsilon_2\alpha_4 - \alpha_2)|}{2\pi\sigma_3(\sigma_2^2 + \beta_1^2\sigma_1^2)^{1/2}} \exp\left\{ -\frac{1}{2}\left[\left(\frac{\sigma_2^2 + \beta_1^2\sigma_1^2}{\sigma_1^2\sigma_2^2}\right) \times (B_{2t} - B_{1t}^2) + A_t\right]\right\} \times \left[1 - \Phi\left(\frac{q_t + B_{1t}}{\sigma_1\sigma_2/(\sigma_2^2 + \beta_1^2\sigma_1^2)^{1/2}}\right)\right]$$

where

$$\begin{aligned} w_{1t} &= \alpha_2 y_{2t} - \alpha_4 y_{3t} - z_{1t} \\ w_{2t} &= z_{4t} y_{2t} - z_{2t} \\ A_t &= \frac{(-\varepsilon_2 y_{2t} + y_{3t} - z_{3t})^2}{\sigma_3^2} \end{aligned}$$

$$B_{1t} = \frac{\sigma_2^2 w_{1t} - \sigma_1^2 \beta_1 w_{2t}}{\sigma_2^2 + \beta_1^2 \sigma_1^2}$$

$$B_{2t} = \frac{\sigma_2^2 w_{1t}^2 + \sigma_1^2 w_{2t}^2}{\sigma_2^2 + \beta_1^2 \sigma_1^2}$$

Inspection reveals that the necessary condition of Proposition 2 is satisfied in this case.

6.3.5 Testing the chronic excess demand hypothesis

Finally, it is important to recognize that the notion of chronic excess demand is only a hypothesis, even in CPEs. Excess demands or shortages are not directly observable and their presence is often inferred from indices of shortage intensity (Kornai, 1982). It is therefore relevant to ask whether one can test the null hypothesis than only excess demand has occurred in a given sample.

There are two key issues to be noted. The first one is that the various proxies for shortages (e.g. length of queues for consumer goods, levels of business inventories), while relevant and interesting quantities, cannot by themselves settle the issue of whether there is excess demand in the aggregate or not, whether shortages outweigh slacks or not. They cannot settle the issue by reference either to the proxies' levels or to their departures from some normal levels. They cannot do so because we do not observe aggregate demand (supply) except in equilibrium and thus cannot know a priori what degree of disequilibrium corresponds to a given level of a shortage indicator. In fact, two economies with identical aggregate demands and supplies may exhibit different levels of shortage indicators, simply because markets are aggregated differently in the two economies or because the distribution of goods is organized differently. Under these circumstances a person who encounters an out-of-stock situation may be counted as rationed or queueing in one case and not in the other.

The second point is that the chronic excess demand hypothesis should not be interpreted to mean that excess supply could not occur under any circumstances; such an assertion would be incompatible with most reasonable assumptions about the stochastic specification of models. Rather, the hypothesis should be interpreted to mean that only excess demand has occurred in the sample at hand. Thus given a particular pattern of excess demands and supplies estimated from a given model, could that pattern plausibly have occurred by choice, if in fact the hypothesis is true? (The accuracy of estimates of the excess demand probabilities is discussed in section 6.4).

A test suitable for this hypothesis has been introduced by Rogers (1984) and has been applied to the US labour market by Quandt and Rosen (1985) and to aggregate consumption in Poland by Portes, Quandt and Yeo (1985).

For purposes of the test we posit a latent variable y^* (here $D-S$) and some exogenous variables x; their joint density may be written as $f^*(y^*|x;\theta)h(x)$, where $h(x)$ is usually not specified explicitly. The econometrician wishes to test hypotheses about the probability that y^* belongs to some subset Y_S of the space Y (here $D - S > 0$); this probability depends only on x (and θ) and is denoted by $p(x; \theta)$. In the simple disequilibrium model, for example, we can easily obtain the density $g(D - S|x; \theta)$, and then

$$\Pr\{D - S > 0|x; \theta\} = \int_0^{\infty} g(D - S|x; \theta)\,\mathrm{d}(D - S) = p(x; \theta)$$

and

$$\Pr\{D - S > 0|\theta\} = \int_x p(x; \theta)h(x)\,\mathrm{d}x = E(p(x; \theta)).$$

Rogers tests the hypothesis $E(p(x; \theta)) \geqq c$ for some constant c. The test statistic is

$$M_T = T^{-1/2} \sum_t [p(x_t; \hat{\theta}) - c]/\hat{v}$$

where $\hat{\theta}$ is the maximum likelihood estimator of θ and where

$$\hat{v}^2 = T^{-1} \sum_t (p(x_t; \hat{\theta}) - c)^2 + \hat{G}'\hat{\Omega}\hat{G}$$

$$\hat{G} = T^{-1} \sum_t \frac{\partial}{\partial \theta} p(x_t; \hat{\theta})$$

$$\hat{\Omega} = -T^{-1} \sum_t \frac{\partial^2}{\partial \theta \partial \theta'} \log f(Q_t|x_t; \hat{\theta})$$

where $f(\;\;)$ is the density of $Q_t = \min(D_t, S_t)$. Then if c is chosen to be near 1 (representing the null hypothesis of almost certain excess demand) and if the hypothesis $E(p(x; \theta)) = c$ is true, M_T is asymptotically distributed as $N(0, 1)$.

Both applications reject the hypothesis that the excess demand (or supply) is chronic in the sense indicated. While this may not seem too surprising for the US labour market, it does seem to fly in the face of orthodoxy concerning CPEs. However, the relevance of the finding is underscored by qualitatively similar findings in Burkett (1986), which he obtains by drastically different techniques. It is difficult to escape the conclusion that the chronic excess demand hypothesis should be subjected to systematic testing.

6.4 SOME COMPUTATIONAL RESULTS

The computation of maximum likelihood estimates for disequilibrium models is not as straightforward as many other econometric computations that are routinely performed. The principal reason is that in many interesting disequilibrium models the maximum likelihood estimates can be obtained only via numerical optimization. This certainly applies to the simple disequilibrium model without an adjustment equation, to the disequilibrium model with an adjustment equation containing an error term, and specifically applies to the CPE-oriented models of

Portes and Winter (1980), Portes, Quandt, Winter and Yeo (1987), and Charemza, Gronicki and Quandt (1988). Since numerical optimization algorithms are not as routinely available as some other algorithms, since they are often difficult to use or to tune, and since they often produce uncertain results, investigators have occassionally shied away from using them and have estimated second-best specifications with more standard procedures.

The present section undertakes some small-scale Monte Carlo experiments to investigate two specific questions: (1) What are the comparative performance characteristics of several numerical optimization algorithms when applied to the likelihood function for a simple disequilibrium model? (2) What is the behaviour of the asymptotic variance estimates of the parameters as the proportion of excess demand observations is varied?

6.4.1 Design of the experiment

Comparative performance evaluation of optimization algorithms often employs classic test functions, often of low dimensionality, such as Rosenbrock's function and others (Goldfeld and Quandt, 1972; Lootsma, 1972). Here we take a simple disequilibrium model defined by

$$D_t = \beta_{10} + \beta_{11}x_{1t} + \beta_{12}x_{2t} + u_{1t}$$
$$S_t = \beta_{20} + \beta_{21}x_{1t} + \beta_{22}x_{3t} + u_{2t}$$
$$Q_t = \min (D_t, S_t)$$

where u_{1t} and u_{2t} are distributed as $N(0, \sigma_1^2)$, $N(0, \sigma_2^2)$, independently of one another and over time. The x's were drawn once-and-for-all from uniform distributions, with x_1 from (55,105), x_2 from (50,150), and x_3 from (0,100). The parameter values were fixed at $\beta_{10} = 100.0$, $\beta_{11} = -1.0$, $\beta_{12} = 1.0$, $\beta_{20} = -40.$, $-20.$, $20.$, $40.$, $\beta_{21} = 1.0$, $\beta_{23} = 1.0$, $\sigma_1^2 = 250.$, 1000.0, 2000.0 and $\sigma_2^2 = 250.0$, 1000.0, 2000.0. The variances were varied in some of the experiments between the given values and β_{20} was varied in order to change the fraction of observations coming from the excess demand regime. The generated sample sizes were $N = 30$, 60.

(a) *Optimization algorithms*

A total of nine optimization algorithms were employed representing four classes of algorithms. We describe them briefly; for detail the reader is referred to Quandt (1983).[10]

(1) THE QUADRATIC HILL CLIMBING ALGORITHM GRADX

This algorithm is a generalization of Newton's method and thus requires first and second partial derivatives (which were evaluated by numerical differencing in the

[10] All the algorithms are incorporated in GQOPT, a general purpose optimization package available from the author.

present experiments). The generalization is designed to cope with the problem that Newton's method encounters if it strays into a region (in a maximization problem) in which the Hessian matrix is not negative definite. If we wish to maximize $F(x)$, with gradient $g(x)$ and Hessian $G(x)$, a Newton step to compute the $(k+1)$th iterate of x is given by

$$x^{k+1} = x^k - G(x^k)^{-1} g(x^k)$$

The first version of GRADX, GRADX1, replaces $G(x^k)^{-1}$ by $(G(x^k) - \alpha I)^{-1}$ where α is chosen so as to make $G(x^k) - \alpha I$ negative definite. The algorithm is then equivalent to maximizing the function at each step on a hypersphere, the radius of which is a decreasing function of α. GRADX2 differs from GRADX1 by maximizing at each step on a suitably oriented ellipsoid rather than sphere, and takes a step according to $x^{k+1} = x^k - (G(x^k) - \alpha B)^{-1} g(x^k)$, where B is a suitable positive definite matrix.

(2) QUASI-NEWTON METHODS

These include the Davidon–Fletcher–Powell algorithm and its variants. DFP2 is the classical Davidon–Fletcher–Powell algorithm. DFP3 is the Broyden modification. DFP4 gives the rank 1 correction formula. DFP1 chooses between DFP2 and DFP3 at each step. This class also contains the algorithm SSVM, the self-scaling variable metric method of Oren and Luenberger. For each of these methods, only first partial derivatives need to be calculated; an approximation to the Hessian is obtained from the x^k, x^{k-1}, $g(x^k)$, $g(x^{k-1})$ and the current approximation to $G(x^k)$.

(3) POWELL'S CONJUGATE GRADIENT METHOD

This algorithm employs no derivatives at all and is based in an n-variable problem on sequences of n linear searches in n linearly independent mutually conjugate directions.

(4) RANDOM SEARCH ALGORITHMS RNSRCH

This algorithm again employs no derivative calculation. It generates trial points randomly within a hypersphere. If a function improvement is found, the centre of the sphere is moved and the search continues. If no improvement is found within a set number of trials, the radius of the sphere is shrunk and search continues. Search terminates when the radius has met the specified termination criterion.

Numerous other algorithms could have been employed but were not. Thus, one could have investigated Mukai conjugate gradient method, the Nelder–Mead simplex method and the Hooke and Jeeves pattern search method. One could also have investigated Newton–Raphson or other methods for solving the first order conditions. Also, in the algorithms that were investigated, a number of fine-tuning variations could have been considered but were not. Thus, for example, all methods other than random search employed linear searches at each iteration; a number of alternatives are available in this respect that were not exploited. The results of the experiments are therefore only suggestive.

6.4.2 Results of optimization experiments

Each of the experiments was replicated 10 times on an IBM 3081 computer. For each generated data set, each of the algorithms was employed to maximize the loglikelihood. We investigated β_{20} values of -20.0 and 40.0 and σ_1^2, σ_2^2 values of 250.0, 1000.0 and 2000.0. The resulting average percentages of excess supply are given in Table 6.1.

Crude measures of performance may be given by the number of function evaluations or by the actual time required for optimization. Tables 6.2–6.5 contain the results.

At first glance, Tables 6.2 and 6.4 suggest that GRADX1, GRADX2 and POWELL are both absolutely more expensive than the other methods and also may become more expensive as the error variances increase; this latter feature presumably flattens the likelihood and makes convergence more difficult. Although this observation is correct as far as it goes, it misses the important observation that not all the methods terminate at the same function value. In fact only these three methods *ever* achieve the best function values, as shown in Table 6.6.

This is a little misleading since the four versions of DFP (but not SSVM and RNSRCH) frequently miss by only a hair.[11] Not to belabour the point, we illustrate this with a single example: in replication 1, case 1, GRADX2 is best with a function value of -118.374; the four DFP methods yield -118.378, -118.378, -118.378, -119.638. A perhaps more revealing measure is the average Euclidean distance over the replications in the space of parameters of the endpoint for each method from the endpoint of the method with the smallest function value.[12] We display these mean distances in Table 6.7 based on the first six parameters only.

Table 6.1 Excess supply percentages

Case	β_{20}	σ_1^2, σ_2^2	*Mean excess supply percentage*	*Standard deviation of excess supply percentage*
1	-20.0	250.0	0.4600	0.0467
2	40.0	250.0	0.8300	0.0277
3	40.0	1000.0	0.7600	0.0389
4	40.0	2000.0	0.6867	0.0499

[11]It is interesting to note that SSVM and RNSRCH frequently converge to nearly identical function values (that are different from those GRADX and POWELL converge to). This seems to suggest that SSVM and RNSRCH, for reasons that are not clear, converge to a second local optimum.

[12]Since no method is uniformly best, this measure will be strictly positive for each method.

Table 6.2 Mean function evaluations

Case	*Algorithm* GRADX1	DFP1	POWELL	SSVM	RNSRCH	GRADX2	DFP2	DFP3	DFP4
1	663	344	1580	218	392	575	346	346	230
2	777	270	1879	257	360	734	270	271	220
3	856	210	1957	259	398	745	209	209	212
4	1043	200	2472	252	393	925	199	202	191

Table 6.3 Standard deviation of function evaluations

Case	*Algorithm* GRADX1	DFP1	POWELL	SSVM	RNSRCH	GRADX2	DFP2	DFP3	DFP4
1	58	113	508	29	56	102	102	116	76
2	151	104	649	17	113	137	104	104	105
3	118	59	571	32	92	132	58	58	71
4	113	91	1428	59	116	114	90	93	76

Table 6.4 Mean timings (seconds of CPU time)

Case	*Algorithm* *GRADX1*	*DFP1*	*POWELL*	*SSVM*	*RNSRCH*	*GRADX2*	*DFP2*	*DFP3*	*DFP4*
1	1.79	0.87	3.97	0.56	1.03	1.61	0.88	0.87	0.58
2	2.07	0.70	4.53	0.68	0.96	2.02	0.70	0.70	0.57
3	2.27	0.53	4.72	0.66	1.04	2.04	0.52	0.52	0.53
4	2.73	0.48	5.60	0.62	0.99	2.50	0.48	0.48	0.46

Table 6.5 Standard deviation of timings (seconds of CPU time)

Case	*Algorithm* *GRADX1*	*DFP1*	*POWELL*	*SSVM*	*RNSRCH*	*GRADX2*	*DFP2*	*DFP3*	*DFP4*
1	0.15	0.29	1.27	0.07	0.15	0.29	0.26	0.29	0.19
2	0.37	0.26	1.60	0.04	0.30	0.37	0.26	0.26	0.27
3	0.32	0.15	1.16	0.08	0.25	0.35	0.15	0.15	0.18
4	0.30	0.23	2.89	0.14	0.29	0.31	0.22	0.23	0.18

Table 6.6 Number of best function values achieved

Case	*GRADX1*	*GRADX2*	*POWELL*
1	5	4	1
2	5	5	0
3	2	3	5
4	3	4	3

This confirms quite dramatically the superiority of GRADX1, GRADX2 and POWELL as well as the rapid deterioration of all methods when σ_1^2, σ_2^2 are large.

A final insight is given by the count of computational failures. These may occur for a variety of reasons – the simplest one being the failure to converge within a set number of iterations (200). Whenever an algorithm failed, the entire replication was discarded and a new one generated so that a matched sample of replications was always available. Only GRADX1 and GRADX2 ever failed, the former 0, 1, 8, 5 times and the latter 1, 2, 5, 4 times in the four cases.

The upshot is fairly clear and can be summarized as follows. (1) In terms of accuracy, GRADX1, GRADX2 and POWELL are dominant. (2) The GRADX algorithms have a non-negligible failure rate. (3) The POWELL algorithm is the most expensive. (4) All methods tend to deteriorate as error variances rise.

Two further points should be noted in conclusion. First, it goes without saying that there is no such thing as a best algorithm. Secondly, the results of Monte Carlo experiments are only suggestive and are in fact conservative. In any concrete optimization problem, the investigator is well advised to do some fine tuning of the algorithm in question which involves both alterations of the algorithm parameters and of starting points. While this was not practical in these experiments, in concrete problems such steps would have certainly improved the performance of all methods.

6.4.3 Probability predictions and asymptotic variance estimates

The first question we investigate is how the asymptotic variance estimates, obtained as the diagonal elements of the negative inverse of the Hessian matrix of the loglikelihood, depend on the split of the data points between excess demand and excess supply periods. In the limit, we know from section 6.3, that if there is only excess demand, the demand function parameters are not estimable and hence their implied variance is infinite. The question is how these variances actually vary as the proportion of excess supply points changes. In Tables 6.8 and 6.9 we display the means of the asymptotic variances over 30 replications and their standard deviations for six cases that have different β_{20} values, which induce the average excess supply percentages shown in column 1. (σ_1^1 and σ_2^2 are 250.0 and the sample size is 30 in these experiments.)

Table 6.7 Mean distances from optimum

Case	*GRADX1*	*DFP1*	*POWELL*	*SSVM*	*RNSRCH*	*GRADX2*	*DFP2*	*DFP3*	*DFP4*
1	6.34	19.58	7.58	52.78	52.92	6.37	22.71	19.58	43.40
2	0.04	60.96	12.55	87.36	87.09	0.33	64.26	60.96	65.60
3	38.54	97.58	7.71	102.52	102.24	37.69	97.57	97.57	101.67
4	101.44	209.06	107.65	226.34	227.41	105.62	209.06	209.06	226.30

Table 6.8 Mean asymptotic variance estimates

Mean fraction of excess S	β_{10}	β_{11}	β_{12}	σ_1^2	β_{20}	β_{21}	β_{22}	σ_2^2
0.8333	10.277	0.226	0.140	66.741	43.083	0.611	0.239	65.680
0.7811	20.802	0.244	0.146	67.526	47.389	0.664	0.244	73.972
0.7256	21.263	0.263	0.152	68.829	45.978	0.634	0.258	83.041
0.4544	27.537	0.349	0.195	85.180	28.161	0.332	0.176	76.000
0.3967	31.017	0.389	0.212	85.248	26.077	0.299	0.179	76.489
0.3378	27.285	0.352	0.217	73.132	24.620	0.280	0.166	75.932

It is immediately evident that the mean asymptotic variances are nearly monotone in the excess supply percentage, and in the predicted direction: as the excess supply percentage declines, the supply function parameters are estimated with increasing, and the demand function parameters with decreasing precision. It is further true that the asymptotic variance estimates themselves behave in similar fashion. It is also interesting to see what happens for a larger sample size, e.g. $n = 60$. We approximately reproduced the last row of Tables 6.8 and 6.9.[13] The mean asymptotic variance estimates for β_{10}, β_{11}, β_{12} changed relatively little, whereas those for β_{20}, β_{21}, β_{22} declined substantially. We conclude that, as has informally always been conjectured, the excess demand/supply percentage causes the maximum likelihood procedure to behave as if the actual number of demand (supply) points were equal to the actual sample size multiplied by the excess supply (demand) percentage.

A second point of interest is the behaviour of the excess demand/supply predictions. Questions that have arisen are just how good the predictions are and what the relationship is between the unconditional probabilities $\Pr\{D_t < S_t\}$ and

Table 6.9 Standard deviation of asymptotic variance estimates

Mean fraction of excess S	β_{10}	β_{11}	β_{12}	σ_1^2	β_{20}	β_{21}	β_{22}	σ_2^2
0.8333	3.430	0.050	0.041	20.887	24.716	0.368	0.129	44.856
0.7811	3.871	0.061	0.041	21.349	27.955	0.412	0.132	42.985
0.7256	4.594	0.070	0.038	22.273	25.192	0.360	0.167	49.526
0.4544	10.184	0.124	0.060	39.984	7.062	0.085	0.066	27.338
0.3967	16.788	0.189	0.100	53.436	6.283	0.067	0.143	29.150
0.3378	12.898	0.160	0.126	57.148	6.905	0.075	0.148	34.288

[13]The mean fraction of excess supply here was 0.3383.

Table 6.10 Probabilities of excess supply

	Based on estimates $F\{S>D\}$	$\Pr\{S>D\}$	$\Pr\{S>D\|Q\}$	*Based on truth* $F\{S>D\}$	$\Pr\{S>D\}$	$\Pr\{S>D\|Q\}$
$N=30$						
Mean	0.7778	0.7682	0.7841	0.8667	0.8073	0.8251
Std Dev.	0.0952	0.0928	0.0988	0	0	0.0077
$N=60$						
Mean	0.8150	0.7861	0.8014	0.8333	0.8002	0.8138
Std Dev.	0.0395	0.0479	0.0516	0	0	0.0045

the conditional probabilities $\Pr\{D_t < S_t | Q_t\}$ (Gersovitz, 1980; Burkett, 1981). For each of the thirty replications we computed the average for that replication of these probabilities using the estimated parameter values as well as the true parameter values. The results for the case when $\beta_{20} = 40$ and $\sigma_1^2 = \sigma_2^2 = 250.0$ are in Table 6.10 (where $F\{S > D\}$ denotes the relative frequency that S exceeds D).

The conditional and unconditional probabilities are close to one another which is not surprising in the light of the fact that $\sigma_1^2 = \sigma_2^2$. More importantly, the estimates of the relative frequency of excess supply and of the excess supply probabilities are close to their true values and this closeness improves with increasing sample size. We conclude that correctly specified disequilibrium models are capable of producing good estimates of these crucial quantities.

6.5 SUMMARY AND CONCLUSIONS

We considered some general problems of specification and estimation in modelling CPE with the aid of disequilibrium econometric models. In the area of the specification we concentrated on the derivation of possible adjustment equations in models of CPEs and suggested that it makes good sense to attempt to derive these from loss function minimization. In the course of this we alluded to the coherency problems that may arise in the construction of more realistic disequilibrium models, particularly in the presence of threshold effects.

We then turned to discussing general problems of estimability in so-called all-excess-demand models. Necessary (but not sufficient) conditions were found for such models to be estimable. Finally, we undertook a Monte Carlo investigation of the computability of maximum likelihood estimates in disequilibrium models with various optimization algorithms and of the behaviour of the asymptotic variance estimates in a variety of circumstances. In general, the results were encouraging for the numerical solution of the likelihood maximization problem.

The problems to which we have pointed provide a reasonable research agenda. Thus we shall have to sharpen our methodology concerning the estimation of disequilibrium models that (a) exhibit chronic excess demand as a matter of original specification, (b) contain price or wage adjustment, (c) contain more than a single sector. We have taken some small steps in the direction of solving these problems, but much remains to be done. But we have established that formal disequilibrium modelling of CPEs and the estimation of these models by proper econometric techniques is not only feasible, but a going concern. In particular, computational difficulties are disappearing rapidly and even the multiple numerical integration problems inherent in likelihood maximization for multisector models are becoming routinely manageable (Plant and Quandt, 1987). The one problem that is likely to remain serious in the short run is the problem of data availability, particularly of quarterly data. We can only hope that CPEs will decide to emulate Western Europe and North America in this particular respect.

ACKNOWLEDGEMENT

Support from the National Science Foundation is gratefully acknowledged.

REFERENCES

Barro, R. J. and Grossman, H. I. (1971) A general disequilibrium model of income and employment. *American Economic Review*, **61**, 82–93.

Burkett, J. P. (1981) Marginal and conditional probabilities of excess demand. *Economics Letters*, **8**, 159–62.

Burkett, J. P. (1986) Slack. Shortage and discouraged consumers in Eastern Europe: estimates based on smoothing by aggregation, University of Rhode Island (*mimeo*).

Charemza, W. (1987) Maximum likelihood methods of estimation for disequilibrium models in a centrally planned economy. *Economics of Planning*, **21**, 449–67.

Charemza, W. and Dlouhy, V. (1987) Two alternative disequilibrium models of a planned economy, Discussion Paper No. 65, Department of Economics, University of Leicester.

Charemza, W., Gronicki, M. and Quandt, R. E. (1988) Modelling parallel markets in centrally planned economies: the case of automobiles in Poland. *European Economic Review*, **32**, 861–83.

Charemza, W. and Quandt, R. E. (1982) Models and estimation for centrally planned markets in disequilibrium. *Review of Economic Studies*, **49**, 109–16.

Chow, G. C. (1983) *Econometrics*, McGraw-Hill, New York.

Eaton, J. and Quandt, R. E. (1983) A model of rationing and labor supply: theory and estimation. *Economica*, **50**, 221–34.

Gersovitz, M. (1980) On classification probabilities for the disequilibrium model. *Journal of Econometrics*, **14**, 239–46.

Goldfeld, S. M. and Quandt, R. E. (1972) *Nonlinear Methods in Econometrics*, North-Holland, Amsterdam.

Goldfeld, S. M. and Quandt, R. E. (1975) Estimation in a disequilibrium model and the value of information. *Journal of Econometrics*, **3**, 325–48.

Goldfeld, S. M. and Quandt, R. E. (1978) Some properties of the simple disequilibrium model with covariance. *Economics Letters*, **1**, 343–6.

Goldfeld, S. M. and Quandt, R. E. (1988) Budget constraints, bailouts and the firm under central planning, *Journal of Econometrics*.

Gourieroux, C., Laffont, J.-J. and Monfort, A. (1980) Coherency conditions in simultaneous linear equation models with endogenous switching regimes. *Econometrica*, **48**, 675–96.

Hare, P. Theoretical foundations: the economics of shortage, Chapter 3, this volume.

Hartley, M. J. and Mennemeyer, S. T. (1974) The estimation of markets in excess demand with incomplete data: the market for physicians' services, Discussion Paper No. 317, Department of Economics, SUNY, Buffalo.

Kornai, J. (1979) Resource-constrained versus demand-constrained systems. *Econometrica*, **47**, 801–20.

Kornai, J. (1980) *The Economics of Shortage*, North-Holland, Amsterdam.

Kornai, J. (1982) *Growth, Shortage and Efficiency*, Blackwell, Oxford.

Laffont, J.-J. and Monfort, A. (1976) Econometrie des model les d'equilibre avec rationnement. *Annales de'l'INSEE*, **24**, 3–40.

Lootsma, F. A. (ed.) (1972) *Numerical Methods of Non-linear Optimization*, Academic Press, New York.

Maddala, G. S. and Nelson, F. D. (1974) Maximum likelihood methods for models of markets in disequilibrium. *Econometrica*, **42**, 1013–30.

Plant, M. W. and Quandt, R. E. (1987) On the cost and accuracy of multiple integration in several variables, Research Memo No. 75, Financial Research Center, Princeton University.

Portes, R., Quandt, R. E., Winter, D. and Yeo, S. (1987) Macroeconomic planning and disequilibrium: estimates for Poland, 1955–1980. *Econometrica*, **55**, 19–42.

Portes, R., Quandt, R. E. and Yeo, S. (1988) Tests of the chronic shortage hypothesis: the case of Poland, *Reviews of Economics and Statistics*, **70**, 288–95.

Portes, R. and Winter, D. (1980) Disequilibrium estimates for consumption goods markets in centrally planned economies. *Review of Economic Studies*, **47**, 137–59.

Quandt, R. E. (1978) Tests of the equilibrium vs. disequilibrium hypothesis. *International Economic Review*, **19**, 435–52.

Quandt, R. E. (1983) Computational problems and methods, *Handbook of Econometrics*, Vol. 1, North-Holland, Amsterdam.

Quandt, R. E. and Rosen, H. S. (1985) Is there chronic excess supply of labor? *Economics Letters*, **19**, 193–7.

Rogers, A. J. (1984) Tests of some hypotheses in latent variable models with stochastic exogenous variables, Research Memo No. 49, Financial Research Center, Princeton University.

Rosen, H. S. and Quandt, R. E. (1978) Estimation of a disequilibrium aggregate labor market, *Review of Economics and Statistics*, **60**, 371–9.

PART TWO

Macroeconomic disequilibrium models of centrally planned economies

7

Disequilibrium models of the Czechoslovak economy

VLADIMIR DLOUHY

7.1 MACROECONOMIC DISEQUILIBRIUM IN THE CZECHOSLOVAK ECONOMY

In this chapter an attempt is made to include the problems of shortage and disequilibrium within the framework of a macroeconometric model of the Czechoslovak economy, in the belief that macroeconometric modelling of centrally planned economies (CPEs) is feasible and can be based on technological and behavioural regularities, well mirrored in the data. The author is strongly convinced of the need to base the whole approach to econometric modelling on the appropriate macroeconomic theory for centrally planned economies. This is a very ambitious goal (which is unlikely to be achieved in the present chapter).

On basic strategy the author concurs with the view of Portes that, in the absence of a macrotheory, when building a CPE macromodel '... one chooses a minimal coherent set [of assumptions] for one's objectives' (Portes, 1986, p. 18). The basic objective here is the theoretical formalization of the forces determining the development of basic macroeconomic aggregates such as output, personal and social consumption, investment, exports and imports, employment and increments of capital and stocks. The strategy adopted by the author is that of more or less traditional macroeconometric modelling with the emphasis on concentrated theory and its empirical verification, without resort to building a large econometric model of type W5 or W7 (Welfe, 1985) or CEM (Šujan, 1985).

The author does not deny the existence of excess demand as a persistent feature in CPEs, certainly not in the state productive sector at least. The economic

International Studies in Economic Modelling Series.
Models of Disequilibrium and Shortage in Centrally Planned Economies.
Edited by C. Davis and W. Charemza.
Published in 1989 by Chapman and Hall Ltd, 11 New Fetter Lane, London EC4P 4EE.

functioning of CPEs is well known (Kornai, 1970, 1972, 1979, 1980) and is now part of a conventional wisdom concerning CPEs. The theory is very attractive, even for CPEs where economic development has been relatively stable and successful, mainly because it can explain the long-term indicators in CPE performance (low and falling productivity in all production inputs, very high energy and material intensity, declining trend in terms of trade, etc.). It represents a significant advance in understanding how CPEs actually work; the same is true of other theoretical and empirical investigations, which seek system-specific signs within the production sector and to explain past dynamics, while the discussion constantly returns to the nature of industrialization, closed loops of production for the sake of production, excessive hoarding of stocks, etc. – see, for example, Winiecki (1986a, 1986b).

Kornai's analysis of CPEs implicitly rejects the possibility of macroeconometric modelling in the traditional, neoclassical sense. Others arguing along the same lines object primarily to the models with quantity rationing, based on the Barro-Grossman disequilibrium theory, appropriately adapted to the CPE context (Portes, 1979, 1981, 1986). Having contributed to the discussion on the Portes model (Dlouhý, 1984a, 1984b; Dlouhý and Dyba, 1985), the author doesn't intend here specifically to defend this type of model, because it clearly doesn't represent the only possible theoretical outcome. He intends simply to contrast past Czechoslovak economic reality with the general a priori assertion of those who – in accordance with Kornai's theory – refute the macroeconometrics of CPEs on the grounds that shortages decisively determine the socialist economy, are chronic and widespread, and represent a normal or equilibrium state (or dynamic equilibrium trajectory), given the definition of equilibrium as the state of rest of the system.

Kornai has himself on occasions presented his arguments in a form directly at odds with the data. Portes (1986) argues with contrapositives of Kornia's assertions and gives selected examples of reasonably successful results for the various econometric models mentioned above. Others besides Kornai, *inter alia* Winiecki (1985, 1986a), Kemme and Winiecki (1984), Pindák (1983) and Podkaminer (1986) challenge the macroeconometrics of CPEs on the basis of more empirical arguments. Before contrasting their assertions with the Czechoslovak reality, we take a brief look at the possible sources of shortages and disequilibria in that economy.

The author agrees unreservedly with the assertion that the permanence of excess demand in the state productive sector is a matter of fact in the case of Czechoslovakia and that a certain level of excess demand can be considered a normal state. Consistent with this normal state is a level of output, consumption, investment and other macroeconomic aggregates that is an equilibrium level (in the sense of equality of demand and supply) related to the behaviour of planners and households. If we try to systematize the main sources of disequilibria, one possible outcome would be the following (given that every classification is somewhat arbitrary):

1. exogenous sources, namely changes in terms of trade, bad harvests, consecutive

losses of export markets, increasing irregularities in decisions concerning imports and random fluctuations in demand for consumption goods;
2. endogenous sources, namely investment rate, noninvestment state expenditure, dynamics of the money income of the population and growth rate of the national income.

While it is beyond the scope of the present chapter to defend this classification it should be made clear that it stems from a belief in the feasibility of macroeconomic regulation in CPEs, for example that planners are able to return the economy to an equilibrium state when any disturbance appears. At the same time, it does not preclude the existence of permanent excess demand in the state productive sector. Whether this excess demand spills over into the consumption goods market is nevertheless completely open to question, and the answer apparently differs for particular CPEs. Furthermore, unlike Kornai and others, the author believes that the existence of this spillover is to a very high degree dependent on the behaviour of the planning centre.

This fact is crucial for the feasibility of macroeconometrics in CPEs. Indeed, Winiecki (1985, 1986b) claims that permanent excess demand and soft budget constraints of firms create a spillover from the state productive sector into the consumption goods market. On the demand side, the suction effect of labour inputs and permanent overemployment leads to chronic overfulfilment of wage plans, while on the supply side the factor responsible is the underfulfilment of planned deliveries of consumption goods because the preferences of the centre are directed towards the means of production. The next step is to check with the data.

There are presented in Table 7.1 the planned and real average growth rates for

Table 7.1 Average growth in nominal money income (%)

	61–65	*66–70*	*71–75*	*76–80*	*81–85*
Total money income					
Plan	3.61	–	5.70	4.32	3.03
Real	3.56	8.16	5.23	4.09	3.14
Money wages					
Plan	4.17	–	4.78	4.11	2.57
Real	4.15	7.50	4.66	3.94	2.69
Co-op(agriculture) income					
Plan	–	–	–	0.38	5.36
Real	3.59	7.07	3.10	1.54	3.80
Social security income					
Plan	–	–	–	7.85	2.63
Real	5.15	10.68	5.12	5.69	4.25
Other income					
Plan	–	–	–	1.84	5.21
Real	11.58	9.45	10.81	3.83	3.41

Source: real values – *Statistical Yearbook of Czechoslovakia*, various issues.

Table 7.2 Total money income and its components (current prices, milliard korunas)

	Total money income			*Money wage*			*Co-op (agriculture) income*			*Social security income*			*Other income*		
	Plan	*Real*		*Plan*	*Real*		*Plan*	*Real*		*Plan*	*Real*		*Plan*	*Real*	
1975	292.1	292.1	0.0	186.5	186.7	+0.2	20.7	20.4	−0.3	51.4	51.7	+0.3	33.5	33.3	−0.2
1976	307.5	306.4	−1.1	195.2	194.3	−0.9	21.1	20.2	−0.9	56.8	56.3	−0.5	34.4	35.6	+1.2
1977	317.9	320.0	+2.1	202.4	203.3	+0.9	20.6	20.3	−0.3	58.8	58.2	−0.6	36.1	38.2	+2.1
1978	332.3	331.4	−0.9	210.3	211.9	+1.6	20.6	20.5	−0.1	65.7	59.7	−6.0	35.7	39.3	+3.6
1979	345.0	343.2	−1.8	219.5	219.2	−0.3	20.7	20.9	0.2	70.1	63.1	−7.0	34.7	40.0	5.3
1980	360.9	356.9	−4.0	228.1	226.5	−1.6	21.1	22.1	1.0	75.0	68.2	−6.8	36.7	40.1	3.4
1981	369.7	366.1	−3.6	233.5	232.4	−1.1	21.7	22.6	0.9	70.3	69.5	−0.8	44.2	41.6	−2.6
1982	380.9	381.8	0.9	240.8	239.8	−1.0	22.9	24.0	1.1	75.7	75.2	−0.5	41.5	42.8	1.3
1983	387.4	393.5	6.1	242.5	246.3	3.8	23.4	25.5	2.1	78.2	77.9	−0.3	43.3	43.8	0.5
1984	400.7	403.8	3.1	249.2	252.4	3.2	25.8	26.5	0.7	80.4	79.8	−0.6	45.3	45.1	−0.2
1985	419.0	416.6	−2.4	258.9	258.7	−0.2	27.4	26.6	−0.8	85.4	84.0	−1.4	47.3	47.3	0.0

Source: real values – *Statistical Yearbook of Czechoslovakia*, various issues.

nominal money income of the population and its main components. In the Czechoslovak case one can't speak of a stable overfulfilment of money wage plans or of plans for total money income. As can be seen from Table 7.1, real growth in money income in the five-year plan periods never significantly exceeded planned expectations and in no case can we speak of a trend or normal state.

Table 7.2 gives more detailed information on the deviation of plans from the reality in the last two five-year plan periods. Certain overfulfilment of planned figures is observed only in the period 1982–84 with a maximal deviation of 6.1 milliard korunas (kčs) in 1983. This is a consequence of two facts:

1. After a period of stagnation in 1981–82, the real growth rate in national income reached approximately 3%. As the planners are a very conservative group, it is very probable that they planned a growth of money income consistent with expected further stagnation. Furthermore, the excess of 6.1 milliards is divided between a 3.8 milliard excess of money wages and a 2.1 milliard excess of income in co-operatives – and there was an exceptionally good harvest in 1983.
2. There was an unprecedented rise in some retail prices in 1982 and a growth in the index of living costs of 4.1% in the same year; planners were probably willing to allow for some compensations, above all in some groups of the population. The figures for 1985, however, establish that the period 1982–84 was more the exception than the rule.

So-called deliveries to the markets fund are the best approximation for the aggregate supply of consumption goods. Planned and real values together with the deviations are presented in Table 7.3. Again, the data do not confirm the assertion regarding the stable underfulfilment of planned supply; the opposite is in fact true.

Table 7.3 Deliveries to the market funds (current prices, milliard kčs)

	Planned	*Real*	*Deviations*
1975	221.4	222.5	+1.1
1976	227.9	225.1	−2.8
1977	232.2	231.3	−0.9
1978	245.0	245.4	+0.4
1979	258.8	259.1	+0.3
1980	266.0	266.9	+0.9
1981	266.1	268.5	+2.4
1982	277.4	279.8	+2.4
1983	280.5	287.2	+6.7
1984	291.9	297.5	+5.6
1985	308.1	311.0	+2.9

Source: real values – *Statistical Yearbook of Czechoslovakia*, various issues.

Table 7.4 Average growth rates of official price indices (five-year averages) (%)

	61–65	*66–70*	*71–75*	*76–80*	*81–85*
Wholesale prices	0.68	2.13[a]	−0.02	1.60	4.37
Purchasing prices of agricultural products	–	–	0.72	2.87	2.90
Retail prices	0.25	1.84	0.19	2.11	2.00
Living costs index	0.58	1.64	0.06	2.03	1.76

[a] Average for 1968–70 only because of the change in the construction of the index in 1967.
Source: *Statistical Yearbook of Czechoslovakia*, various issues.

Obviously, this doesn't reveal anything concerning the equilibria in the markets: plans may be inconsistent, there is competition in the consumption goods market between households and firms (although this is less important) and the picture is further complicated by deliveries from so-called state natural reserves – but here data are not available. However, the main lesson for future discussion on the sources of disequilibria in CPEs, drawn from Tables 7.1–7.3, is that the experience of one country cannot serve as a basis for generalization.

Pindák (1983, p. 125) claims that '. . . even if the exact level of forced savings remains unclear', in spite of some simplication, traditional monetary indicators such as the average propensity to save, the incremental ratio of household money to income and the income velocity of household money can serve as the best measures of repressed inflation. Inflationary pressure can express itself in the form of official open inflation and/or non-official open inflation (hidden inflation) and/or repressed inflation. For the sake of clarity the official data are presented in Table 7.4.

The official inflation remains low indeed, with some more significant acceleration in the last period only, especially for wholesale prices. The only striking increase in retail prices occurred in 1982, when the total index rose 5.04%, and the index for foodstuffs rose 9.40%.

There is no doubt that hidden inflation is present in the Czechoslovak economy and that it probably expresses itself more often through quality deterioration than through inappropriate weights in the consumption goods basket (although this factor exists as well). A comparison of purchasing power parities, pioneered in this context by Askanas and Laski (1985), and applied for Czechoslovakia by Havlik (1985), gives a reasonable estimate of hidden inflation in Czechoslovakia. Implicit growth in retail prices, based on a PPP comparison between Austria and Czechoslovakia in 1964 and 1980, is 169% (3.3% p. a.), while officially admitted growth is 122.7% in the same period (1.29% p.a.). Hence, hidden inflation in this period is estimated at a level of 2.02% p.a. Cautious observation reveals that this is probably an accurate figure; at the same time, acceleration in hidden inflation after 1980 is a question of fact.

In Table 7.5, Pindák's (1983, p. 127) table, with household money figures, allegedly measuring repressed inflation, is presented, extended and slightly revised on the basis of the Czechoslovak official data.

Table 7.5 Household money in Czechoslovakia: incremental ratio to income, average propensity to save and income velocity

	Incremental ratio to income	*Average propensity to save*	*Income velocity*
1960	1.39	2.15	4.03
1961	3.79	2.51	3.70
1962	2.33	2.28	3.51
1963	2.20	2.14	3.33
1964	3.50	2.97	3.10
1965	3.88	3.51	2.91
1966	3.82	2.84	2.73
1967	4.71	3.84	2.56
1968	4.17	2.41	2.55
1969	3.20	2.74	2.59
1970	5.43	5.08	2.37
1971	5.96	5.18	2.17
1972	6.45	5.59	2.00
1973	6.88	5.87	1.85
1974	4.26	3.75	1.79
1975	4.35	3.50	1.71
1976	5.06	4.07	1.64
1977	5.41	4.11	1.57
1978	2.98	2.29	1.54
1979	2.64	1.64	1.52
1980	3.56	2.78	1.50
1981	3.71	3.15	1.45
1982	4.99	4.01	1.41
1983	5.03	4.08	1.35
1984	4.78	3.97	1.30
1985	4.41	4.11	1.26

Source: *Statistical Yearbook of Czechoslovakia*, various issues. Own computations.

The development of liquid balances among the population is a misleading measure of repressed inflation (as a rate of change of excess demand). For general discussion see Nuti (1985); here the author stresses mainly the fact that falling revealed income velocity is due to the increasing importance of the second economy, together with an increase in required cash holdings, desired for speculative purposes. However, it may in addition be an expression of a general trend, where an increase in the standard of living generates a higher level of savings – this effect may be multiplied by the diverted Czechoslovak price structure, where consumer durables (houses in private possession, cars, consumer electronics, etc.) are much higher priced compared with those in market economies. Moreover, the average propensity to save is modest and again does not differ from its counterparts in other developed economies.

The problem of repressed inflation is returned to in the next part of the chapter.

To conclude here, it is worth repeating that for Czechoslovakia we have no convincing information for spillover of disequilibria from the state productive sector into the consumption goods market nor for persistent repressed inflation. All monetary and other indicators can be very reliably explained on the basis of a hypothesis of global equilibrium in time in the Czechoslovak consumption goods market. It is evident, however, that this doesn't exclude frequent structural disequilibria and shortages in particular markets.

7.2 DISEQUILIBRIUM MODELS IN THE CZECHOSLOVAK ECONOMY

7.2.1 Consumption goods market

We start with the demand side of the market. Given the assumption of equilibrium, most studies (see *inter alia* Janáček, 1975; Goldman *et al.*, 1978; Klaus and Rudlovčák, 1977; Klaus, 1979, Klaus and Rudlovčák, 1982) have applied the Houthakker–Taylor consumption function with very acceptable empirical results.

Statistically, this is not surprising. Looking at Table 7.6, where five-year average growth rates are presented for disposable income and consumption together with

Table 7.6 Growth rates for real disposable income and consumption (average growth rates and growth in particular years)

	Disposable income	*Consumption*
1961–65	4.27	3.31
1966–70	5.65	5.69
1971–75	4.87	5.11
1976–80	1.60	2.02
1981–85	1.01	0.60
1962	1.45	2.98
1963	2.84	2.97
1964	5.36	2.25
1968	9.85	10.48
1969	6.56	7.64
1974	3.84	6.76
1978	1.27	3.88
1979	−0.64	−0.29
1980	0.91	1.23
1981	1.42	0.06
1982	−0.47	−1:78
1983	1.86	1.79
1984	1.56	1.83
1985	0.82	1.21

Source: own calculations.

Table 7.7 Change in savings deposits and cash of households in the 60s (million kčs in 1977 proces)

1961	*1962*	*1963*	*1964*	*1965*	*1966*	*1967*	*1968*	*1969*	*1970*
4823	3011	2927	4904	5777	5921	7643	7424	6072	10635

Source: *Statistical Yearbook of Czechoslovakia*, various issues. Own calculations.

single growth rates for particular years, we see how consumption followed the development of disposable income through the whole period. As such, this fact doesn't prove very much about equilibrium or disequilibrium; however, data for the years of particular interest, displayed in Table 7.6, reveal the ability of the centre to influence at least partially real consumption via real disposable income.

Exceptional years are 1964, 1968–69, 1974 and 1978. The period 1968–69 was characterized by an acceleration in the income of population, and the gap between change in income and consumption was not as wide as is commonly believed. There was probably very strong structural disequilibrium: the point of saturation with basic consumption goods was reached in the first half of the 60s and the increase in income materialized mainly in a demand for consumer durables. Purchases in these years were of course very irregular, influenced by non-economic events and by expectation; total instability headed for a decrease in savings, but not significantly so, as the conventional wisdom indicates (Table 7.7.).

Both in 1974 and 1978 changes in consumption were strikingly higher than changes in income. This is a consequence of several intervals of turbulent purchases in these years. Even if one admits the influence of autonomous stimuli, the main explanation seems to lie in the somewhat cyclical nature of the replacement of consumer durables. Klaus (1979) analysed the residuals of the regression between changes in purchases and changes in disposable income and discovered that a high positive residual (an excess of true purchases over fitted purchases, hence an excess above the normal level) is preceded by a cumulation of negative residuals and vice versa. The estimation period is 1956–77 and the forecasting performance (given the reality of turbulent purchases in 1978) is admirable.

The cyclical nature of purchases of consumer durables is not inconsistent with a hypothesis of global equilibrium, accompanied by structural disequilibria. Shortages and slacks in micromarkets, random breakdowns in supply, etc. generate some degree of uncertainty; a single event or item of information (an increase in the world price of raw materials, signs of acute shortage in neighbouring countries, an increase in the price of selected commodities – gasoline in 1974, some foodstuffs in 1977, etc.) then provides a strong impetus for irrational reactions, and purchases of consumer durables which would normally be spread over a period are concentrated into a single interval.

Hence, 1964 remains the only year where we accept a priori the existence of a strong global disequilibrium; this is connected with the overall slowdown in economic activity in the period 1962–63 combined with unfavourable exogenous conditions (a severe winter and a poor harvest).

The Houthakker–Taylor (H–T) consumption function in its general, unrestricted form doesn't perform specially well for the period 1960–85:[1]

$$\underset{(1.330)}{CO = 3776.000} + \underset{(3.199)}{0.695DIN_{-1}} + \underset{(4.779)}{0.699\Delta DIN} + \underset{(1.142)}{0.266CO_{-1}} \qquad (7.1)$$

$$\bar{R}^2 = 0.998, \quad s = 2257.823, \quad \xi^2 = 0.935, \quad \xi^3 = 3.175, \quad \xi^4 = 8.735$$

Here (and in all following estimated equations) R^{-2} is the coefficient of determination adjusted for degrees of freedom, s is the estimated standard error of the regression, ξ^2 is a linearity test, ξ^3 is a homoscedasticity test, and finally, ξ^4 is the F form of the LM test for residual autocorrelation up to second order (for detailed discussion see Spanos, 1986).

In spite of a high $\bar{R}^2$ value the estimates are weakly determined and misspecification diagnostics indicate serious statistical problems. Inspection of the residuals reveals that although the model tracks the development in the 60s well (including 1968–69), it performs relatively worse at the end of the estimation period when compared with the following restricted form.

This form of the H–T consumption is presented by Portes and Winter (1978, 1980) and is a derivate of the H–T savings function (where depreciation rate $\delta = 0$ in the basic equation of the H–T model). For the period 1961–85 we obtained the following result:[2]

$$CO = -\underset{(7.521)}{0.452SAV_{-1}} + \underset{(6.312)}{0.848\Delta DIN} + \underset{(12.894)}{0.978DIN_{-1}} \qquad (7.2)$$

$$\bar{R}^2 = 0.998, s = 2162.701, \quad \xi^2 = 0.873, \quad \xi^3 = 2.52, \quad \xi^4 = 3.108$$

Here all parameters are statistically significant, the fit is again quite good and the misspecification diagnostics especially are significantly improved *vis-à-vis* the previous estimate. All estimates are in accordance with the a priori specified hypothesis: $-1 < \alpha_1 > -\frac{1}{3}, 0 < \alpha_2 < 1, \alpha_3 = 1$ (Portes and Winter, 1980). In tracking the turning points the model now performs better, and is able to track the period 1968–69, the slowdown in consumption in 1980–82 and a slight acceleration in 1984–85; the only problem remains the year 1978, where the model predicts an increase in consumption of 3.2 milliards against roughly a 9.8 milliards real increase. It should be noted that there is no dummy for 1968–69 in both estimates as compared with the majority of the other models.

Klaus and Rudlovčák (1982) challenge the general H–T specification and present their consumption function where variables are measured in nominal values and the retail price index is included among the explanatory variables. For the same

[1] Notation: CO personal consumption; DIN disposable income; Δ first difference operator.
[2] Notation: SAV savings

period as above the following results were obtained:[3]

$$\underset{}{CO} = \underset{(-0.005)}{-64.001} + \underset{(41.017)}{0.950DIN} - \underset{(-1.396)}{0.221\Delta DIN} + \underset{(0.209)}{3568.005PMC} \tag{7.3}$$

$$\bar{R}^2 = 0.969, \quad s = 2543.062, \quad \xi^2 = 1.017, \quad \xi^3 = 5.819, \quad \xi^4 = 5.237.$$

When compared with Klaus and Rudlovčák's (1982) estimate (which was for 1965–81), this is a very poor result. The inclusion of a dummy for 1968–69 improved only the fit, but not the significance of parameters, and not the problem of heteroscedasticity and autocorrelation. Given this result, we accept the restricted H–T model as a basis for further experiment.

For the supply side, existing estimated relations are mostly of an *ad hoc* nature (see Klaus and Rudlovčák's discussion on their surrogate supply function). Portes *et al.* (1984) include planned consumption in the supply function and build a theory supporting their formalization of the process of plan formation. Not having a time-series of planned data we resorted to a specification à la Portes and Winter (1977), with some modification:[4]

$$CO = \underset{(395.752)}{0.998COT} + \underset{(4.897)}{0.264YX} + \underset{(3.936)}{0.428MONX} + \underset{(1.725)}{1.548AGX} \tag{7.4}$$

$$\bar{R}^2 = 0.997, \quad s = 2696.537, \quad \xi^2 = 0.537, \quad \xi^3 = 1.901, \quad \xi^4 = 2.015.$$

Here the supply depends on a trend value of past consumption (as a proxy for plan) and on the deviations from trend values of national income, financial assets of households and agricultural production. Financial holdings are included here because their level is apparently taken into account by the centre when determining the supply, even in the current period (Portes and Winter, 1977; Portes and Santorum, 1986). The empirical performance is quite satisfactory: all estimates satisfy the a priori hypothesis of $\beta_i > 0$, 0 and the fit is very good as are other diagnostics, with the exception of autocorrelation. The residuals behave satisfactorily and for that reason insignificant AGX was retained in the model.

Allowing for disequilibrium in the consumption goods market we estimated a basic model (a canonical model following the taxonomy of Quandt, 1982) with demand and supply functions and the 'min' condition. For general discussion see Portes and Winter (1980): in spite of obvious deficiencies the author still considers this exercise a legitimate test for the hypothesis of repressed inflation.

Recently, Burkett (1986) proposed an alternative method of estimation for the canonical model, taking into account Kornai's (1980) criticism of the discrete switching model. Burkett's approach allows for both excess demand and supply at the microlevel – for applications see Burkett (1986) and Portes and Santorum (1986).

[3]Notation: PMC retail price index.

[4]COT trend value of CO; Y net output; YX deviations from quadratic trend of Y; MON household money (cash plus cumulated savings); AG agricultural production; MONX, AGX deviations from quadratic trends.

Table 7.8 Estimation results for discrete switching and the Burkett model

Discrete switching:

$$CD = \alpha_1 SAV_{-1} + \alpha_2 \Delta DIN + \alpha_3 DIN_{-1} + u_1$$
$$CS = \beta_1 COT + \beta_2 YX + \beta_3 MONX + \beta_4 AGX + u_2$$
$$C = \min(CD, CS)$$

Birkett:

$$c = 0.5(\alpha^T x + \beta^T y) - 0.5/(\alpha^T x - \beta^T y)^2 + 4\gamma^2(\alpha^T x)(\beta^T y)/\tfrac{1}{2} + u$$
$$\alpha = (\alpha_1, \alpha_2, \alpha_3)^T, x = (SAV_{-1}, \Delta DIN, DIN_{-1})^T$$
$$\beta = (\beta_1, \beta_2, \beta_3, \beta_4)^T, y = (COT, YX, MONX, AGX)^T$$

	Discrete switching	*Burkett*
α_1	− 0.491	− 0.223)
	(0.073)	(0.001)
α_2	0.892	1.007)
	(0.347)	(0.500)
α_3	1.000	1.000
s_1^2	4.692×10^6	–
β_1	1.000	1.000
β_2	0.315	0.273
	(0.006)	(0.010)
β_3	0.400	0.419
	(0.397)	(0.057)
β_4	1.511	1.523
	(0.392)	(0.005)
s_2^2	4.004×10^6	–
g	–	0.001
$\log L$	− 61.23	− 71.28

Note: Asymptotic standard errors in parentheses. GRAD × routine from GQOPT pachage used for estimation. To obtain convergence, restrictions in α_3 and β_1 were essential.

The results of estimation for both discrete switching and Burkett's model are presented in Table 7.8. At first sight the results seem very good: optimum was reached in both cases (although only when restrictions on both α_3 and β_1 were imposed), estimates of parameters are mostly significant and a priori expectations for signs mostly fulfilled. Notice that in both cases there are no significant differences from ordinary least squares (OLS) estimations. The results are rather surprising when compared with Burkett (1986), where the estimate for Czechoslovakia gave a weaker outcome (unacceptable signs etc.).

All the results indicate that we should accept the hypothesis of global equilibrium in the consumption goods market in Czechoslovakia. The only contradictory results we obtain are from comparison of Burkett's index of relative shortage and the percentage of excess demand for the discrete switching model. For the whole period of the 60s and 70s both are very similar, while for the last period they differ significantly: while the discrete switching model indicates a state of equilibrium in the market, Burkett's model strongly suggests that 1979–85 were the only years of prevailing shortage in the Czechoslovak consumption goods market.

Disequilibrium estimates in Table 7.8 should be considered preliminary unless they are confirmed by independent estimate using alternative computer routines.

Nuti (1985) criticizes the use of sophisticated econometric methods when estimating excess demand in CPEs. Although his reservations may be challenged, the author agrees with his explanation why we obtain estimates of low excess demand even for CPEs where all other evidence suggests a high level. Nuti's argument reverts to Grossman's (1977) idea, that the second economy is flexible enough to generate a price, which equilibrates the overall demand and supply for consumption goods in the economy. Hence when using official data on consumption, we estimate the demand on the state consumption goods market, regressing it on the explanatory variables (disposable income etc.) that influence the overall demand (including that in the second economy). Hence, we may obtain a biased picture, indicating global equilibrium (which, considering the second economy, is indeed true), while in the state sector large excess demand may prevail. This calls for the separation of the state market and the second economy in the data, and in the main for alternative specifications of demand and supply functions (see Chapter 11). In spite of the many useful results we obtained from the aggregate disequilibrium model, it seems to be functioning at its information limit. However, reliable information on the second economy, on the process of formation of demand and supply in that economy, and on the pricing process, etc. is a major unknown for us at the present stage of research.

7.2.2 Investment and disequilibria

Modelling of the investment process within the framework of the macroeconometric model differs substantially from the models of demand and supply in the consumption goods market. Investment is (and has been since the early years of existence of the socialist economy) a core for generating shortages and disequilibria. Institutional features allowed for both micro and macroeconomic conditions which stimulated the accelerated growth of investment compared with consumption. The accumulation rate was boosted to maximal level within a short period while, at the same time, the logic of central planning allowed for maximal concentration of investment resources and their allocation.

The stability of this process in the long run was governed by the separation of the state productive and household sectors, and regulation of wages and money incomes (see section 1) was essential – for a general discussion see Kalecki (1965). Savings (either forced or intended) gradually represented an important investment resource, together with a two-level price system (cheap investment was thus created), the centralization of depreciation, a relationship between the state budget and credit plans, etc.

The process of formation of investment plans in CPEs has recently been described by several authors, and its inner logic is well known (Kornai, 1980; Bauer, 1978; Winiecki, 1982, 1985, etc.). Firstly, disequilibria appear at the point when firms put forward their claims for new investment, which are compared with the total investment budget of the centre. Second, the bargaining character of the formation of the investment plan leads to tension between the total investment budget and the investment project selected for realization; see in particular Winiecki's (1985) account of how equal cuts to everybody undeniably result in disequilibrium at this point, when the demand for investment is created. Finally, disequilibria exist between the demand for investment and real capacities in the economy (supply).

The socialist economy has adapted to the existing tensions and disequilibria. The existence of fixed prices and other institutional characteristics has generated a quantitative adjustment of a special nature: this is manifested mainly in prolongation of the time for completion of projects and the high number of projects currently under realization.

The above discussion opens up the possibility of modelling the investment process having regard to the sources of possible disequilibria. Klaus (1985) specifies a small model with three equations where total investment depends on the past growth rate of national income and past balance of trade; expected costs for newly started projects are determined by current investment and by the residual expected costs of projects currently under realization; the ratio of costs of currently completed projects and the volume of unfinished projects (of various vintages) is determined by the degree of disequilibrium, e.g. by the tension between the demand for and supply of investment outlays in the current period.

Klaus's model offers the possibility of modelling the disequilibria stemming from various sources. The full description of the contributions and flaws of this model is outside the scope of this chapter (see Kupka (1986) for general comments); given the data problem, we do not present empirical results here.

If we discontinue the effort of modelling the disequilibria at various stages of the investment process, we are left with a requirement to track the development of the total investment only. This brings us to the theory of investment cycles (Bauer, 1978; Winiecki, 1982), although in the case of Czechoslovakia once again the data do not seem to be very consistent with theory. Because of this, we opted for *ad hoc* specification, where investment is seen as a process of muddling through. The main barriers consist of the balance of payments and a cyclically appearing tension, which expresses itself as an increase in the completion time of projects.

Looking at the first differences in investment outlays (solid line in Fig. 7.1), we observe a changing pattern where large increases are followed by small ones. Furthermore, given the existing structure of the Czechoslovak economy with a high share of heavy industries, the level (and change) of investments depends on past development of gross output; a significant part of investment is devoted to the industries producing intermediates. Finally, the balance of trade with non-socialist countries became an important factor influencing planners' decisions on total investment outlays in the current period. The following equation represents one

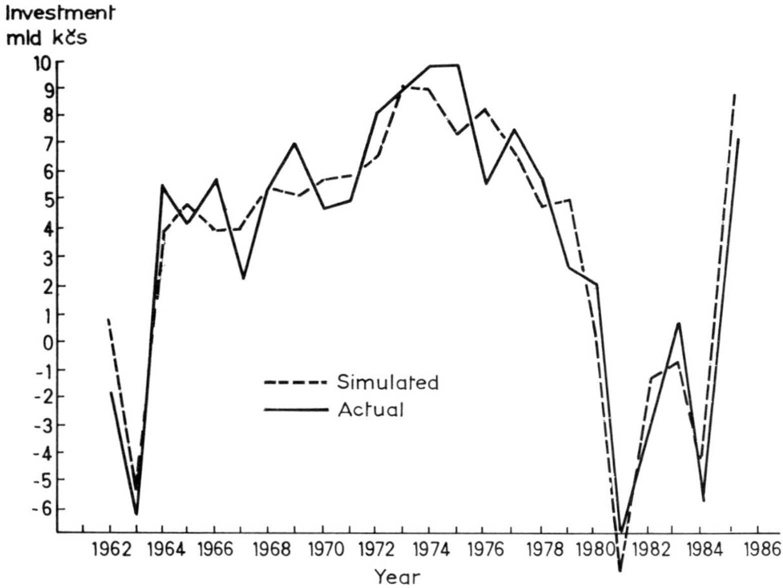

Fig. 7.1 Annual increase of investment in Czechoslovakia.

possible theoretical specification.[5]

$$\text{IN} = \alpha_0 + \underset{(+)}{\alpha_1\text{OUT}_{-1}} + \underset{(-)}{\alpha_2\text{IN}_{-1}} + \underset{(+)}{\alpha_3\Delta\text{IN}_{-1}} + \underset{(4)}{\alpha_4\text{NE}} \tag{7.5}$$

Here $\alpha_3 > 0$ represents a tendency towards an investment drive, while the achieved level of past investment IN_{-1} operates more as a break in the investment process ($\alpha_2 < 0$). Note that stability requires

$$0 < 1 + \alpha_2 + \alpha_3 < 1, \quad \text{hence} \quad -1 < \alpha_2 + \alpha_3 < 0.$$

The empirical performance is quite satisfactory for the period 1962–85:

$$\begin{aligned} \text{IN} = \; & \underset{(-2.178)}{-3564.013} + \underset{(6.325)}{0.080\text{OUT}_{-1}} - \underset{(-6.709)}{0.625\text{IN}_{-1}} + \underset{(3.179)}{0.327\Delta\text{IN}_{-1}} \\ & \underset{(-5.125)}{-1.823\text{NE}} - \underset{(-5.501)}{7141.057\text{D}134'} \end{aligned} \tag{7.6}$$

$$R^2 = 0.918, \; s = 2083.005, \; \xi^2 = 0.901, \; \xi^3 = 3.174, \; \xi^4 = 2.718.$$

Actual and fitted figures are given in Fig. 7.17; the inclusion of dummy D314 (with values equal to unity in 1963, 1981 and 1984) was necessary to track the very unusual negative changes in investment for these particular years. Apart from this, the results are good: all coefficients are significant, with signs as expected, and there are no signals of misspecification.

[5] Notation: IN investment outlays; OUT gross output, NE net export.

7.3 MACROECONOMETRIC MODEL OF THE CZECHOSLOVAK ECONOMY

After our closer look at the problem of equilibrium and disequilibrium in the consumption goods market and investment sector, an aggregate econometric model of the whole economy is presented in the present section. As underlined above, the emphasis is directed here towards empirical verification of the hypothesis on the development of the main macroeconomic aggregates. This approach calls primarily for identification of a basic theoretical macroeconomic skeleton of the socialist economy. Given the absence of an established theory the author attempted to generalize the results of empirical macromodelling on the basis of existing models (Dlouhý, 1986). This generalization reveals several theoretical links that exist in more or less all econometric models of CPEs as well as market economies: production functions, determining the level of gross or net output, aggregate consumption function, investment function, export and import functions, the chain from investment outlays to gross capital formation, increase in capacities and increase in output, inflationary feedback, etc. Given the system-specific features of CPEs, there are also specific blocks or at least equations. Following the taxonomy of Welfe (1985) one can specify a closed loop, bottleneck multiplier, describing the spread of scarcities in supplies in certain industries to the whole economy, the existence of shortages and disequilibria calling for the construction of disequilibria indicators, taking into account excess demand or supply, etc. In foreign trade a separation of the trade with socialist and non-socialist countries is inevitable, as well as a separation among the most important commodity groups.

System-specific features of CPEs have created new concepts, such as the supply accelerator, bottleneck multipliers, adjustment for shortages, capacity utilization and foreign exchange constraints. Growth is, in the long run, engendered primarily by increasing capacities; there is no clear idea how to model technological change, and the shortages in material inputs are taken into account only indirectly, which (implicitly) reflects the possibilities for modelling structural changes with these tools.

However, not only have these models largely contributed to (or even proved) the existence of some regularities in the functioning of CPEs, but they have also enlarged our knowledge of particular regularities in both a qualitative and quantitative sense. All theoretically postulated links, multipliers and accelerators were actually estimated for independent sets of data for various countries, and the conclusion can be drawn that we possess a general idea of the macroeconomics of CPEs or at least of the macroeconomics during a rather turbulent period at the end of the 70s and beginning of the 80s.

Nevertheless, the general structure of large macroeconometric models doesn't contain an endogenous, explicitly built-in equilibrating mechanism. At the same time, all existing models tend to stress the dominance of the supply side, and growth is governed essentially by a (sometimes modified) supply accelerator. Several years ago, this led Portes to the conclusion that '... a better basis for a structural model will be the more symmetrical quantity rationing framework, appropriately adapted to the CPE context' (Portes, 1978).

Within this framework, a better possibility exists for explicit specification of the equilibrating mechanism in the model, i.e. the feedback from excess demand or supply to planners' behaviour and endogenous quantity adjustment. The theoretical model is much simpler: it has only two markets (for consumption goods and labour) and two agents, households and planners; an explicit treatment of spillover between the two markets is allowed for and the whole specification of the model leads to a standard classification of short-run outcomes: Keynesian unemployment, repressed inflation, classical unemployment and underconsumption in the spirit of the Barro–Grossman theory. However, for the reasons outlined above the author does not expect the effort in specifying the model of the whole economy strictly à la Barro–Grossman to bring successful results.

The question remains whether and if so how we should include the disequilibria in our model. As a logical outcome of the conclusions from the previous section the model below doesn't allow for any disequilibria. This fact – given the persistence of excess demand and shortages in the state productive sector – clearly shows where the key lies to the solution of today's problems not only in Czechoslovakia but also in all centrally planned economies: in breaking down system-specific features leading to excess demand and shortages, i.e. improving today's inefficient system of planning and managing the economy. At the same time, this doesn't diminish the gains from macroeconomic analysis that we endeavour to set up here – CPEs are

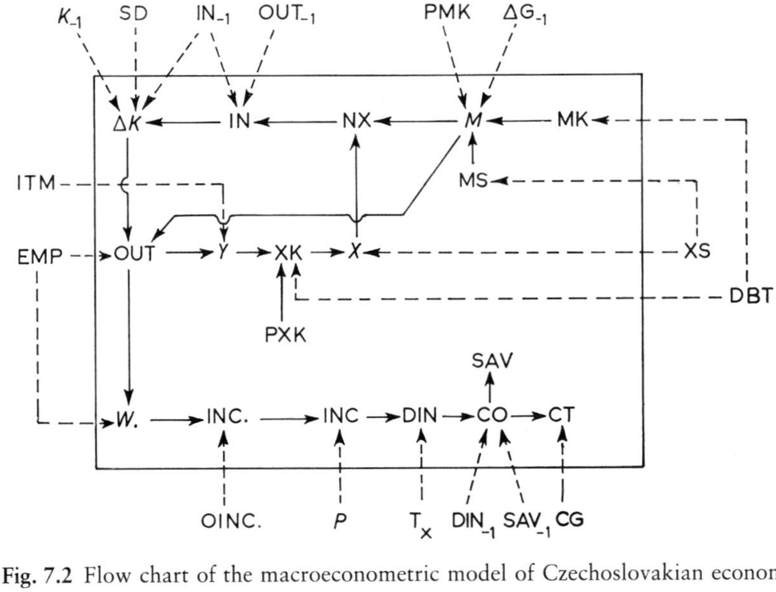

Fig. 7.2 Flow chart of the macroeconometric model of Czechoslovakian economy.

unsound through inefficiency and in medicine autopsy of parts of the diseased body represents a most powerful analytical method.

A flow-chart of the model, presented in Fig. 7.2, contains only the main connections and links. The first important link goes from gross output: the centre sets (with regard to productivity) the money wage, which in turn represents the main part of money income INC.[6] This is deflated by the retail price index, and, given the tax-rate TX, disposable income DIN is obtained. Past savings and disposable income determine personal consumption and savings and with exogenous social consumption CG we arrive at the total consumption CT. Hence, this is a completely demand-determined link, assuming global equilibrium in the consumption goods market and without any spillover from (possibly constrained) consumption to the labour market.

The second basic link stems from the conditions in foreign trade. Given net output (national income) *Y*, the centre sets the supply of exports to NSC, and given world economic activity, demand sets the price PXK. Foreign debt conditions DBT determine both the supply of exports and demand for imports *vis-à-vis* NSC. Exogenously determined demand for Czechoslovak exports in socialist countries, XS, allows for imports MS. Total exports *X* and total imports *M* give net exports NE and, in addition, total imports *M* are a factor of growth of gross output OUT.

The third link starts from predetermined variables of past gross output OUT_{-1} and past investment IN_{-1}, which together with current net exports set investment IN. This and past investment (and depreciation rate and physical liquidation) give the increase in stock of capital which enters the production function of gross output. Exogenous intermediate consumption ITM then transforms gross into net output. This is obviously the most controversial link, which hides and at the same time preserves the main sources of both disequilibria and long-term falls in productivity and efficiency. However, we aim here to track the development of the main macroeconomic variables outside the state productive sector, given the long-term conditions (i.e. the long-term, existing disequilibria, taut planning, shortages, planners' and firms' behaviour, etc.) in this sector and that these long-term conditions are well covered by the production function.

For the main equations of the model see the Appendix; a full description, including diagnostics and detailed explanation, is beyond the scope of this chapter. Base run simulations for personal consumption, investments, total exports and imports and net output are presented in Figs. 7.1 and 7.3–7.5.

This macroskeleton of the Czechoslovak economy is based on several theoretical assumptions, stemming from the analysis of past planners' behaviour and

[6]All variables are assumed to be expressed in real terms; an asterisk after the symbol means the expression is in current prices. Symbols not explained in the text: *W* money wage; INC total money income; OINC = INC-*W*, *P* retail price index; *K* stock of capital; EMP employment; XK exports to NSC; XS exports to SC; *M* imports; MK imports from NSC; MS imports from SC; PMK price of imports; PXK price of exports; SD depreciation and scrappings, AG agricultural production; WA total world imports; PW price index of WA; SAD scrappings and depreciation.

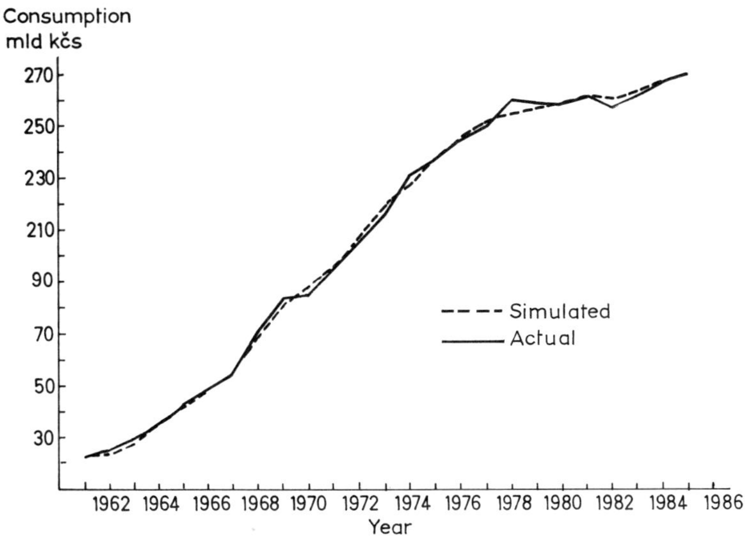

Fig. 7.3 Comsumption in Czechoslovakia.

supported by careful inspection of the data. These assumptions are: the separability of the production and consumption sectors of the economy, planners' ability to control the money wage and retail prices and the ability to meet the resulting demand for consumption goods and services, the determination of exports by foreign conditions and the dependence of imports on exports and finally, the

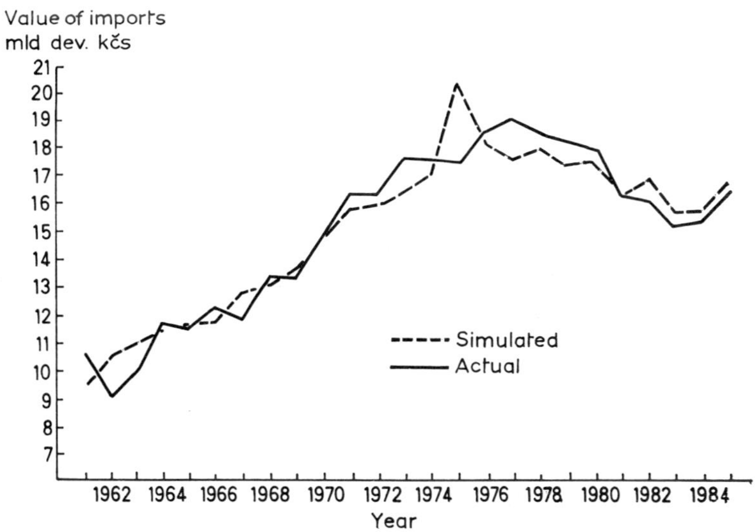

Fig. 7.4 Czechoslovak imports from non-socialist countries.

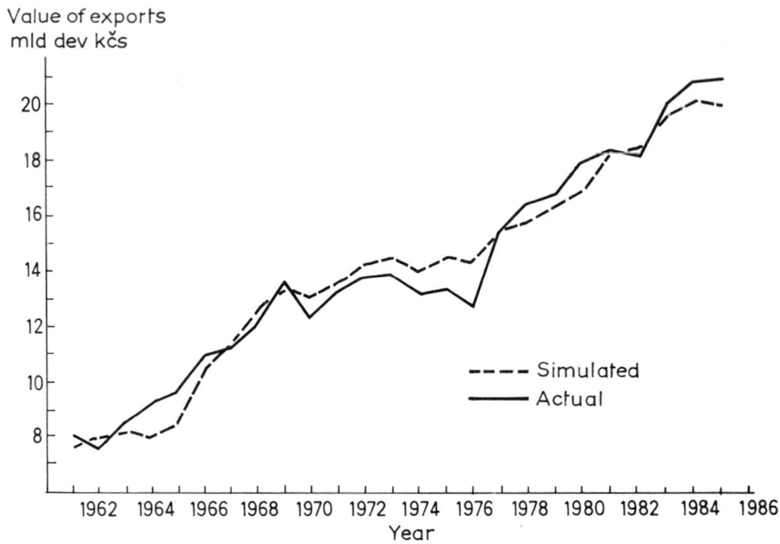

Fig. 7.5 Czechoslovak exports to non-socialist countries.

somewhat autonomous development of investments, discussed in the previous section.

There is no doubt that this is not a very fleshy skeleton; indeed, the model itself is much more detailed. However, we have achieved partial progress along two main lines: we have a theoretical description of the behaviour of both planners and households and exogenous factors are clearly identified, and we have set up a model which tracks the data with reasonable accuracy.

There is a certain equilibrating mechanism in this approach as well. Exogenous disturbances (a fall in world activity, a bad harvest, irregularities in traditional supplies from socialist countries, etc.) appear in net exports NE and in fall of total imports M; the latter variable effects rate of output both directly and indirectly, through investments and lag structure in forming an increase in capital. Changes in output generate changes in wage rate and in the aggregate price index (this is true for the last decade only and is not yet covered by the model and also in the adjustment of supplies. Inconsistencies between the structure of supply and the structure of consumers' preferences are solved partly by forced substitution and partly in the second economy, and global equilibrium is always achieved.

Equilibrating reactions to endogenous disturbances are only approximated. Ruling out global disequilibria in the consumption goods market and taking into account non-investment expenditures of the state as given, the only two main endogenous sources are the rate of investment and rate of growth of output. Strictly speaking, this calls for equations specifying planned investment and planned output as well as for a model block, to display how these plans together with system-specific features of CPEs gradually generate disequilibria and call for strict restrictions in investment with consequences for output etc. This is related to the investment cycles in CPEs, discussed above.

To conclude, the model presented here and the related discussion in previous sections can be considered only as a first step in the theoretical specification of planners' and consumers' behaviour at the macroeconomic level in the Czechoslovak economy. The problem is, in general, that both the assumptions and the structure of the model may not fit the economic experience in other CPEs. In spite of this, the author believes that further development of the model is necessary mainly for two reasons: first, it can provide a deeper insight into the macroeconomics of the Czechoslovak economy in the immediate past and it can at least provide us with short-run forecasts; second, it helps in clarifying the bounds of macroeconomics in CPE performance both formalized and nonformalized.

The principal aim of this chapter has been to discuss the questions of macroeconomic equilibrium and disequlibrium in the Czechoslovak economy. By means of traditional analysis of the statistical data and by means of econometric analysis the following basic problem has been examined: in the past, has the planning centre in Czechoslovakia been able to separate the persistent disequilibrium (given system-specific features) in the state productive sector from the situation in the consumption goods market? As should be clear from the results above, the answer to this question is basically yes. On the basis of this result the author has presented an outline of an econometric model, heavily based on the assumption of equilibrium in the whole economy. The author does not deny that the analysis here is incomplete and lacks profundity. However, he is convinced of the robustness of the answer to the basic question. Nevertheless, at the same time, he believes that substantial progress is needed in research if we are not to be stuck with the rather simple framework of Portes' model, which was set up almost ten years ago.

In this respect, Charemza's, Nuti's and Brada and King's contributions to the present volume represent substantial progress, while going some way to reconciling the disequilibrium macroeconomics of CPEs with Kornai's theory of shortage. This opens up possibilities of many interesting directions for future research, but we should still be aware of the most important fact: that the main source of disequilibrium is in the state-productive sector and that this is system determined. If we are to draw at least some policy-oriented conclusions, we must direct our efforts along these lines as well.

REFERENCES AND FURTHER READING

Askanas, B. and Laski, B. (1985) Consumer prices and private consumption in Poland and Austria. *Journal of Comparative Economics*, **9**, No. 2 June, 164–77.

Bauer, T. (1978) Investment cycles in planned economies. *Acta Oeconomica*, **21**, No. 3.

Burkett, J. (1986) Slack, shortage and discouraged consumer in Eastern Europe: estimates based on smoothing by aggregation, University of Rhode Island, New York.

Dlouhý, V. (1984a) Macroeconomic disequilibrium model of centrally planned economy. *Ekonomicko-matematický Obzor*, **20**, 374–87.

Dlouhý, V. (1984b) On the problem of macroeconomic equilibrium in centrally planned economies, Paper presented to Econometric Society European meeting, Madrid.

Dlouhý, V. (1986) On the problem of macroeconometric modelling in centrally planned economies. *Ekonomicko-matematický Obzor*, **22**, 361–77.

Dlouhý, V. and Dyba, K. (1985) *Econometric Model of the Czechoslovak Balance of Trade* (in Czech), Academia, Praha.

Dlouhý, V. and Dyba, K. (1985). Modelling Czechoslovak foreign trade flows with non-socialist countries (equilibrium and some disequilibrium estimation). *Czechoslovak Economic Papers* **23**, pp. 91–110.

Grossman, G. (1977) The second economy of the USSR. *Problems of Communism*, September/October, pp. 25–40.

Goldman *et al.* (1978) *An Introduction to Macroeconomic Analysis*, Academia, Prague (in Czech).

Havlík, P. (1985) A comparison of purchasing power parity and consumption levels in Austria and Czechoslovakia. *Journal of Comparative Economics*, **9**, 178–90.

Janáček, K. (1975) *Macroeconomic Consumption Function*, Acadmmia, Prague (in Czech).

Kalecki, M. (1965) *Introduction to the Theory of Growth in a Socialist Economy*, Oxford University Press, Oxford (in Czech).

Kemme, D. M. and Winiecki J. (1985) Disequilibrium in centrally planned economies, Working papers in Economics, Center for Applied Research, University of North Carolina at Greensboro, August.

Klaus, V. (1979) Irregularities in household consumption expenditures in the 70s. *Politická Ekonomie*, **27**, 455–67 (in Czech).

Klaus, V. (1985) Unfinished constructions, model and empirical analysis. *Finance a úvěr*, **35**, 152–62 (in Czech).

Klaus, V. and Rudlovčák, V. (1977) Dynamic model as a possible basis for judging growth of savings in Czechoslovakia. *Politická Ekonomie*, **25**, 717–22 (in Czech).

Klaus, V. and Rudlovčák, V. (1982) Savings function as an inverse problem of disequilibrium consumption modelling: a case study of Czechoslovakia, Paper presented to International Conference on Applied Econometrics, Budapest.

Kornai, J. (1970) *Anti-equilibrium*, North-Holland, Amsterdam.

Kornai, J. (1972) *Rush versus Harmonic Growth*, North-Holland, Amsterdam.

Kornai, J. (1979) Resource-constrained versus demand–constrained systems. *Econometrica* **47**, No. 4.

Kornai, J. (1980) *Economics of Shortage*, North-Holland, Amsterdam.

Kornai, J. (1982) *Growth, Shortage and Efficiency*, Blackwell, Oxford.

Kupka, V. (1986) The investment tension and unfinished construction. *Finance a úvěr*, **36**, 236–42 (in Czech).

Nuti, D. M. (1985) Hiddlen and repressed inflation in soviet-type economies: definitions, measurement and stabilization. *EUI Working Papers*, No. 85 200, Florence.

Pindak, F. (1983) Inflation under central planning. *Jahrbuch für Wirtschaft-Osteuropas*, **10/2**.

Podkaminer, L. (1986) Measurement of macroeconomic disequilibrium in centrally planned economies: a critical review of Professor Portes' results paper presented to the IIASA task force meeting on Applied General Equilibrium Modelling, Laxenburg.

Portes, R. (1978) Macroeconomic modelling of centrally planned economies: thoughts on SOVMOD I, Discussion Paper No. 621, Harvard Institute of Economic Research.

Portes, R. (1979) Internal and external balance in a centrally planned economy. *Journal of Comparative Economics*, 3, 325–45.

Portes, R. (1980) Effects of the world economic crisis on the east European economies. *The World Economy*, **3**, 13–52.

Portes, R. (1981) Macroeconomic equilibrium and disequilibrium in centrally planned economies. *Economic Inquiry*, **19**, 559–78.

Portes, R. (1983) Central planning and monetarism: fellow travellers? in *Marxism, Central Planning and the Soviet Economy* (ed. P. Desai), MIT Press, Cambridge, Mass., pp. 149–65.

Portes, R. (1986) The theory and measurement of macroeconomic disequilibrium in centrally planned economies, Discussion Paper No. 91, CEPR.

Portes, R., Quandt, R., Winter, D. and Yeo, S. (1984) Planning the consumption goods market: preliminary disequilibrium estimates for Poland 1955–80, in *Contemporary Macroeconomic Modelling* (eds P. Malgrange and P.-A. Muet), Blackwell, Oxford, pp. 254–71.

Portes, R. and Santorum, A. (1986) Money and the consumption goods market in China, Discussion Paper No. 193, Birkbeck College, London.

Portes, R. and Winter, D. (1977) The supply of consumption goods in centrally planned economies, *Journal of Comparative Economics*, **1**, 351–65.

Portes, R. and Winter, D. (1978) The demand for money and for consumption goods in centrally planned economies. *Review of Economics and Statistics*, **60**, 8–18.

Portes, R. and Winter, D. (1980) Disequilibrium estimates for consumption goods markets in centrally planned economies. *Review of Economic Studies*, **47**, 137–59.

Portes, R., Winter, D. and Burkett, J. (1980) Macroeconomic adjustment and foreign trade of centrally planned economics, Paper presented at the Econometric Society World Congress, Aix-en-Provence.

Quandt, R. (1982) Econometric disequilibrium models. *Econometric Reviews*, 1–63.

Spanos, A. (1986) *Statistical Foundations of Econometric Modelling*, Cambridge University Press, Cambridge.

Šujan, I. (1985) The use of econometric macromodels under changing conditions in centrally planned economies, Paper presented at the Ninth International Conference on Models and Forecasts, Bratislava.

Welfe, W. (1985) Econometric macromodels of unbalanced growth: the Polish experience, Paper presented at the Fifth World Congress of the Econometric Society, Cambridge, Mass.

Winiecki, J. (1982) Investment cycles and an excess demand inflation in planned economies: sources and processes. *Acta Oeconomica*, **28** (1–2).

Winiecki, J. (1985) Portes ante portas or a revisionist interpretation of inflaton under central planning: a critique. *Comparative Economic Studies* No. 2.

Winiecki, J. (1986a) Distorted macroeconomics of central planning. *Banca Nazionale del Lavoro Quarterly Review*, **157**, 197–223.

Winiecki, J. (1986b) The share of industry in the national income of centrally planned economies. *Ekonomista* No. 1, 109–33 (in Polish).

APPENDIX

Equations of the model (basic version, identities omitted).

Net output:

$$Y = 0.963e^{0.018t - 0.001t^2}/0.977K^{-3.337} + 0.023\text{EMP}^{-3.337}/ - 0.299 \times e^{-0.067\,\text{D624}}$$

D624 – dummy for 1962–64

Gross output:

$$\frac{\mathrm{OUT}}{\mathrm{Y}} = \left(\frac{M}{\mathrm{Y}}\right)^{0.205} \times e^{0.675}$$

Consumption function:

$$\mathrm{CO} = -0.452\mathrm{SAV}_{-1} + 0.848\Delta\mathrm{DIN} + 0.978\mathrm{DIN}_{-1}$$

Investment function:

$$\begin{aligned}\mathrm{IN} = &-3564.013 + 0.080\mathrm{OUT}_{-1} - 0.625\mathrm{IN}_{-1} + 0.327\Delta\mathrm{IN}_{-1}\\ &- 1.823\mathrm{NE} - 7141.057\mathrm{D314}\end{aligned}$$

Foreign trade:

$$\mathrm{MK} = 4551.625 - 12343.831\mathrm{PMK} + 10826.171\mathrm{PXK} + 0.047\mathrm{XKPW}^{7}$$

$$M = \mathrm{MK} + \mathrm{MS}$$

$$\mathrm{XK} = -367.472 + 0.055Y - 5016.426\mathrm{PXK} - 0.070\mathrm{INX} - 0.397\mathrm{XSX}$$

$$\mathrm{PXK} = 3.271 + 0.816\mathrm{WA} - 0.853\mathrm{XK} + 0.629\mathrm{PW}$$

$$X = \mathrm{XK} + \mathrm{XS}$$

[7] XKPW proxy for the level of foreign debt.

8

Macroeconomic disequilibrium models of Poland

WLADYSLAW WELFE

8.1 INTRODUCTION

The early multisectoral models of the Polish economy of the 70s: KP1 and KP2 (Maciejewski, 1976, 1980) and W1, W3 (Welfe 1973; Czyzewski, Debski and Welfe, 1976) assumed that the national economy, its main sectors and markets are, in principle, balanced.[1] The imbalances were temporary and thus did not require special treatment (the use of dummies was advocated by Pawlowski, 1960 and Kolupa, 1965). This was based on the belief that the planners' activities at the stage of plan construction are consistent, i.e. that they bring the requirements and resources to a balance and that the central management system using commands can ensure – mainly through quantitative adjustment – balanced implementation of the plans.

One of the major potential applications of these models was – by means of appropriate simulation exercises – to discover the potential imbalances and analyse the efficiency of policy instruments aimed at preventing disequilibria.[2] To achieve these goals the models had to be complete, i.e. to include equations determining both demand and supply of commodities and labour and define the excess demand (supply). This rule was, for the first time, applied in a systematic

[1]We have tried to avoid using the term 'equilibrium' in a similar sense, but without much success, however. Nevertheless, where we do use the term it means merely that the equality of demand and supply, which might be temporary or stable, frequently approximates. It does not necessarily indicate that the state of the system in question is normal, natural or optimal.
[2]The problem setting here shows some similarities with the computable general equilibrium (CGE) models. However, the econometric macromodels have been stressing quantitative adjustments made mainly by the central planner, compared with price adjustments within a market system as in the CGE models.

International Studies in Economic Modelling Series.
Models of Disequilibrium and Shortage in Centrally Planned Economies.
Edited by C. Davis and W. Charemza.
Published in 1989 by Chapman and Hall Ltd, 11 New Fetter Lane, London EC4P 4EE.

way in the W1 model, which explicitly determined excess demand; model KP2 defined demand and supply of commodities for the major sectors, i.e. at the households' and producers' levels.[3]

Growing disequilibria (demand pressures, shortages in supplies) in the second half of the 70s, which found their momentum in the economic crises in Poland in the early 80s (associated with a decline in imports and overall eonomic activities by *c.* 25%), brought new attempts in modelling efforts. The model builders had to acknowledge the existence of excess demand (supply shortages) and associated, directly unobservable phenomena. These were: departure of effective demand from its notional content (due to forced substitution, intertemporal spillovers, expectations, etc.). The problem of their measurement had to be solved as well. The most interesting and difficult question was how to model the impact of disequilibria in various sectors and markets, especially in the production sector (bottlenecks and growing underutilization of capacities due to shortages of production factors), assuming price rigidity.

The first reaction was to develop supply determined models such as WE3S (Juszczak and Welfe, 1983) and then W7S (W. Welfe, 1986b) that completely neglected the determination of demand and concentrated instead on generating output (supply) and its allocation. They introduced additional constraining factors (i.e. material inputs) into the production functions, however, without identifying conditions under which the material inputs really constrain the use of production capacities, i.e. are in short supply.

The attempts to model explicitly the disequilibria at the macrolevel were undertaken in the early 80s in various research centres. The so-called unbalanced version of the W3 model (Czerwinski and Welfe, 1982; Welfe, 1983) was rebuilt and extended and a large, multisectoral annual W5 model, which is being used regularly in medium-term forecasting, plan evaluation and other policy simulations, was constructed (Welfe, 1985a, b, c). It includes, in the majority of sectors and markets, the equations determining both supply (output and its allocation) and demand, as well as excess demand (supply).[4] It puts special emphasis on the production sector, trying to explain the impact not only of disequilibria in labour markets but also in foreign trade and shortages in supplies of energy and materials on the production process (utilization of capacities). The impact of disequilibria in the consumer goods markets on price formation and productivity, in the labour market on wage increase and investment activities, etc. has been studied only recently. This happens both in the framework of model respecification and in a project aimed at construction of a small model W6, which is an aggregated version of the large

[3]The use of these models was conceptually difficult. To ensure a unique solution equalizing demand and supply, a set of originally exogenous variables (quantitative adjustment instruments) should be defined and endogenized. On this assumption it would have to be done for a sample period too, but it has never been systematically performed. Otherwise, the assumption that both demand and supply are observable (being equal transcated values) would not hold, and the use of conventional estimation techniques would not be justified.

[4]In the specification of consumers' demand equations, the construction of specific disequilibrium indicators, the model builders borrowed from other authors, especially from A. Welfe (1984, 1985).

model W5 (Romanski and Welfe, 1985, 1986). As this research has not yet been completed, we refer below mainly to the findings related to the W5 model.

Studies of the properties and methods used in the classical econometric disequilibrium models led W. Charemza and M. Gronicki (see Charemza, 1981) to the construction of sectoral disequilibrium-type models. In the early 80s, they built a comprehensive, small (one-sectoral) disequilibrium model for the Polish economy and thoroughly analysed its properties (Charemza and Gronicki, 1985). The model defines both demand and supply and tries to introduce broadly the likely effects of disequilibria in consumer goods markets, labour supply, imports demand, spillovers from labour markets into consumers' demand, etc. Being short-term oriented it also introduces expectations (including those for disequilibrium intensity), and endogenizes the production plan. The model was applied to study the alternatives of market disequilibria in the consumer goods and labour market by means of historical simulations (for the sample period 1960–80) and the chances of reducing disequilibria using optimal control techniques. It is the most comprehensive model of this type built for a centrally planned economy (compare Dlouhý, 1984 for CSRS).

In this chapter a comparison of the main properties of the two macroeconomic models will be made. Both models attempt to describe the functioning of the Polish economy during a similar period, and make similar assumptions regarding the management and planning system, with the same goal – to discover the disequilibria, analyse their likely implications and provide policy instruments to minimize their magnitude. However, they show important differences in their objectives, underlying hypotheses, specification, size and methodology used.

The basic assumptions regarding the behaviour of economic agents, their interrelations and the economic mechanisms are discussed first. This is followed by a discussion of the role of disequilibria and modifications in the behaviour of economic units and economic mechanisms under disequilibria. The specification of particular blocks of equations will be compared, i.e. production, employment and fixed capital formation, foreign trade, investment and consumption, wages and prices. The final remarks contain some important methodological issues.

8.2 BASIC ASSUMPTIONS

The models describe the activities of the following economic units: producers (mainly socialized enterprises), government (institutions of the non-material services sector), households, foreign agents and the central planner, controlling these activities. The Charemza–Gronicki (CH–G) model treats the above-mentioned units and their activities globally after allowing for suitable aggregation. Model W5 partitions the economic agents by sectors and industries and decomposes their activities. Let us summarize the basic assumptions about their behaviour, disregarding the sectoral differences and ignoring in this preliminary section the existence of disequilibria.

8.2.1 Producers

Producers are involved in the production process, i.e. transformation of production factors – fixed capital (K), employment (N), materials and energy (QM) using prescribed (changing in time) technology (TO) – into production.[5] This process is described by a technological production function, which has the general form

$$X = f(K, N, \mathrm{QM}, \mathrm{TO}) \tag{8.1}$$

where X is net output. The introduction of the use of materials and energy is rather new and unconventional. In practice the energy input was used (since the energy crises). As at the macrolevel the material input reduces to the imports of the intermediate commodities – imports of these commodities if available or total imports were used as additional factors in production functions, especially in the models for CMEA and developing countries (for instance in the UNCTAD SIGMA model, see United Nations, 1987).

There is a technological relationship linking net output with gross output (Q):

$$Q = X + \mathrm{QM} + A \tag{8.2}$$

where A is depreciation.

The use of production factors is determined by the technology described by the production function (Equation 8.1). We have for the use of fixed capital

$$K = f^{-1}(X, .) \tag{8.3}$$

labour

$$N = f^{-1}(X, .), \tag{8.4}$$

and materials

$$\mathrm{QM} = f^{-1}(X, .) = h(Q, .) \tag{8.5}$$

At the macrolevel the material-inputs of domestic origin cancel out and the last term reduces to the use of imported materials and energy. If Equation 8.5 is linear, then for a disaggregated system we have a Leontief-type system of equations explaining the intermediate use of commodities.

The capital stock formation is related to investments put into operation (I) and scrappings:

$$K - K_{-1} = I + dK_{-1}, \tag{8.6}$$

where d is the rate of scrapping.

Investments put into operation can be decomposed into the respective portions of investment outlays (J) realized in current and previous periods:

$$I = \sum_{i=0}^{m} v_i J_{-i}, \tag{8.7}$$

[5] We use here the notation applied in the W model, explained in the text and the tables that follow. It differs to some extent from that used in the Charemza–Gronicki (CH–G) model.

where $v_i > 0$ are the parameters of the lag distribution and m is the length of completion period.

Investment outlays are composed of investment goods either domestically produced or imported:

$$J = v_k X + v_m M \tag{8.8}$$

where M is imports and v_k, v_m are allocation coefficients. Equations 8.1, 8.6, 8.7 and 8.8 determine a fundamental dynamic feedback between output and investment outlays that is a supply-type accelerator which we call the supply investment–output multiplier (Welfe, 1983, p. 215). In modified forms it is available in both models.

The financial side is summarized by defining the components of expenditure and surplus which add up to the value of gross output. We have the following identities.[6] Value of net output (XP):

$$\mathrm{XP} = \mathrm{ZP} \cdot N + \mathrm{AFP} \tag{8.9}$$

where ZP is average wages, AFP is financial accumulation (gross surplus) and the value of gross output (QP) is:

$$\mathrm{QP} = \mathrm{QMP} + \mathrm{AP} + \mathrm{XP} \tag{8.10}$$

Dividing both sides of Equation 8.10 by Q we obtain an equation defining producers' prices (PQ):

$$\mathrm{PQ} = (\mathrm{QMP} + \mathrm{AP})/Q + \mathrm{ZP}/(Q/N) + \mathrm{AFP}/Q, \tag{8.11}$$

where the particular components mean: unit material costs, labour cost and mark-up (including taxes). The price equations are defined in the W5 model only, whereas the CH–G model assumes that prices are exogenous.

The average wages are either policy instruments or they change as a result of negotiations taking into account the expected productivity increase and inflation rate (which is adopted in the W5 model only). Thus

$$\mathrm{ZP} = z(X/N, \mathrm{PC}), \tag{8.12}$$

where PC is the prices of consumer goods.

If we establish an equation linking prices of consumer goods (PC) with producers' prices (PQ), then Equations 8.11 and 8.12 will define the inflationary loop (absent in the CH–G model). The financial accumulation can be decomposed in net surplus, taxes and subsidies being financial policy instruments. These variables are defined and used in the W5 model only.

Under the centralized management system which prevailed in the past (in the sample period 1960–80) producers were basically motivated to maximize their output (initially gross, then net, output). However, because of the existence of

[6] We use the following notation: P at the end of a symbol indicates a variable expressed in current prices (its nominal value).

other, competing goals, it was rather difficult to formalize their (optimizing) behaviour. The dependence of producers' decisions on the central plan has led several authors, including Charemza and Gronicki, to assume that producers' activities can be represented at the macrolevel by central planners' activities (where the national economy is treated as an immense factory). This simplification also goes too far in a highly centralized economy. However, it is not too confusing provided there is no reason to believe that the current producers' decisions are no longer in line with that of the central planners.

As this is obviously true with regard to investment decisions the CH–G model treats producers and central planner separately, the first representing demand, the latter controlling supply. The W5 model also consistently distinguishes between producers' behaviour and central planner's decisions with regard to the generation of requirements for remaining production factors (including imports) and of wages.

A further important common assumption is that producers show low sensitivity to (relative) price changes, including prices of primary production factors. The economic reforms introduced in the 80s, which tend to change producers' motivation towards profit maximizing, do not make it possible to abandon the above assumption as long as producers are constrained, especially with regard to the material inputs. This explains why not much use could be made of the established, formalized theory of producers' behaviour in market economies.

On these assumptions the supply of production and demand for production factors can be defined in a general way only.[7] Supply can be determined either in terms of net output from Equation 8.1 or in terms of gross output from Equation 8.2, after allowing for changes in inventories of products (which typically are negligible). Demand for production factors can be determined using respectively Equations 8.3, 8.4 and 8.5 after allowing for changes in stocks and lags of adjustments.[8] To arrive at demand for investment outlays the system of Equations 8.3, 8.6 and 8.7 must be solved for J.

8.2.2 Public institutions

The activities of public institutions consist of providing nonmaterial services (educational, social, medical, cultural and general, including administrative, defence, etc.). In principle, they can be described by a specific production function and the derived functions for the use of production factors, including material inputs (being equal to the so-called social consumption G). They are financed – together with transfers to households and producers – from the state budget, its basic source being income from taxes.

The above activities were either treated as completely exogenous (central planner instruments) as in the CH–G model or partly assumed to depend on the

[7] We use the following notation: S at the end of a symbol means supply, D means demand.
[8] Modifications due to temporary and chronic disequilibrium will be discussed below.

availability of resources (capital formation) and also their use in other sectors, i.e. as residuals (labour) as in the W5 model.

8.2.3 Households

Households are assumed to provide the labour force for production and services sectors; they accumulate the wealth and use it for current consumption. They receive income basically from employment and transfers. Personal income (YP) is obtained as a sum:

$$\mathrm{YP} = \mathrm{ZP} \cdot N + \mathrm{YIP} \tag{8.13}$$

where YIP is other incomes.

Personal income is spent or saved:

$$\mathrm{YP} = \mathrm{CP} + \mathrm{DOZP} \tag{8.14}$$

where CP is consumer expenditures and $\mathrm{DOZP} = \mathrm{OZP} - \mathrm{OZP}_{-1}$ is current savings, equal to the increase in cash and bank deposits.

Households are assumed to behave rationally. In the CH–G model a utility function is explicitly employed and demand, savings and labour supply functions are derived. The W5 model uses the conventional specifications of the above functions, which are not inconsistent with previous more demanding approach. In general the above functions can be specified as follows.

Consumer demand (CD) will basically depend on personal income and relative prices. For total demand we have:

$$\mathrm{CD} = C(\mathrm{YP/PC}) \tag{8.15}$$

Personal savings (DOZP) will depend on personal income and changes in motivation to save:

$$\mathrm{DOZP} = O(\mathrm{YP}, r\%) \tag{8.16}$$

where $r\%$ is the real interest rate on deposits, labour supply (NS) will be determined by demographic factors and labour participation changes negatively with the ratio of transfers (pensions, etc.) to wages:

$$\mathrm{NS} = N(\mathrm{LPR}, (\mathrm{YIP}/L)/\mathrm{ZP}) \tag{8.17}$$

where L is population and LPR is population of productive age.

8.2.4 Foreign trade

Foreign trade plays a significant role: exports provide foreign currency to finance imports, extend the participation in the results of world division of labour, increase the efficiency and quality of domestic output, etc. The foreign demand for exports depends on factors determining foreign countries' imports; domestic supplies for export depend on the production capacities and attractiveness of exports compared with domestic uses.

Imports are basically complementary, serving either current production (materials, energy) or fixed capital formation (technology and machinery equipment) or consumption. They are specified as demand functions. Because of differences in planning and financing foreign trade both models distinguish between trade with socialist and nonsocialist countries.

The simplest description of the functioning of the national economy as a system of interlined units exchanging their products and services can be given *ex post* – by the following balance equation. It says that total resources coming from production and imports are equal to their use:

$$M + Q = \mathrm{QM} + H + E \qquad (8.18)$$

where M is imports, E is exports, H is final domestic use, $H = C + G + J + \mathrm{DR}$, where J is investment outlays and DR is increase in stocks of inventories. The above material balance identity is equivalent to the national income accounting identity:

$$X = \mathrm{XG} + (E - M) \qquad (8.19)$$

where $\mathrm{XG} = C + G + \mathrm{JA} + \mathrm{DR}$ is national income distributed and $\mathrm{JA} = J - A$ is net investment.

It has an equivalent in financial terms, i.e. current prices: $\mathrm{XP} + X \cdot \mathrm{PX}$. Substituting from Equation 8.11 and solving for deflators of commodities entering national income (PX) we have:

$$\mathrm{PX} = m(\mathrm{PM}) + \mathrm{z}(\mathrm{ZP} \cdot N/X + \mathrm{AFP}/X) \qquad (8.20)$$

where PM is imports prices. This equation establishes the (reduced) links between prices and wages (given the labour productivity) and financial accumulation–output ratio (mark-up), the latter being partly dependent on tax rates.

8.2.5 Central planner

The role of the central planner clearly depends on the prevailing management system. Nevertheless, the planner's responsibilities without any doubt cover control over public expenditures and partly over investment outlays in order to ensure the assumed rates of growth. The planner influences the changes in reserves and inventories and to some extent prices and wages in order to prevent domestic disequilibria and inflation. The planner controls the foreign trade balance and frequently imports. With economic reform direct control measures are being substituted by indirect financial measures. It is associated with a decline in formal rationing of the markets for intermediate and investment goods.

8.3 MODELLING DISEQUILIBRIA: THE PRODUCTION PROCESS

The economic situation in the last decades has frequently shown serious departures from assumed balanced growth. Producers, investors and consumers experienced numerous constraints in realizing their plans and especially their demands; prices were either repressed or adjusted too slowly to clear the markets.[9] On the other hand, quantitative adjustments quickly approached their ceilings, i.e. in the case of an increase in demand buffer stocks were insufficient, production factors were constrained, etc. and the result was excess demand (shortages).

In general, producers' supplies may be either (a) insufficient to meet demand: $XS < XD$ and $XS = X$, i.e. supplied quantities are transacted or (b) in excess: $XS > XD = X$, i.e. demanded quantities will be transacted. We can thus generally assume that the minimum condition holds:

$$X = \min(XS, XD). \tag{8.21}$$

We shall say that in case (a) the economic system is supply determined, and in case (b) demand determined. Let us analyse more carefully the supply-determined system, which many scholars believe was characteristic of the centralized management system that prevailed in Eastern Europe in the 60s and 70s.

The shortages of supplies at producers' level are due to constraints in the availability of at least one of the production factors: fixed capital (bottlenecks), labour or material inputs, which at the macro-level (after solution for intermediate supplies) translate into imports of materials. However, excess demand at producers' level means that at least one of the users suffers from short supplies. This provides a broad variety of potential situations (regimes, states) which the model builders have to consider.

The disequilibrium theory, developed for the market economies, initially put the emphasis on simplified situations, assuming that in the short term either labour or fixed capital (adjusted to the level which ensures profitability) are the binding constraints and that the consumer goods market is only rationed (intermediate demand was assumed to be met, investment and exports exogenous). This led to a broadly accepted classification of economic regimes,[10] which for the supply-determined system distinguishes (a) the repressed inflation regime, where labour is constrained, i.e. $ND > NS$, (b) classical unemployment, where there is excess supply of labour, i.e. $ND < NS$ and the production is constrained by fixed capital (due to profit maximization). We are opposed to using this classification for a description of the centrally planned economies (CPEs) as it is incomplete and does not accentuate the most important types of constraint occurring in the real world. Secondly, the names are misleading as they suggest the existence in the CPEs of economic mechanisms specific to the capitalist economies, as is obviously the case with regard to the classical unemployment regime. Again, the source of the

[9]Let us remember here that producers have shown low sensitivity to price changes, i.e. their response in terms of supply increases or by substituting production factors has been weak.

[10]It assumes symmetrically for demand-determined systems: Keynesian unemployment if excess supply in commodity markets is associated with unemployment ($ND < NS$), and underconsumption, if it is associated with excess demand for labour ($ND > NS$) (see Barro and Grossman, 1976; Malinvaud, 1977). The latter regime is sometimes regarded as a realistic alternative to the repressed inflation system for the CPEs (see Charemza and Gronicki, 1985, p. 161). See also Welfe (1987).

repressed inflation regime is not necessarily economic policy repressing prices and allowing unconstrained wage increases. Charemza and Gronicki (1985) believe that consumers' pressure on wages, based on their erroneous expectations, form a mechanism (money illusion) that explains the growing excess demand.[11]

Kornai's (1982) argument seems more general and convincing, i.e. that the basic source of disequilibria in all other markets is also the absence of hard budget constraint in the producer's sector (being an institutional factor). Given the tendency to maximize current or future output the producers are not effectively constrained in their current or investment expenditures. This might explain the existence of chronic (normal) disequilibria. However, to explain the changes in their intensity we need more – an explanation of changes in constraints for the particular production factors and their role in the rationing process (regime changes). Kalecki's theory of growth barriers (1963) seems to provide the arguments which explain how, in the process of accelerated growth available resources are exhausted and barriers appear: (a) fixed capital, which in the long run can be increased until the investment level reaches its ceilings determined by the assumed minimum level of consumption, (b) the available labour force, (c) imports, which depend on exports (its increase being typically lagged in the case where imports lead growth) and available foreign financing.

In line with the above arguments we suggest extending the classification of the supply-determined system into basically (a) fixed capital (K) constrained, i.e. such that $\mathrm{KD} > \mathrm{KS}$, (b) labour ($N$) constrained, i.e. such that $\mathrm{ND} > \mathrm{NS}$ and (c) material input (QM) constrained, $\mathrm{QMD} > \mathrm{QMS}$, which at the macrolevel translates into $\mathrm{MQMD} > \mathrm{MQMS}$, where MQM is imports of energy and materials.[12]

This calls for a further extension of the notions of variables used in the analysis – their potential values should define not only offers or requirements (planned or expected) but also their expected value for a given state (regime) of the economy. Thus, by adding at the end of a variable the symbols K, N, QM we indicate that the variable belongs to the respective regime. Using this notation we write that the supplies at the producers' level can be determined from the following minimum condition

$$\mathrm{XS} = \min(\mathrm{XK}, \mathrm{XN}, \mathrm{XQM}) \tag{8.22}$$

The (equipment) capacity constrained output is given by:

$$\mathrm{XK} = f(\mathrm{KS}, \mathrm{NK}, \mathrm{QMK}) \tag{8.23}$$

where $\mathrm{NK} \leqslant \mathrm{NS}$, $\mathrm{QMK} \leqslant \mathrm{QMS}$ are respectively employment and material inputs sufficient to utilize fully the existing productive equipment. Notice that in this case full employment is not excluded. The values of NK and QMK can be obtained from

[11]We comment on their hypotheses further on, discussing the disequilibrium indicators concept.

[12]This constraint must not be identified with the foreign trade barrier. If the latter arrives (for instance because of limitation of foreign financing opportunities),imports of consumer goods may decline first, and imports of machinery equipment (investment) may be reduced before the supplies of raw material will be cut off and affect the current producers' activities.

demand functions 8.4 and 8.5. The system of Equations 8.23, 8.4 and 8.5 can thus be solved for XK, yielding the reduced form production function:

$$\mathrm{XK} = r(\mathrm{KS}). \tag{8.24}$$

Labour force constrained output (XN) is given by:

$$\mathrm{XN} = f(\mathrm{KN}, \mathrm{NS}, \mathrm{QMN}) \tag{8.25}$$

where $\mathrm{KN} \leqslant \mathrm{KS}$, $\mathrm{QMN} \leqslant \mathrm{QMS}$ are productive equipment and material input sufficient to ensure full employment; their values can be derived from demand functions 8.3 and 8.5. Then $\mathrm{XN} = r(\mathrm{NS})$. Material inputs constrained output (XQM) is defined from:

$$\mathrm{XQM} = f(\mathrm{KQM}, \mathrm{NQM}, \mathrm{QMS}) \tag{8.26}$$

where $\mathrm{KQM} \leqslant \mathrm{KS}$, $\mathrm{NQM} \leqslant \mathrm{NS}$, i.e. productive equipment and employment are in excess. Again, their required values will be determined from Equations 8.3 and 8.4 and after substituting in Equation 8.26 the reduced form is derived: $\mathrm{XQM} = r(\mathrm{QMS})$, suggesting that the impact of fixed capital and employment might be ignored.

The results obtained so far are valid for a long-term type of analysis, i.e. if we assume that the substitution process between production factors can be completed and that the adjustments in the level of the production factors to that being constrained (bottleneck) have taken place. For short-term analysis none of these assumptions is justified.

Shocks in supplies of the constraining factors bring immediate, quasi-proportional changes in output as no factor substitution is feasible. For instance, electric energy constraints imposed on the cement industry reduce its output proportionally. Thus, to reflect this impact properly, we can generally assume that

$$\mathrm{XS} = \mathrm{XZ}(1) \cdot \mathrm{ZS}/\mathrm{ZD} \tag{8.27}$$

where $Z = K, N$, QM is the constraining factor in the current period and $Z(1)$ is in the previous period. The equation reads that the current output will be obtained after adjusting the potential output defined for the regime prevailing before the shock took place – using a simple shortage indicator, relating supply of factor Z to the demand for it, ZD. It can be interpreted as a coefficient of utilization of potential output as the value of ZD is obtained referring to the potential output XZ(1). Returning to the previous example, let us assume that the cement industry was originally constrained by available equipment. A decline in electricity supplies would lead to an immediate decline in cement production, which can be calculated from:

$$\mathrm{XS} = \mathrm{XK} \cdot \mathrm{QMS}/\mathrm{QMK}$$

where QM stands for electricity supplies, and the ratio QMS/QMK estimates the rate of capacity utilization.

It is important in this context to take into account that in general the demand for constraining factor ZD cannot be simply derived from one of the previously defined demand functions 8.3, 8.4 or 8.5 which determine the notional demand. The effective demand may differ from it because of (a) producers' expectations with regard to the intensity of shortage and (b) likely forced substitution within the constrained factor. If the producers expect growing shortages of raw materials, say, they will tend to extend their stocks (stock hoarding) and thus their demand. If they can substitute imported raw materials for domestic materials of inferior quality, they will have to increase their total demand too. It means that we have

$$\mathrm{ZD} = h(f^{-1}(\mathrm{XZ}(1)), (\mathrm{ZS}/\mathrm{ZD})^{\mathrm{E}}, S) \tag{8.28}$$

where superscript E means expectations, and S stands for a parameter expressing the results of forced substitution, which may assume certain lags in forming effective demand.

The second objection raised in the context of short-term analysis rests on the observation that, if a shock in supplies of a constraining factor occurs, the other factors do not adjust immediately, i.e. their values cannot be established using demand functions 8.3, 8.4 and 8.5 respectively as has been claimed before.

Suppose there is a shock in imports of raw materials – a serious decline such as happened in Poland in 1981–82. The respective decline in industrial output brought a less than proportional decrease in employment and a still increasing fixed capital. Of course, the working time both of employees and machines declined respectively, and the productivity of labour and especially that of equipment decreased dramatically. The lags of adjustments of employment can be explained in a similar way to the phenomenon of labour hoarding in the market economies during recessions. There were optimistic expectations of producers and expected costs of firing and hiring experienced personnel that prevented firms from making immediate downward adjustments of employment. The opposition of the firms' managers to adjusting capacities downwards by means of increased scrapping (eliminating old, inefficient equipment) has similar roots, additionally strengthened by pessimistic expectations with regard to new investments (shortages in equipment supplies). Thus, the simple adjustment schemes applied in macromodels (for instance that used by Sneeśens, 1981 to allow for labour hoarding) may be insufficient. Anyhow, to deal with these phenomena more information is necessary to distinguish between flows of labour and fixed capital inputs and stocks of these variables. Flows (measured in hours worked, say) quickly adjust to output changes, whereas stocks, or to put it more properly their utilization, need additional explanation in terms of lags and expectations of disequilibria in respective markets.

It must be emphasized that in econometric modelling only a few of the potential variables analysed so far are directly observable. The values of others must be estimated, which poses a series of technical problems. Some of them will be discussed in the next sections.

The analysed models of the Polish economy are in the above broad context rather dissimilar. Both models leave open the question whether the economic system is supply or demand determined. However, the authors of the CH–G model assume that over the sample period the economic system was supply determined. It follows indirectly from the assumption that there is excess demand for consumer goods and from special treatment of investments in the process of aggregation discussed below, which ensures excess demand over the whole sample period 1960–80. The simulation results confirm this global assumption yielding an average 3%–4% excess demand for GDP (see Charemza and Gronicki, 1985, pp. 138–40). As no one of the above assumptions was positively verified, and this is documented below, the question remains open.

The equations explaining the production process in the CH–G model are shown in Table 8.1. The formulation of the equations determining production plans (C1), production (C2) and demanded employment (C3) implies that the model implicitly assumes that the production is equipment capacity constrained. Indeed, the production function enters not the labour supply but demand for labour[13] which in turn depends on (planned) output. Hence, specification 8.23 is assumed (material inputs are ignored). Thus, there is a contradiction in the model's assumptions, as at the same time the authors assume that there is generally a labour shortage, i.e. ND > NS, which would suggest the production function specification given by Equation 8.25. The model does not include any constraints in material inputs nor does it analyse any distortions due to disequilibria in intermediate commodity markets or foreign trade. The production function (C2) contains *G*, government expenditure as an explanatory variable, as a proxy for capacity utilization, following the Barro–Grossman specification. This might be justified only in the case where underutilization of capacities was due to insufficient demand (represented by *G*). It does not apply, however, for the Polish economy, where it was dependent on shortages of material supplies and/or labour force. The production function discussed above does not allow for technical progress either, and uses the linear approximation, which qualifies it for short-term analysis only.

The W5 model contains a specification which allows definition of a variety of economic regimes and switching from one state to another. No definite results can be demonstrated for the sample period yet. However, there is no doubt that from the late 70s the supply-determined regime prevailed. As for potential (or factual) output XS the model does not distinguish the three states defined above using Equations 8.23, 8.25 and 8.26. This is because of hesitation in explicit use of the long-term specification and also because the model builders have not succeeded in defining labour supply by industries. Instead, the short-term specification outlined above was used. The equipment capacity constrained regime was assumed as a

[13]The authors follow the Barro–Grossman specification, claiming that the demand for labour should be understood as a sum of labour supply and excess demand. The rationale behind the use of the latter variable is that excess demand for labour stimulates measures to increase productivity of labour (Charemza and Gronicki, 1985, pp. 62–3). The problem is, however, that the above-mentioned variables do not enter the production function separately, neglecting the fact that their likely effects are different. Moreover, the excess demand for labour as an implicit component of labour demand would depend on the planned output which is determined by the (expected) production and positvely by the excess demand for labour again. Because of the above specifications the suggested interpretation cannot be accepted as it leads to logic inconsistency.

Table 8.1 Model CH–G equations*

A. Production sector

PLANNED GDP

$$X^{P}_{+1} = 0.88X^{E}_{+1} + 352(\text{ND} - \text{NS}) \tag{C1}$$

where superscript P means plan, E means expectations and subscript + 1 indicates that plans (expectations) which were formed in the current period (0) apply to the next period (1)

TOTAL GDP

$$X = -1235 + 0.20K + 80\text{ND} + 1.73G - 3.5Z4 \tag{C2}$$

where $Z4 = 1$ in 1980.

DEMAND FOR LABOUR

$$\text{ND} = 10 + 0.0034(0.6X^{P} + 0.4X^{P}_{+1}) - 1.05Z3 + e_{\text{ND}} \tag{C3}$$

where $Z3 = 1$ in 1976–80.

AVERAGE WAGES

$$\text{ZP} = \text{ZP}_{-2} + 0.17(\text{CPD} - \text{CPS}) \tag{C4}$$

B. Investment

SUPPLIES OF INVESTMENT GOODS

$$\text{JS} = -152 + 0.36X^{P} + e_{\text{IS}} \tag{C5}$$

DEMAND FOR INVESTMENT OUTLAYS

$$\text{JD} = 0.088(\text{KD} - \text{K}_{-1}) \tag{C6}$$

REQUIRED FIXED CAPITAL

$$\text{KD} = 288 + 5.90(0.6X^{P} + 0.4X^{P}_{+1}) + 697Z2 + e_{\text{KD}} \tag{C7}$$

where $Z2 = 1$ in 1978–80.

FIXED CAPITAL

$$K = (1 - 0.08)K_{-1} + I \tag{C8}$$

INITIAL GROSS FIXED CAPITAL FORMATION

$$\tilde{I} = 15.0 + 0.57J + 0.54J_{-1} + 0.18J_{-2} - 0.21J_{-3} - 0.37J_{-4} + e_1 \tag{C9}$$

GROSS FIXED CAPITAL FORMATION

$$I = \tilde{I} - 2.8\max(0, (\text{JD} - \text{JS})_{-2}) \tag{C10}$$

INVESTMENT OUTLAYS

$$J = \min(\text{JD}, \text{JS}) \tag{C11}$$

(contd.)

Table 8.1 Contd.

INVESTMENT OUTLAYS IN SOCIALIZED SECTOR

$$\mathrm{JU} + J = \mathrm{JN} \tag{C12}$$

where JN is private investment.

PLANNED INVESTMENT OUTLAYS

$$\mathrm{JP} = J_{-1} = \begin{cases} 2.40\ (\mathrm{JD} - \mathrm{JS}) \text{ if } \mathrm{JD} - \mathrm{JS} > 0 \\ 1.35\ (\mathrm{JD} - \mathrm{JS}) \text{ of } \mathrm{JD} - \mathrm{JS} \leqslant 0 \end{cases} \tag{C13}$$

* Estimates of structural parameters are shown; they were obtained using OLS and TSLS to respecified equations. Symbol *e* stands for estimates of disturbance terms.

point of reference. The potential output XK was obtained from the production function 8.23 which entered employment as being demand determined. Then using Equation 8.27 the labour constraint was superimposed. The ratio NS/NK defining the rate of capacity utilization was approximated using shifts coefficients.[14]

Material input was omitted in the production function 8.23 as its impact, also in the form of imports of materials, was not meaningful and was absorbed by the technical progress indicator defined below. It is interesting to know that different studies show that the isolated long-term impact of the imports-input is about 0.2 in terms of elasticities of output with regard to this variable. However, for the periods characterized by shortages in supplies of materials, especially from imports, these shortages must be additionally superimposed. Special shortage indicators, proxies for the ratio QMS/QMN, were constructed and applied using Equation 8.27 again. The specification of the above production function and related equations is shown in Table 8.2.

More specifically the two-factor Cobb–Douglas production function, extended to allow for technical progress, was used to determine the capacity output (W1) and next the indicators of availability of working time and material input were added to explain its utilization and thus to arrive at potential output (supply) (W2). The estimates of respective elasticities are shown in Table 8.3. The link to potential gross output was established (W3). This variable was compared with total demand at producers' level composed of intermediate and final demand (W4). The empirical results for selected years of the sample period are shown in Table 8.4.

The equation explaining requirements for employment and working time allowing for disturbances due to material-input shortages are given in (W7) and (W8). The empirical results showing how similar (between industries) was the response of labour demand to output changes are demonstrated in Table 8.3.

[14] By definition they relate employment to the available number of jobs.

Table 8.2 Model W5 equations (disturbance terms omitted)

A. Production sector (by industries)

CAPACITY OUTPUT

$$\ln XK = b_0 + b_1 \ln (K \cdot WZM) + (1 - b_1) \ln HK + b_2 \ln (WK \cdot WZM) + b_3 \ln (M7/JV) \quad (W1)$$

where WZM is the maximum number of shifts, HK is the working time ensuring full use of capacity output, $WK = WK_{-1} + K$ is the embodied technical progress indicator, $M7$ is the import of machinery equipment, JV is the investment of machinery equipment.

NET OUTPUT

$$\ln XS = b_0 + b_1 \ln (K \cdot WZ) + (1 - b_1) \ln H + b_2 \ln (WK \cdot WZ) + b_3 \ln (M7/JV) + b_4 (\ln QMI_{-1}) UM_{-1} \quad (W2)$$

where WZ is the number of shifts, H the hours worked, $QMI = \tilde{Q}MS/\tilde{Q}MD$ is the material input shortage ratio, $UM = 1$ if $QMI < QMI^* =$ equilibrium level.

GROSS OUTPUT

$$QS = XS/(1 - WKM) \quad (W3)$$

where WKM is the material input–gross output ratio.

DEMAND FOR GROSS OUTPUT

$$QD = QQD + QHD \quad (W4)$$

where QQD is the intermediate demand to be determined from input–output framework or its stochastic approximation, QHD is the final demand: $QHD = CD + GD + JD + DRD + (ED - MD)$.

GROSS OUTPUT PRODUCED

$$Q = \min (QS, QD) \quad (W5)$$

NET OUTPUT PRODUCED

$$X = Q(1 - WKM) \quad (W6)$$

Notice that values of output in current prices are obtained using respective deflators.

EMPLOYMENT REQUIRED

$$ND = a_0 + a_1 (X_{-1}/WK_{-1}) \quad (W7)$$

HOURS WORKED

$$H = c + c_1 ND - c_2 QMI_{-1} UM_{-1} \quad (W8)$$

SHIFTS WORKED

$$WZ = a_0 + a_1 K + a_2 ND \quad (W9)$$

Table 8.3 Elasticities for production and employment functions

Industry	*Elasticities of net output X with regard to*			*Average elasticity of employment N with respect to net output X*
	Fixed assets K	*Embodied technical progress* WK	*Share of imported machinery equipment* $(M7/JV)_{-1}$	
Fuel and energy	0.5	0.97	0.66	0.43
Metallurgic, chemical	0.6	0.90	0.20	0.34
Electroengineering	0.6	0.74	0.03	0.69
Light, paper, others	0.5	0.98	—	0.55
Foodstuffs processing	0.5	0.13	0.10	0.55
Building	0.4	0.28	—	0.57
Transportation and communication	0.5	1.00	0.11	0.49

All $R^2 > 0.95$ and DW > 1.5.

Table 8.4 Rates of utilization of fixed assets: determined capacities (WX) in %

Industry	*1965*	*1970*	*1975*	*1978*	*1979*	*1980*	*1981*	*1982*
Fuel and power	100.2	100.0	96.9	102.9	93.1	96.8	44.8[a]	40.7[a]
Metallurgic, chemical	98.5	89.8	99.2	96.3	88.4	84.2	79.4	81.1
Electroengineering	96.6	97.6	98.7	91.7	96.5	94.8	83.4	87.4
Light	95.2	97.7	98.0	91.1	87.7	85.2	81.8	76.0
Food processing	97.0	97.1	96.4	92.0	78.3	75.2	77.4	67.9
Agriculture	93.3	97.1	98.1	94.3	92.1	87.9	91.0	92.5
Building	98.9	101.4	104.4	93.8	82.9	88.0	70.7	47.6
Transportation	102.7	96.5	102.5	100.4	100.0	95.0	85.3	77.6

[a] Underestimated due to the change in the system of constant prices.

The equations explaining the fixed capital formation investment process as well as financial variables including wages are discussed below.

8.4 THE MODELLING OF FINAL USERS' BEHAVIOUR UNDER DISEQUILIBRIA

The existence of excess demand at the producers' level raises the question of how it is allocated between potential users. In general, the problem can be formalized in a simple way. Let us assume that total supply and demand are given and QD > QS = Q. The allocation of excess demand QE = QD-QS can be demonstrated using the following equation system which determines the transacted values for particular users:

$$\begin{aligned} Q &= \mathrm{QMD} - q\cdot \mathrm{QE} \\ H &= \mathrm{HD} - h\cdot \mathrm{QE} \\ E &= \mathrm{ED} - e\cdot \mathrm{QE} \\ M &= \mathrm{MD} + m\cdot \mathrm{QE} \\ \hline Q &= \mathrm{QD} - \mathrm{QE} \end{aligned} \tag{8.29}$$

It can easily be found that the sum of allocation coefficients equals unity: $q + h + e + m = 1$. If we assume that the producers are not material input constrained, i.e. $q = 0$, then the above system can be reduced to show the allocation of excess demand for distributed national income $\mathrm{XE} = \mathrm{XD} - \mathrm{XS}$; $\mathrm{XS} = \mathrm{X}$. We have:

$$\begin{aligned} \mathrm{C} &= \mathrm{CD} - c\cdot \mathrm{XE} \\ \mathrm{G} &= \mathrm{GD} - g\cdot \mathrm{XE} \\ \mathrm{JA} &= \mathrm{JAD} - j\cdot \mathrm{XE} \\ \mathrm{DR} &= \mathrm{DRD} - d\cdot \mathrm{XE} \\ E &= \mathrm{ED} - e\cdot \mathrm{XE} \\ M &= \mathrm{MD} + m\cdot \mathrm{XE} \\ \hline \mathrm{X} &= \mathrm{XD} - \mathrm{XE} \end{aligned} \tag{8.30}$$

Again, $(c + g + j + d) + (e + m) = 1$.

If the markets for particular users are not separated (common pool approach) spillovers between markets can take place freely. In an extreme case if no user is preferred, the allocation coefficients will be proportional to the users' demand shares and the initial demand will be cut off by the same average percentage rate.[15] For instance, for consumer goods markets we have $c = \mathrm{CD/XD}$ and

$$\mathrm{C} = \mathrm{CD}(1 - \mathrm{XE/XD}) \tag{8.31}$$

If there are no constraints in the balance of foreign trade, the shortages of domestic products can be fully compensated by additional supplies from imports. We have $m = 1$ and all other coefficients equal zero.

Typically, however, we have to assume that the imports are constrained, that preferences exist and that markets are separated (this being an institutional measure to ensure the implementation of central planners' preferences).

The imports constraint means that there is excess demand for foreign currencies and the initial imports demand has to be adjusted downwards so as not to exceed the level of foreign currencies supply. Thus $M < \mathrm{MD}$ and $m < 0$. In this case

[15] This rule was also used in model W5 to allocate excess demand of investment outlays between industries.

Table 8.5 Model CH–G equations

C. Foreign trade

NATIONAL INCOME DISTRIBUTED

$$XG = X + BT \tag{C14}$$

DIFFERENCE BETWEEN NATIONAL INCOME DISTRIBUTED AND GDP

$$BT = 13 + 39(MD - ES) + e_{BT} \tag{C15}$$

TOTAL IMPORTS (DEMAND)

$$MD = MND + MNS \tag{C16}$$

TOTAL EXPORTS (SUPPLY)

$$ES = ENS + ESS \tag{C17}$$

IMPORTS FROM NON-SOCIALIST COUNTRIES (DEMAND)

$$\begin{aligned} MND = & -1.23 + 0.0021(0.6X^{P} + 0.4X^{P}_{+1}) \\ & + 0.029(0.75(CD - CS)^{E} + 0.25(CD - CS)^{E}_{+1}) \\ & + 0.64(ZK - ZK_{-1}) + 1.18Z4 + e_{MN} \end{aligned} \tag{C18}$$

where ZK is total debt in hard currencies, end of period.

EXPORTS TO NON-SOCIALIST COUNTRIES (SUPPLY)

$$\begin{aligned} ENS = & -0.72 + 0.0033(0.6X^{P} + 0.4X^{P}_{+1}) \\ & - 0.013(CD_{-1} - CS_{-1}) + 0.63Z4 + e_{EN} \end{aligned} \tag{C19}$$

where $Z4 = 1$ in 1980.

IMPORTS FROM SOCIALIST COUNTRIES (DEMAND)

$$\begin{aligned} MSD = & -1.11 + 0.00096(0.6X^{P} + 0.4X^{P}_{+1}) + 0.04RW_{-1} \\ & + 0.35ESS_{-1} + 0.33Z5 + e_{MS} \end{aligned} \tag{C20}$$

where RW is total imports of CMEA countries (Poland excluded), $Z5 = 1$ in 1971–75.

EXPORTS TO SOCIALIST COUNTRIES (SUPPLY)

$$\begin{aligned} ESS = & -2.29 + 0.0047X^{P}_{+1} - 0.031 \max(0, (JD - JS)) \\ & - 0.25(ZK - ZK_{-1}) + e_{ES} \end{aligned} \tag{C21}$$

imports will be obtained from an identity: the right-hand side will be equal to the deflated sum of receipts from exports (EP) and net foreign financing (equal to the assumed balance of trade EMP with interchanged sign):

$$M = (EP - EMP)/PM \tag{8.32}$$

This is the specification applied in the W5 model, which assumes imports constraint over the sample period. This global constraint is then used to adjust the

Table 8.6 Model W5 equations

B. Foreign trade

IMPORTS (BY GROUPS OF COUNTRIES AND COMMODITIES)

INITIAL DEMAND

$$MD = a_0 + a_1Q + a_2J + a_3C + a_4PM/PQ \quad (W9)$$

where PM = PM$·WZLD and PM$ is the imports deflator in US$, WZLD is the exchange rate zloty/US$, PQ is the gross output deflator.

EFFECTIVE DEMAND = TRANSACTED VALUES

$$M = a\,MD \quad (W10)$$

where a is the ratio of total imports demand to total available imports fund (deflated) obtained by adding up exports plus net foreign financing (Equation 8.32).

EXPORTS (BY GROUPS OF COUNTRIES AND COMMODITIES) DEMAND

$$ED = b_0 + b_1EHS + b_2PE\$/PEHS\$ \quad (W11)$$

where PE$ = PE/WZLD and PE is the exports deflator in zlotys, PEHS$ is the world exports deflator (in US$), and EHS is world trade (for specific regions).

EXCESS DEMAND

$$EE = b_3IE \quad (W12)$$

where IE is the disequilibrium indicator for exports.

TRANSACTED VALUES

$$E = ED - EE \quad (W13)$$

Values of imports and exports in current prices in zlotys and in US$ are obtained using respective deflators.

C. Investment

INVESTMENT OUTLAYS, INITIAL DEMAND (BY INDUSTRIES)

$$JD = a_0 + a_1J_{-1} + a_2XD + a_3WZ_{-1} \quad (W14)$$

INVESTMENT OUTLAYS (BY INDUSTRIES)

$$J = a\,JD \quad (W15)$$

where a is the ratio of total demand for investment outlays to total available (deflated) investment fund JC.

SUPPLIES OF INVESTMENT GOODS (BY COMMODITY GROUPS)

$$JS = b_0 + b_1QS + b_2M7 \quad (W16)$$

where $M7$ is imports of machinery equipment.

(contd.)

Table 8.6 Contd.

GROSS FIXED CAPITAL FORMATION (BY INDUSTRIES)

$$I = a_0 + a_1 \sum_{i=0}^{m} W_i J_{-i} \quad \text{(W17)}$$

FIXED ASSETS (BY INDUSTRIES)

$$K = K_{-1} + I - \text{KA} \quad \text{(W18)}$$

SCRAPPINGS (BY INDUSTRIES)

$$\text{KA} = d_0 + d_1 K_{-1} + d_2 I \quad \text{(W19)}$$

demand by commodity groups using an allocation scheme similar to Equations 8.29 and 8.31. The imports demand equations are shown in Table 8.6. They relate the imports of energy and raw materials to the gross output of respective industries (using the information on input–output coefficients) and relative prices. The imports of investment and consumer goods are complementary and depend respectively on the level of investment outlays for machinery equipment and private consumption. The major empirical results are reported in Table 8.7.

The CH–G model does not constrain directly the demand for imported goods. However, imports from non-socialist countries positively depend on an increase in foreign debt which for the 70s was associated with an increase in net foreign financing. This specification cannot be applied for the period of debt repayment as a further increase in foreign debt is associated with repayment of the interest rate, i.e. negative net foreign financing (growing $\text{EMP} > 0$). For details see Table 8.5, Equations C18, C20.

With regard to exports it can be assumed that because of their high position on the central planners' preference list, foreign demand will be constrained only exceptionally ($e = 0$). This is explicitly assumed in the W5 model. It defines the foreign demand allowing for shortages in domestic supplies in the form of disequilibrium indicators only. The lack of supply equations limits the possibilities

Table 8.7 Average elasticities in foreign trade

	SITC commodity groups				
Groups of countries	*0–1*	*2, 4*	*3*	*7*	Other I
Elasticities of exports with regard to total world exports					
Socialist	0.52	0.90	0.40	0.85	0.96
Non-socialist	1.85	2.21	2.37	2.75	1.02
Elasticities of imports with regard to domestic activity variables					
Socialist	0.53	3.00	1.00	0.77	1.34
Non-socialist	2.03	0.85	1.06	1.36	1.40

of an analysis of the sources of changes in excess demand. Details are given in Table 8.6, whereas the empirical results are in Table 8.7.

The CH–G model takes a different position assuming – as was common in the past – that exports mainly provide the means for necessary imports and should be specified in the form of supply functions only. It allows for spillovers from domestic markets, such that the excess demand in the consumer goods markets forces a decline in exports to non-socialist countries.[16] (See Table 8.5, C19.) On the other hand, positive excess domestic demand of investment goods induces a decrease in supplies for exports to socialist countries (C21). It is an interesting contribution; it seems, however, that it should have been extended to the case where a (positive) excess supply of investment goods leads to an increase in supplies for exports. As the demand functions have not been defined it is not possible to estimate directly the excess demand in exports markets. It might be obtained as a residual from the allocation system (Equation 8.30). Government expenditures (*G*) are either exogenous or as in the W5 model it can be assumed that the funds are limited to such an extent as to reduce if not eliminate likely excess demand.

The remaining components of final use are personal consumption and investment outlays. In this case there exists a natural market separation. It can, however, be relaxed via foreign trade (e.g. excess machines exchanged for grain lacking) or via shifts of materials and labour from industries producing consumer goods to industries producing investment goods or vice versa. Thus it is unlikely to experience growing excess demand of consumer goods and excess supply of investment goods simultaneously for a long period of time.[17] In these conditions the total excess demand estimated from the users' side (as a sum of positive excess demand in consumer and investment goods markets) may be lower than the excess demand estimated at the producers' level (XD – XS) where the users' demand is being added up and compared with global supply.[18]

The models do not explicitly assume spillovers between the two markets. They allow for shifts in supplies using national income identity (Equation 8.19). However, the likely shifts are not related to the intensity of disequilibria in these markets. In the CH–G model an increase in investment outlays (C33) induces an equivalent decrease in supplies of consumer goods (being residual). In the W5 model the direction of change is exactly opposite – an increase in supplies of consumer goods is followed by a decline in investment outlays (being residual). The latter assumption seems to be more realistic for the reasons listed below.

[16]This assumption does not seem to be realistic enough. It is not easy to test it in the framework of the analysed model because the estimated excess demand for consumer goods is upward biased as follows from our further remarks.

[17]We therefore disagree with the simulation results obtained by Charemza and Gronicki (1985, pp. 138–9) showing the above tendencies in the late 70s.

[18]It has to be mentioned that Charemza and Gronicki use erroneous methodology in arriving at total GDP excess demand. For periods characterized by (positive) excess supply of investment goods, they add this variable to the excess demand for consumer goods (see pp. 137–8). This is based on an inappropriate interpretation of excess supply of investment goods as an equivalent of capacity shortages. Obviously, it means the opposite – the excess of capacities of industries producing investment goods.

The investment outlays description in analysed models is highly advanced compared with previous models of the CPEs. Both models define demand and supply functions.[19] There are minor differences in the specification of particular equations, which will be discussed later. The determination of investment in the system of model equations shows noticeable differences, however. The CH–G model determines investment outlays from the minimum condition

$$J = \min(\text{JD}, \text{JS}) \tag{8.33}$$

Supply is determined as a function of planned GDP (C5). It is interpreted as a central planners' estimate of available investment goods. Demand is proportional to the difference between desired and initial fixed capital and is believed to represent the producers' estimate (C6). However, the desired fixed capital depends on planned GDP gain (C7), which – let us recall – reflects the central planner expectations based on supplied (generated via production function) and not demanded output (C1 and C2). This raises serious doubts whether the model is really able to generate the appropriate values for investment demand and excess demand.

The W5 model introduces the concept of central planners' desired investment level JC, which can differ from the supplies of investment goods JS, but cannot exceed them ($\text{JC} \leqslant \text{JS}$) and which is used to depress the demand, i.e. $\text{JC} \leqslant \text{JD}$. It is assumed that JC is realized. Hence

$$\text{JC} = J \leqslant \min(\text{JS}, \text{JD}) \tag{8.34}$$

The desired value JC is obtained in the model as a residual from the national income accounting identity (Equation 8.19). It reflects the view that investment programmes can be expanded if there are either high rates of growth or foreign lending available and they have to be squeezed if the rates of growth decline (or output declines) and/or the foreign debt service is rising. As the condition 8.34 is not automatically fulfilled in the model – in case it is not met – the value of JC has to be modified. This can be done by adjusting other components of the balance identity (inventory increase, government expenditures or household expenditures). In this way shifts of commodity supplies can follow. Let us mention that supplies of investment goods depend on the imports and gross output of industries producing investment goods (Table 8.6, W16), whereas demand (by industries) depends on net output (demand)[20] and lagged investment outlays and capacity utilization (W14). There is a feedback of accelerator type in the model as demanded net output is codetermined by the investment outlays.

[19] Only the CH–G model applies disequilibrium indicators in the process of their estimation. The W5 model estimates can be biased as no serious disequilibria are assumed for the sample period in the estimation process. None of these models has yet been able to take into account the modifications of effective demand (compared with notional demand) due to forced substitution, postponement in its realization, etc.

[20] In the earlier versions of the model the required net output was dependent on supplies and not on demand for consumer goods. This specification erroneously modified the demand for investment outlays.

Table 8.8 Lag distributions in investment process

	Lag in years					
Industry	*0*	*1*	*2*	*3*	*4*	*5*
Agriculture	0.096	0.774	0.130	–	–	–
Fuel and energy	0.330	0.349	0.264	0.057	–	–
Metallurgic, chemical	0.382	0.205	0.179	0.174	0.060	–
Electroengineering	0.276	0.343	0.268	0.113	–	–
Light	0.645	0.290	0.065	–	–	–
Foodstuffs processing	0.305	0.411	0.238	0.046	–	–
Building	0.445	0.240	0.128	0.079	0.062	0.046
Transportation	0.289	0.445	0.195	0.007	0.021	0.043

Both models contain equations linking investment outlays with gross capital formation, defining lag distributions (C9 and W17). The weights for the W5 model are shown in Table 8.8. The weights estimated for the CH–G model are given in C9 (Table 8.1). Some of them have negative signs, which is a likely result of collinearity. By definition they must be positive. The distortion of the investment process may change their values (their average will increase) but not their signs as the authors believe. On the other hand, the effects of likely distortions are in an interesting way superimposed on the value of gross fixed capital formation (C10).

The demand and supplies of consumer goods are specified in both models. The minimum condition is formally applied: $C = \min(\text{CD}, \text{CS})$. However, the CH–G model explicitly assumes positive excess demand for the whole sample period: $\text{CD} > \text{CS} = C$. In the W5 model the less rigid assumption was made that excess demand prevailed in the sample period, but in a few periods the markets for particular commodities were balanced, i.e. $\text{CD} \geqslant \text{CS} = C$.[21]

The authors of the models distinguish between notional and effective demand. As they differ considerably in the concepts used let us first examine this problem. The CH–G model defines Walrasian notional demand CD^{W} and effective demand CD. In this specification the notional demand changes proportionally to the potential income from labour obtained by multiplying the potential labour force (LPR, in original notation T) by the wage rate plus nonlabour income and initial money balances (OZP_{-1}):

$$\text{CD}^{\text{W}} \cdot \text{PC} = a(\text{OZP}_{-1} + \text{ZP.LPR} + \text{YIP}). \tag{8.35}$$

This is an abstract concept suitable for long-term analysis as it assumes that the households when planning expenditures take as a point of reference the income from potential employment of all family members and the total available money balance. In the short term we would expect that the point of reference will be the

[21] In the process of estimation disequilibrium indicators were used, of completely different types, however. They are discussed below.

potential income calculated for the notional supply of labour NS^W, $NS^W < LPR$, with no reference to initial money balance. We must not ignore the fact that a certain portion of the population of a productive age deliberately stays out of the labour force (housewives, students, etc.), which on the other hand is accepted by the authors in defining the notional supply of labour NS^W. As the households are not rationed in the labour market ($NS^W < ND$ by assumption) there are no direct spillovers from this market to the consumer goods market.[22] However, in the case when the consumer market is rationed ($CD^W > CS = C$) there is excess demand in this market. The authors believe, following orthodox theory, that this causes a decline in the labour supply, i.e. a spillover from the consumer goods market into the labour market. Simplified, it can be written as

$$NS = NS^W - f(CD^W - C) \tag{8.36}$$

Of course, from $CD^W - C > 0$ it follows that $NS < NS^W$ (and also $NS < ND$). This will cause a spillover back to the consumer markets. Lower potential income will induce lower demand. Thus the effective demand will be given by[23]

$$CD = CD^W - h(NS^W - NS) \tag{8.37}$$

Obviously, from $NS^W - NS > 0$ it follows that $CD < CD^W$. This adjustment process, however, is not being continued (until it would reach the convergence point). Thus the so-called supply multiplier (Barro–Grossman, 1976) is being assumed in the model in a restrictive form only (no impact of declining labour supply on output and consumer goods supply is introduced either).

This specification seems at first glance to confirm fully the demanding postulates of the theory. However, it can hardly be accepted for the following reasons. Firstly, the effective demand is not defined in an appropriate manner as the above specification allows for the shift from potential labour force (LPR) used in determining notional demand (Equation 8.37) to the notional supply NS^W, only under restrictive assumptions as we will demonstrate below. From this point of view the use of current income from labour (assuming NS^W instead of LPR) in the consumer demand function would be appreciated; the resulting demand might be called initial demand (CD^I). Thus we prefer the following specification:

$$CD^I = f((ZP \cdot NS^W + YIP)/PC) \tag{8.38}$$

Secondly, it is frequently argued that the spillovers from the consumer goods to the labour market do not take the form of a decline in labour supply (and employment). The shortages in the consumer goods markets can be evaluated by households as temporary. If they become chronic, the institutional factors (privileges associated with employment) will prevent employees from leaving the

[22] In theory, it is assumed that the expected constraints in the labour market $N = ND < NS$ (i.e. potential unemployment) leading to expected decline in income bring about a decrease in consumer effective demand compared with its notional value. This is by assumption excluded in the model postulating full employment ($ND \geqslant NS$).

[23] We ignore for the moment other sources of discrepancies between the notional and effective demand.

Table 8.9 Model CH–G equations

D. Households and disequilibria

NOTIONAL DEMAND FOR CONSUMER GOODS

$$CD^{W}P = 0.58(OZP_{-1} + ZP \cdot LRP + YIP - JNP) \qquad (C22)$$

where CDP = CD·PC, YIP is other income and JNP are investments in private sector.

NOTIONAL DEMAND ADJUSTED FOR EXCESS DEMAND EXPECTATIONS

$$\tilde{C}D^{W}P = CD^{W}P + 0.56(CDP - CSP)^{E}_{+1} \qquad (C23)$$

EFFECTIVE DEMAND FOR CONSUMER GOODS

$$CDP = \tilde{C}D^{W}P + 68 + 0.69ZP(N - NS^{W}) + e_{CD} \qquad (C24)$$

NOTIONAL SUPPLY OF LABOUR

$$NS^{W} = 0.84LPR - 0.16(OZP_{-1} + YIP - JNP)/ZP \qquad (C25)$$

NOTIONAL SUPPLY OF LABOUR ADJUSTED FOR PRICE EXPECTATIONS

$$\tilde{N}S^{W} = NS^{W} + 1379(0.3(P - P_{-1})^{E} + 0.7(P - P_{-1})^{E}_{+1})P/ZP \qquad (C26)$$

EFFECTIVE SUPPLY OF LABOUR

$$NS = \tilde{N}S^{W} - 52 + 0.35(CP - CD^{W}P)/ZP + 30Z1H + e_{NS} \qquad (C27)$$

where $Z1 = 1$ in 1971–76.

NOTIONAL MONEY BALANCES

$$OZD^{W}P = 0.26(OZP_{-1} + ZP \cdot LPR + YIP - JNP) \qquad (C28)$$

NOTIONAL MONEY BALANCES ADJUSTED FOR EXPECTATIONS OF DISEQUILIBRIA IN CONSUMER GOODS MARKETS

$$\tilde{O}ZD^{W}P = OZD^{W}P + 0.92(0.75(CDP - CSP)^{E} + 0.25(CDP - CSP)^{E}_{+1}) \qquad (C29)$$

EFFECTIVE MONEY BALANCES

$$OZDP = \tilde{O}Z^{W}P - 0.78P + 0.65(CD^{W}P - CP) - 38P \cdot Z2 + e_{OD} \qquad (C30)$$

where $Z2 = 1$ in 1978–80

TOTAL MONEY WEALTH OF HOUSEHOLDS

$$OZCP = OZDP + ZLP \qquad (C31)$$

TOTAL DEBT OF HOUSEHOLDS

$$ZLP = ZLP_{-1} + KRP - SP \qquad (C32)$$

where KRP is consumer credit, SP is repayment of consumer credit.

TOTAL SUPPLIES OF CONSUMER GOODS TO HOUSEHOLDS

$$CS = XG - G - DR - J \qquad (C33)$$

where DR is increase in stocks of inventories.

(Contd.)

Table 8.9 Contd.

TOTAL PERSONAL CONSUMPTION		
	$C = \min(\mathrm{CD}, \mathrm{CS})$	(C34)
TOTAL EMPLOYMENT		
	$N = \min(\mathrm{ND}, \mathrm{NS})$	(C35)

labour force (an important factor of inertia). The most important seems that, in actual terms, shortages are experienced in socialized trade only. There exist private trade or second economy markets where potentially the consumer demand can be met (at higher prices).[24] However, a spillover effect exists in the form of a decline in utilization rates of the employed labour force, i.e. a decrease in the working time and its intensity (causing a labour productivity decline).[25]

For the above reasons, in the equation defining effective demand, we would prefer to insert effective labour supply $\mathrm{NS} = N$ instead of its notional value NS^{W} as in Equation 8.38. Thus we are assuming a linear approximation:

$$\mathrm{CD}\cdot\mathrm{PC} = c_0(\mathrm{ZP}\cdot N + \mathrm{YIP}) + c_1 = c_0(\mathrm{ZP}\cdot\mathrm{NS}^{\mathrm{W}} + \mathrm{YIP}) + c_0\mathrm{ZP}(\mathrm{NSNS}^{\mathrm{W}}) + c_1 \quad (8.39)$$

if the existence of spillovers from the labour market (Equation 8.36) is believed. After substituting from Equations 8.36 and 8.38, using the initial demand concept (CD^{I}) we have

$$\mathrm{CP}\cdot\mathrm{PC} = \mathrm{CD}^{\mathrm{I}}\cdot\mathrm{PC} + c_0\mathrm{ZP}f(\mathrm{CD}^{\mathrm{I}} - \mathrm{C}) + c_1 \quad (8.40)$$

where the second term distinguishes the effects from a feedback showing how the initial excess demand in consumer goods markets reduces the supply of labour and this, in turn, reduces initial demand and excess demand for consumer goods.

It can easily be verified that, only after accepting very restrictive assumptions, the first term of Equation 8.40 equals the notional demand $\mathrm{CD}^{\mathrm{W}}\cdot\mathrm{PC}$ as postulated in the CH–G model. We have:

$$a(\mathrm{ZP}\cdot\mathrm{LPR} + \mathrm{YIP}) = c_0(\mathrm{ZP}\cdot\mathrm{NS}^{\mathrm{W}} + \mathrm{YIP}) \quad (8.41)$$

only if the notional labour supply equation has the following specific form:

$$\mathrm{NS}^{\mathrm{W}} = b\mathrm{LPR} - (1 - b)\mathrm{YIP}/\mathrm{ZP} \quad (8.42)$$

and $a = c_0 b$, i.e. if the average propensity to consume a (determined for income associated with potential labour force) is equal to the propensity to consume c_0

[24] The results obtained by Howard (1976) for the USSR do not positively verify the above hypothesis. The empirical results brought by the Charemza–Gronicki model positively support the above questioned hypothesis. However, they cannot be used as a convincing test because of the use of disequilibrium indicators yielding biased estimates of the effective demand.

[25] For details see Winiecki (1985) and Welfe (1985c).

(defined for the income associated with notional employment) adjusted for labour participation ratio b. These are the assumptions being made in the CH–G model, which have led to the debatable specification[26]

$$\mathrm{CD{\cdot}PC} = a(\mathrm{OZP{\cdot}LPR} + \mathrm{YIP}) + c_0\mathrm{ZP}(\mathrm{NS{\cdot}NS^W}) + c_1, \tag{8.43}$$

The model CH–G introduces the following innovation, which has to be welcomed. The effect of consumer's expectations with regard to the future disequilibria in consumer markets, further deterioration of the consumer's position was defined.[27] The equations defining notional demand (C22), adjusted for excess demand expectations (C23) and effective demand (C24), are shown in Table 8.9. The propensity to consume measured for notional demand is rather low $a = 0.58$; as it is measured for effective demand is $c_0 = 0.84$ and labour participation ratio $b = 0.69$. In Table 8.9 are listed the equations explaining money deposits and the above discussed notional and effective labour supply functions. The adjustments of labour supply for price expectations (C28) seem arbitrary, however. The spillover effect from consumer goods markets to the labour market in C27 is clearly significant but rather small. This result is not convincing as the values of notional demand $\mathrm{CD^WP}$ were obtained using an inappropriate disequilibrium indicator in the consumer goods market, as is demonstrated in the next section.

The above specifications do not exhaust the list of desired adjustments in the first line in the consumer demand function. It should include as an additional explanatory variable forced savings, i.e. the remainder of the purchasing fund (minus normal savings) from the previous period which households were not able to spend and are still willing to spend.[28] This adjustment could not be observed in the CH–G model as it assumes all accumulated savings (including forced savings) are the components of the households' purchasing fund.

In models which decompose consumer demand by commodity groups (markets) further adjustments are necessary:

1. those which allow for intertempora spillovers, i.e. postponed demand (for durables mainly) and
2. those which allow for demand transfer (spillovers) between commodity groups – due to forced substitution – towards those commodities which are in excess supply.[28]

The model W5 uses the concept of initial demand outlined above. As it rejects the hypothesis of spillovers from consumer goods markets to the labour market (for the

[26] The derivation of the demand function from a constrained utility function was based on a misleading assumption. It was assumed that there exists a constraint in the labour market, i.e. that $N = \bar{N}$. However, introduction of this constraint ($\bar{N}$) makes sense only if it is interpreted as labour demand and not if it is understood as the labour supply after spillovers from consumer goods markets have taken place.

[27] Again, the empirical results are not convincing because of the use of the disequilibrium indicator yielding questionable values of excess demand.

[28] For details see Chapter 12 by A. Welfe.

Table 8.10 Model W5 equations

D. Households
CONSUMER DEMAND (BY COMMODITY GROUPS)
$$CD = a_0 + a_1(Y + OZE_{-1}) + a_2PC/PY \qquad (W20)$$
where Y is real income, OZE is real forced savings, PC is consumer prices, PY is real income deflator.
SUPPLIES OF CONSUMER GOODS (BY INDUSTRIES TO THE WHOLESALE TRADE AND BY COMMODITY GROUPS TO THE RETAIL TRADE AND HOUSEHOLDS)
$$CS = b_0 + b_1(QS -) + b_2M \qquad (W21)$$
CONSUMPTION (TRANSACTED VALUES)
$$C = \min(CD, CS) \qquad (W22)$$
The values of consumer expenditures in current prices are obtained using respective deflators.
LABOUR SUPPLY
$$NS = b_0 + b_1LPR - b_2(YIP/L)/ZP \qquad (W23)$$
where LPR is active population, YIP is other income, ZP is average wages.

reasons given above) no feedback adjustment enters the demand function. However, effective demand differs from initial demand as current income was being adjusted by adding an estimate of forced savings. It partly allows for shifts in expenditures due to postponed demand and demand transfers. We have

$$CD_i = c[(Y + OZEP_{-1}/PC), PC_i/PC] \qquad (8.44)$$

Table 8.11 Marginal propensity to consume, income and price elasticities

Commodity group	*Marginal propensity to consume in %*	*Average elasticities of demand with regard to*	
		Income	*Relative prices*
Foodstuffs	29.0	0.77	−0.20
Alcoholic beverages	17.0	1.27	−0.38
Tobacco and cigarettes	1.6	0.68	−0.26
Textiles, apparel and shoes	8.5	0.78	−0.04
Durables	9.1	1.06	−0.21
Other non-durables	19.5	1.28	−0.08
Services	3.7	1.01	−0.01

Table 8.12 Model W5 equations

E. National income, its distribution

NATIONAL INCOME DISTRIBUTED (DEMAND)

$$XGD = CD + GD + JAD + DRD \qquad (W24)$$

GOVERNMENT (OTHER) CONSUMPTION

$$GD = a_0 + a_1CD \qquad (W25)$$

NET INVESTMENT OUTLAYS

$$JAD = a_0 + a_1JD \qquad (W26)$$

INCREASE IN INVENTORIES

$$DRD = a_0 + a_1DQD + a_2R_{-1} \qquad (W27)$$

F. Wages, incomes, savings, other financial processes

AVERAGE WAGES

$$ZP = c_0 + c_1(X/N) + c_2PY + c_3(ND - NS) \qquad (W28)$$

PERSONAL INCOME

$$YP = a(ZP \cdot N, YRP, YIP) \qquad (W29)$$

where: YRP is the income of private farmers, YIP is other income.

SAVINGS, DEMAND

$$DOZDP = c_0 + c_1YP + c_2DYP + c_3(r/PY) \qquad (W30)$$

where r is the interest rate.

(ACCUMULATED) FORCED SAVINGS

$$OZEP = (CD - CS)PY \qquad (W31)$$

TOTAL MONEY DEPOSITS

$$OZP = OZDP + OZEP \qquad (W32)$$

STATE BUDGET REVENUES (BY ITEMS)

$$BYP = a_0 + a_1QP + a_2XP \qquad (W33)$$

STATE BUDGET EXPENDITURES (BY ITEMS)

$$BCP = b_0 + b_1GP \qquad (W34)$$

G. Prices

PRODUCERS' PRICES (BY INDUSTRIES)

$$PQ = a_0 + a_1PQM + a_2PM + a_3ZP \cdot N/Q + a_4AFP_{-1}/Q_{-1} \qquad (W35)$$

where: APF/Q is the markup, PQM is the deflator of expenditures on material inputs (an input-output framework is being used).

(Contd.)

Table 8.12 Contd.

CONSUMER GOODS PRICES (BY COMMODITIES)

$$PC = b_0 + b_1 PQ + b_2 PM + b_3 (CD - CS) \quad (W36)$$

PERSONAL INCOME DEFLATOR

$$PY = a_0 + a_1 PC \quad (W37)$$

INVESTMENT OUTLAY DEFLATORS (BY INDUSTRIES AND COMMODITY GROUPS)

$$PJ = b_0 + b_1 PQ + b_3 PM \quad (W38)$$

EXPORTS DEFLATORS (BY COMMODITY GROUP)

$$PE = b_0 + b_1 (PE\$ \cdot WZLD) + b_2 PQ \quad (W39)$$

where i is the commodity group (see Table 8.10, W20). The empirical estimates of propensities to consume and elasticities are given in Table 8.11.

The supplies of consumer goods are determined taking into account the available domestic resources, i.e. total domestic output plus imports minus exports. The CH–G model assumes that they have the lowest position on the preference list and thus are determined as residual from the national account identity (Equation 8.19), whereas the W5 model assumes that supplies of consumer goods, which depend on the gross output, are only modified by changes in foreign trade (W21). It means that they are not affected by changes in supplies for investment purposes or the services sector. It seems that the latter specification was more realistic, especially for the period of the 70s and the 80s when the policy of protecting consumption was successfully applied. Table 8.13 shows the estimates of excess demand for the sample period obtained from the model W5.

No reference will be made to the detailed results obtained for total consumption from the CH–G model as the consumer excess demand was generally overes-

Table 8.13 Effective demand in % of real expenditures of households

Commodity groups	*1965*	*1970*	*1978*	*1979*	*1980*	*1981*	*1982*
Foodstuffs	102.7	100.0	103.0	101.2	102.2	110.3	111.4
Alcoholic beverages	100.0	100.0	100.0	100.0	100.0	146.7	117.9
Textiles, apparel, shoes	119.4	106.2	114.5	115.8	113.0	112.7	141.3
Durables	104.3	116.3	100.0	104.2	103.9	102.7	130.2
Other nondurables	103.2	100.0	102.5	101.1	100.0	101.0	104.2
Services	100.0	102.7	120.7	119.2	117.4	122.9	130.4

timated for the reasons given below. It declined from 6% in 1960 to 3% in 1969, then increased rapidly (as wages increased) to 10% in 1974 (which contradicts the other evidence), further declining until 1978 (6.5%) and started growing again (7.7% in 1980). The estimates of excess demand for labour are still less reliable, and were already 6% in 1973; they reached 13% in 1980(!) (see Charemza and Gronicki, 1985, p. 139).

The excess demand for consumer goods plays a significant role in market economies as a signal in explaining price changes. In fixed price disequilibrium models price changes are frequently ignored, i.e. prices are treated exogenously, as is the case in the CH–G model. W5 tries to explain price movements taking into account, first of all, the changes in fabrication costs (see Table 8.12; W35). Recent studies have shown that in the past price changes were sensitive only to global excess demand and this applies first of all to the prices of alcoholic beverages and cigarettes (where indirect tax changes reflect the intensity of the general disequilibrium – W36).

8.5 EXPECTATIONS AND PLANS: MEASUREMENT OF DISEQUILIBRIA

A problem frequently discussed in recent years in modelling CPEs was how and to what extent the expectations and plans of economic units should explicitly enter the equations, and whether and how to model the central planners' policies and interventions.

The traditional approach was to assume that expectations can be represented by an autoregressive process. It was broadly applied in the W5 model, especially in the derivation of demand functions for investment outlays. Their expected value was assumed to depend on the anticipated value of production or its increase which, in turn, was generated from an autoregressive scheme. The CH–G model defines the expectations explicitly and tries to use past information more broadly, applying the concept of rational expectations. It is done first of all with respect to the total output, which we find encouraging (see C1, Table 8.1). In this case it seems fully justified to assume that the central planners in forming their expectations have a good knowledge of the rules (exemplified by the model) by means of which the output is being generated. This does not apply, however, to other applications, which deal with the expectations of households with regard to the intensity of disequilibria used in consumer demand functions (C23) and equations explaining money balances (C29) and also imports from non-socialist countries (C18). There are no good grounds for assuming that households may have more than an incomplete, subjective picture of the forces driving disequilibria – far from that which the model offers in explanation. The same applies to the price change expectations used in the labour supply functions (C26). Here the use of the rational expectations concept is not justified at all as the CH–G model does not contain any explanation of price movements. In this case, there is also no serious argument available to introduce any kind of price expectations to explain the changes in labour supply.

Central plans are only recently being treated in macromodels as endogenous variables, which affect the activities of the economic units. The prevailing approach followed by the W5 model was to assume that the (annual) plans are typically close to their fulfilment and their explanation is embedded in the explanation of variables showing the plans' realizations. The reason for splitting the plans from the realizations may lie in the fact that they differ significantly from each other.

From this, we would expect selection for more comprehensive analysis of plans and their realization in the area of investment and wage formation rather than total output as was assumed in the CH–G model. The concept of forming the production plan in this model is as follows. It depends on central planners' expectations (based on the underlying production function), which we find interesting and on excess demand for labour:

$$X^{P} = f(X^{E}, \mathrm{ND} - \mathrm{NS}) \tag{8.45}$$

The incorporation of the last term seems superfluous and the underlying assumptions inappropriate. It is assumed that if there is excess demand for labour (higher than normal) the planners will increase the production plan against its expected value. The opposite seems to be true – assuming labour supply constraint, production plans would be adjusted downwards.

The appearance of significant differences between the values of plans and their fulfilment may, in general, suggest the existence of disequilibria. As one of the major short-term tasks of the central planner is to restore equilibria, the plans will thus be adjusted allowing for observed disequilibria. This gave rise to the construction by Charemza (1981) of a class of disequilibrium indicators called plan indicators discussed below (see also Charemza and Quandt, 1982). They were applied in the CH–G model in explaining the excess demand for investment outlays and in modified form in explaining the excess demand for labour.

The variables representing demand and excess demand, sometimes also supply, are, as a rule, not directly observable (one noticeable exception is the labour market in the market economies, where the data on unemployment are available; it must be admitted, however, that many students question their reliability).

The approach that has been used in both models rests on the concept of deterministic disequilibrium indicators. This concept uses the condensed form of demand (or supply) equation making use of the identity:

$$\mathrm{YE} = \mathrm{YD} - \mathrm{YS} \tag{8.46}$$

where Y is the variable in question and YE is excess demand. Thus the condensed form will be, using the demand function as a point of reference:

$$Y = \mathrm{YD} - \mathrm{YE} \cdot U, \tag{8.47}$$

where $U = 1$ if $\mathrm{YE} > 0$, $U = 0$ otherwise and using the supply function:

$$Y = \mathrm{YS} + \mathrm{YE}(1 - U). \tag{8.47'}$$

Let us define the disequilibrium indicator IE as a variable (single variable or a combination of variables), which is strongly associated with excess demand YE. In the deterministic case we have:

$$\mathrm{YE} = a \cdot \mathrm{IE} \tag{8.48}$$

Two types of these indicators can be distinguished. The first uses the causal relationship established between the variable (IE) and excess demand YE. The classical example is the price determination formula, assuming that $\mathrm{IE} = \Delta P$, i.e. that price increase depends on excess demand. We thus postulate the existence of a relationship:

$$\mathrm{IE} = 1/a \cdot \mathrm{YE} \tag{8.49}$$

and the value of disequilibrium intensity will be derived from Equation 8.48. Frequently the appropriate specification of the relationship in question requires the incorporation of further factors determining variable IE. In the classical example to explain price changes we would also need variables showing the changes in fabrication costs.

The above method was applied in the CH–G model in defining the disequilibrium indicators – explaining excess demand for consumer goods, for investment goods and in the labour market. The first plays an important role in the model as it is used not only to determine the value of excess demand but also indirectly to study the impact of disequilibria in consumer goods markets, labour supply, money balances and imports.

As the price changes were only weakly sensitive to intensity of disequilibria, they were eliminated as candidates for disequilibria indicators. The authors have formulated the hypothesis that under disequilibria, households – given the supply – will be ready to pay higher prices than administered prices (in a second economy market, say) and will therefore execute pressures for a wage increase. The authorities will be reluctant to agree to it. The result will be an increase in wages and further increase in excess demand. They call it an effect of money illusion (Charemza and Gronicki, 1985, p. 75). This wage-forming mechanism destabilizes the model's behaviour.

The logic of the above hypothesis raises certain doubts. It implies a very restrictive assumption, especially that households have to abstain from making any expenditures at higher prices in the second economy market. Otherwise, the effective price increase would be the basic argument in wage negotiations in the next period. If we agree that the hypothesis in question is valid, it must not be overlooked that a wage increase in the past was simultaneously determined by many other factors. The students of this problem agree that these were firstly productivity of labour increase, which was used as an argument to justify the government's decisions (in the 70s) and in wage negotiations, secondly the cost of living increase (in the late 70s and 80s) when the rate of inflation became two-digit. In the last period the rising pressures in the labour market were the additional factor of wage increase. We

Table 8.14 Average elasticities of average wages

Industry	*With regard to* Productivity of labour X/N	Living cost index PY
Fuel and energy	0.48	1.26
Metallurgic, chemical	0.46	0.62
Electroengineering	0.53	0.59
Light, paper, others	0.51	0.62
Foodstuffs processing	0.79	0.87
Transportation and communication	0.69	0.75

have, in fact:

$$\Delta ZP = f(\Delta X/N, \Delta PC, CD\ CS, ND - NS) \tag{8.50}$$

The above specification was applied in the W5 model (the last term was added only recently: W28 in Table 8.12). The addition of this last term can be interpreted as an extension of the hypothesis known as the Phillips curve phenomenon to the regions where the unemployment rate is negative (W. Welfe, 1988). The earlier empirical results showing the average elasticities of wages with regard to productivity of labour and inflation rates are shown in Table 8.14. For an empirical verification of the impact of tensions in the labour market see Debski (1987).

In the CH–G model the disequilibrium indicator was defined from

$$ZP - ZP_{-2} = a(CD - CS) \tag{8.51}$$

and we have:

$$CE = CD - CS = 1/a(ZP - ZP_{-2}) \tag{8.52}$$

From the above argument it follows that because other significant factors have been omitted from Equation 8.51 the estimates of excess demand will be upward biased as wages were increasing over the whole sample period. To obtain an unbiased estimate it would be necessary to eliminate from the wage increase the effects of changes in other explanatory variables.

The practical implications of the above misspecification are far reaching. Overestimation of excess demand for consumer goods influences the estimates of expectations and spillovers. It overstates the differences between effective and notional demand, etc. In fact, no meaningful conclusions about the intensity of disequilibria in the above area can be reached for the sample period as was mentioned above in the preceding sections.

The plan disequilibrium indicators used in the model seem to have a firmer justification. The indicator related to the excess demand for investment outlays implies that the plan is increased if there is excess demand and reduced in the opposite case:

$$J - J_{-1} = \begin{cases} a(\mathrm{JD} - \mathrm{JS}) & \text{if } J^{\mathrm{P}} - J > 0 \\ b(\mathrm{JD} - \mathrm{JS}) & \text{if } J^{\mathrm{P}} - J \leqslant 0 \end{cases} \tag{8.53}$$

where superscript P means plan value.

The above assumption roughly conforms to the observations. However, it would seem to be more realistic to interchange the role of variables, i.e. to assume that because of lags in the adjustment process, the (planned) increase of investment resulted in short supplies of investment goods and (planned) decrease in excess capacities in industrices producing these goods.

The disequilibrium indicator assumed to measure the intensity of disequilibria in the labour market is based on a comparison of planned and expected output which follows from the equation determining planned output:

$$X^{\mathrm{P}} = \mathrm{XE}^{\mathrm{E}} + n(\mathrm{ND} - \mathrm{NS}) \quad n > 0 \tag{8.54}$$

As has been argued before, the assumed direction of change (i.e. the sign of the coefficient) should be reverse. The excess demand for labour should adjust the planned output against the expected output downwards and not upwards.

The second type of disequilibrium indicator treats it as a suitable approximation of excess demand as in Equation 8.46. By definition, excess demand is determined by the factors influencing demand and supply, i.e.

$$\mathrm{YE} = \mathrm{YD}(\mathbf{XD}) - \mathrm{YS}(\mathbf{XS}) \tag{8.55}$$

where **XD** and **XS** are vectors of variables explaining respectively demand and supply. YE can be approximated by a function of selected variables from the above sets of variables:

$$\mathrm{YE} = y(\tilde{\mathrm{X}}\mathrm{D}, \tilde{\mathrm{X}}\mathrm{S}) \tag{8.56}$$

where $\tilde{\mathrm{X}}\mathrm{D}$, $\tilde{\mathrm{X}}\mathrm{S}$ are known, single variables or subsets of variables determining respectively YD and YS, being closely correlated with them. Thus $\tilde{\mathrm{X}}\mathrm{D}$, $\tilde{\mathrm{X}}\mathrm{S}$ can be interpreted as instrumental variables. For instance, in early econometric investigations the excess demand for consumer goods entering the price equation was approximated by a function whose arguments were total nominal income and total output adjusted for foreign trade balance.

In order to summarize the above combination of factors in one variable, A. Welfe proposed the following definition based on a simple transformation of the basic formula (Equation 8.46). We have

$$\begin{aligned} \mathrm{YE} &= \mathrm{YS}[(\mathrm{YD}/\mathrm{YS} - \mathrm{YD}_0/\mathrm{YS}_0) + \mathrm{YE}_0/\mathrm{YS}_0] \\ &= a\,\mathrm{YS}(\mathrm{YD}/\mathrm{YS} - \mathrm{YD}_0/\mathrm{YS}_0) \end{aligned} \tag{8.57}$$

as $\mathrm{YE} \cong 0$, i.e. for the period 0 a situation close to equilibrium is assumed.

The disequilibrium indicator will thus be defined from the following approximation:

$$IE = \tilde{X}S(\tilde{X}D/\tilde{X}S - \tilde{X}D_0/\tilde{X}S_0) \tag{8.58}$$

where $\tilde{X}D$ and $\tilde{X}S$ are single variables or linear combinations of variables. The practical problem is the choice of instrumental variables. It can be solved if the demand and supply functions are well specified (in the theoretical plane). Secondly, it is the choice of period for which it is assumed that $YE_0 \cong 0$. The practical advice is to accept as such those periods for which the value of the disequilibrium indicator was the lowest in the sample period. This rule might be misleading as it might qualify all periods but a few in the past as periods characterized by disequilibria. Extraneous information would then be necessary.

This approach was broadly used in the W5 model to define the disequilibrium indicators as proxies of excess demand

1. in consumer goods markets, where personal income and supplies to the households were used as instrumental variables
2. in exports, where world trade and domestic gross output were used
3. in the labour market, where we used the number of vacancies and men and women searching for work as instruments related to the changes in excess demand for labour
4. in the production process – an inverse concept of the material input shortage indicator (in non-linear form) was used, where the lagged gross output stood for a determinant of demand for materials and an estimate of material supplies based on input–output allocation coefficients was used.

Use of the above disequilibrium indicators poses special estimation problems if the system of equations to be estimated including condensed form equations is simultaneous and non-linear. They are discussed by W. Charemza in his monograph (1981). It has to be emphasized that, after the parameters of the condensed functions *Y* have been estimated, the equations explaining (or defining) variables *Y* are ignored in the model version used for simulations and forecasting. The model after suitable respecification returns to its original structural form and to the variables defining demand (YD), supply (YS) and excess demand (YE) – from identity 8.46. The transcated values *Y* are obtained from the minimum condition again. The most difficult practical problem is to ensure sufficient accuracy of excess demand in the simulations estimates as it may accumulate errors of demand and supply estimates. Otherwise, the systems of equations can be used operationally as in the case of model W5.

8.6 COMPARATIVE DESCRIPTION OF OTHER PROPERTIES OF THE MODELS

The basic characteristics of the models analysed are summarised in Table 8.15. Both models use annual data based on official Polish statistics. The W5 model

Table 8.15 Basic characteristics of macromodels of the Polish economy CH–G (disequilibrium model by Charemza and Gronicki, 1985) and W5 (macroeconometric model by W. Welfe, 1985c)

	CH–G	W5
1. *Equations and variables*		
Number of exogenous variables:		
= number of equations	35	1076
among them		
Stochastic	13	436
Deterministic	22	640
Behavioural	11	287
Balance identities	8	286
Minimum conditions	3	67
Number of exogenous variables		
Except for dummies	11	52
Dummies	5	0
2. *Composition*		
Blocks of equations	3	15
Disaggregation	One sector	Industries: 11 MP sectors 2 services sectors; commodity groups 3–7
3. *Structural relationships*		
Equations	Restrictions upon parameters, nonlinear with respect to parameters, dynamic	Parameters unrestricted nonlinear, linear with respect to parameters, dynamic
Interrelationships	simultaneous	simultaneous
Methods of estimation	searching procedures OLS, TSLS and alternatively Bayesian methods	OLS, TSLS
4. *Data and use*		
Data used for estimation	1960–1980	1963–1982
Constant prices of the year	1980	1982
Applications	Historical simulations and use of optimal control for the period 1960–1980	Forecasts twice a year, 5 and 10 years horizon, policy simulations

applies the concepts defined in the MP system to be found in the official national income statistics. The CH–G model uses the concepts close to the NA systems: special attempts were made by the authors to calculate independently GDP and government expenditures, also the concept of private income is slightly different from that of personal income used in the statistics. Therefore certain results of estimation and simulation exercises are not directly comparable between the two models.

The CH–G model is small but highly sophisticated. The structural equations are well elaborated. They rest on established economic theory, where only that is available. They also reflect the authors' specific hypotheses with regard to the functioning of a socialist economy (under the management system that prevailed in the 60s and 70s), some of them being debatable. It is a fashionable disequilibrium-type model. It not only distinguishes demand, supply and excess demand but also broadly defines the impacts of disequilibria in various areas introducing numerous spillover effects. It makes use of expectations and plan figures.

This is the reason why the equations have a formally complicated structure. They are non-linear, parameter values are frequently restricted, the rational expectations hypothesis is used, several variables are not directly observable, etc. The variables are strongly interdependent and many simultaneous and dynamic feedbacks are defined in the model. It all calls for the use of elaborate computational procedures. First of all, the structural equations need to be respecified in order to eliminate the unobservables. This is achieved making use of disequilibrium indicators (of the first type) and restrictions imposed upon parameters. A variety of advanced estimation techniques (parallely classical including OLS, TSLS, search procedures and Bayesian methods) were then used. The results of estimation were carefully checked for economic consistency (signs, etc.) and statistical significance (conventional statistical tests were applied with R^2 higher than 0.98 for all but one equation, DW showing low serial correlation, etc.). The authors used special packages of programs elaborated for WANG minicomputers.

The empirical result are, however, in general disappointing. This is mainly because of the weaknesses of the concepts and techniques applied to measure the intensity of disequilibria, especially with regard to consumer goods markets. For this reason, the majority of estimates of the disequilibria in these markets, their impact on other areas, and the results of counterfactual simulations including the use of optimal control techniques are unconvincing.

The specification of the model accentuates the phenomena observed in the 60s and 70s ignoring the constraints in the production sector due to disequilibria in foreign trade and intermediate commodity markets. This explains why it could not be applied directly to analyse the Polish economy of the 80s.

The W5 model is large. This is due to disaggregation of the economic activities (by industries) and markets of commodities.[29] It gives an insight into the supply side of the economy, allows for identification of sources of disequilibria (and inflation) and specific policy measures that make the model suitable for a variety of

[29] Its stylized replica, model W6, will be much smaller, directly comparable to the CH–G model. As we mentioned earlier in the introduction it is not yet operational and for that reason was not directly the subject of our analysis.

applications. This is achieved at the cost of several simplifications. The model makes use of the established economic theory of a socialist economy. However, the authors tried to avoid the use of too restrictive assumptions. The model distinguishes between general demand, supply and excess demand (in a few sectors supply can only be indirectly determined). However, in its reported version the impact of disequilibria is only infrequently fed back into the system.[30]

The model is non-linear (linear in parameters, however), dynamic but does not explicitly use variables representing expectations or plans. It also contains unobservables whose values are approximated using disequilibrium indicators of the second type. Extraneous information is broadly used. As a result, the formal structure of equations tends, in general, to be simple and thus operational. On the other hand, the variables are highly interdependent with many feedbacks, simultaneous and dynamic, built in. OLS and TSLS estimation techniques were used and Gauss–Seidel procedures of simulation applied.[31] Validation of the equations and equation system follows the conventional rules. It took a long period of time for convergence to be reached and satisfactory empirical results obtained.

The model is updated and revised once every 3–4 years (the new version prepared rests on a sample period extended until 1985). It is used in regular service for medium and long-term forecasting (also for the world economy model of the PROJECT LINK) and policy simulations preceding the five year and annual central plan elaboration.

REFERENCES AND FURTHER READING

Artus, P., Laroque, G. and Michel, G. (1984) Estimation of quarterly macroeconomic model with quantity rationing. *Econometrica*, **52**, 1387–414.

Barro, R. J. and Grossman, H. J. (1976) *Money, Employment and Inflation*, Cambridge University Press, Cambridge.

Charemza, W. (1981) *Ekonometryczne modele nierownowagi. Problemy specyfikacji i estymacji* (Econometric disequilibrium models. Specification and estimation problems), Wyd. Uniwersytetu Gdanskiego, Gdansk.

Charemza, W. and Quandt, R. E. (1982) Models and estimation of disequilibrium for centrally planned economies. *Review of Economic Studies*, **49**, 109–16.

Charemza, W. and Gronicki, M. (1985) *Ekonometryczna analiza nierownowagi gospodarczej Polski* (Econometric analysis of economic disequilibrium in Poland), PWN, Warsaw.

Czerwinski, Z. and Welfe, W. (1982) Macroeconometric modeling based on econometric and simulation models for the Polish economy, in *Advances in Econometrics* (ed. W. Hildebrand), Cambridge University Press, Cambridge, pp. 285–301.

Czyzewski, A. B., Debski, W. and Welfe, W. (1976) Dynamic simulations based on Revised W-1 model and model W-3 of the Polish economy. *Prace IEiS UL*, **26**, Lodz.

Debski, W. (1987) *Ekonometryzne modele dochodów osobistych bidności Polski* (Econometric models of personal incomes for Poland), PWE, Warsaw.

[30] New versions of the W5 model as well as the W6 model will largely extend the list of spillover effects and the impact of excess demand.

[31] The package STS for estimation and LINK for a simulation are basically used.

Dlouhy, V. (1984) Macroeconometric disequilibrium model of a centrally planned economy, *Economicki-matematicky obzor*, **20**, 374–87.

Garbioz, M., Kuszewski, T., Mazur, K., Maciag, W. and Pawilno-Pacewicz, J. (1985) Prognostyczny model gospodarki narodowej SAPO (Forecasting model of the national Economy SAPO). *Gospodarka Planowa*, **1**, 34–40.

Green, D. and Higgins, C. I. (1977) *SOVMOD I: A Macroeconometric Model of the Soviet Union*, Academic Press, New York.

Howard, D. (1976) The disequilibrium model in a controlled economy: an empirical test of the Barro–Grossman model. *American Economic Review*, **66**, 871–9.

Jaramenko, J. W. (ed.) (1984) *Modelirowanie mezotraslevych wzaimodejstwij* (Modelling inter-industry interactions), Izd. Nauka Moskwa (in Russian).

Juszczak, W. and Welfe, W. (1983) A conditional forecast for development of the Polish economy in the period 1983–1985 (forecast: June 83). *Prace IEiS UL*, **44**, Lodz.

Kalecki, M. (1963) *Zarys teorii wzrostu gospodarki socjalistycznej* (On the theory of growth of the socialist economy), PWN, Warsaw.

Kooiman, P. (1986) Some empirical models for markets in disequilibrium. PhD dissertation, Erasmus University, Rotterdam (*mimeo*).

Kooiman, P. and Kloek, T. (1980) The specification of spillovers in the empirical disequilibrium models, Erasmus University, Rotterdam (*mimeo*).

Kolupa, M. (1965) *Badanie popytu w warunkach niedostatecznej podazy* (Demand analysis under short supplies), PWE, Warsaw.

Kornai, J. (1982) *Growth, Shortage and Efficiency. A Macrodynamic Model of the Socialist Economy*, Basil Blackwell, Oxford.

Maciejewski, W. (1976) *Zastosowania ekonometrycznych modeli rozwoju gospodarki narodowej* (Applications of the econometic models of the national economy development), PWE, Warsaw.

Maciejewski, W. (1980) *Ekonometria stosowana. Analiza porownawcza* (Applied econometrics. A comparative analysis), PWE, Warsaw.

Malinvaud, E. (1977) *The Theory of Unemployment Reconsidered*, Basil Blackwell, Oxford.

Pawlowski, Z. (1960) Problem nieobciazonosci parametrow popytu a przypadki przejsciowej niedostatecznej podazy dobr na rynku (The problem of unbiasedness of estimators of demand functions in case of temporary supply shortages). *Preglad Statystyczny*, **7**, (1), 41–50.

Portes, R. and Winter, D. (1980) Disequilibrium estimates for consumption goods markets in centrally planned economies. *Review of Economic Studies*, **47**, 137–59.

Portes, R., Quandt, R. E., Winter, D. and Yeo, S. (1983) Macroeconomic planning and disequilibrium: estimates for Poland, 1955–1980, Discussion Paper No. 139 (revised), Birkbeck College, London.

Romanski, J. and Welfe, W. (1985) Simulation minimodel W-6 of the Polish economy under disequilibrium, Paper presented to the Conference on Macromodels 85, Szczyrk Bila (*mimeo*).

Romanski, J. and Welfe, W. (1986) An alternative variant of the minimodel of the Polish economy under disequilibrium, Paper presented to the Conference on Macromodels 86, 3–5 December 1986, Szczyrk Bila (*mimeo*).

Sneessens, H. (1981) *Theory and Estimation of Macroeconomic Rationing Models*, Springer Verlag, Berlin.

Staruch, D. (1985) Main features of the macroeconometric model of the Czechoslovak economy (Version CEM – 3.2), Paper presented to the Conference on Macromodels 85, Szczyrk Bila, 10–13 December 1985, Bratislava (*mimeo*).

Suchecki, B. and Welfe, A.(1988) *Popyt i rynek w warunkach nierownowagi* (Demand and markets under disequilibrium), PWE, Warsaw.

Sujan, I. (1986) Econometric models of Czochoslovakia. A survey, in *Macromodels and forecasts of national economy. Proceedings of Models and Forecasts 83 and Macromodels 83* (eds W. Milo and M. Miszezynski), PWN, Warsaw, pp. 171–84.

Szakolczai, Gy., Marjanek, K. and Sisak-Feket, Zs. (1985) Adaptation of the Hungarian economy to change on the world markets. An analysis based on econometric model, Paper presented to the Econometric Society Fifth World Congress, 17–24 August 1985, Cambridge, Mass. (*mimeo*).

Tomaszewicz, L. (1983) *Zintegrowane modele gospodarki narodowej* (Integrated models of the national economy), PWE, Warsaw.

United Nations (1987) *Trade and Development Report*, UNCTAD, Geneva, New York.

Vilares, J. M. (1986) *Structural Change in Macroeconometric Models. Theory and Estimation*, Martinus Nijhoff, Dordrecht.

Welfe, A. (1984) Analiza popytu w warunkach nierownowagi (Demand analysis in conditions of disequilibrium). *Ekonomista*. No. 5, 1045–64.

Welfe, A. (1985) The system of demand equation in a state of disequilibrium, Collaborative Papers, CP-85-3p, International Institute for Applied Systems Analysis, Laxenburg.

Welfe, A. (1986) Intensity of disequilibrium and changes in inventories, Collaborative Papers, CP-86-7, International Institute for Applied Systems Analysis, Laxenburg.

Welfe, A. and Welfe, W. (1986) Modele i prognozy rynku w warunkach nierownowagi, (Market models and forecasts under disequilibrium). *Ekonomista*, No. 2, 279–314.

Welfe, W. (1973) A medium-term econometric model of the Polish economy. *Prace IEiS UL*, **8**, Poland.

Welfe, W. (1976) The unbalanced econometric macromodels, in *Computing Equilibria: How and Why* (eds J. Los and M. W. Los), North-Holland, Amsterdam.

Welfe, W. (1983) Models of the socialist economy, in *Lectures in Econometrics* (L. R. Klein), North-Holland, Amsterdam, pp. 197–227.

Welfe, W. (1985a) *Model W-5 the Polish economy. General assumptions and first empirical results*, Netherlands' Economic Institute, Rotterdam, No. 2.

Welfe, W. (1985b) Moglichkeiten und Probleme der okonometrischen Modelierung sozialistischer Volkswirtschaften: das Modell W-5 der polnischen Wirtschaft. *Mitteilungen Rheinisch-Westfalisches Institut für Wirtschaftsforschung Essen*, **36**, (1) 1–21.

Welfe, W. (1985c) Econometric macromodels of unbalanced growth. *Prace IEiS*, **52**, Lodz, 1–64.

Welfe, W. (1986a) Model W-5 of the Polish economy. General assumptions and the first empirical results, (revised version of (1985a) paper), in *Problems of Building and Estimation of Large Econometric Models*) eds W. Welfe and W. Juszczak), Acta Universitatis Lodziensis, Folia Oeconomica **54**, Uniwersytet Lodzki, Lodz, pp. 93–124.

Welfe, W. (1986b) The econometric models of the Polish economy, in *Macromodels and Forecasts of National Economy. Proceedings of Models and Forecasts 83 and Macromodels 83*, (eds W. Milo and M. Miszczynski), PWN, Warsaw, pp. 185–213.

Welfe, W. (1987) New version of W-models for Poland. Patterns of constrained growth, disequilibria and mixed government policies under economic reform. In *Models and Forecasts 84 and Macromodels 84*, (eds W. Welfe and K. Markowski), Ossolineum, Wroclaw, pp. 101–11.

Welfe, W. (1988) Econometric models of rhe centrally planned economies in the state of disequilibrium. *Ekonomicko–Matematicky Obzor*, **24**, (2), 121–34.

Winiecki, J. (1985) Portes ante Portas: a critique of the revisionist interpretation of inflation under central planning. *Comparative Economic Studies*, **27**, (2), 25–52.

9

Macroeconomic disequilibrium model of Hungary

KATALIN HULYÁK

9.1 MOTIVATION AND AIMS OF THE MODEL

The model presented in this chapter forms part of the macroeconometric model building activity which began many years ago in Hungary (Halabuk and Shapiro, 1976). The project is motivated by a strong belief in the existence of stable behavioural economic relations in a centrally planned economy and thus in the need for a macroeconomic quantitative approach in economic analysis. The disequilibrium character of econometric models appeared as rather a new feature in our model building processes. Until the 1980s econometric models focused in detail on the supply side of production and on its allocation by items of final use. Demand side and the feedback from demand to supply have been neglected by model builders. Equilibrium is obtained by a one-way connection and is based on the formal equilibrium of a data system derived from the national accounting system.

While macroeconomic models reflect and assume equilibrium, concrete economic policy and planning permanently struggle with disequilibrium problems. The introduction of a disequilibrium technique in modelling practice seems to bridge this gap and facilitates the use of econometric models in policy-making and planning.

On the other side, the possibilities of disequilibrium modelling emerged from a definite line of theoretical and empirical research work in the field of disequilibrium theory and applications to analysing markets in disequilibrium (Barro and Grossman, 1976; Quandt, 1982). Among these works numerous studies have been devoted to the disequilibrium phenomenon in centrally planned economies. The alternatives of disequilibrium models for planned economies are generally based on

International Studies in Economic Modelling Series.
Models of Disequilibrium and Shortage in Centrally Planned Economies.
Edited by C. Davis and W. Charemza.
Published in 1989 by Chapman and Hall Ltd, 11 New Fetter Lane, London EC4P 4EE.

Barro–Grossman theory and evaluate the evidence for repressed inflation in these economies (Charemza and Gronicki, 1984; Portes, 1986). A model framework has been developed in which the disequilibrium states and spillover effects of different markets can be formalized.

In a situation of repressed inflation the two or three markets of consumption goods, labour and money are investigated respectively and models highlight the spillovers of these markets. Our model does not follow this approach completely. After some elaboration of single market disequilibrium models (Hulyák, 1985; Csunderlik, 1985), the stimulus of our economic situation led us to build a multimarket disequilibrium macromodel which aimed at analysis of foreign and domestic disequilibrium states of the Hungarian economy.

On the economy's equilibrium, state planners refer first of all to the balance of foreign trade and to the balance of payments of the country. At the same time, we suppose that the improvement in imbalances in foreign resources has led to deterioration in the equilibrium states of domestic markets, for example with respect to both consumption goods and investment goods. Thus we think that an important feedback effect in our economy appears in the latter response, that is in the interdependence of foreign and domestic disequilibria in the economy. We hope that our model provides a tool for analysing the interdependence of foreign and domestic markets and hence the consequences of economic policy aimed, first of all, at improving foreign imbalances. Believing in the potential importance of this question we used a disequilibrium modelling framework to test the direction and size of these responses.

This task naturally requires elaboration and verification of a multimarket disequilibrium model. Accordingly, the model contains equations for the consumption goods market, for the investment goods market, for the labour market and for exports and imports. Using these equations we derived the main items of the supply and demand sides of total gross domestic products. In spite of our rather ambitious objectives the structure of the model is relatively simple.

While we emphasize its experimental character, the complete model can be considered as one of the necessary first steps in building a multimarket disequilibrium macromodel for a centrally planned economy.

9.2 SPECIFICATION OF THE MODEL

In the next section we consider the essential points in the specification of the model. The structure of the model presented here is a slightly modified version of the original macromodel (Hulyák, 1985) which is based on the period 1965–81. In 1986 the model's observation period was extended to the years 1982–84 and respecification of some equations seemed necessary. The modifications performed did not alter the basic structure of the model. It may be regarded as the same disequilibrium multimarket econometric macromodel with the same assumptions and problems.

Only the block of consumption goods and that of investment goods form disequilibrium models in the classical sense and instead of price adjustment equations the model has adjustment equations of special types. Maintaining an a priori hypothesis of excess demand, in some cases, these adjustment mechanisms express an increasing or decreasing intensity of excess demand. Before presenting the complete model with estimation results we run through the specification of the main blocks.

Beginning with the consumption goods market (C) our approach is based on the results of earlier studies (Portes and Winter, 1978; Hulyák, 1982). We have found that total consumer expenditure in Hungary can be characterized by an 'almost-equilibrium state' with only small size movements in the direction of excess demand. The demand and supply functions are complemented by a savings adjustment equation verifying the interdependence of the intensity of excess demand and irregular changes in the second differences of savings of the population. While the first differences in the stocks of savings (the balance of savings and credits) have increased year by year, rather irregular fluctuations appeared in the second differences in the stocks of savings of the population. That is why we have tried to relate the intensity of excess demand in the consumption goods market to the second differences of savings stock supposing that the influence of forced savings is behind these irregular changes.

Nevertheless, this hypothesis can be argued from several angles. First from the theoretical standpoint we cannot be sure that excess demand in the consumption goods market leads to forced savings and that the irregular changes in savings flows express only the influence of forced savings. In addition to these actual theoretical problems we have to admit that using savings differences as disequilibrium indicators implies very serious consequences for the model.

In the case of our consumption goods market, while we can accept the phenomenon of forced savings as a disequilibrium indicator, still we cannot be satisfied with the methodological treatment of this question assuming that all increases in savings should be identified with forced savings. Thus we cannot consider the savings adjustment equation in this deterministic form as an integral part of the model since it is used only for estimation purposes. The only justification for this solution is the relatively good estimation results.

Our use of the savings adjustment equation does not reveal an excess supply regime because the decreases in savings (forced savings) do not signify an excess supply. In accordance with our a priori assumption a savings adjustment was only used for detecting excess demand regimes with increasing intensity.

The consumption demand function fits well in the classical Houthakker–Taylor form stating that consumption demand is a function of consumption habits (C_{t-1}) and income (J_t) factors. The supply of consumption goods to the population depends on domestic production (Y_t) and on the situation and possibilities in our foreign resources in previous years. We assume that during the observation period the supply of consumption goods has been determined by gross domestic product but it has permanently been adjusted by planners with respect to the cumulated foreign trade deficit (BT_t) and the deficit in other items of the balance of payments

(BP_t) in the previous two years:

$$DC_t = \alpha_0 + \alpha_1 C_{t-1} + \alpha_2 \Delta J_t + \alpha_3 J_{t-1} \tag{9.1}$$

$$SC_t = \beta_0 + \beta_1 Y_t + \beta_2(BT_{t-1} + BT_{t-2}) + \beta_3(BP_{t-1} + BP_{t-2}) \tag{9.2}$$

The second differences in stocks of savings ($\Delta^2 S_t$) is a function of assumed excess demands:

$$\Delta^2 S_t = \gamma_C(DC_t - SC_t) \tag{9.3}$$

$$C_t \leqq DC_t \tag{9.4}$$

Whilst we suppose that the consumption goods market in Hungary may be characterized by an 'almost-equilibrium state' as a consequence of the protection of the standard of living of the population, the situation is quite different in the market of investment goods (I). This seems to be a consequence of the practice of allocating resources. As a result of restricted resources for investment supply and the lack of institutional restrictions on demand for investment goods, during the 70s and 80s our investment goods market has suffered from permanent excess demand. We have no grounds for doubting this hypothesis. We hope that by estimating the changing intensity of excess demand we will be able to connect with the analysis of the formation of cycles in investment activity. The investment block contains a disequilibrium model complemented by a plan adjustment mechanism (Charemza and Quandt, 1982).

The demand for investment goods is described by a classical-type demand function which is derived from the following three functions:

$$\Delta K_t = (1 - \lambda)(K_t^* - K_{t-1}) \tag{9.5}$$

assuming that the change in capital stocks (ΔK_t) is determined by partial adjustment to the difference of the desired capital stocks (K_t^*) and the capital stocks of the previous year.

The desired value of capital stocks is a function of the planned production value (Y_t^P):

$$K_t^* = \omega Y_t^P \tag{9.6}$$

The demand for investment expenditure of the year (DI_t) must cover ΔK_t and the depreciated part of K_{t-1} i.e. δK_{t-1} (where δ is the rate of depreciation):

$$DI_t = \Delta K_t + \delta K_{t-1} \tag{9.7}$$

finally, giving the demand function:

$$DI_t = (1 - \lambda)\omega Y_t^P - (1 - \lambda)(1 - \delta)\omega Y_{t-1}^P + \lambda DI_{t-1} \tag{9.8}$$

The supply of investment goods is determined in a similar way as in the case of consumption goods. Domestic production (Y_t), foreign trade deficit (BT_t) and the deficits of balance of payments of the previous two years (BP_{t-1}, BP_{t-2}) are the explanatory variables:

$$SI_t = \beta_0 + \beta_1 Y_t + \beta_2 BT_t + \beta_3 BP_{t-1} + \beta_4 BP_{t-2} \quad (9.9)$$

Assuming that:

$$DI_t > SI_t = I_t \quad (9.10)$$

the demand and supply functions are complemented by a plan adjustment mechanism supposing that planning activity intends to decrease the tension caused by the increasing intensity of excess demand in the sphere of investments:

$$I^P_{t+1} = I^P_t + \gamma_I(DI_t - SI_t) \quad (9.11)$$

The specification of our investment block is not based on a belief in the CPEs 'explosive investment demand' and does not suggest that investment hunger is almost insatiable in a centrally planned economy. We do not think that an investment demand of infinite type is a reasonable category for modelling at macrolevel.

We intend to specify and measure the deviations of a classical demand category (the claims to desired capital stocks for planned production) from investment supply, which in our model is a function of material and financial sources for investment expenditures. We thought that this solution could also throw some light on the excess demand pressure on the investment goods market.

Investments put into operation are expressed by

$$IU_t = \theta_0 + \theta_1 I_{t-1} + \theta_2 I_{t-2} + \theta_3 BT_{t-1} \quad (9.12)$$

Finally, an identity connects investments with capital stocks:

$$K_t = K_{t-1} + IU_t - Z_t \quad (9.13)$$

After specifying the excess demands in the two main markets of final domestic use we are able to consider the supply and demand sides of total domestic production. As a result of the a priori hypothesis of excess demands in domestic markets, total domestic production is determined by the supply side. In our model supply is represented by a Cobb–Douglas production function in which production value (Y_t) is explained by the utilized (U_t) stocks of fixed assets (K_t) and actual employment (L_t)[1]

$$\log Y_t = \log B(t) + \beta_1 \log (U_t K_t) + \beta_2 \log L_t \quad (9.14)$$

Following a priori assumptions we derived the aggregate variable of domestic disequilibrium (DDE_t):

$$DDE_t = DC_t - SC_t + DI_t - SI_t \quad (9.15)$$

[1](a) U_t is a Wharton-type utilization index constructed from the ratios of Y_{mt} to K_{mt}, where Y_{mt} and K_{mt} are production outputs and capital stocks in material branches, respectively.
(b) t is a time trend
(c) instead of K_t and L_t in the course of the estimation $\bar{K}_t$ and $\bar{L}_t$ the mid-year stocks were used.

For the demand side of total production we assume that:

$$DY_t = Y_t + DDE_t \tag{9.16}$$

In the function of inventory accumulation (IN_t) we tested its possible role as a buffer to absorb shocks. We could reasonably expect that the influence of foreign and domestic disequilibria of the year and those of the previous years appear in the function of the change of inventory accumulation (ΔIN_t):

$$\Delta IN_t = \eta_0 + \eta_1 \Delta IN_{t-1} + \eta_2 BT_t + \eta_3 (BT_{t-1} + BT_{t-2}) + \eta_4 DDE_{t-1} \tag{9.17}$$

As a first step, we have chosen a rather simple variant for the foreign trade block. It contains only three functions: one for total imports, one for total exports and an identity defining the main item of foreign disequilibrium by the foreign trade deficit (import surplus).

Total imports deflated by import prices (M_t/PM_t) are a function of the planned value of production (Y_t^P), the deficits of our trade balance (BT_t), other items of balance of payments (BP_t), and the domestic disequilibrium of the previous year (DDE_{t-1}):

$$M_t/PM_t = \alpha_0 + \alpha_1 Y_t^P + \alpha_2 BT_{t-2} + \alpha_3 (BP_{t-1} + BP_{t-2}) + \alpha_4 DDE_{t-1} \tag{9.18}$$

Total exports deflated by export prices (E_t/PE_t) are determined in a similar manner:

$$E_t/PE_t = \beta_0 + \beta_1 Y_t + \beta_2 BT_{t-2} + \beta_3 BP_{t-1} + \beta_4 BP_{t-2} \tag{9.19}$$

Foreign trade deficit is derived from

$$BT_t = M_t - E_t \tag{9.20}$$

This simple treatment of foreign trade is a weakness of the model. We think that by proper disequilibrium modelling of foreign trade, the problems in our foreign trade could be measured by more sophisticated methods than by simple imbalances, that is by import surpluses and other deficits in the balance of payments.

The labour market connected with the consumption goods market has a crucial importance in models which are based on Barro–Grossman theory. The spillover effect of the two markets is responsible for the supply multiplier in the regime of repressed inflation. However, we have found that the disequilibrium states of our labour market have a special character. Naturally, the labour market is not independent of the consumption goods market but the situation in our labour market is very complex and we could not connect it with the small-size excess demands in our consumption goods market. Wage adjustment cannot be observed either.

Instead of using classical solutions for disequilibrium modelling we assume that in the first part of the observation period the labour market was dominated by excess supply and during the second part by excess demand. This assumption seems to be supported by the size and direction of deviations of actual employment from the planned level of it. Considering the second part of the observation period the planned level of employment (L_t^P) generally exceeds the actual number of employees (L_t). Both demand function and supply function have been estimated for the whole observation period using the variable $(L_t^{\mathrm{P}} - L_t)$ only in the estimation of the demand function.

The demand for labour (DL_t) is a function of employment in the previous year (L_{t-1}) and the planned growth rate of gross production value $(\Delta Y_t^{\mathrm{P}})$:

$$\mathrm{DL}_t = \alpha_0 + \alpha_1 L_{t-1} + \alpha_2 \Delta Y_t^{\mathrm{P}} \tag{9.21}$$

while the supply of labour is explained by the potential labour force (LF_t) and real wages (RB_t):

$$\mathrm{SL}_t = \beta_0 + \beta_1 \mathrm{LF}_t + \beta_2 \mathrm{RB}_t \tag{9.22}$$

We assume that

$$L_t = \begin{cases} \mathrm{DL}_t & 1965\text{—}72, \quad \text{when} \quad (L_t^{\mathrm{P}} - L_t) \sim 0 \\ \mathrm{SL}_t & 1972\text{—}84, \quad \text{when} \quad (L_t^{\mathrm{P}} - L_t) > 0 \end{cases} \tag{9.23}$$

Finally, we have a rather formal equation specifying the total disposable income of the population (J_t), as a function of the income of the previous year, real wages (RB_t) and financial and natural benefits of the population (TJ_t):

$$J_t = \xi_0 + \xi_1 J_{t-1} + \xi_2 \mathrm{RB}_t + \xi \mathrm{TJ}_t \tag{9.24}$$

9.3 EMPIRICAL RESULTS OF THE MODEL

The estimation of the final version of the model is based on the annual time series of variables from 1965 to 1984. The primary sources of data are official Hungarian statistical publications as regards the data of the national accounting system. In a few cases supplementary calculations have been made on the basis of original official data. Data on the balance of payments are from the IMF Yearbooks (International Financial Statistics). The list of variables is presented after the complete model with estimation results.

The model was estimated in several steps. The OLS estimation of all equations in the equilibrium framework seemed to be a necessary first step. Following that, the model was estimated in a disequilibrium framework by a 2SLS version of the quantitative method of Fair and Jaffee (1972). From the results of the latter step the aggregate variable of domestic disequilibrium was derived using the estimated excess demand of the consumption goods market and that of the investment goods market. As a last step some equations needed re-estimation with this created domestic disequilibrium variable.

Structural equations and estimation results of the model

(a) *Consumption goods market*

$$\underset{}{\mathrm{DC}_t} = C_t + \underset{(2.1)}{0.634}\Delta^2 S_t = \underset{(3.2)}{23.180} + \underset{(1.5)}{0.274}C_{t-1} + \underset{(7.9)}{0.698}\Delta J_t + \underset{(4.1)}{0.629}J_{t-1}$$

$$\bar{R}^2 = 0.9991 \qquad \mathrm{DW} = 1.99$$

$$SC_t = \underset{(9.7)}{56.352} + \underset{(49.9)}{0.555Y_t} - \underset{(-2.4)}{0.0565}(BT_{t-1} + BT_{t-2}) - \underset{(-2.7)}{0.0870}(BP_{t-1} + BP_{t-2})$$

$$\bar{R}^2 = 0.9968 \qquad DW = 1.58$$

$$C_t \leqslant DC_t$$

$$\Delta^2 S_t = \frac{1}{0.634}(DC_t - SC_t)$$

(b) *Investment goods market*

$$DI_t = I_t + \underset{(1.8)}{0.438\Delta I^P_{t+1}} = \underset{(0.9)}{1.453} + \underset{(2.6)}{2.575Y^P_t} - \underset{(-2.9)}{2.419Y^P_{t-1}} + \underset{(3.6)}{0.723I_{t-1}}$$

$$\bar{R}^2 = 0.9663 \qquad DW = 1.89$$

$$SI_t = \underset{(-5.5)}{-47.252} + \underset{(21.6)}{0.365Y_t} + \underset{(3.0)}{0.303BT_t} - \underset{(-1.6)}{0.155BP_{t-1}} - \underset{(-3.9)}{0.425BP_{t-2}}$$

$$\bar{R}^2 = 0.9795 \qquad DW = 2.22$$

$$DI_t > SI_t = I_t$$

$$\Delta I^P_{t+1} = \frac{1}{0.438}(DI_t - SI_t)$$

$$IU_t = \underset{(2.1)}{6.635} + \underset{(5.3)}{1.138I_{t-1}} - \underset{(-1.9)}{0.189I_{t-2}} - \underset{(-2.3)}{0.285BT_{t-1}}$$

$$\bar{R}^2 = 0.9710 \qquad DW = 2.48$$

$$K_t = K_{t-1} + IU_t - Z_t$$

(c) *Supply and demand sides of total GDP*

$$\log Y_t = \underset{(-4.6)}{-4.317} + \underset{(5.2)}{0.012t} + \underset{(11.4)}{0.873} \log (U_t\bar{K}_t) + \underset{(2.7)}{0.438} \log \bar{L}_t$$

$$\bar{R}^2 = 0.9992 \qquad DW = 2.04$$

$$SY_t = Y_t$$

$$DDE_t = DC_t - SC_t + DI_t - SI_t$$

$$DY_t = Y_t + DDE_t$$

(d) *Changes in inventory stocks*

$$\Delta IN_t = \underset{(5.7)}{17.317} - \underset{(-1.5)}{0.217\Delta IN_{t-1}} + \underset{(7.4)}{0.639BT_t} - \underset{(-1.6)}{0.103}(BT_{t-1} + BT_{t-2}) - \underset{(-1.3)}{0.845DDE_{t-1}}$$

$$\bar{R}^2 = 0.8049 \qquad DW = 1.87$$

(e) *Foreign trade*

$$M_t/\mathrm{PM}_t = -62.448 + 1.516Y_t^{\mathrm{P}} - 0.078\mathrm{BT}_{t-2} - 0.267(\mathrm{BP}_{t-1} + \mathrm{BP}_{t-2}) + 0.755\mathrm{DDE}_{t-1}$$
$$(-4.3)\quad (18.4)\qquad (-1.4)\qquad (-3.3)\qquad\qquad (1.6)$$
$$\bar{R}^2 = 0.9735 \qquad \mathrm{DW} = 2.19$$

$$E_t/\mathrm{PE}_t = -89.704 + 0.477Y_t + 0.119\mathrm{BT}_{t-2} + 0.024\mathrm{BP}_{t-1} + 0.568\mathrm{BP}_{t-2} - 0.0653\mathrm{DDE}_t$$
$$(-8.0)\quad (22.4)\qquad (1.7)\qquad (1.2)\qquad (4.1)\qquad (-1.6)$$
$$\bar{R}^2 = 0.9896 \qquad \mathrm{DW} = 1.93$$

$$\mathrm{BT}_t = M_t - E_t$$

(f) *Labour market*

$$\mathrm{DL}_t = L_t + 38.862(L_t^{\mathrm{P}} - L_t^*) = 1039.32 + 0.797L_{t-1} + 2.522\Delta Y^{\mathrm{P}}$$
$$(3.8)\qquad\qquad (2.0)\qquad (7.6)\qquad (1.4)$$
$$\bar{R}^2 = 0.9483 \qquad \mathrm{DW} = 2.34$$

$$\mathrm{SL}_t = 2230.12 + 0.352\mathrm{LF}_t + 3.362\mathrm{RB}_t$$
$$(11.7)\qquad (14.6)\qquad (8.9)$$
$$\bar{R}^2 = 0.9473 \qquad \mathrm{DW} = 1.75$$

$$L_t = \begin{cases} \mathrm{DL}_t & 1965\text{–}72, \quad \text{when} \quad (L_t^{\mathrm{P}} - L_t^*) \sim 0 \\ \mathrm{SL}_t & 1972\text{–}84, \quad \text{when} \quad (L_t^{\mathrm{P}} - L_t^*) > 0 \end{cases}$$

(g) *Total disposable income of the population*

$$J_t = -25.658 + 0.644J_{t-1} + 1.003\mathrm{RB}_t + 0.374\mathrm{TJ}_t$$
$$(-1.2)\quad (3.1)\qquad (2.9)\qquad (0.9)$$
$$\bar{R}^2 = 0.9954 \qquad \mathrm{DW} = 1.41$$

List of variables

Endogenous variables

DC_t	Demand for consumption goods at constant prices in milliard forints
SC_t	Supply of consumption goods at constant prices in milliard forints
C_t	Actual consumption of the population at constant prices in milliard forints
DI_t	Demand for investment goods at constant prices in milliard forints
SI_t	Supply of investment goods at constant prices in milliard forints
I_t	Actual investments at constant prices in milliard forints
$\mathrm{I}_t^{\mathrm{P}}$	Planned value of investments (1962 = 100)
IU_t	Investments put into operation at constant prices in milliard forints
K_t	Fixed assets at constant prices in milliard forints ($\bar{K}_t = (K_{t-1} + K_t)/2$)

Y_t	Total value added at constant prices in milliard forints
SY_t	Supply of total value added at constant prices in milliard forints
DY_t	Demand for total value added at constant prices in milliard forints
DDE_t	Excess demand in the domestic markets at constant prices in milliard forints
IN_t	Inventory stocks at constant prices in milliard forints
M_t	Total imports at current prices in milliard forints
E_t	Total exports at current prices in milliard forints
BT_t	Balance of foreign trade (import surplus) at current prices in milliard forints
DL_t	Demand for labour force, thousand
SL_t	Supply of labour force, thousand
L_t	Actual employment, thousand, $\bar{L}_t = (L_{t-1} + L_t)/2$, $L_t^* = L_t/L_{1972}$
S_t	Stocks of savings of the population at constant prices in milliard forints
J_t	Total income of the population at constant prices in milliard forints

Exogenous variables

Y_t^P	Planned value of total value added (1962 = 100)
L_t^P	Planned level of employment (1972 = 100)
Z_t	Scrapping and decay of fixed assets at constant prices in milliard forints
U_t	Capacity utilization of fixed assets (1970 = 1000)
LF_t	Sources of labour force, thousand
RB_t	Real wage index (1960 = 100)
TJ_t	Social benefits at constant prices in milliard forints
BP_t	Current account balance of payments, excluding the trade balance (the balance on other goods, services and incomes; private and official unrequired transfers), hundred thousands of US dollars
PM_t	Import price index (1981 = 1.000)
PE_t	Export price index (1981 = 1.000)
t	Time

9.4 CONCLUSIONS AND PROSPECTS

The specification and estimation of the experimental disequilibrium macromodel has aimed at analysis of the last twenty years of the Hungarian economy. The results of the work support our underlying hypothesis that there has been a strong relationship between internal and external disequilibrium states of the economy and that the problem can be analyzed by an econometric disequilibrium technique. The following main conclusions can be drawn from this study.

1. During the period 1965 to 1984 there was a stable relationship between the foreign and domestic disequilibrium states of the economy. It can be observed

that a disadvantageous foreign trade balance and balance of payments were followed by restrictions in the supply of investment goods and consumption goods in the next one or two years, leading to increasing domestic disequilibria. On the other side these excess demands in the domestic markets yield immediate changes in inventory stocks and responses in future exports and imports. These fluctuating responses in foreign trade have generated further import surpluses leading again to deficits in the balance of trade and in the balance of payments, and thus it has begun all over again.

The direction and size of these responses referring to SC_t, SI_t, ΔIN_t, M_t and E_t are presented in thousand million forints in Table 9.1.

Table 9.1 Relation between foreign and domestic imbalances in some macrocategories of the economy

	BT_t	BT_{t-1}	BT_{t-2}	BP_{t-1}	BP_{t-2}	DDE_{t-1}
SC_t		−0.057	−0.057	−0.087	−0.087	
SI_t	+0.303			−0.155	−0.425	
ΔIN_t	+0.639	−0.103	−0.103			−0.845
M_t/PM_t			−0.078	−0.267	−0.267	+0.755
E_t/PE_t			+0.119	+0.024	+0.568	−0.653

Note: Based on the structural parameters of the model, in 1981 thousand million forints.

2. The above lag structure in responses provides some explanation for the existence of inventory cycles and investment cycles. As Fig. 9.1 shows, cycles in inventory stocks have taken a symmetrical form for cycles in import surpluses and an asymmetrical form for cycles in domestic excess demand. In addition, our results support the theory of investment cycles, stating the investment cycles in the 70s have been asymmetric as compared with changes in foreign trade balance (for the 80s see the next point). Figure 9.1 shows fluctuations in foreign and domestic disequilibrium and in inventory stocks and investments during the observation period.
3. By comparing the results of the two versions of the model we can draw a final conclusion about our recent economic processes, namely, with respect to the foreign debt situation. In the first version (Hulyák, 1985), which was estimated for the period 1965–81, the economy's foreign disequilibrium situation was described by only one factor, i.e. by the foreign trade deficit. After the actual economic events of the early 1980s we were interested in the model's forecasting performance. We would have liked to know whether the cyclical economic movements revealed still existed after 1981. In 1984 *ex post* forecasts were made for a set of endogenous variables for the years 1981–84 (Hulyák, 1985). Inspection of the forecast errors then indicated that the nature of the interdependence of foreign and domestic excess demands had changed in the

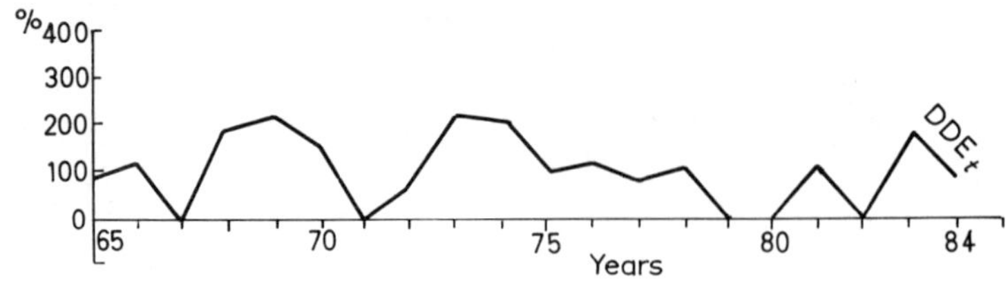

(a) Excess demand in domestic markets (DDE_t)

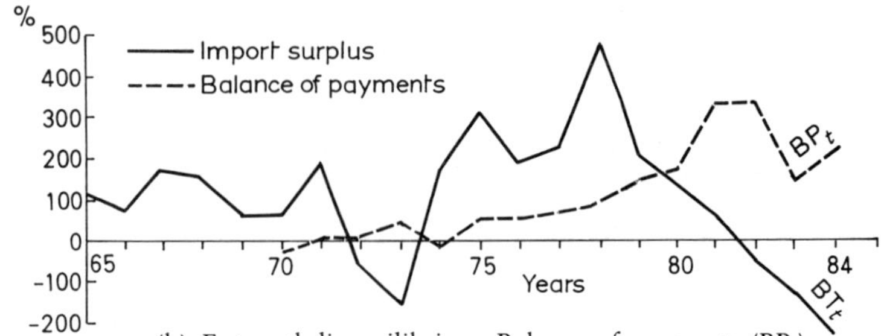

(b) External disequilibrium: Balance of payments (BP_t)

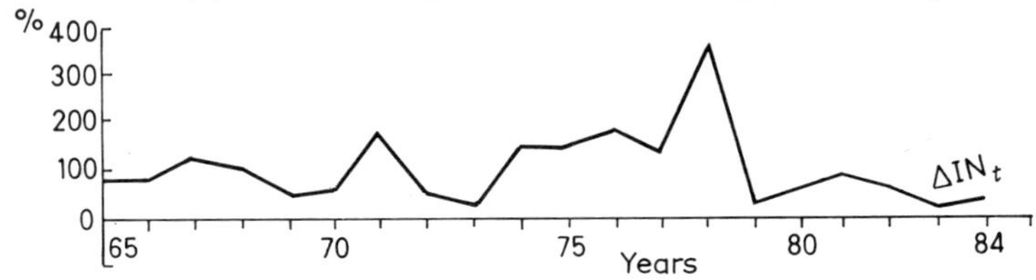

(c) Annual increases in inventories (ΔIN_t)

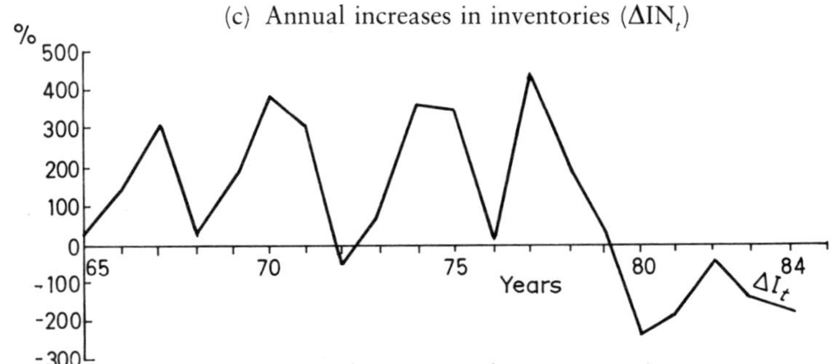

(d) Annual increases of investment (ΔI_t)

Fig. 9.1 Disequilibria and cycles in the Hungarian economy (percentage of the average). Note that DDE_t, BP_t, BT_t, ΔIN_t and ΔI_t are transformed to standardized index numbers: each variable is divided by its arithmetical mean and multiplied by 100.

early 1980s. That is why economic policy has been permanently focused on improving the foreign trade balance and there has been no interaction to improve disequilibrium in the domestic markets.

In the second version of the model not only the foreign trade deficits but also the deficits of other items of the current balance of payments appear in the model as factors of similar importance characterizing the foreign debt.

Re-estimation of the modified version of the model for the period 1965–84 reflects the view that the interdependence of foreign and domestic excess demands still existed in the eighties, but that this connection has become more complex. In spite of the improvement in the foreign trade balance, the deficits of other items in the balance of payments have permanently and significantly increased in later years (Fig. 9.1). Excess demands in domestic markets have been influenced by foreign deficits of both kinds. We can state that while in the 70s trade deficits caused the main problem, in the 80s the whole foreign debt situation must be considered in the structure of our economic processes.

4. The results of the model support our underlying hypothesis that the disequilibrium technique gives new possibilities in empirical economic analysis. At the same time we emphasize that we are at the very beginning of this work. For future research work we consider that the two main directions are the improvement and extension of the theoretical model and the widening of the data base and estimation framework.

The following are some new possibilities for extending disequilibrium model-building in order to develop macromodel-building activity in Hungary.

The main weakness of our model appears to lie in the simplified treatment of foreign trade processes and the labour market. We intend to disaggregate and improve the foreign trade block and to build a more sophisticated block for the labour market. In addition to this we should not completely exclude the state budget sector, price-formation and inflation measures and other monetary processes from a disequilibrium macromodel. In the case of Hungary, especially since the late 1970s, we have to investigate monetary activity too. In order to complete the macromodel we have to extend disequilibrium analysis to the production sphere by sectoral demand functions and supply functions as well.

Our model has proved inadequate for explaining planning behaviour in our economy. Using the achievements of research works in this field, we should follow up this line too.

Also important for extending the range of the modelling framework would be to pay explicit attention to the second and third economies of Hungary.

To sum up the prospects of all these future tasks, we hope that macro-disequilibrium modelling activity will contribute to the elaboration of a more specific and more realistic macroeconomic theory of centrally planned economies.

REFERENCES AND FURTHER READING

Barro, R. J. and Grossman, H. I. (1971) A general disequilibrium model of income and employment. *American Economic Review*, **61**, 82–93.

Barro, R. J. and Grossman, H. I. (1976) *Money, Employment and Inflation*, Cambridge University Press, Cambridge.

Benassy, J. P. (1982) *The Economics of Market Disequilibrium*, Academic Press, New York.

Charemza, W. and Quandt, R. E. (1982) Models and estimation of disequilibrium for centrally planned economies. *Review of Economic Studies*, **49**, 109–16.

Charemza, W. and Gronicki, M. (1984) Plans and investment disequilibrium in a partially commanded economy: an econometric investigation for Poland, 1960–1980, Paper presented at Models and Forecasts 1984, Warsaw, 1984.

Charemza, W. and Gronicki, M. (1984) Simultaneous consumption–labour–money analysis in a disequilibrium framework: Poland 1960–1980, Series RC/B20, CREES Discussion Paper, University of Birmingham.

Csunderlik, C. (1985) The interrelationship of the private housing expenditures and personal savings: an econometric disequilibrium analysis of the Hungarian housing market, paper presented at the World Congress of the Econometric Society, Cambridge, MA.

Fair, R. C. and Jaffee, D. M. (1972) Methods of estimation for markets in disequilibrium. *Econometrica*, **40**, 497–514.

Halabuk, L. and Shapiro, H. T. (1976) Macroeconometric model building in socialist and non-socialist countries: a comparative study. *International Economic Review*, **17**, 529–65.

Howard, D. H. (1976) The disequilibrium model in a controlled economy: an empirical test of the Barro–Grossman model. *American Economic Review*, **66**, 871–9.

Hulyák, K. (1982) disequilibrium models for consumption goods in Hungary, Paper presented at the 9th International Conference on Applied Econometrics, Budapest.

Hulyák, K. (1985) An econometric disequilibrium macromodel for Hungary, Paper presented at the World Congress of the Econometric Society, Cambridge, Mass.

Malinvaud, E. (1982) An econometric model for macro-disequilibrium analysis, in *Current Developments in the Interface: Economics, Econometrics, Mathematics* (eds M. Hazewinkel and A. H. G. Binnooy Kan), Reidel, Dordrecht.

Portes, R. (1977) Effective demand and spillovers in empirical two-market disequilibrium models, Discussion Paper No. 595, Harvard University, Harvard Institute of Economic Research, December.

Portes, R. and Winter, D. (1978) Disequilibrium estimates for consumption goods markets in centrally planned economies, Discussion Paper No. 612, Harvard University, Harvard Institute of Economic Research, March.

Portes, R. (1966) The theory and measurement of macroeconomic disequilibrium in centrally planned economies, Paper No. 91, CEPR, London.

Quandt, R. E. (1982) Econometric disequilibrium models. *Econometric Reviews*, **1**, 1–63.

Quandt, R. E. (1982) The structure of disequilibrium models, Econometric Research Program, Research Memo. No. 295, Princeton University.

PART THREE

Sectoral shortage and disequilibrium models of centrally planned economies

10

Sectoral shortage models in Hungary

MÁRIA LACKÓ

10.1 INTRODUCTION

In Hungary, a descriptive–explanatory theory of the socialist economy and a family of models connected with it have gradually been developing since the early 1970s and have preserved their relative independence despite the emergence of the disequilibrium models. The main fields of interest in this school are the behavioural rules originating from the institutional system of the socialist economy, the different forms of appearance of shortage and its permanent reproduction mechanisms. The school was started by Kornai's book *Anti-Equilibrium* (1971), and a series of verbal works, theoretical–mathematical and econometric models followed.

1. Verbal works, which do not or only rarely apply formalization or quantification are: Kornai (1980, 1984, 1985, 1986), Kornai–Matits (1984), Bauer (1978, 1981), Soós (1975, 1984, 1986), Gács (1981, 1982, 1983a), Gács and Lackó (1973), Laki (1980, 1984), Farkas (1983) Fazekas–Köllõ (1985), Kapitány, Kornai and Szabó (1983), Fábri (1982), Lukács (1984), Varga (1984), Major (1982), Chikán (1981), Dániel (1984), Dániel and Semjén (1987).
2. Theoretical–mathematical models, which investigate the functioning and behavioural rules of different systems in the framework of rigid formulized models are: Kornai (1982), Kornai–Martos (1971, 1981), Kornai–Simonovits (1977, 1985), Kornai–Weibull (1978, 1983), Martos (1985), Simonovits (1981, 1982, 1985), Dániel (1983), Dániel–Kornai–Weibull (1981).
3. Econometric models, which aim to test the hypotheses propounded in the verbal

International Studies in Economic Modelling Series.
Models of Disequilibrium and Shortage in Centrally Planned Economies.
Edited by C. Davis and W. Charemza.
Published in 1989 by Chapman and Hall Ltd, 11 New Fetter Lane, London EC4P 4EE.

works and theoretical–mathematical models on the bases of statistical data are: Gács (1983b), Hewett (1980), Hunyadi (1984), Hunyadi *et al.* (1980), Lackó (1976, 1980, 1984a, 1986), Marrese (1977, 1981), Simon (1977, 1980, 1981), Budavári–Náray (1982), Riecke (1982), Vanous (1980).

The fact is that econometric investigations accomplished so far constitute a vulnerable point for this research direction: in a series of very important fields there are no formulized econometric models. A model which would embrace the special behavioural rules of all spheres of the shortage economy has not yet been worked out. There are no econometric investigations into such macrolevel problems as shortage and inflation, the macro-consequences of the functioning of the private sector and the second economy in centrally planned economies, since the models describe only the behavioural reactions of the state sector.

This chapter summarizes the results of those macrolevel econometric models which have been elaborated for the Hungarian economy and are related to this school. In section 10.2 the major features of the models are summarized, in sections 10.3, 10.4 and 10.5 the investment, foreign trade and consumption sphere models are scrutinized, and in section 10.6 the tasks for the future are defined.

10.2 THE MAIN CHARACTERISTICS OF MODELS OF THE SHORTAGE ECONOMY

10.2.1 The cardinal role of institutional conditions

The models of the shortage economy are intended to show the specific rules of behaviour of the socialist economy and their empirical verification. This feature of the models is based on a firm belief that the regulation of the socialist economy has its own rules and laws which are in many respects different from the rules and laws of market economies. The behaviour of the actors is powerfully influenced by the institutional conditions of the economy.

In the dominant state sector of the socialist economy, enterprises and institutions are positioned in vertical subordination, more or less irrespectively of the direct or indirect systems of control. The budget constraints of these institutions and enterprises are soft. As a general implication, incomes and expenditures are in loose relationships, the frequent expenditures in excess of receipts being covered by an institution, mostly by the state, and so survival of the firm is warranted. The behaviour of the institution or enterprise is determined by this perpetual opportunity as follows:

1. The general price sensitivity of enterprises is low;
2. Efficiency is an unimportant consideration in decision-making;
3. Artificial indicators of success and efficiency as well as budgets are used, usually abortively, by the central administration in the hope of controlling enterprise behaviour;

4. There is a strong tendency for the demand for current inputs and investments to grow boundlessly.

The budget constraint of the consumer may also be stated to be hard in the socialist economy, incomes and expenditures being closely correlated. As a result of the institutional conditions existing in the other, state-owned sphere of the economy, however, the consumer's behaviour is strongly affected by chronic shortages in the economy.

10.2.2 Demand and supply functions are not separated in the models

The permanent bargaining process between the higher and lower levels of the economic hierarchy is an outcome of the softness of the budget constraint of microunits and the consequential enterprise behaviour.

In this bargaining process and because of the tensions shown in the functioning of the economy, the higher authorities feed positive plan directives and more market-like instruments along with a variety of other signals down to the enterprises, and the enterprises take this into account in stating their claims for various inputs and investments. The enterprises try to control the conflicts that might develop in their relations with higher administration and at the microlevel. Consequently they adjust their claims to the possibilities and the supply. In these economies forced substitution is a frequent instrument of adjusting demand to supply. That is why, unlike disequilibrium models, the shortage economy models do not formulate discrete functions of demand and supply, and do not attempt to state the degree of shortage on the basis of the difference between the two. Instead of separating these functions the shortage economy models are devoted to presenting and modelling the development of strains and measurable variables and the responses to them.

At the same time the formulation of demand functions at macrolevel and shortage modelling are thwarted by the difficulty of aggregation. In this economic system aggregation of shortages and surpluses presents a problem. Although surplus is the opposite to shortage, in many instances it necessarily accompanies shortage, while surplus does not induce behavioural responses opposite to those provoked by shortage. It follows *inter alia* from these features that shortage is a notion, the occurrence of which at the microlevel, when simply aggregated, cannot offer a shortage indicator of macroeconomic validity.

In the consumer sphere, despite the toughness of the budget constraint and the significant bearing that incomes and prices have on consumer behaviour, owing to the chronic shortage of certain commodities and services, the consumers, too, must adjust their demand and savings to supply and must regularly revert to forced substitution. Demand and supply are therefore not distinguished by the shortage economy models for this sphere either. Instead the models give a description of the actual processes and their regulation upon feedback of various strains and shortages.

10.2.3 The important role of non-price signals among explanatory variables of the models

Owing to the institutional features outlined above, the drives, criteria of policy-making and tensions that would control the behaviour of capitalist economies, are largely missing from the conduct of socialist economies with central planning. Tensions, however, do not vanish but assert themselves in other forms and in other spheres of the economy. It is a typical tension transformation when they are pushed up to macrolevel, having cumulated in the meanwhile, and become the main state variables of macro control. Tensions like the regular above-plan implementation of inputs, investment outlays and imports are recorded at the level of the national economy while the outputs and convertible exports are regularly underfulfilled at the national economy level. These tensions powerfully affect the behaviour of the higher policy-makers and plan drafters.

It is a noteworthy feature of shortage economy models that various macrotensions and shortages are presented as signals. On the basis of tension signals, endogenous regulation takes place in the models, whereupon some strains are eliminated but others are recovered.

The traditional explanatory variables, namely incomes and prices, play an important role in consumer behaviour because of the hardness of the budget constraint. These latter variables are not independent of the macrotensions noted above. That is, consumer behaviour is under the influence of the robust central control of a few major macrocategories (wages, prices) and shows the typical features of traditional responses along with hard budget constraints and adjustment to the chronic state of shortage.

Among the quantitative non-price signals, prominent importance is attributed to norms. In the bargain process system, striving after preserving the status quo is often the characteristic conduct of the actors in the socialist economy. In shortage economy models this sympton is described as control by norm taking place around various settled norms established by historical behaviour.

10.2.4 Data problems

The shortage economy approach has achieved only a few econometric results mostly because of a lack of relevant data. There are two main groups of data problem.

The first group is connected with the fact that due to the bargaining processes generated by the institutional setup of the system some variables are unobservable. This economy can be characterized much more by the adjustment of demand to supply than by the adjustment of supply to demand. Accordingly it is very difficult to quantify the original magnitude of the demand, i.e. the kind of demand that has been drawn up independently of the pressing supply situation. Of course at one point in time in the bargaining process a certain kind of demand is outlined, e.g. the firms' investment credit requests can be interpreted as an indicator of investment demand. However, it should not be forgotten that this is already the outcome of a reconciliation between the applicants and the allocators.

The second group of data problems is connected with the practice of the official statistical publications. There are some important phenomena for which statistical observations are not organized. For another set of variables data are either collected and published irregularly or the content of the statistical notion varies from time to time. This drawback of the data makes time series too short to construct reliable econometric models.

Thus not only theoretical considerations but also data problems prompt model builders of the shortage economy to use such tension and shortage indicators, as well as deviations from given norms, as can be derived from observable data.

An evolution has been experienced recently in the methodology of using unobservable variables in econometric models, so there is hope for the development of more advanced econometric models of the shortage economy in the future.

10.3 INVESTMENT MODELS

In this sphere, the shortage economy models express, in one way or another, the theoretical statement that in socialist economies the regulation of investments is based on signals of tensions added up at the macrolevel, for no or hardly any profitability constraint asserts itself at the microlevel.

Most of the motive powers, decision criteria, and tensions regulating the investment behaviour of capitalist firms are missing from the behaviour of enterprise in the centrally planned socialist economies. Before a decision is made on investment, the requirement that the present value of the future profits of the new establishment or production unit should cover investment and other costs, as well as the mass of the normal profit is barely considered. Instead, great importance is attached to the physical objectives the enterprise uses to justify its capacity-increasing ambitions. Moreover when a new project has been put into operation, tension in sales and efficiency rarely arises to threaten the enterprise in its existence. In the socialist economy, the budget constraint of enterprises and institutions is soft,[1] which creates an almost insatiable investment demand at all hierarchical levels of the economy. The applicants spend almost all their energy in acquiring the permissions necessary for starting the investments. The allocators, i.e. decision-makers, change the approval coefficient[2] of the permissions following changes in the tensions adding up at the macrolevel: as tensions grow, the number of permissions granted is lowered; whereas if tensions lessen, the high demand for new investments need not be so strictly screened. Yet regulation based on macrotensions is not as simple as it seems. Namely, the different kinds of macrotensions behave, as

[1]The softness of the enterprises' budget constraint, and empirical analysis of the resulting microlevel behavioural regularities are not covered within the framework of the present chapter. The subject was worked out by Kornai–Matits (1983).

[2]This concept was introduced by Bauer (1981). Approval in the strict sense ceased after 1968.

follows from their nature, in a conflicting manner in the course of regulation; resolved to ease one tension by their decision, planners may involuntarily increase another.

Macrotensions as signals are guided towards micro-units in several different ways, by means of direct plan instructions, certain indirect means, more like those of a market economy, and informal control. The application of the latter is dissolved in bargaining processes in the socialist economy, and the outcome is embodied in consensuses, expressed or implied. The large investment demand – owing to institutional causes – of enterprises is moderated by strict portioning, the screening of demands, bargaining, and in many cases the consensus, and this is also what renders the concept of investment demand in the socialist economy quite incomprehensible. The measuring of demand is further made impossible by forced substitution, a frequently occurring phenomenon in the socialist economies.

It follows from the preceding paragraphs that the models of this family do not set up demand and supply functions separately, but try to grasp the origin of the tensions and the reaction to them. In the following paragraphs, Lackó's (1986) model will be presented, with short references to other econometric models containing a description of behavioural regularities.

Lackó's (1986) model, a new version of her earlier one (1980), describes the regulation of investments by macrolevel tensions, and the reproduction of tensions. The model relies, on the one hand, on the comprehensive empirical study of Bauer (1981), and on the other, on the theoretical statements of Bauer (1981) and Kornai (1980) concerning the mechanism of investment cycles in the socialist countries.

In the model, special roles are assigned to two tension indicators:

1. The carrying out of investments is an activity well prolonged in time; several years after it has been started, an investment represents almost certain demand in the form of commitment. The rate of macrolevel commitment to GDP is, owing to the institutional causes outlined above, always higher than the level which could provide for a fast and smooth implementation of investments, without shortages and forced substitutions (Bauer, 1981). The divergence of the commitment ratio from that level is considered the internal tension of the investment process.
2. The relative net exports, i.e. the rate of balance of foreign trade to imports may be considered an indicator of external tension: an increase in relative net exports indicates an easing of external tension, its decrease indicates its rising.

Model variables

- $s(t)$ Investment start ratio to GDP: total cost of investment started during t per GDP
- $i(t)$ Investment outlays ratio to GDP: actual outlays on investments effectuated during t per GDP
- $d(t)$ Investment commitment ratio: outstanding outlays (commitment) to investments effectuated during t per GDP

$k(t)$ Relative net export: macroeconomic export–import differential per import
$q(t)$ Stockpiling ratio: stockpiling during t per GDP
$r(t)$ Growth rate of GDP
Endogenous: $s(t), i(t), d(t), k(t)$
Exogenous: $r(t), q(t)$

The model

$$\Delta s(t) = \alpha_d \Delta d(t-1) + \alpha_k k(t-1) + \alpha_c \qquad \alpha_d < 0 \qquad \alpha_k > 0 \tag{10.1}$$

The first term of the right-hand side of the behaviour equation (Equation 10.1) expresses the concept that if, *ceteris paribus*, internal tension grows which is, as is known, an increasing function of the commitment ratio, for the next period planners will reduce the increment of the start ratio. And, in the opposite case, if the increment of the commitment ratio decreases, planners understand it as a relative release of investment capacities, so that they can increase the increment of the start ratio.

The other factor taken into account in determining the start ratio is the relative net exports which are considered, from the aspect of the investment process, as an indicator of external tension. The parameter $\alpha_k > 0$ expresses the concept that if, *ceteris paribus*, relative net exports decrease, planners will also ease external tension by means of cutting down the start ratio; whereas the reverse case clearly allows them to increase the investment ratio.

$$s(t) = s(t-1) + \Delta s(t) \tag{10.2}$$

$$i(t) = \beta_s s(t) + \beta_d d(t-1) + \beta_k (t-1) + \beta_c \qquad \beta_s > 0 \quad \beta_d > 0 \qquad \beta_k > 0 \tag{10.3}$$

Equation 10.3 of the model explains the investment outlays ratio by the start ratio of the same period, the commitment ratio inherited from the preceding period, and the earlier net exports: the higher the start ratio, and the higher the commitment ratio, the greater the pressure, *ceteris paribus*, on planners to further raise the investment outlays ratio.

$$d(t) = \frac{d(t-1)}{r(t)} + s(t) - i(t) \tag{10.4}$$

$$\Delta d(t) = d(t) - d(t-1) \tag{10.5}$$

Equation 10.4 of the model formulates the reproduction of the commitment ratio.

$$k(t) = \omega_i i(t) + \omega_q q(t) + \omega_c \qquad \omega_i < 0 \quad \omega_q < 0 \tag{10.6}$$

The behaviour equation (Equation 10.6) represents the reproduction of external tension: it explains the changes in relative net exports by the investment outlays ratio and by the stockpiling ratio considered to be exogenous; their growth reduces, *ceteris paribus*, the value of relative net exports.[3]

The recursive pattern of the model (Equations 10.1–10.6) is as follows:

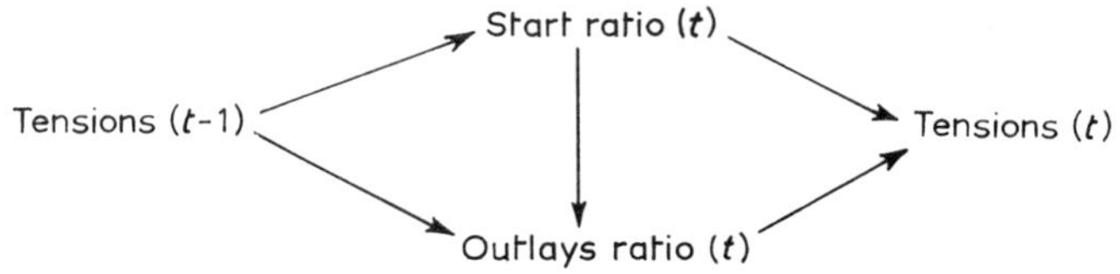

Fig. 10.1 Recursive pattern of the shortage model of investment.

The result of the estimate is shown in Table 10.1. Below estimated parameters, the value of the t-statistics is shown in parentheses.

The signs of the parameters developed as had been expected: their values seem to be plausible. In the course of making the estimate, the variable $k(t-1)$ was in close parallel movement with the variable $d(t-1)$ and, because of multicollinearity, it did not prove to be significant in the estimate. $k(t-1)$ was replaced by $k(t-2)$. The

Table 10.1 The estimated[4] equations

Explanatory variables	*(10.1)* $\Delta s(t)$	*(10.3)* $i(t)$	*(10.6)* $k(t)$
$\Delta d(t-1)$	−2.055 (10.57)		
$d(t-1)$		0.1864 (3.74)	
$s(t)$		0.1546 (4.99)	
$k(t-2)$	0.2273 (1.96)	0.0241 (1.47)	
$i(t)$			−3.662 (4.59)
Constant	−0.034 (2.95)	0.1880 (5.13)	0.8743 (4.69)
rho	0.6807 (3.08)	0.9599 (11.36)	
aR^2	0.79	0.83	0.78
DW	1.28	1.67	1.67
SE	0.043	0.006159	0.0484
Mean of the dependent variable	−0.01852	0.2389	−0.07017
Standard deviation of the dependent variable	0.0939	0.01489	0.10397

[3] Since the mid-1970s, changes in relative net exports have been considerably influenced by changes in the terms of trade; however, they have not proved a significant variable for the whole period.

[4] The model estimates were hindered by the lack of data mentioned in section 10.2.4. In the first stage the missing data were provided then the estimations were made by the two-stage least squares method. Modifications by first degree serial correlated error were occasionally carried out by the Cochran–Orcutt (CORC) method. The estimation is presented in detail by Lackó (1986).

significance of the estimates is to be treated with reservation, because of the data generation.

With the aid of the model, we tried to describe the cyclical mechanism by which planners and decision-makers shape the investment process relying on tension indicators. Higher tension, the increased commitment arising from investments started in the past, and decreasing relative net exports compel the planners to reduce the number and costs of new investments, and to spend capacities on fulfilling the commitments of investments under way. The investment outlays thus effected do not influence tensions all in the same way: they reduce the commitment, and thereby also the internal tension, whereas they further increase the external tension measured by the relative net exports. As a result of tensions easing, central planners make way for the enterprises' investment demand, as they see the increase in investments themselves as a condition of economic growth, and, by raising the approval coefficient of new investments, they contribute to the reproduction of tensions.

In the following we give an outline (not in detail) of investment models similar to the model of Lackó (1980, 1986) that describe the regulatory role of macrotensions and the reproduction of tensions.

In his model, Marrese (1977) explained the deviation of the Hungarian investment volume from the trend by means of various tension indicators: building material imports as an indicator of supply shortage, the deviation of consumption from the trend in the past which is expressive of unbalanced living standards, the deviation of imports of machinery and instruments against dollars from the trend, which is, according to the author, an indicator of repressed excessive demand, and the deviation of uncompleted investments from the trend in the past. Marrese's model consists of isolated equations; it does not cover the entire process of endogenous regulation.

The investment equations of Hewett's (1980) macromodel produce endogenously the magnitude of the yearly plan of investment outlays. The investment plan is determined by the difference plan – the facts for the foregoing period, yearly net exports (if the plan figure is not known, the facts), and the plan of the national income, and the fulfilment is determined by the plan, the real GDP, and net exports. Real investment outlays have a certain action on net exports. In this model, the difference between the plan and the facts may be considered a kind of indicator of internal tension: overfulfilment of the investment plan is an indication to the planners of runaway investments, to which they react by reducing the next yearly plan.

Simon's (1981) model describing endogenous regulation is a special Cobweb model, in which the two-year lag value of the balance of foreign trade accounted in dollars affects the investment outlays (positive effect), which – together with the terms of trade – determines the balance of foreign trade accounted in dollars (negative effect). Investments decided on the enterprise level are explained by changes in the credit stock which is endogenously regulated: by the time-lag value of the GDP and of the balance of foreign trade accounted in dollars, and by the current value of the price.

In his model, Hunyadi (1984) takes apart centrally decided investments and those decided on the enterprise level, and separately examines the large sectorial aggregates. Enterprise investments are explained by amortization and the time-lag values of the enterprise's development funds, as well as by the centrally planned accumulation ratio.

Centrally decided investments are examined separately in industry, the building industry, and in communal sectors. In the author's conception, the various kinds of tension exert an influence on the starts of investments, and the investment outlays can be described by the distributed lags of the starts. Thus he describes the investment outlays by the distributed lags of the various tensions, in this way avoiding the difficulties arising from the lack of data on starts. In industry and the building industry, investments are explained by the distributed lags of the balance of foreign trade accounted in dollars and by the outstanding stock of investments, while those in the communal sector are explained by the latter, and by the distributed lags of the deviations of consumption from the trend.

Lackó's study (1986) covers an examination of the investments of the individual sectors of the national economy (industry, the building industry, agriculture, trade, transport and news communication, communal branches). In her hypothesis sectorial investments are not only influenced by the above-mentioned macrotensions affecting all investments, but also a certain role is played by the sectorial tensions. She demonstrated that, in the infrastructural branches (trade, transport and news communication, communal branches), besides the effect of changes in total investment a kind of control by norms operates, where norms are the long-term average of the investment rate in the branches;[5] hence the importance of the earlier position of the branch in question: whether it received an investment rate smaller or greater than the norm. In these branches, an investment rate lower than the norm leads to an increased growth rate of investments in the next period (and conversely) if the growth rate of the total amount of investments does not change. It was quantified in the model that, in changes in public health and welfare investments, a relevant factor is represented, besides total investment, by a shortage signal that reflects well, from a certain point of view, the tensions in the functioning of the sector: a deviation in the supply of hospital beds from the ordinary supply in the foregoing period. Relative shortage, i.e. with the supply indicator below the long-term trend, leads to an increased growth rate of investments granted to the sector for the next period.

10.4 SHORTAGE ECONOMY MODELS OF FOREIGN TRADE[6]

For the foreign trade sector description and analysis by econometric models are elaborated in much less detail than for investments. As in the investment models,

[5] In these models norms are identified with the time trend of the respective variable. Although it is evident that a norm can be better characterized by a normal zone than by a selective line, we opted for this indicator to make the calculations easy to handle.

[6] In writing this chapter, I relied on Gács's (1983) study.

macrotensions play the main part in regulating this sphere, too, and though for different reasons, this is so for foreign trade with the capitalist as well as the socialist economies.

The majority of the models make no attempt at all to use relative prices as an explanatory factor, as it is known that in the planned economies foreign trade decisions and behaviours are not primarily guided by efficiency requirements and by the need to select the cheaper source of supply and the most profitable channel of sales, but by the central determination that the physical balances should be in a state of equilibrium. Nor do those who examine the explicit effect of relative prices believe that foreign trade processes develop as a result of the millions of decisions made by decentralized decision-makers taking into account relative prices as well; however, they do assume that planners who basically influence changes in trade also have an eye to changes in relative prices (Neither Vanous, 1980 nor Simon, 1980 obtained a convincing result for the role of prices.)

In trade with the socialist countries, parties strive for a bilateral balance of trade at the macrolevel, in compliance with the system of clearing accounts; moreover as the transferable rouble and the prices expressed in it are inapt for measuring the different degrees of hardness of the different kinds of goods, the parties also further strive to keep a bilateral balance of trade in the major groups of products.

In their model Hunyadi *et al.* (1980) present a simultaneous balancing of total imports accounted in roubles: the growth of exports accounted in roubles is the condition of the growth of imports accounted in roubles.

In his equation set up for changes in exports accounted in roubles, Hewett (1980) indicates factors of several years' cumulated changes in imports: the growth of exports is so much the greater, the greater were the imports accounted in roubles one, two, or three years before.

In more disaggregate models, there are a few examples to show that efforts are made at reaching equilibrium in respect of groups of products also. In Simon's (1980) model, material exports to socialist countries are a function of simultaneous material imports from socialist countries and machinery exports from socialist countries are referred, among other things, to the lagged value of machinery imports from socialist countries. In the model of Hunyadi *et al.* the exports of consumer articles accounted in roubles also depend on the imports of consumer articles accounted in roubles. At the same time, on examining the foreign trade of several CMEA countries, Vanous (1980) did not find any adequate quantitative proof of a balancing by groups of products.

In trade with the capitalist countries, at the macrolevel, efforts are not directed at reaching perfect equilibrium, but at a given level of the balance of trade accounted in hard currencies.

In his model Hewett (1980) describes the changes in exports accounted in dollars as a negative reaction, among other things, to the change in the dollar balance of two years earlier: an improving balance leads to decreasing exports, and a deteriorating balance leads to increasing exports, and besides the deviation of investments from the trend as a market indicator, the imports accounted in dollars are explained by a positive reaction to the one-year lag dollar balance: an improving balance leads to increasing imports, and a deteriorating balance leads to a setback in imports.

The foreign trade turnover both with capitalist and socialist economies taken together can also be described by the reaction to their aggregate balance: on examining foreign trade plans (by rough methods), Gács-Lackó (1973) found a disposition where a deterioration in the foreign trade balance led to a raising of export plan targets and a lowering of import plan targets, whereas improvement in the foreign trade balance led to a lowering of export plan targets and a raising of import plan targets in Hungary.[7]

The macrotensions of the foreign trade with capitalist countries flowing into the microsphere, and the behaviour of the microsphere, were analysed by Gács (1981). In this sphere, as well as in investments, the almost insatiable demand must be mitigated, in this case the demand for imports against hard currencies which present a threat to the foreign trade balance. Here, too, direct methods such as tightening of permissions are completed by bargaining and informal regulation (consensus). In this way the mechanism hinders development of a bad shortage situation, whereas it prompts the enterprises in need of imports to resort to forced substitutions over a wide field. The signs of forced substitution are also found at the macrolevel: in his model Hewett (1980) demonstrates that the volume of Hungarian imports against roubles not only affects the value of the rouble trade balance of a year earlier in the positive direction, but also affects the timelag value of the dollar trade balance, and in a negative direction. This latter correlation indicates that in the period under examination the import burdens of the deteriorating dollar trade balance were gradually taken over by socialist imports.

In the family of shortage economy models, an important question is that of the switchover between internal and external shortages. A phase of this, i.e. the effect of the internal tension of investment on foreign trade and the regulatory role of the latter has already been discussed in the treatment of investment models. In the following paragraphs, we focus on another switchover mechanism, i.e. the relation between internal stocks and foreign trade.

The stock model of Budavári-Náray (1982) investigates the motives of stockpiling and tries to reveal a relation between the stockbuilding cycles and foreign trade processes. Their hypothesis is that in the years following fast growth in investments, enterprises tie up their money reserves in stocks in advance, fearing central restrictive measures – which is again a consequence of the soft budget constraint of the socialist enterprise, a lack of the force of profitability, and the possibility of shifting on of costs. This enables enterprises to apply a restraint to the actual taxation perspectives of the central organs and they expect that in the emerging difficult situation of payments the Hungarian National Bank will be compelled to bear the burdens of a high stock level. Also, in the year of an expected restriction, accumulated material stock facilitates a reduction in material purchases, which makes it easier to bear the pressure exerted by the financial

[7] On this subject, see also the study of Suba (1984).

regulators. Stock fluctuation at the macrolevel is almost the mirror image of the foreign trade balance accounted in dollars. The relatively weak link with the balance of foreign trade with socialist countries can be explained by the rouble trade being fixed in long-term agreements.

Since stocks are more closely related to imports against dollars than to exports accounted in dollars, it is reasonable to assume that stocks are more likely to be a result of deliberate reserve-building than the consequences of sales difficulties.

Riecke (1982) examined stocks in Hungary be means of a simultaneous model.[8] He represents stockpiling as a negative reaction to earlier stocks and balances of foreign trade with the capitalist countries, besides GDP, thus demonstrating the behaviour of planners and economic policy-makers in shaping the balance of foreign trade (expressive of external tension), as already put forward in respect of the investment sphere. Together with the GDP, investments and consumption, stockpiling completed in this way shapes the balance of foreign trade with the capitalist countries. It is an interesting result of the author's computations that stockpiling in Hungary follows a constant cycle of 4.4 years.

The models briefly presented above all bear the marks that have been enumerated as characteristic of shortage economy models. As also in the investment sphere, macrotensions play the primary regulatory role in both foreign trade relations, because of the lack of profitability considerations at the microlevel. In the capitalist relation, the soft budget constraint of the enterprises generates an almost insatiable demand, the mitigation of which is largely left to informal regulation, and the extent to which it can be satisfied is regulated by the balance of foreign trade with the capitalist countries as a tension indicator. Moreover, in the socialist relation, largely apart from price effects and efficiency criteria, trade is pursued within the framework of long-term agreements, in which outstanding importance is attached to maintaining, in the short term, an even balance of foreign trade with the socialist countries.

In addition, the mechanisms of the various different tensions, mutually substitutive to a certain extent (dollar and rouble balance) and mutually intensifying (investments, stocks, balance of foreign trade with the capitalist countries), were also formulated in the models.

10.5 THE SHORTAGE MODELS OF CONSUMPTION

In the socialist economy, the investment and foreign trade spheres are characterized, as a consequence of the enterprises' soft budget constraint, by an extremely high demand for investments and imports, which is kept within limits by the consensus reached in the bargaining mechanism functioning between the enterprise and the central organs. In the consumption sphere, the budget constraint of households can be called hard: purchasing decisions are definitely influenced by

[8] Forerunners of W. Riecke's study: Chikán (1981, 1982) and Abel–Riecke (1981).

income and prices. In addition to these factors, however, shortages in different consumer goods, which weaken the effect of incomes and prices, are also of decisive importance. A good example is provided by the study of Dániel–Semjén (1987) investigating the effect of a shortage of state-owned dwellings at the microlevel demand for tenement dwellings. In their investigation these authors found a striking difference between two groups of consumers concerning their response to price rises. It emerged that the consumers unsatisfied with their housing position (i.e. those experiencing a strong perception of shortage), would respond to price rises in a different way from those satisfied with their housing position, the two groups being identical in other respects (income, family size etc.). From this phenomenon the authors came to the conclusion that where shortage exists the functional relationship between the price of and demand for flats is not uniform: because of the strong perception of shortage that can be experienced by a wide range of consumers, prices can increase over a long period without having an effect on demand.

Shortage influences demand in different ways: it obliges consumers to queue, whereas it also turns them away from the queue and decreases demand. Depending on the rates of mixture of the two opposing effects in the case of a given product, demand may increase or decrease, as shortage grows more intense. At the same time, if shortage grows too fast, it will also affect supply: economic policy-makers and planners will make efforts to ease the shortage. If, because of intensifying shortage of a certain product, demand decreases (i.e. the effect of turning consumers away from the queue is stronger than the effect of attracting them to it), and supply grows, this dual effect acts towards reducing shortage, which in turn leads to growing demand and a relative decrease in supply, whereupon shortage is reproduced. This mechanism was demonstrated by Lackó (1986) in respect of the demand for, and supply of, state-owned tenement-dwellings in Hungary.

Moreover, in the car market where the effect of shortage attracting consumers to the queue is strong, the deviation of the shortage from the usual extent functions as a feedback signal for both demand and supply. (See the study of Kapitány–Kornai–Szabó, 1982 in Chapter 15 of this volume.)

It follows from the preceding that the consumer demand function of the socialist economy undoubtedly changes compared with the traditional demand function. As was mentioned in the areas discussed previously, another modifying factor connected with shortage is the forced substitution that frequently occurs. Where shortage exists, if consumers cannot find the product they want on the market they have the following alternatives: (1) either they can spend their money on some other, similar product (forced substitution), or (2) they can save their money (forced saving). When considering the latter alternative, several researchers of the socialist economy apply a traditional macrocategory of saving for determining shortage accompanying consumption (Portes–Winter, 1980; Hulyák, 1985). Whereas in the shortage economy models the following hypotheses are latent:

1. Shortage may well be the cause of money left unspent (forced saving), but it does

not necessarily have this consequence, if forced substitution and forced spending are frequently occurring phenomena in the socialist economy.

2. The magnitude of the saving is shaped by different considerations, traditional and related to shortage (transactional, voluntary advance saving, obligatory advance saving, consumers in readiness, the building of safety reserves), which are almost impossible to separate, and it is just as impossible to demonstrate, at the macrolevel, forced saying owing to shortage.

It follows from these fundamental hypotheses that all that the shortage economy models can undertake is to show which factors – besides the role of incomes and prices – influence substitutions between the different large groups of products that are the subjects of consumption, and the changes in savings. These questions are treated in the studies of Lackó (1976) and Simon (1977); the latter is presented in detail below. Simon's (1977) econometric investigations led to the main conclusion that, besides incomes and prices, supply limits also play a part in shaping household consumption and savings in Hungary.

Firstly, the study examined the consumption of groups of products affected by a shortage in the supply of cars – a situation clearly prevailing in the period under examination. Changes in the intensity of the car shortage were approximated by a ratio, the numerator of which features private car supply (=car purchases of households), and the denominator the exp $(-36.79+4.325 \log \text{income})$ function. The author's assumption is that where prices are unchanged, the demand for cars grows at a rate proportional to incomes. According to his computations, the income elasticity of this demand is 4.325. If actual car purchases are greater than those computed according to this function, the car supply is relatively great, and free market prices fall; it they are lower, supply is relatively scarce, queues are, longer, and free market prices higher. On quantifying the function, he found that, of the groups of consumer goods, it was first of all in consumer durables that purchases reflected missed car purchases:

$$\begin{aligned}\log \text{TARTN} = & -\underset{(4.33)}{2.58} + \underset{(20.19)}{1.05}\,\text{JÖV} - \underset{(9.64)}{2.56}\,\text{PTARTN/PFOGY} \\ & -\underset{(4.25)}{0.22} \log \frac{\text{AUTO}}{\text{AUTOB}} \end{aligned} \qquad (10.7)$$

$$R^2 = 0.972 \qquad \text{SE} = 0.039 \qquad \text{DW} = 1.71$$

in which

TARTN = consumption of consumer durables at unchanged price
JÖV = real income
PTARTN/PFOGY = relative price shift in consumer durables
AUTO/AUTOB = households' car purchases (exp $-36.79 + 4.325 \log \text{JÖV}$)

According to the function, by 1975, for example, the substitution between consumer durables and cars had covered 20–40% of the fluctuations in the car supply.

Secondly, on examining household spendings on housing accumulation, credits, the car shortage indicator, and an indirect shortage indicator of private home building were taken into consideration besides the role of incomes and prices. The latter indicator is the increment of total national economic investment, in accordance with the hypothesis that national economic investments siphon off building materials and capacities from private home building activities to a different extent each year. As opposed to the similar indicator of market economies, changes in the credit supply are not dependent on changes in the interest rate, but are directly regulated by economic policy based on considerations of equilibrium.

The computations showed that, besides the credit and price variables, fluctuations in car shortage also cause changes in housing accumulation (by 1975, for example, about 40% of the 1975 shortage had been spent on housing accumulation), and furthermore, the hypothesis was confirmed that the changes in housing accumulation necessary for private home building move contrary to the increment in total national economic investment.

Thirdly, Simon (1977) also pointed out that housing accumulation costs are not replaced by consumer goods, but appear 100% in savings. Savings largely depend on households' housing accumulation, and thus indicate changes in the home building shortage. The estimated function is as follows:

$$\begin{aligned} \text{MEGTAKF} = \; & \underset{(4.98)}{0.33}(\text{JÖVF} - \text{JÖVF}_{-1}) + \underset{(5.61)}{0.62}\text{MEGTAKF}_{-1} \\ & - \underset{(3.51)}{3.55}\text{LAKVÁLT} - \underset{(1.24)}{51} \end{aligned} \tag{10.8}$$

$$R^2 = 0.963 \qquad \text{SE} = 0.73 \qquad \text{DW} = 1.099$$

in which

MEGTAKF = savings at current prices

JÖVF = income at current price

LAKVALT = the weighted average of the explanatory variables of housing accumulation, in which weights are the coefficients that figure in the equation of housing accumulation

This is in good agreement with the econometric results of Lackó (1984b) establishing the negative influence of fluctuating intensity in private home building, *ceteris paribus*, on savings.

She also referred the variations in savings to the annual growth rate of building material imports, an assumed building material shortage indicator. She quantified the consequential increase in savings upon a growing building material shortage. The forced savings content of savings is verified by the results of both Lackó and Simon.

10.6 TASKS FOR THE FUTURE

From this short review of the above-mentioned sectoral shortage models in Hungary the reader can clearly see that from the relationships given verbally, only a fraction could be verified by econometric methods as yet.

The emerging role of the second economy and the private sector are absolutely not covered in the models, nor are the resolution mechanisms induced by the said spheres to relieve shortages in the state sector. With respect to the state sector the presentation is limited to the connections of the development and reproduction of a few strains or shortages.

It must also be admitted that the models are not really advanced methodologically. This may be attributed partly to the theoretical considerations resulting from the specific features and institutional conditions of the socialist economy which stand against the simple and mechanical adoption of market economy models. Furthermore, econometric analysis of the functioning of the socialist economy is historically much younger than that of market economies.

Consequently, the tasks for the future are as follows:

1. Quantifying the strains and shortages that exist in the state sphere but are not quite or not at all specified, stating the interdependence of strains and shortages (namely, which are contingent upon one another and which are mutually substitutive); constructing the models of the aforesaid.
2. Including the secondary economy and the private sphere in the models.
3. Using acknowledged, advanced econometric methods for the model estimates and improving on them.

REFERENCES AND FURTHER READING

Abel, I. and Riecke, W. (1981) A model of aggregate inventory behaviour in the Hungarian economy, in *Economics and Management of Inventories* (ed. A. Chikán), Akadémiai Kiadó, Budapest.

Bauer, T. (1978) Investment cycles in planned economies. *Acta Oeconomica*, **21** (3), 243–60.

Bauer, T. (1981) *Tervgazdaság, beruházás, ciklusok*, Közgazdasági és Jogi Kiadó, Budapest.

Budavári, P. and Náray, L. (1982) A készletfelhalmozás és növekedés néhány összefüggése a magyar gazdaságban, Discussion Paper, Országos Terhivatal, Budapest.

Chikán, A. (1981) *The Economics and Management of Inventories*, Akadémiai Kiadó, Budapest.

Chikán, A. (1982) Inventory fluctuations in the Hungarian economy, Paper presented at the Second International Symposium of Inventories, Budapest.

Dániel, Zs. (1983) Modell a lakásváltoztatási szándékok és lehetőségek elemzésére. *Szigma*, **16** (1–2), 31–60.

Dániel, Zs. (1984) A lakbérek és a bérlakások iránti kereslet. *Közgazdasági Szemle*, **31** (3), 290–305.

Dániel, Zs. and Semjén, A. (1987) Housing shortage and rents: the Hungarian experience. *Economics of Planning* (to be published).

Dániel, Zs., Kornai, J. and Weibull, J. (1981) A Model Framework for Analysis and

Simulation of Housing Markets (*mimeo*), Department of Mathematics, Royal Institute of Technology, Stockholm.
Fabri, E. (1982) Superficial changes and deep tendencies in the inventory process in Hungary. *Acta Oeconomica*, **28** (1–2), 133–46.
Farkas, K. (1983) The changing role of enterprise inventories. *Eastern European Economics*, **21** (3–4), 49–66.
Fazekas, K. and Köllõ, J. (1985) Fluctuations of labour shortage and state intervention after 1968, in *Labour Market and Second Economy in Hungary* (eds. P. Galasi, and Gy. Sziráczky), Campus Verlag, Frankfurt/New York.
Gács, J. (1981) Importkorlátok, hiányjelenségek és a vállalati alkalmazkodás, Dissertatin, Budapest.
Gács, J. (1982) Passive purchasing behaviour and possibilities of adjustment in the Hungarian industry. *Acta Oeconomica*, **28** (3–4), 337–49.
Gács, J. (1983a) Planning and adjustment policy changes in the development of the building materials industry. *European Economics*, **21** (3), 105–25.
Gács, J. (1983b) A központi tervezésü gazdaságok külkereskedelmi magatartásának irodalnáról (*mimeo*), Konjunktura-és Piackutató Intézet, Budapest.
Gács, J. and Lackó, M. (1973) A study of planning behaviour at national economic level. *Economics of Planning*, **13**, 91–119.
Hewett, E. (1980) A macroeconometric model of a centrally planned economy with endogenous plans: the Hungarian case, Discussion Paper, University of Texas, Austin.
Hulyák, K. (1985) An econometric disequilibrium macromodel for Hungary, Paper presented at the World Congress of the Econometric Society, Cambridge, Mass.
Hunyadi, L. (1984) Osztott késleltetésü modellek elmélete és alkalmazása, Dissertation, Budapest.
Hunyadi, L. *et al.* (1980) *A rövidtávu tervezés ökonometriai modellje*, Közgazdasági és Jogi Kiadó, Budapest.
Kapitány, Zs., Kornai, J. and Szabó, J. (1983) The reproduction of shortage on the Hungarian car market. *Soviet Studies*, **36**, 236–56.
Kornai, J. (1971)*Anti-Equilibrium*, North-Holland, Amsterdam.
Kornai, J. (1980) *Economics of Shortage*, North-Holland, Amsterdam.
Kornai, J. (1982) *Growth, Shortage and Efficiency*, Basil Blackwell, Oxford.
Kornai, J. (1983) A lakosság fogyasztásának vizsgálata Almon ökonometriai modelljével (*mimeo*), Konjunkturaés Piackutató Intézet, Budapest.
Kornai, J. (1984) Descriptive–explanatory theoretical models of the Socialist economy: review of a research direction. *System Research*, **1** (2), 135–43.
Kornai, J. (1985) *Contradictions and Dilemmas*, Corvina, Budapest.
Kornai, J. (1986) The soft budget constraint. *Kyklos*, **39** (1), 3–30.
Kornai, J. and Martos, B. (1971) Gazdasági rendszerenk vegetative mukodese, *Szigma*, **4**, 34–50.
Kornai, J. and Martos, B. (1981) *Non-Price Control*, Akadémiai Kiadó, Budapest.
Kornai, J. and Matits, A. (1983) A koltsegvetesi korlat puhasagarol vallalati adatok alapjan, *Sazdasag*, **16**, 3–30.
Kornai, J. and Matits, Á. (1984) Softness of the budget constraint: an analysis relying on data firms. *Acta Oeconomica*, **32** (3–4), 223–49.
Kornai, J. and Simonovits, A. (1977) Decentralized control problems in Neumann economies. *Journal of Economic Theory*, **14**, 44–67.
Kornai, J. and Simonovits, A. (1985) Investment efficiency and shortage: a macro-growth model. *Matekon*, **22**, 3–29.
Kornai, J. and Weibull, J. (1978) The normal state of the market in a shortage economy: a queue model. *Scandinavian Journal of Economics*, **80**, 375–98.
Kornai, J. and Weibull, J. (1983) Paternalism, buyers' and sellers' market, *Mathematical Social Sciences*, **6**, 153–69.

Lackó, M. (1976) Consumer savings and the supply situation. *Acta Oeconomica*, **15** (3–4), 365–84.
Lackó, M. (1980) Cumulating and easing of tensions. *Acta Oeconomica*, **23** (3–4), 357–77.
Lackó, M. (1984a) Behavioural rules in the distribution of sectoral investments in Hungary. *Journal of Comparative Economics*, **8**, 290–300.
Lackó, M. (1984b) A hátrány megőrzése. *Közgazdasági Szemle*, **31** (3), 305–20.
Lackó, M. (1986) Feszültségek és normák szerepe and beruházások szabályozásában (*mimeo*), Közgazdasági Intézet, Budapest.
Laki, M. (1980) End-year rush-work in Hungarian industry and foreign trade. *Acta Oeconomica*, **25** (1–2), 37–65.
Laki, M. (1984) The enterprise crises. *Acta Oeconomica*, **32** (1–2), 113–24.
Lukács, P. (1984) Oktatástervezés és hiány (*mimeo*), Oktatáskutató Intézet, Budapest.
Major, I. (1982) Reproduction of tensions in transport in Hungary. *Economics of Planning*, **18**, 85–100.
Marrese, M. (1977) Hungarian investment fluctuations; a theoretical and econometric study of hierarchical decision-making, PhD dissertation, University of Pennsylvania.
Marrese, M. (1981) The bureaucratic response to economic fluctuation: an econometric investigation of Hungarian investment policies. *Journal of Policy Modelling*, **3**, 221–43.
Martos, B. (1985) Gazdasági szabályozási strukturák és müködésük: a nem walrasi eset, Ph. D. dissertation, Közgazdaságtudományi Intézet, Budapest.
Portes, R. and Winter, D. (1980) Disequilibrium estimates for consumption goods markets in centrally planned economies. *Review of Economic Studies*, **47** (1) 137–59.
Riecke, W. (1982) A simultaneous model of inventory accumulation and foreign trade in the Hungarian economy, Paper presented at Second International Symposium on Inventories, Budapest.
Simon, A. (1974) A magyar külkereskedelem vizsgálata ökonometriai módszerekkel. *Szigma*, **3**, 166–76.
Simon, A. (1977) A lakosság kiadásainak és megtakaritásainak vizsgálata ökonometriai módszerekkel. *Szigma*, **10**, pp. 249–64.
Simon, A. (1980) An econometric model of the Hungarian economy, Discussion Paper, Institute for Economic and Market Research, Budapest.
Simon, A. (1981) A magyarországi beruházások ciklikusságának egy modellje. *Közgazdasági Szemle*, **28**, 293–302.
Simonovits, A. (1981) A teljesen decentralizált szabályozás, Dissertation, Közgazdaságtudományi Intézet, Budapest.
Simonovits, A. (1982) Buffer stocks and naive expectations in a non-Walrasian dynamic macromodel: stability, cyclicity and chaos. *Scandinavian Journal of Economics*, **84** (4), 570–81.
Simonovits, A. (1985) Growth and allocation of tensions in an open Socialist economy: theoretical simulation, Discussion Paper, Institute of Economics of the Hungarian Academy of Sciences, Budapest.
Soós, K. A. (1975) Causes of investment fluctuations. *Eastern European Economics*, **14**, 25–36.
Soós, K. A. (1984) A propos the explanation of the shortage phenomenon: volume of demand and structural inelasticity. *Acta Oeconomica*, **33**, 305–20.
Soós, K. A. (1986) *Terv, kampány, pénz*, Kossuth Kiadó, Budapest.
Suba, J. (1984) Egy alkalmazkodási mechanizmus a külkereskedelemben. Forgalmi ciklusok Magyarországon, (*mimeo*), Közgazdaságtudományi Intézet, Budapest.
Vanous, J. (1980) An econometric model of intra-CMEA foreign trade, in *East-European Integration and East–West Trade* (eds P. Marer and J. M. Montias), Indiana University Press, Bloomington.
Varga, J. (1984) Tanteremhiány az általános iskolákban 1950–1980. *Valóság*, **4**, 15–30.

11

Disequilibrium modelling of consumption in the centrally planned economy

WOJCIECH CHAREMZA

1.1 INTRODUCTION

Consumption market modelling is, without doubt, the best developed area of disequilibrium analysis of centrally planned economies (CPEs). There are many more papers that estimate unbalanced consumption demand/supply equations and excess demand than there are studies of other sectors of a planned economy which frequently operate in a disequilibrium state, such as labour or investment. This special attention devoted to the consumer goods market by econometric modellers can be attributed to several factors. First, consumer markets in CPEs provide clear and ample evidence of disequilibrium, such as queues, formal or informal rationing, waiting lists, forced substitution, forced savings, etc. This suggests that the traditional equilibrium modelling is not of great use, especially for some chronically unbalanced markets like housing and meat. Second, statistical data for consumption analysis are relatively abundant, and of good quality, by CPE standards (but see section 11.4 for an assessment of the completeness and bias of consumption data). Finally, the practical importance of thorough empirical analysis of consumption (which includes the disequilibrium approach) cannot be overestimated. If decision-makers knew the extent to which demand exceeded or fell below the actual consumption level, then they could more easily develop a realistic policy of dealing with the problems of excess demand or supply. Given this, there always has been a strong demand for realistic information concerning imbalance in the consumption markets.

International Studies in Economic Modelling Series.
Models of Disequilibrium and Shortage in Centrally Planned Economies.
Edited by C. Davis and W. Charemza.
Published in 1989 by Chapman and Hall Ltd, 11 New Fetter Lane, London EC4P 4EE.

During the last decade, empirical papers on disequilibrium modelling of consumption markets in Czechoslovakia, Hungary and Poland have mushroomed. In the same period, however, the number of controversies in this field has increased. This is hardly surprising; at the early stage of development of a new research technique, there are usually premature attempts to apply it and to skip or avoid some of the unresolved problems. The controversies in disequilibrium consumption modelling have been due not only to overenthusiasm, but also to the uneven developments in economic theory and econometric computation techniques in recent years. A decade ago tight and unrealistic assumptions were often imposed on a model because of computational limitations. Today, technologically determined restrictions are less severe. Despite this progress, though, it appears that contemporary models developed and estimated in CMEA countries usually incorporate more restrictive assumptions than those built in countries where computer technology is advanced.

Disputes in disequilibrium consumption modelling are caused as well by differences in theoretical approaches (Chapters 2, 3 and 4 of this volume). Authors of some empirical papers claim that a state of permanent excess demand exists in the aggregate consumption markets in European CMEA countries. This assumption of 'chronic excess demand' or 'known excess demand' has been generally challenged by other analysts, who believe in 'testable excess demand'. Inevitably, empirical results generated by models based on these two approaches often appeared to be in conflict.

This chapter evaluates the econometric disequilibrium models of aggregate consumption markets in CPEs that reflect these two trends. In section 11.2 'chronic excess demand' models of consumption are reviewed and criticized. Section 11.3 provides a similar assessment of the 'testable excess demand' models. Finally, a new approach to disequilibrium consumption analysis is suggested in section 11.4, which involves the modelling of the interrelations between official and 'parallel' private (legal and underground) markets in CPEs. A model of the consumption sector in Poland is estimated and used to evaluate consumption excess demand in the period 1956–86.

11.2 FIRST DECADE: CHRONIC EXCESS DEMAND (DISEQUILIBRIUM INDICATORS) MODELS

Consumption disequilibrium models, for which a known (usually permanently positive) sign of excess demand is a priori assumed, have been developed primarily in the Eastern European countries. These models chronologically preceded the 'testable excess demand' disequilibrium consumption models, in which no a priori assumption about the excess demand was made.

Econometricians in Eastern Europe realized quite early that equilibrium type models were inadequate for analysing the consumption markets in CPEs. The first papers to tackle the problem of bias of consumption demand parameters in the case of insufficient supply were published in Poland almost thirty years ago (e.g. Pawlowski (1959), Kolupa (1961, 1965)). The main contributions of these initial studies were to identify the problem of bias and to point out the inappropriateness of the widespread equilibrium approach. However, the proposed estimation methods were of rather limited utility since they were based on strict assumptions about time invariant excess demand.

After these pioneering works, the disequilibrium issue was neglected for over a decade. In the 1970s the problem of modelling consumption demand—supply imbalances was reconsidered by W. Welfe (1973, 1976) and especially by Dyba (1973), but only modest progress was made in solving it. Also, consumption market disequilibrium was explicitly examined in the SOVMOD model of the CPEs built in the USA (Green and Higgins (1977)). The publication in *Econometrica* of Fair and Jaffee (1972) opened new perspectives. In this paper the canonical disequilibrium model with the minimum condition was introduced for the first time (Chapter 6). In subsequent years both the Fair and Jaffee (1972) quantitative disequilibrium model and related estimation techniques were intensively developed in Eastern Europe and energetically applied in consumption modelling. In its original version, the Fair–Jaffee quantitative method involves combining the deterministic price-adjustment equation for sticky prices (Chapter 6, section 6.2)

$$p_t = p_{t-1} + \alpha(Q_t^d - Q_t^s), \quad \alpha > 0 \tag{11.1}$$

(where p_t stands for an endogenous price in time t and Q_t^d, Q_t^s denote quantities demanded and supplied on the market), with the demand and supply equations. Following the notation used in Chapter 6, demand and supply equations can be formulated as:

$$\begin{aligned} Q_t^d &= \alpha_1 p_t + \beta_1' \mathbf{X}_{1t} + u_{1t} \\ Q_t^s &= \alpha_2 p_t + \beta_2' \mathbf{X}_{2t} + u_{2t} \end{aligned} \tag{11.2}$$

where $\mathbf{X}_{1t}, \mathbf{X}_{2t}$ are vectors of exogenous variables with structural parameters β_1, β_2, and u_{1t}, u_{2t} are error terms. Since the sign of excess demand is assumed to be known, the above three equations, together with the minimum condition

$$Q_t = \min\{Q_t^d, Q_t^s\} \tag{11.3}$$

can be written in a condensed form as

$$Q_t = \alpha_1 p_t + \beta_1' \mathbf{X}_{1t} - \frac{1}{\alpha} \Delta p_t^+ + u_{1t} \tag{11.4}$$

$$Q_t = \alpha_2 p_t + \beta_2' \mathbf{X}_{2t} + \frac{1}{\alpha} \Delta p_t^- + u_{2t} \tag{11.5}$$

where

$$\begin{aligned} \Delta p_t^+ &= \max\{0; p_t - p_{t-1}\} \\ \Delta p_t^- &= \min\{0; p_t - p_{t-1}\} \end{aligned}$$

In Equations 11.4 and 11.5 all the variables are observable and the model can be directly estimated by standard econometric techniques (for a Monte Carlo comparative evaluation of the various methods used to estimate a quantitative disequilibrium model, see Charemza (1981)).

The Fair–Jaffee 1972 article stimulated considerable further theoretical discussion. In particular Amemiya (1974) proposed several estimation methods for the quantitative model, Laffont and Garcia (1977) modified the model by allowing an asymmetric price adjustment mechanism, and Bowden (1978) proposed a model that was similar in its estimation properties, but better grounded in the neoclassical macroeconomic theory.

It is clear that the above model, with endogenous, freely negotiated prices and perfect information cannot be explicitly used for most of the CPE consumption markets(see the critical discussion of the model's more general form in section 6.2 of Chapter 6). Consequently, during the subsequent decade, efforts in consumption disequilibrium modelling in Eastern Europe were concentrated on the problem of finding appropriate observable variables which could be used in Equation 11.1 in place of the price.

The usually adopted solution to this problem was given in the first quantitative disequilibrium consumption model developed by Charemza and Gierusz (1978) (this is still the only such model for CPEs which uses quarterly data). It illustrates, in a clear way, the advantages and drawbacks of applying the Fair–Jaffee method in the analysis of CPEs. After some reformulation, the model can be presented in a three equation form, describing demand, supply, and excess demand in the retail sales market:

$$Q_t^d = Q^d(Q_{t-1}, Y_t) + u_{1t} \tag{11.6}$$

$$Q_t^s = Q^s(Q_{t-1}, X_i, IM_t) + u_{2t} \tag{11.7}$$

$$N_t = \gamma \frac{\Delta S_t}{\overline{Y}_t^{(2)}}, \quad \gamma > 0 \tag{11.8}$$

where Y_t is the households' money income, Q_t is the quantity transacted (sold and bought), X_t is the industrial output, IM_t stands for imports, ΔS_t is the increase in personal savings during a quarter ($\Delta S_t = S_t - S_{t-1}$), and $\overline{Y}_t$ is the average income of households over the last two quarters (i.e. $\overline{Y}_t = 1/2(Y_t + Y_{t-1})$). N_t is by definition equal to consumption excess demand. An additional assumption imposed is that demand cannot be smaller than the observed supply, i.e. $N_t \geqslant 0$ and $Q_t^s = Q_t$ for all t.

Provided that the demand and supply equations are linear, the rough estimation of the above model is straightforward. Equation (11.8) can be used to solve out the unobservable Q_t^d variable through the substitution $N_t = Q_t^d - Q_t^s$. This leads to

$$Q_t = Q^d(Q_{t-1}, Y_t) - \gamma \frac{\Delta S_t}{\overline{Y}_t^{(2)}} + u_{1t} \tag{11.9}$$

In Charemza and Gierusz (1978) Equations 11.7 and 11.9 were estimated by ordinary least squares using 56 observations from the period 1963–76. Statistically significant student t-ratios were obtained for all the estimated parameters and the determination coefficient had a high value.

Despite its computational attractiveness, the above procedure used has numerous flaws. First, the introduction of the 'variable' N_t directly to the model (i.e. putting $Q_t^d - Q_t^s$ on the left-hand side of Equation 11.8) makes at least one of the variables on its right-hand side endogenous, because changes in ΔS_t and/or $\bar{Y}_t$ are caused by variations in u_{1t} and/or u_{2t}. Secondly, the assumption of a deterministic excess demand equation is evidently too tight for the needs of empirical analysis. The model would generate estimates of consumption excess demand that are exactly proportional to $\Delta S_t/\bar{Y}_t$, and would reduce the problem to estimating the parameter γ.

A more suitable alternative to Equation 11.8 would for example be the following savings equation

$$S_t = S_{t-1} + \alpha(Q_t^d - Q_t^s)\bar{Y}_t + u_t \tag{11.10}$$

where $\alpha > 0$ and u_t is an error term. This would make the restrictive and disputable chronic excess demand assumption unnecessary, since the right-hand side of Equation 11.6 can now be explicitly substituted to Equation 11.10. Furthermore, the economic justification for the stochastic-type Equation 11.10 is stronger than for the deterministic Equation 11.8. If this alternative model is adopted, the estimated Equations 11.6, 11.7 and 11.10 can be utilized to simulate excess demand from a reduced form using all the parameters and variables.

It should be recognized, however, that when the original model was built ten years ago, estimation of such an alternative was not feasible, because the only estimation method available in Poland for models with more than 30 observations was ordinary least squares. The endogenous nature of S_t and the nonlinear form of Equation 11.10 mean that non-linear two stage least squares is the simplest technique that can be used to estimate the model consistently. Moreover, the problem of identification of the error terms variances remains, since the error term of the savings equation to be estimated would be equal to $u_t + \alpha\bar{Y}u_{1t}$. In such a situation pretending that $u_t \equiv 0$ is probably the worst solution, for it leaves us with a deterministic formula for consumption excess demand. The more reasonable approach consists of finding a proportional coefficient (for example the ratio of S_t to Q_t) that can be used *ex-post* to retrieve the unobservable u_{1t} and u_t from the compound residual of the estimated equation.

A number of problems with the Charemza–Gierusz retail sales model were caused mainly by the limitations of available computational capabilities. In subsequent years developments in computer hardware and software loosened these constraints. Despite this, many subsequent studies adhered to this increasingly outmoded approach.

W. Welfe (1978) introduced the concept paper in which the disequilibrium indicators explicitly introduced macroeconometric modelling of CPEs. It was claimed that it is usually possible to define a set of variables which directly determine (or are directly determined by) consumption excess demand. These variables (and often just one variable) are called *disequilibrium indicators* and can be used to solve out the unobservable demand (and/or possibly supply) variable.[1] If one uses this terminology, then the disequilibrium indicator in the original Fair–Jaffee model is the first difference of prices and in the Charemza–Gierusz model the ratio $\Delta S_t / \bar{Y}_t$.

In Charemza (1981) the theory of disequilibrium indicators was consolidated and generalized. Several types of disequilibrium indicators were distinguished, both deterministic and stochastic. Unfortunately, insufficient attention was paid to difficulties associated with this approach. In particular, problems of an exogenous or endogenous nature of the disequilibrium indicators remained undecided (see the review of that book: Klaus and Rudlovčák (1982a)). This especially resulted in a number of misunderstandings and confusions in subsequent works which utilized the known excess demand approach to disequilibrium consumption modelling.

The number of empirical studies that use deterministic disequilibrium indicators in the modelling of consumption excess demand has grown significantly. Michalak and Starzynska (1979) developed a disequilibrium model of the Polish food market in which excess demand was assumed to be proportional to the difference between the free market and official state prices for food (further developments of this model are presented in Starzynska (1986; forthcoming)). Gronicki and Szreder (1981, 1986) constructed a model of the television set market in Poland that expressed the excess of demand over supply as a function of changes in television inventories. Deterministic disequilibrium indicators have also been extensively used in formulating consumption blocks in large scale econometric models of CPEs (e.g. W. Welfe (1985), Romanski and Welfe (1986) and also Chapter 8). Czerwinski and Jurek (1984) appear to have extended the disequilibrium indicators concept to its limit by making the assumption that consumer excess demand could be fully explained by an exponential time trend.

For Hungary, Hulyak (1982) in a model of the car market assumed that the difference between demand and supply for cars was proportional to changes in the number of people on waiting lists. Her model of the entire durable goods market in Hungary used a Fair–Jaffee disequilibrium indicator of the first difference of prices. The subsequent model of the Hungarian economy (Hulyak, 1985), used a modified disequilibrium indicator of the first difference of total personal savings.

The common feature of the disequilibrium consumption models listed above is that they possess relatively good *ex-post* statistical diagnostics (high values of the coefficients of determination, good student *t*-ratios for the estimated parameters, etc). None of them, however, has been used for *ex-ante* forecasting or simulation exercises. This can be explained by the way of modelling excess demand adopted

[1]The phrase 'disequilibrium indicator' could, in its present context, confuse those familiar with the latent variable literature, where an 'indicator' stands for a proxy of effects rather than causes. Perhaps the better expression would be 'correlate' rather than 'indicator'. Nevertheless, to avoid more confusion, throughout this chapter the 'disequilibrium indicator' is used in its originally defined sense.

in these models. Models with deterministic, extraneously driven disequilibrium indicator do not permit well-defined *ex-ante* computation of excess demand. Their utility in the *ex-post* simulation of excess demand is also questionable. What is the point of building and estimating a complicated disequilibrium model if it is recognized in advance that consumption excess demand is entirely determined by one known variable?

The most recent significant application of the deterministic disequilibrium indicators approach is made in the consumption block of the simulation minimodel W6 of the Polish economy (Romanski and Welfe, 1986). This block consists of the following demand and supply equations (original notation has been changed):

$$Q_t^d = Q^d(YR_t + FSR_t, \Delta r_t) + u_{1t} \tag{11.11}$$

$$Q_t^s = Q_t = Q^s(X_t) + u_{2t} \tag{11.12}$$

where YR_t is real income, FSR_t stands for forced savings in real terms, (evaluated outside the model and assumed to be exogenous), Δr_t is a first difference of interest rate for the households' savings deposits (with the expected negative first derivative of demand with respect to interest) and X_t is industrial output.

The actual formulation of the consumption model using these equations is interesting in several respects. First, the function $Q^s(\cdot)$ uses various linear and multiplicative dummy variables, and therefore is no longer linear. In fact the supply equation becomes a switching parameters function. These switches are exogenously imposed on both the structural parameters and the constant term. Secondly, the specification of the consumption demand function is innovative, by adding forced savings to the total disposable real money income and by successfully applying interest on households' deposits as an explanatory variable. Nevertheless, such specification also has some disputable points. Description of the model does not indicate whether the interest rate is nominal or is in real terms. If the interest rate is real, and there are no forced savings, the consumption demand function is essentially that analysed by Gapinski (1982, pp 144–6), where introduction of the interest rate is derived from an intertemporal households' preference function; households could refrain from consumption at present, expecting consumption in the future due to the attractive interest rate. This might be the case only if they do not expect limitations of their future consumption. If the interest rate is nominal rather than real, additional restriction appears; the expected future open inflation rate has to be smaller than the interest rate. Consumption demand definition in the Romanski–Welfe model is, as in all disequilibrium models, in the Walras and not in the Dreze sense; it is not corrected downwards in case of an appearance of shortages. Consequently, if persistent shortages are expected by households for the future, a potential consumer is normally facing a dilemma: is it better to have goods at present and little money in the future (when there will be less goods to buy), or to have a lot of money in the future and few goods at present (and in future, due to the prospective rationing) instead? It is puzzling why he or she is choosing the latter. It seems that the presence of the interest rate in the consumption demand equation can be justified only in the rather unlikely case where consumption shortages are expected to decline and the interest rate is greater than the expected open inflation rate.[2]

The fact of adding forced savings to the households' real incomes gives rise to another query. If no savings at all are added to YR_t, one is left either with the Keynesian-type absolute income hypothesis (see Portes and Winter (1978) for its application to CPEs), with all its known limitations or – if the real interest rate is added as an explanatory variable – with the Gapinski-type intertemporal consumption demand function. If, on the other hand, stock of all personal savings from the previous period is added to the real income, we have a wealth-oriented consumption demand function (analysed originally by Ball and Drake (1964) and frequently used in disequilibrium econometrics, see for example Ito (1980), Charemza and Gronicki (1988)), pooled with the intertemporal interest rate approach. It is not, however, clear why one part of the total savings, namely forced savings, should affect consumption demand. This implies that the cumulative non-forced 'voluntary' savings cannot be regarded as a cause for intended perspective consumption according to the Romanski–Welfe model; people are not saving (at present) to consume (in future). Consequently, it is not clear why households keep voluntary savings. Using forced savings in the consumption demand function seems also to be in contradiction to the presence of the interest rate variable – there is no reason to refrain from consumption at present and saving money, even with attractive interest, if the saved and accumulated stock of money does not stimulate future consumpton demand.

If there is positive excess demand in the consumption market, demand function (11.11) is unidentified, since, according to the minimum condition, all the data lie on the supply curve. Romanski and Welfe (1986) have estimated the demand equation using an extraneously formulated disequilibrium indicator of the form

$$I_t = \left[\frac{\tilde{Q}_t^{\mathrm{d}}}{Q_t} - \frac{\tilde{Q}_{t'}^{\mathrm{d}}}{Q_{t'}}\right] Q_t \qquad (11.13)$$

where

$$\tilde{Q}_t^{\mathrm{d}} = \mathrm{YR}_t + \mathrm{FSR}_t)^{\alpha}$$

is an approximation of the consumption demand, t' stands for such time period t, for which the ratio $\tilde{D}_t^{\mathrm{d}}/Q_t$ was closest to unity, and α is a constant equal to 0.98. It has been shown by A. Welfe (1984) and A. Welfe and W. Welfe (1986) that if $\tilde{Q}_t^{\mathrm{d}}$ is a close approximation to Q_t^{d}, I_t is a close approximation of the consumption excess demand.

The above solution raises some additional difficulties in the context of the Romanski–Welfe model. First, it is not clear whether Equation 11.13 is a part of the model together with Equations 11.11 and 11.12 or not. From the original Romanski and Welfe (1986) paper, it can be deduced that Equation 11.13 is to be separated from the consumption demand and supply equations, nevertheless in the more

[2]In the model of the Polish economy presented in Chapter 8, W. Welfe treats the interest rate on savings deposits in a slightly different way.

general description of their attitude to disequilibrium modelling, A. Welfe and W. Welfe (1986) explicitly pooled this relation with the consumption demand equation. If Equation 11.13 is a part of the consumption model, then one of the variables in this equation – either YR_t or FSR_t – must be jointly dependent with Q_t^d, Q_t^s, since I_t is directly proportional to $Q_t^d - Q_t^s$. Despite being inconsistent in estimation (the estimation method applied by Romanski and Welfe (1986) was ordinary least squares), and, what is more important, diversive in *ex-ante* simulation, the entire model in which the relation in Equation 11.13 is included is also non-stationary, if there is even a slight autocorrelation in error terms. To explain this, let us start at time $t = 1$; error terms u_{11}, u_{21} are correlated with the disequilibrium indicator arguments (say, with FSR_1) and, in the case of autocorrelation, with the lagged error terms u_{10}, u_{20} and consequently with FSR_0. In time $t = 2$, u_{12}, u_{22} are correlated with FSR_2, FSR_1 and FSR_0, etc. Consequently covariance between u_{1t}, u_{2t} is not a function of a phase, as in conventional econometric models, but is a function of the time index itself.

Another query is connected with the applied definition of aggregate consumption demand, and especially excess demand. In spite of whether the I_t equation is included in the model or excluded, it seems that consumption demand is defined inconsistently since, according to Equation 11.11, $Q_t^d = Q^d(\cdot) + u_{1t}$, and simultaneously, following Equation 11.13, $Q_t^d = \tilde{Q}_t^d + \text{approximation error}$, while $Q_t^d = (YR_t + FSR_t)^{0.98}$. Evidently, the values of $Q_t^d(\cdot)$ can be far away from the values of $\tilde{Q}_t^d$. It is also not clear what is the definition of consumption excess demand and its relation to forced savings; usually forced savings are defined as a sum of inflationary overhang (Polish; nawis inflacyjny) and inflationary gap (Polish; luka inflacyjna) – see for example Herer and Sadowski (1981). If this is the case, FSR_t is nothing but the entire consumption excess demand, expressed in real terms (Nuti, 1986).

Despite the above reservations, the consumption models with the disequilibrium indicators of the type (11.13) have an important practical advantage; they gave sensible *ex-post* estimates of consumption excess demand for Poland (A. Welfe and W. Welfe (1986)).

In recent years some disequilibrium indicators of various types have been developed for the consumption goods markets. For analysing of the Polish economy in the period 1960–80, Charemza and Gronicki 1988) proposed a wage illusion mechanism, according to which workers are additionally paid for disutilities caused by the existence of consumption excess demand; this eventually leads to the wage equation

$$w_t = w(w_{t-\tau}, Q_t^d - Q_t^s) \tag{11.14}$$

where w_t is the nominal wage and τ is a time lag. The above equation has received a substantial critique, mainly stressing an absence in the equation of some relevant explanatory variables, especially labour productivity and cost of living index – see Chapter 8 and also A. Welfe and W. Welfe (1986).

The argument that wages might be a function of consumption prices and labour productivity is perfectly correct. Nevertheless, the question is whether this was really the case in Poland in the sixties and seventies. As far as labour productivity is concerned, there is little evidence that its increase had been substantial enough to be regarded as a statistically significant cause for wage increases. Empirical results are in that case not conclusive. However, there is much stronger support for the low (even zero or negative) aggregate labour productivity growth hypothesis in Poland, 1960–80, (e.g. Gomulka, 1986, pp. 96–7; Kemme, 1984; Klacek and Nesporova, 1985; Nasilowski, 1980) than for its alternative (Krzysztofiak, 1982).

Moreover, even in periods when an increase in labour productivity really took place, it was likely to be highly correlated with the consumption excess demand and thus to be statistically untestable. For instance, it would be interesting to consider a situation where increase of labour productivity would raise production mainly in the investment sector, leaving the level of consumption supply, expressed in real terms, virtually unchanged. Hence, wages would rise and consumption excess demand would immediately rise by a similar amount. This seems to be the likely case for Poland in the seventies, where excessive heavy industry, metallurgy and chemical plants projects had been undertaken. Obviously, long-run effects of such a policy on consumption excess demand depends on reality and the effectiveness of the completed projects; as we have learned from the recent economic history of Poland, realism was not a prime cause in that case.

The problem of the consumers' price index as the second potentially omitted variable in Equation 11.14, is more complex. Let us assume that, other things being constant, wages can be expressed as a function of either prices, or consumption excess demand, or both. In the first case, where price increases enforce some kind of wage indexation, the dynamic tendency is towards a long-run disequilibrium; increase of nominal wages causes increase of consumption excess demand, preceded by its decrease, due to substitution of the repressed inflation by open inflation. If there is a feedback from wages to prices in a long-run, there is a tendency to stabilize consumption disequilibrium over time, leaving us with an analogy to Kornai's concept of normal shortage.

In the second case, where the quantity constraints on the consumption market cause wage changes (i.e. where the wage illusion effect takes place), a causal relation from wages to prices has a stabilizing effect, since repressed inflation is substituted in time by open inflation. Hence, consumption disequilibrium would not be unstable, at least in the medium term. If, however, there is an inverse causality, from prices to wages, the wage illusion mechanism is essentially unstable.

Thirdly, wage changes might be jointly affected by consumption excess demand and consumption prices (as is pointed out by A. Welfe and W. Welfe (1986)). In that case there is by definition a causal relation from prices to wages (wage indexation); the entire mechanism of wage formation is unstable, unless there is a feedback from wages to prices. Besides, the concept of the wage equation with both consumption excess demand and prices as explanatory variables seems to be statistically awkward; a positive effect of both variables on wages is in immediate contradiction with the negative trade-off between repressed and open inflation.

It can be concluded from the above, that the nature of causal relations between consumers' prices and wages could reside in formulating the one- or two-factor wage equation. This would also help in answering the question of its stability. Generally, a causal relation from prices to wages is destabilizing, and from wages to prices is stabilizing for the consumption market. In particular, if prices do cause changes in wages and there is a wage illusion effect, the system is unstable; if wages do cause changes in consumption prices and there is a wage illusion, the market is in the state of a stable disequilibrium, being an analogy to normal shortage.

To find out the nature of price–wage relations in Poland, an appropriate causality test has been computed. The Granger definition of causality has been used; a variable y is understood to be a cause for x (i.e. $y \rightarrow x$), if present x can be predicted with better accuracy with the use of past y's, rather than by not doing so (e.g. Harvey, 1981, p. 302). A test for Granger causality has been run for various possible definitions of wages and prices; in particular wage means alternatively nominal official wages in state sector (ws), net labour earnings per employee in the state and private sector (wa); price means the official consumers' price index (ps) and an aggregate index for the official and secondary markets (pa), calculated by pooling the official price index and index for black market price of US dollars (for further details of its calculation see Charemza (in preparation), and also section 11.4). The applied test is the original Granger (1969) one, which in a number of Monte Carlo experiments has been found as relatively most powerful against a number of alternatives (Guilkey and Salemi, 1982). The procedure consists in testing for $y_{t-\tau}$, $\tau = 1, 2, \ldots$ as for the omitted variables in an ordinary least squares regression of x_t on the lagged x's and a time trend. As a diagnostic, the Lagrange multiplier test has been used in its two forms: as a modified x^2 statistic ($LM^\dagger$), having $x^2(2)$ distribution under the null hypothesis that there is no Granger causality from y to x, where one- and two-year lags of y's have been used (Harvey, 1981, p. 174), and as an F statistic (LMF, see Kiviet 1986), with $F(2, T-6)$ distribution for the same null hypothesis, where T is the sample size. The test has been repeatedly run for Poland, for the period 1960–80, i.e. for which Charemza and Gronicki (1988) claimed the wage illusion effect took place, and for 1960–86. The results are given in Table 11.1; the statistics significant at the 0.01 significance level are marked by ***, those significant at the 0.05 level by ** and at 0.10 level by *.

It can be seen that a dynamic feedback is clearly detected for wages earned in the state sector and the official consumer's prices. It would have been the evidence against a wage illusion mechanism, if the official consumption market only were considered and wages were indeed equal to the basic pay in the state sector. In that case it seems that in the medium and long run trade-off between prices and wages could have had the role of keeping the consumption disequilibrium close to its normal level, without leading to a substantial wage illusion effect. Nevertheless, in the Charemza and Gronicki (1988) model, total (officially reported) earnings in both private and state sectors have been used in defining wages. Hence, for the period 1960–80, data in Table 11.1 shows strong support for the Granger causality of the type: wage $\rightarrow$ price rather than price $\rightarrow$ wage. This seems to be consistent

Table 11.1 Granger causality test for consumer's prices and wages in Poland

	1960–80		1960–86	
	$LM^{\dagger}$	LMF	$LM^{\dagger}$	LMF
ws → ps	12.53 ***	4.48 **	33.89 ***	13.18 ***
wa → ps	27.02 ***	9.72 ***	36.16 ***	14.06 ***
wa → pa	16.73 ***	5.98 **	5.17 *	2.01
ps → ws	26.59 ***	9.50 ***	18.36 ***	7.14 ***
ps → wa	4.28	1.53	20.15 ***	7.83 ***
pa → wa	1.54	0.55	2.34	0.91

with the Charemza–Gronicki consumption model: during the sixties and seventies excess demand in the consumption market caused the increase in labour earnings not by increasing its basic level, but rather by adding numerous supplements to it, which in turn allowed for spilling the monetary overhang, accumulated by households, over the private and black (grey) markets. It must be stressed that for the entire 1960–86 period this type of causality has not been confirmed. It seems for the post-1980 era that the clear picture of the wage illusion effect has been broken; huge open inflation on the state market associated with substantial basic pay remuneration speaks in favour of an almost immediate price–wage feedback.

The answer to the question of future prospects for the disequilibrium indicators approach to consumption modelling in CPEs is not straightforward. On one side, the a priori admitted assumptions about mechanisms generating excess demand are particularly tight and not easily testable. In addition, for the models already estimated some other assumptions usually have to be made before any simulation and forecasting experiments can be done (Charemza, 1987). Also, controversy may easily arise between a realistic formulation of a disequilibrium indicator and its awkward stochastic structure. On the other side, however, it seems that even in the present era of rapidly growing computational possibilities the only feasible way of introducing consumption disequilibrium to large econometric models would be through disequilibrium indicators.

It is expected that for the next decade disequilibrium indicators will not be generally forgotten. Their future would be likely to be associated with the development of an expectations approach to disequilibrium modelling, in which the rigorous assumption about the proportionality between consumption excess demand and a given set of observable variables would be replaced by a more flexible mechanism of adjusting households' expectations towards the demand restrictions.

11.3 FIRST DECADE: TESTABLE EXCESS DEMAND MODELS

Development of the consumption disequilibrium models, in which no explicit or hidden assumptions about the sign or size of excess demand has been a priori assumed, took place at a similar time as the development of disequilibrium indicator models. Nevertheless, there is an interesting geographical difference – the former have been elaborated in Czechoslovakia, Great Britain and the USA, the latter in Hungary and Poland. This seems clearly to be caused by computer software availability; non-linear maximum likelihood estimation packages required for estimation of the models with unknown excess demand were for a long time available only in Western Europe and Czechoslovakia.

While disequilibrium indicator models originated from the quantitative Fair and Jaffee (1972) model, unknown excess demand models benefited mainly from Maddala and Nelson (1975) and Goldfeld and Quandt (1975) in which foundations for estimating a canonical disequilibrium model were formulated. As shown in Chapter 6, the canonical model (often called in the literature a generalized model) consists of demand and supply equations together with the minimum condition, without any additional information about excess demand. Such a model can be sensibly estimated by a non-linear maximum likelihood method, where the unobserved parts of demand or supply (whichever is the greater) are integrated out.

For modelling consumption markets in CPEs, the above method was first applied by Portes and Winter (1980). They constructed a consumption disequilibrium model with a Houthakker–Taylor (1970, pp. 13–17) type of demand equation, a supply equation, where consumption supply is a non-linear function of material product, households' financial assets, agricultural output and investment plus defence expenditures, and a minimum condition. The model was estimated for Czechoslovakia, GDR, Hungary and Poland by the Maddala–Nelson maximum likelihood method. This was the first model explicitly used for estimation of the consumption excess demand for the period from mid- to late fifties (varying for different countries) to 1975. The numerical results were quite surprising; the estimated amount of consumption excess demand, measured in percentage of the quantity transacted, was never higher than 3% and often reached negative levels, as low as −9%. These findings sharply contradicted numerous theories of positive consumption excess demand prevailing in CPE markets (see for example Beksiak and Libura, 1974; Barro and Grossman, 1974; Bauer, 1978; Kornai, 1971, 1980; Zwass, 1979; for a more detailed discussion and review see Pindak, 1983). Consequently, the findings of Portes and Winter were disputed, their estimates of excess demand were questioned and described as unrealistic (see Kornai, 1982, p. 35; Kemme and Winiecki, 1984; Winiecki, 1985; Nuti, 1986; Podkaminer, 1986). Portes (1986) response to these critiques is presented in detail in Chapter 2 of this volume.

Moving away from the discussion about the theoretical formulations of the Portes theory (see Charemza and Gronicki (1988) and Charemza and Dlouhý (1987) in which the consistency and theoretical validity of its foundations have been confirmed), it seems that surprisingly low consumption excess demand estimates resulted from an imperfect econometric tool applied for estimation and, what is clearly related to this, from using an equilibrium type demand function, unidentified in the process of maximum likelihood estimation from excess demand conditions. As far as the estimation technique is concerned it is easy to show the following shortcomings of the original Maddala–Nelson maximum likelihood method:

1. The likelihood function is unbounded, which necessitates the implementation of additional restrictions for the parameters (see Goldfeld and Quandt, 1975; Maddala, 1986, p. 1643).
2. In the case of permanent excess demand the model is not identified with respect to parameters (see Chapter 6).
3. If the probability of positive (or negative) excess demand is equal to unity, the conventional assumptions for the error terms as being normally distributed cannot be maintained (the conditional probability distributions for demand and supply are truncated).

The above reservations are not, in practice, absolutely essential and are relevant only in extreme cases. It seems that the more important shortcoming of the Portes and Winter model is that the Houthakker–Taylor consumption demand function is of an equilibrium (or excess supply) type and its application in the case of positive excess demand develops a problem of structural equivalence. It can be shown that the Houthakker–Taylor consumption demand function has, in a state of positive excess demand, a structure similar to that of an intertemporal budget constraint, being a linearized accounting identity (for a full definition of the accounting identity in a context of equilibrium relationships see Spanos (1982)). Let us rewrite the Portes and Winter consumption demand equation in the form:

$$Q_t^{\mathrm{d}} = \alpha_1 \Delta S_{t-1} + \alpha_2 \Delta Y_t + \alpha_3 Y_{t-1} + u_{1t} \tag{11.15}$$

where the symbols have the same meanings as those used in section 11.2, especially S_t standing for personal savings and Y_t for personal incomes. It is expected that $-1 < \alpha_1 < -1/3, 0 < \alpha_2 < 1, \alpha_3 = 1$.

Suppose that Equation 11.15 is a true relation, without misspecification error and the only problem consists in estimating its structural parameters. Independently from Equation 11.15, a budget constraint

$$Q_t = Y_t - \Delta S_t,$$

where Q_t stands for the observed consumption expenditures, can be represented in its intertemporal form as:

$$Q_t = A_{1t} \Delta S_{t-1} + A_{2t} \Delta Y_t + Y_{t-1}$$

where

$$A_{1t} = -\frac{\Delta S_t - \Delta S_{t-1}}{\Delta S_{t-1}}, \quad A_{2t} = \frac{\Delta Y_t - \Delta S_{t-1}}{\Delta Y_t}$$

As the accounting identity derived from the budget constraint the above relation is always true, no matter whether Q_t is endogenous or exogenous, or whether savings are forced or voluntary. It can also be expressed in an error-in parameters form, i.e.

$$Q_t = a_1 \Delta S_{t-1} + a_2 \Delta Y_t + Y_{t-1} + \text{error} \tag{11.16}$$

and $A_{1t} = a_1 + \xi_t$, $A_{2t} = a_2 + \eta_t$, where ξ_t, η_t are approximation errors. For relatively smooth data a_{1t} is likely to be negative and not smaller than -1, while a_2 is likely to be positive and not greater than 1. Hence, any reasonable method of fitting the constants a_1 and a_2 in (11.16), for example ordinary least squares, could give a high goodness of fit and significant student's t-ratios for the estimated constants. Generally, the relation 11.16 has nothing in common with a behavioural econometric model; it is just a linearized budget constraint. Nevertheless, if for some t the analysed market is in equilibrium or in an excess supply regime, i.e. $Q_t^d = Q_t$, Equation 11.16 becomes a Houthakker–Taylor consumption demand function, which can easily be interpreted and the estimates of a_1, a_2 express some kind of household behaviour. If, however, for some other t, $Q_t^d > Q_t$, the relation 11.16 still holds and is easily estimable but its parameters have nothing in common with a consumption demand function (Equation 11.15). If one attempts in this situation to estimate as part of a disequilibrium model, its structure cannot be identified from the structure of Equation 11.16. In other words, the estimation of a model consisting of Equation 11.15 and a supply function may lead to numerically equivalent results as estimation of a model containing the same supply function and the linearized intertemporal budget constraint (Equation 11.16). This provides an explanation for the surprisingly low estimates of positive consumption excess demand obtained by Portes and Winter; i.e. what was estimated in the case where $Q_t^d > Q_t$ was Equation 11.16 rather than 11.15.

It is worth noting that prospects for identification of a disequilibrium Houthakker–Taylor structure are improved if some type of disequilibrium indicator is combined with Equation 11.15. If, for instance, there is an equation for a variable y_t which can be added to the model in the form

$$y_t = \alpha_y (Q_t^d - Q_t^s) + v_t$$

where $\alpha_y \lesseqgtr 0$ and v_t is an error term, then for $Q_t^d > Q_t^s = Q_t$ the condensed equation

$$Q_t = \alpha_1 \Delta S_{-1} + \alpha_2 \Delta Y_t + \alpha_3 Y_{t-1} - \frac{1}{\alpha_y} y_t + u_{1t} - \frac{1}{\alpha_y} v_t$$

has a different structure from Equation 11.16 and is therefore identifiable (apart from the error term variances – see Chapter 6).

The canonical disequilibrium model described above has been used repeatedly for consumption modelling in Czechoslovakia. Dlouhý (1981) published a seminal paper in Eastern European literature on the fundamental ideas of disequilibrium econometrics. He formulated and estimated a canonical-type model for the meat market in Prague, 1960–78. In this model consumption demand was assumed to be a function of disposable income and prices, while the supply was determined by time trend. Subsequently, Klaus and Rudlovčák (1982b) presented estimates of an aggregate canonical demand/supply model for the entire Czechoslovak consumption market. The ideas contained in their consumption demand function are similar to those proposed by Davidson *et al.* (1978), consumption demand being a positive function of current income and negative of its first differences. The main difference is in the quite justifiable dropping of lagged consumption from the set of explanatory variables (which saves the task of dealing with the complicated dynamics of disequilibrium models with lagged endogenous variables) and by using levels rather than differences of the variables in logarithms. The Klaus and Rudlovčák supply function takes as explanatory variables the deliveries for the retail trade increased by a difference between the observed and expected level of inventories (which was calculated outside the model). As in the case of the original Portes and Winter model, it seems that an important shortcoming of the model is in the unidentified structure of the demand equation in the case of positive excess demand. A little algebraic exercise would transfer a budget constraint into a formula similar to that of Equation 11.16, but with current rather than lagged income, which is unidentified in the case of positive excess demand from the Klaus and Rudlovcak demand function. Formally, the Davidson *et al.* consumption model is again of the equilibrium type, which can be interpreted only when all existing savings are voluntary. If they are not voluntary, there is no reason for the households to try to retain the permanent consumption level, so that the increase of income would increase their savings ratio. Nevertheless the use of the equilibrium-type consumption demand function is entirely consistent with the basic assumptions of Klaus and Rudlovčák, who claim that in the period investigated, 1967–81, in the Czechoslovak consumption market there was very little (if any) imbalance. Not surprisingly, their estimates of the probability of positive excess demand are generally closer to zero rather than to unity (estimates of excess demand were not reported). They are also significantly different from those calculated for Czechoslovakia by Portes and Winter.

The next attempt to estimate a canonical-type disequilibrium consumption model for the retail sales market in Czechoslovakia has been done by Viktorinova (1986). In her model the Houthakker–Taylor consumption demand function has been used again, nevertheless she allowed for the Deaton (1978) unanticipatory inflation effect, adding the consumer's price index as the additional explanatory variable. Consumption supply is assumed to be a function of production output, import and lagged retail sales. By adding the price variable, the prospects for structural identification have been improved; nevertheless it seems hard to believe that the difference between consumption demand and the equilibrium structure of the type in Equation 11.16 is caused by consumers refraining from buying goods due to unanticipated inflation. The specially long period of flat consumption prices would substantively bias the estimates of the structural coefficients towards its equilibrium equivalents.

While in Czechoslovakia the original canonical model has been repeatedly estimated, Portes and his associates have made substantial progress in its development. In a series of papers Portes *et al.* (1983, 1984, 1987) have gradually developed a more complex model. Although the Houthakker–Taylor consumption demand has been retained, the consumption supply function has been modified; it has been formulated in the form of deviations from planned consumption rather than of deviations from a second-order exponential trend, as had been the case in the Portes and Winter (1980) model. What is more important, the model has been extended by including a plan-adjustment equation (see Chapter 6), according to which the consumption plan is a function of consumption excess demand and some other variables. Since the unobservable demand, (in the case of positive disequilibrium) appears in this model in an equation for plans, there is no danger of structural unidentification; the reduced form of the demand equation now has nothing in common with the intertemporal budget constraint (Equation 11.16). Nevertheless, the reported estimates of the consumption excess demand for Poland, 1975–80, again look biased downwards and haphazard (for example, the highest estimated positive excess demand is supposed to be equal to 7% in 1972, with negative estimates for 1976–78 and for the crisis year 1980 not exceeding 1.2%). This is more difficult to explain, especially because the statistical diagnostics for the estimated model are satisfactory. Nevertheless, in the case of Poland there is some evidence against using consumption plans as an endogenous variable with respect to consumption output. Many have argued that consumption planning has not been an essential task for planners in Poland, especially in the sixties and seventies. Consumption supply has often been regarded as being residual subject to material production and investment (Green and Higgins, 1977; Libura, 1979; Balicki, 1983; Charemza and Gronicki, 1988), and consequently the consumption plans, although officially announced, are either spurious, which under- or overfulfillment not resulting from a planner's loss function, or caused by investment rather than planning targets.

The above hypothesis seems to be confirmed empirically by results of exogeneity analysis for the consumption plan-output relationships in Poland, 1958–86 (Charemza and Kiraly, 1988). According to their findings Poland can be regarded as strongly exogenous (in the Engle, Hendry and Richard (1983) sense), rather than endogenous, although this finding is not definite due to the detected structural instability of the plans-output relation. Hence, admitting that consumption plans are caused by consumption output, as is the case in the Portes *et al.* (1987) plan-adjustment equation, is questionable.

An additional step forward in the development of the original Portes and Winter (1980) model has been made by Burkett (1988). He relaxed the disputable minimum condition of the canonical disequilibrium model. With demand and supply functions essentially similar to those of Portes and Winter (i.e. without the plan-adjustment), he has replaced the minimum condition with the more realistic assumption that the quantity actually transacted on the aggregate consumption market is lower than the minimum of demand and supply, due to the existence of slacks (unsaleable supplies) simultaneously with positive excess demand. He has used a simple hyperbolic approximation for the relation between slacks and shortages. The Burkett model substantially closes the gap between Kornai's economics of shortage and recent disequilibrium econometrics. The imposed hyperbolic relation between the slack and shortages can be written in the form

$$\frac{\hat{Q}_t^{\mathrm{d}} - Q_t^{\mathrm{a}}}{\hat{Q}_t^{\mathrm{d}}} \; \frac{\hat{Q}_t^{\mathrm{s}} - Q_t^{\mathrm{a}}}{\hat{Q}_t^{\mathrm{s}}} = \gamma' z_t, \tag{11.17}$$

where $\hat{Q}_t^{\mathrm{d}}$, $\hat{Q}_t^{\mathrm{s}}$ stand for linear combinations of explanatory variables of the demand and supply equations with their parameters, i.e.

$$\hat{Q}_t^{\mathrm{d}} = \alpha' X_{1t}, \quad \hat{Q}_t^{\mathrm{s}} = \beta' X_{2t}$$

X_{1t}, X_{2t}, z_t are sets of variables with the corresponding vectors of parameters α, β and γ, and Q_t^{a} is the so-called agreed transaction, equal to the quantity observed Q_t plus an error term u_t. Solving Equation 11.17 with respect to Q_t^{a} and then for Q_t gives

$$Q_t = \tfrac{1}{2}(\hat{Q}_t^{\mathrm{d}} + \hat{Q}_t^{\mathrm{s}}) - \tfrac{1}{2}[(\hat{Q}_t^{\mathrm{d}} - \hat{Q}_t^{\mathrm{s}})^2 + 4\gamma' z_t \hat{Q}_t^{\mathrm{d}} \hat{Q}_t^{\mathrm{s}}]^{1/2} + u_t \tag{11.18}$$

for $Q_t < Q_t^{\mathrm{d}}$ and $Q_t < Q_t^{\mathrm{s}}$ (u_t is an error term).

In the empirical part of the paper Burkett (1988) admitted that $z_t \equiv 1$ and γ is a scalar. Hence, parameters α, β and γ can be easily estimated in Equation 11.18 by non-linear least squares or a maximum likelihood method. With the use of these estimates, aggregate shortages and slacks can easily be computed. In further development of this model, it would be interesting to endogenize z_i, which should reflect asymmetric changes in consumption shortages and slacks. Although the assumption that slacks are proportional to the consumption supply seems reasonable, it is not likely to be the case for shortages. A straightforward modification of Equation 11.17 consists in allowing the right-hand side factor to vary according to changes in the demand/supply ratio. If in Equation 11.17 $\gamma' z_t$ is substituted by $\gamma(\hat{Q}_t^{\mathrm{d}}/\hat{Q}_t^{\mathrm{s}})$, it gives

$$Q_t = \tfrac{1}{2}(\hat{Q}_t^{\mathrm{d}} + \hat{Q}_t^{\mathrm{s}}) - \tfrac{1}{2}[(\hat{Q}_t^{\mathrm{d}} - \hat{Q}_t^{\mathrm{s}})^2 + 4\gamma(\hat{Q}_t^{\mathrm{d}})^2]^{1/2} + u_t,$$

which seems to be numerically as tractable as Equation 11.18.

The Burkett-type modelling immediately reached a new area when Portes and Santorum (1987) applied this for building a disequilibrium model of a consumption market in China. They have applied two models, Portes and Winter (1980) and Burkett (1988), receiving in both cases small estimates of excess demand and shortages (especially the Burkett-type estimates of shortages which were for most of the 1955–83 period smaller than 0.1%).

In the above survey very little has been said about the specification of the consumption supply function. This is mainly due to a tendency to refrain from going deeply into details of the supply side of the market by authors of particular models. Generally, two approaches in the specification of the consumption supply function can be distinguished. The first approach, associated mainly with the disequilibrium indicators mainstream, regards consumption supply either as being exogenously given or derived from a deterministic identity as the difference between material product and investments (taken together with the other components of material product such as government expenditures and foreign trade balance). The second approach uses an explicitly formulated stochastic supply function; this is applied mainly for the testable excess demand type models, which require full stochastic specification of both the demand and supply functions. Nevertheless, some justification has been given for these specifications of consumption supply functions. Almost all the authors stressed that the supply function is *ad-hoc*, preliminary, etc. Bearing this in mind it seems to be too early for a systematic theory of the consumption supply function (Pickersgill, 1980).

11.4 AN ALTERNATIVE: MODEL OF THE OFFICIAL AND SECOND ECONOMY

The last two sections of this chapter indicate that serious problems exist in virtually all the disequilibrium consumption models for CPE's under review, also those of the author. The deficiencies include questionable specification of the stochastic properties of these models and unrealistic estimates of consumption excess demand. In consequence, doubts have arisen about the prospects for disequilibrium econometric modelling of consumption in CPEs.

However, research by Nuti (1986, see also Chapter 5) has revealed a promising new approach to the empirical analysis of the CPE consumption markets. In his paper he pointed out the necessity of distinguishing between two heterogenous components of the consumption market: an official market with exogenously fixed prices and the secondary market, with freely negotiated equilibrium prices. The concepts of second economy and secondary markets are used hereafter in their broad sense of meaning, to describe all types of economic activity, legal and illegal, taking place outside the official state and cooperative sectors. For a more detailed definition and description of the particular markets making up the second economy in CPEs see for example Grossman (1977) and Katsenelinboigen (1977). Empirical analyses of secondary markets in the European CPEs have been carried out by Starzec (1983), Galasi and Sziraczki (1985), and Wisniewski (1985). There are various interrelations between both types of markets. In particular a spillover effect of the unsatisfied demand from the state to the secondary market is likely to appear. This insight provides the foundation for the formulation of a new estimable disequilibrium econometric model of consumption.

To derive a relatively simple model of this type, let us assume that all the households intend, in the first instance, to purchase goods on the official market. The households are divided into those earning in the state sector and directing their consumption demand primarily to the first sector (their income is denoted by Y_{1t} and their money stock at the end of the previous year is M_{1t-1}) and those earning in the secondary economy (with Y_{2t} income and M_{2t-1} money stock respectively), and interested in buying commodities on both the official and secondary markets. For a standard Cobb–Douglas type of household utility function the consumption demand function for the official (state) sector (Q^d_{1t}) can be written as

$$Q^d_{1t} = \underset{(+)}{\alpha} (M_{t-1} + Y_t) + u_{1t} \tag{11.19}$$

where $M_{t-1} = M_{1t-1} + M_{2t-1}$, $Y_t = Y_{1t} + Y_{2t}$, u_{1t} is an error term and a plus in brackets below a structural parameter denotes its expected sign.[3] Since the secondary household also intends to spend money in the secondary market, this gives

$$\tilde{Q}^d_{2t} = \underset{(+)}{\beta_1} (M_{2t-1} + Y_{2t})$$

where $\tilde{Q}^d_{2t}$ is the notional demand in the secondary market. The significant restriction of the above model is that it assumes an institutional barrier which prevents, in case of equilibrium primary households from their entry into the secondary market. This assumption reflects the necessary condition for identification of the parameters.

Next, let us assume that there is a disequilibrium on the primary market, say in the form of positive excess demand. This could spillover from the primary to the secondary market as a forced substitution and open inflation effect (for a discussion of the multiplier effect of the repressed to open inflation see Brus and Laski (1983)). This is specially evident where the price on the secondary market, p_{2t} is a free-market endogenous price and p_{1t}, the official price, is exogenously fixed. This leads to the effective demand equation

$$Q^d_{2t} = \underset{(+)}{\beta_1}(M_{2t-1} + Y_{2t}) + \underset{(+)}{\beta_2}(Q^d_{1t} - Q_{1t}) + \underset{(-)}{\beta_3} p_{2t} + \underset{(+)}{\beta_4} p_{1t} + u_{2t} \tag{11.20}$$

where Q_{1t} is the quantity transacted on the first market and u_{2t} is the error term.[4]

The model is closed by a consumption supply equation for the secondary market. Since the theory of consumption supply functions in CPEs is still relatively underdeveloped (see the discussion in section 11.3) three alternative specifications have been used. These consumption supply functions are solved with respect to p_{2t}, the remaining endogenous variable of the system.

$$p_{2t} = \underset{(+)}{\gamma_1} Q^s_{2t} + \underset{(+,-)}{\gamma_2} p_{1t} + \underset{(-)}{\gamma_3} Q_{2t-1} + u_{3t} \tag{11.21a}$$

$$p_{2t} = \underset{(+)}{\gamma_1} Q^s_{2t} + \underset{(+,-)}{\gamma_2} p_{1t-1} + \underset{(+)}{\gamma_3} p_{2t-1} + y_{3t} \tag{11.21b}$$

$$p_{2t} = \underset{(+)}{\gamma_1} Q^s_{2t} + \underset{(+,-)}{\gamma_2} p_{1t} + \underset{(+)}{\gamma_3} Q^e{}_{1t|t-1} + u_{3t} \tag{11.21c}$$

where $Q^s_{2t} = Q^d_{2t} = Q_{2t}$ (quantity transacted), u_{3t} is the error term and $Q^e_{1t|t-1}$ is the expectations-type variable, expressing households' expectations, taken in time $t-1$ for time t, for the constraints on the official market. The models consisting of Equations 11.19, 11.20 and one of Equations 11.21a, b and c are denoted hereafter,

[3]The variable $M_{t-1} + Y_t$ is a result of aggregation, where the microfunctions do not necessarily have identical parameters. More generally, the right-hand side of Equation (11.9) can be written as $\alpha_1(M_{1t-1} + Y_{1t}) + \alpha_2(M_{2t-1} + Y_{2t})$, where $\bar{\alpha}_1$, $\bar{\alpha}_2$ are microparameters.

[4]Function (11.20) is, in a few respects, analogous to the Howard (1976) private consumption demand function, estimated for the Soviet Union. The essential difference is in the inclusion in Equation (11.20) of the unobserved demand variable Q^d_{1t} from the primary market.

with respect to the third equation chosen, as A1, if 11.21a is selected, A2 with 11.21b, and A3, with 11.21c.

The above consumption supply equations are essentially similar; the consumption supply is assumed to be jointly dependent on p_{2t}, with a relative price effect p_{1t}, delayed in time in the case of A2. There is an uncertainty about the expected sign of γ_2; the relative price effect might be either negative (as a real income and substitution effect) or positive (as a cost effect). The variables standing with the parameter γ_3 in the different variants of the model represent alternative adjustment mechanisms for the consumption supply; A1 quantity autoregressive, A2 price autoregressive and A3 expected forced substitutions.

It has been assumed in models A1, A2 and A3 that state authorities cannot (or do not want to) limit the expansion of the second economy; the second market share is in fact limited only by incomes earned in the second economy and by the repressed inflation in the primary market which evidently means no limitations in a long-run. It seems therefore reasonable to formulate an alternative model of the secondary market, in which it is assumed that consumption supply can be exogenously limited, for example by tight state control, scarcity of raw materials, etc. There is no limit on price; this is perfectly flexible and could effectively reduce demand down to the level of the constrained supply. The secondary market is still in equilibrium but now there are no demand or supply equations; what is left is a price equation, in which demand and supply explanatory variables appear, but not consumption demand and supply themselves. For this model the assumed price equations are alternatively

$$p_{2t} = \underset{(+)}{\delta_1}(M_{2t-1} + Y_{2t}) + \underset{(+)}{\delta_2}(Q^{\mathrm{d}}_{1t} - Q_{1t}) + \underset{(+,-)}{\delta_3}\, p_{1t} + \underset{(-)}{\delta_4} Q_{2t-1} + \underset{(+)}{\delta_5} p_{2t-1} + v_t, \tag{11.22a}$$

or

$$p_{2t} = \underset{(+)}{\delta_1}(M_{2t-1} + Y_{2t}) + \underset{(+)}{\delta_2}(Q^{\mathrm{d}}_{1t} - Q_{1t}) + \underset{(+,-)}{\delta_3}\, p_{1t-1} + \underset{(-)}{\delta_4} Q_{2t-1} + \underset{(+)}{\delta_5} p_{2t-1} + \underset{(+)}{\delta_6} Q^{\mathrm{e}}_{1t|t-1} + v_t \tag{11.22b}$$

The model consisting of 11.19 and 11.22a is herewith denoted as B1, and that of 11.19 and 11.22b as B2. The real difference between 11.22a and 11.22b is in adding to the latter the expectations variable, as in A3. Moreover, in order to avoid multicollinearity the primary market price which appears in 11.22b is lagged. Generally, the variables in the 11.22 equations are those of the demand and supply equations for the 'secondary' market of the models A1, A2 and A3.

The only unobservable variable, apart from the expectations, in all the above models is Q^{d}_{1t} when there is a positive consumption excess demand on the primary market. It can be solved out, leaving all the parameters of the model identifiable, by combining Equation 11.19 and the corresponding equation with the spillover effect. For A1, A2 and A3 it leads to the equation

$$Q^{\mathrm{d}}_{2t} = \beta_1(M_{2t-1} + Y_{2t}) + \beta_2\alpha(M_{t-1} + Y_t) - \beta_2 Q_{1t} + \beta_3 p_{2t} + \beta_4 p_{1t} + (u_{2t} + \beta_2 u_{1t}), \tag{11.23}$$

which reduces each model to a two-equation system (the second equation explains the corresponding consumption supply (11.21a,b or c). For B1 and B2 the solution is even simpler; what is left for estimation is a one-equation model, with p_{2t} on the left-hand side, estimable directly for example by nonlinear least squares method. Apart from straightforward estimation, the advantages of the proposed models are the following:

1. The consumption demand equations are structurally non-equivalent with budget constraints.
2. A sign for the consumption excess demand on the official market is not a priori assumed; it might take either positive or negative values. Note, however, that in the case of persistent excess supply in the official market, a downward bias of the estimated parameter α might appear; since there is no restriction in estimation for the sign of the difference $Q^{d}_{1t} - Q_{1t}$, it might be the case that this sign is negative. Nevertheless, it seems that the existence of persistent consumption excess supply in the official market is not likely.
3. Although negative excess demand is admissible, the models are not in contradiction with the chronic excess demand hypothesis; if there is a permanent state of repressed inflation on the official market, parameters of the models are still identifiable.
4. The number of the estimated parameters is relatively small, which leaves room for further improvement of the model specification. For instance, the Houthakker–Taylor demand function can be applied instead of Equation 11.19, without danger of unidentification.
5. The consumption supply function for the state side, with its most disputable specification, if not formulated does not affect the formulation of the model.

Nevertheless, there is an important obstacle to the estimation of these models, namely obtaining the unconventional data. For the models A1, A2 and A3 data on the second economy prices, quantities, money stocks and personal incomes are necessary. Models B1 and B2 are easier to handle, since the most doubtful data on the second economy consumption expenditure are not needed.

At first sight, it might appear impossible to obtain this statistical information. But through careful reviewing of existing materials one can retrieve time series on black-market hard currency prices, hard currency money stocks in private hands, measures of the underground economic activity, etc. Obviously, the accuracy of this information is unlikely to be great; data are almost persistently subject to measurement error, usually much greater than for the data gathered and published by statistical offices. It is necessary to realize that from the point of view of the empirical analysis of consumption in CPEs the statistical yearbooks' data are always, although not necessarily deliberately, biased; they simply reflect, with better or worse precision, the state of the official economy, ignoring its black counterpart. If the underground economy exists, and it clearly does for Poland as in any other country, the official data are biased when utilized in the evaluation of total (official plus unreported) consumption. Hence, analysts of consumption in

CPEs face a dilemma: to use the data on the first economy that are biased due to their neglect of the second economy, or to employ the full set of consumption data, which is subject to a greater measurement error but less bias. I would suggest the latter course. It is better to attempt to include underground transactions in consumption analysis, even if some of the data are inaccurate, than to pretend that the second economy does not exist.

The models presented above have been estimated using the data reflecting the official and unofficial economic activity of households in Poland, in 1956–86. The detailed description of the methodology and sources for constructing particular data series is given by Charemza (in preparation). All the data have been retrieved or recalculated from the officially published sources; activity on secondary markets is analysed on the basis of official, but not regular, information on private hard currency trade flows and stocks of hard currency in the households' hands. Consumption has been defined as personal money consumption (i.e. where the consumption in kind is excluded) of all the material goods and services and non-material services purchased on the official and unofficial markets. The official consumption, i.e. that corresponding to the legal, private, cooperative and state sectors, has been calculated as:

reported consumption expenditures = official total disposable money income in domestic currency − households' money balance of domestic currency − housing and other expenditures of non-consumption nature

which gives figures close to those published in statistical yearbooks (a minor difference is in the slightly different way of constructing the households' money balance). To include the unreported consumption, two main components of the above balance have been recalculated. Firstly, it has been assumed that total velocity of money in private hands is greater than reported and some velocity multipliers (varying from 1.1 to 1.3 for particular components of the disposable money income) have been applied – for an outline of methodology of the velocity multiplier approach to the second economy problem see Colijn (1985). Secondly, data on hard currency incomes and on money stocks, expressed in US dollars, have been retrieved from the official sources and converted to the domestic currency using black-market dollar prices. The US dollar black-market prices are published officially in the Polish weekly *Veto* (strictly speaking, the free market prices for the dollar vouchers 'bony', are published, which are directly related to the dollar black-market prices). Systematic time series of the black-market dollar prices are also published in *Pick's Currency Yearbook* and its continuation, *World Currency Yearbook*. Recent data on hard currency incomes and bank savings have also been published; for earlier years some interpolation has been made using fragmentary information officially published and data on foreign tourist flows as a proxy for changes in hard currency stocks in private hands. Finally, the total, reported and unreported consumption is estimated as:

total, reported and unreported consumption expenditures = $\sum$(official disposable money income components∗velocity multiplier) + hard currency earnings − households' domestic currency money balance − households' hard currency money balance ($M_{2t} - M_{2t-1}$) − housing and other expenditures of a non-consumption nature

The unreported consumption is the difference between the total and reported consumption expenditures. Next, the reported consumption has been divided into that connected with purchasing of goods and services from the state and co-operative suppliers (which is equal to Q_{1t} in the model notation) and that going to the official private sector. The latter is subsequently added to the estimated unreported consumption giving Q_{2t} as a result.

Finally, the price index for the secondary market, p_{2t}, has been calculated as an aggregate Laspeyres-type index, pooling the official price and black-market hard currency indices. The weights are respectively equal to the shares of the official private and unreported components of Q_{2t} in 1980.

The model has been estimated with a use of annual data for Poland, 1956–86 (for 1986 data are preliminary). Values for $Q^e_{1t|t-1}$ were obtained by the Kalman filter optimal prediction of Q_t given the information on predetermined variables from period $t-1$. Models A1, A2 and A3 have been estimated by the TSP − 4.0 non-linear full-information maximum likelihood (FIML) method, based on the

Table 11.2 Results of FIML estimation for models A1, A2, A3

Parameter	*A1*	*A2*	*A3*
α	0.796	0.906	0.782
	(0.163)	(0.242)	(0.192)
β_1	0.167	0.161	0.183
	(0.040)	(0.031)	(0.052)
β_2	0.127	0.116	0.120
	(0.078)	(0.067)	(0.100)
β_3	−79.46	−19.58	−78.23
	(13.36)	(14.36)	(13.47)
β_4	151.88	71.50	152.14
	(23.09)	(25.41)	(22.03)
α_1	0.00283	0.00130	0.00142
	(0.00100)	(0.00083)	(0.00082)
α_2	1.263	−0.293	1.439
	(0.156)	(0.188)	(0.189)
α_3	−0.00452	1.104	−0.000558
	(0.00110)	(0.209)	(0.000212)
SSE_1	2769.7	2831.3	2702.00
SSE_2	0.696	1.306	.607
AIC	193.02	214.08	193.00

SSE_1 sum of squared errors of the Q^d_{2t} equation.
SSE_2 sum of squared errors of the p_{2t} equation.
AIC Akaike information criterion.
Asymptotic standard errors of the parameter estimates are in brackets

Table 11.3 Results of NLS estimation for the models B1, B2

Parameter	*B1*	*B2*
α	0.708	0.826
	(0.904)	(0.092)
δ_1	0.000657	0.000350
	(0.000352)	(0.000557)
δ_2	0.00134	0.00162
	(0.00054)	(0.00049)
δ_3	0.363	0.504
	(0.125)	(0.122)
δ_4	−0.00643	−0.00525
	(0.00081)	(0.00248)
δ_6	–	0.000609
		(0.000369)
SSE	0.381	0.288
AIC	−38.38	−45.06

For a description of symbols see Table 11.2

Davidon–Fletcher–Powell optimization algorithm. Models B1, B2 were estimated by the TSP − 4.0 non-linear least squares method (NLS), which in this case is equivalent to FIML. Convergence was obtained in each case. Stability of B1 and B2 was difficult to obtain until the parameter δ_5 has been a priori fixed as equal to one. The estimation results are given in Tables 11.2 and 11.3.

Estimates of the stochastic structure of the models seem to be satisfactory. All the parameter estimates have their signs consistent with previous expectations; standard error of estimates are usually (with the exception of δ_1 in B2 and the constant form in B1) much smaller than the absolute values of the corresponding parameters. Nevertheless, the estimated propensity to consumption (parameter α) in the primary market is relatively high, reaching a level of over 0.9 for model A2. This might be a result of overestimation of this parameter due to the fact that Equation 11.19 might overlap in estimation with a similarly specified demand function for both primary and secondary markets.

The possibility of overlapping of the two demand functions has been taken into account in estimation of the consumption excess demand. Firstly, assuming that there was no overlapping and the parameter of Equation 11.19 has been perfectly estimated, Q_{1t}^{d} has been simulated from Equation 11.19. One may notice the unidentification of the variance of the error term in Equation 11.19 – see Chapter 6. In practice, having the compound residuals estimated for Equation 11.23, the residuals of the Equation 11.19 have been retrieved assuming proportionality between u_{2t} and u_{1t} as that between p_{2t} and Q_{1t}. The retrieved residuals have been respectively used in simulating Q_{1t}^{d}.

As was mentioned before, such estimates of Q_{1t}^{d} are unbiased only if there is no overlapping between the demand functions for Q_{1t}^{d} and for the aggregate consumption on both markets. If, however, Equation 11.19 is confused in estimation with the consumption demand function for $Q^d_{1t} + Q^d_{2t}$, what has been simulated in the above procedure is not Q^d_{1t} but rather $Q^d_t = Q^d_{1t} + Q^d_{2t}$. Evidently, in such a case consumption demand in the primary market can be calculated by subtracting Q^d_{2t} from the simulated value of Q^d_t.

In Tables 11.4 and 11.5 the simulated upper and lower limits of consumption excess demand (in percentages of the total observed consumption) are given. The upper limits correspond to the assumption that no overlapping takes place and what was simulated from Equation 11.19 is really Q^d_{1t}; in calculation of the lower limits the full overlapping is assumed – in that case Equation 11.19 generates Q^d_t rather than Q^d_{1t}.

Although significantly different in absolute size, the estimates of consumption excess demand behave similarly in time for the particular models. The estimates for the first two years, 1961–62 seem to be significantly affected by the out-of-sample start of simulation. From 1963, it is a period of relatively stable moderate positive excess demand with some indications of the possibility for the negative excess

Table 11.4 Limits for estimated consumption excess demand for Poland 1961–86, models A1, A2, A3

	A1		*A2*		*A3*	
Year	*Upper*	*Lower*	*Upper*	*Lower*	*Upper*	*Lower*
1961	10.1	−3.1	28.4	15.1	19.4	6.1
1962	−2.0	−14.2	10.7	−1.5	6.6	−5.6
1963	2.9	−8.4	17.5	6.1	9.5	−1.9
1964	5.4	−6.0	19.5	8.1	9.0	−2.4
1965	5.6	−5.3	20.3	9.3	8.3	−2.7
1966	7.4	−3.2	22.0	11.4	8.5	−2.1
1967	9.5	−1.0	22.8	12.3	9.3	−1.2
1968	8.9	−1.2	22.2	12.2	7.8	−2.1
1969	11.6	1.6	29.1	19.1	9.0	−1.0
1970	10.0	0.4	27.4	17.8	7.8	−1.8
1971	15.4	5.7	31.1	21.4	12.2	2.5
1972	20.6	10.9	34.6	24.9	17.3	7.5
1973	27.4	17.2	37.9	27.7	22.6	12.4
1974	30.8	20.3	40.4	30.0	25.3	14.9
1975	34.9	24.2	45.7	35.1	28.8	18.1
1976	37.2	26.2	50.2	39.2	30.7	19.7
1977	43.9	31.7	60.0	47.9	38.0	25.8
1978	46.1	33.1	64.9	52.0	42.5	29.5
1979	46.8	33.6	65.8	52.6	45.1	31.9
1980	53.9	39.2	71.1	56.3	51.6	36.9
1981	53.2	37.3	67.5	51.3	49.0	32.8
1982	53.0	37.1	62.8	46.8	50.5	34.5
1983	45.9	30.6	64.6	49.3	43.0	27.7
1984	47.1	31.3	64.5	48.7	45.6	29.9
1985	56.1	38.7	76.2	58.8	53.7	36.4
1986	54.6	37.2	76.2	58.8	52.6	35.1

Table 11.5 Limits for estimated consumption excess demand for Poland 1961–86 models B1, B2

	B1		*B2*	
Year	*Upper*	*Lower*	*Upper*	*Lower*
1961	– 17.4	– 30.8	– 6.3	– 19.6
1962	37.7	25.5	42.5	30.3
1963	3.4	– 8.0	11.2	– 0.2
1964	6.4	– 5.1	17.6	6.2
1965	7.1	– 3.8	19.2	8.3
1966	2.5	– 8.1	16.5	5.9
1967	11.1	0.6	25.2	14.7
1968	28.4	18.3	39.7	29.7
1969	7.9	– 2.1	24.4	14.4
1970	6.1	– 3.5	22.4	12.8
1971	– 3.2	– 13.0	14.7	5.0
1972	–7.1	– 16.8	11.4	1.7
1973	7.6	– 2.6	26.4	16.2
1974	16.0	5.6	35.3	24.8
1975	24.9	14.2	42.7	32.0
1976	26.7	15.7	45.7	34.7
1977	24.8	12.7	44.7	32.6
1978	26.3	13.4	46.4	33.5
1979	28.0	14.8	45.8	32.5
1980	38.1	23.4	54.7	39.9
1981	58.4	42.2	73.9	57.7
1982	33.2	17.2	55.3	39.3
1983	29.3	14.0	51.2	35.9
1984	36.8	21.1	54.2	38.4
1985	35.8	18.5	56.2	38.9
1986	41.1	23.7	62.2	44.8

demand. The end of this period of moderate disequilibrium is different for particular models – 1969–70 for A1 and A2, 1971 for A3, 1974–75 for B1 and 1973 for B2. These years indicate continuously growing positive excess demand in time; for all models it reaches a maximum in 1980–81. For the eighties there is a period of excess demand stabilized on a high level, usually with a slight fall in 1982–83. It is interesting to note that for all the models but A2 there is a significant drop in 1983. It seems that this can reasonably be explained by a substantial relaxation of state control over the private market resulting in flows of personal incomes from the primary to the secondary economy. After this short period, the consumption excess demand estimates start to rise again, reaching another peak in 1985–86.

It seems that the above estimates are, to a large extent, realistic and interpretable. Estimates derived with the use of B1 and B2 generally have greater variance and a tendency to rapid changes in time, while those obtained with A1, A2 and A3 are smoother. It is difficult to select the best model; the goodness-of-fit and the AIC criterion suggest B1 and B2 to be superior to A1, A2 and A3, nevertheless, excess demand estimates of the former model look more dubious. The models A1, A2, A3 and B1, B2 correspond in fact to different types of state policy towards private economic activity; in the sixties and seventies the policy was to limit its expansion, which corresponds to models B1 and B2, while in the eighties state control was much more relaxed and the economic reality was closer to models A1, A2 and A3.

Another way of evaluating the performance of the models is by checking their *ex-ante* forecasting abilities. As a post-sample period the years 1984–86 were chosen; these were the years of the development of Polish economic reform, generally recognized as based on slightly different economic mechanisms from those implemented in the previous years. All the models were re-estimated for the period 1956–83 and then forecasts of the observable dependent variables were made (for example for Q^{d}_{2t} and p_{2t} in the case of A1, A2, A3 and for p_{2t} in the case of B1, B2) through dynamic simulation on the estimated models, assuming known values of the predetermined variables in the period of prediction.

To evaluate the forecasting properties of the models for the period 1984–86 *ex-ante* projections of the consumption excess demand have also been made. The method of doing this projection was in this case slightly different. Since in all the models the simulated *ex-ante* value of Q^{d}_{1t} depends entirely on one parameter, α, it

Table 11.6 Percentage prediction errors and *ex-ante* projections of excess demand, Poland, 1984–86

Year	Q^{d}_{2t}	p_{2t}	*Excess demand*
Model A1			
1984	−0.53	5.58	54.4
1985	3.24	6.82	60.6
1986	1.10	6.98	63.2
Model A2			
1984	0.97	−2.60	53.4
1985	6.81	−15.94	59.5
1986	6.68	−29.97	62.2
Model A3			
1984	41.39	9.09	55.4
1985	40.05	10.59	61.6
1986	36.70	10.94	64.3
Model B1			
1984		11.52	21.3
1985		−15.14	26.3
1986		−17.80	28.4
Model B2			
1984		2.04	33.2
1985		−15.22	38.6
1986		−9.47	40.9

seems reasonable to apply a heuristic procedure of combining its sample estimate for 1956–83, with a straightforward prior assumption about its value and standard error. The projections of excess demand have been calculated assuming prior knowledge that for all the models the parameter α in the prediction period is normally distributed with mean equal to 0.8, standard deviation equal to 0.15. This prior has been pooled with the sample estimate of α and its estimated standard error, giving as the result a simple Bayesian estimate of α as equal to a weighted average of 0.8 and the sample estimate of α with the weights inversely proportional to the corresponding errors (see for example Freund and Walpole (1987, p. 359)).

The percentage prediction errors for the observable endogenous variables and the projections of excess demand (in percentages of the quantity actually transacted) are given in Table 11.6. For the entire three-year prediction period the best results are those for A1; if only one-step ahead prediction errors are concerned, models A2 and B2 also perform well. The predictions derived from A3 seem to be totally inadmissible, with the percentage prediction error for Q^{d}_{2t} exceeding 40%. Projections of the consumption excess demand are overshot (i.e. they exceed the estimated upper limit) for A1 and A3, and undershot (are below the lower limit) for B2. For the remaining two models, A2 and B2, these projections lie between the estimated upper and lower limits of excess demand.

Summing up, it is difficult to select the best model (if a decision has to be made, the estimates and performance of A1 seem the most convincing). Nevertheless, it does not seem to be necessary; within the general framework of modelling the consumption disequilibrium through secondary markets, there is a lot of room for improvement. The applied functions have deliberately been of a simple nature, with a limited number of parameters and without any extraneously given assumptions, like hints about excess demand or dummy variables. In future investigations a lot of improvements can be made, from applying more sophisticated consumption demand functions to adding subjective a priori information.

ACKNOWLEDGEMENTS

The ESRC Research Grant No. B00230025 is gratefully acknowledged.

I am indebted to John Burkett and Chris Davis for their valuable comments on the earlier draft of this chapter. Neither of them bears any responsibility for the remaining deficiencies.

REFERENCES

Amemiya, T. (1974) A note on a Fair and Jaffee model, *Econometrica*, **42**, 759–62.

Balicki, W. (1983) Theory of disequilibrium in centrally planned economies, *Jahrbuch der Wirtschaft Osteuropas*, **10**, 9–39.

Ball, R. J. and Drake, P. S. (1964) The relationship between aggregate consumption and wealth, *International Economic Review*, **5**, 63–81.

Barro, R. J. and Grossman, H. J. (1974) Suppressed inflation and the supply multiplier, *Review of Economic Studies*, **41**, 97–104.

Bauer, T. (1978) Investment cycles in planned economies, *Acta Oeconomica*, **21**, 243–60.

Beksiak, J. and Libura, U. (1974) *Rownowaga ekonomiczna w socjalizmie*, PWN, Warsaw.

Bowden, R. J. (1978) Specification, estimation and inference for models of markets in disequilibrium *International Economic Review*, **19**, 711–26.

Brus, W. and Laski, K. (1983) Repressed inflation and second economy under central planning, Paper presented at the Conference on the Economics of the Shadow Economy, Bielefeld.

Burkett, J. (1988) Slack, shortage and discouraged consumers in Eastern Europe: estimates based on smoothing by aggregation. *Review of Economic Studies*, **55**, 483–506.

Charemza, W. (1981) *Ekonometryczne modele nierownowagi*, University of Gdansk.

Charemza, W. (1987) Plans, disequilibria and macroeconomic policy simulations, *International Review of Applied Economics*, **1**, pp. 190–208.

Charemza, W. (in preparation) Alternative data for consumption analysis in Hungary and Poland.

Charemza, W. and Dlouhý, V. (1987) Two alternative disequilibrium models for a planned economy, Discussion Paper No. 65, Department of Economics, University of Leicester.

Charemza, W. and Gierusz, B. (1978) The estimation and application of a supply/demand quarterly model of retail sales in Poland, Paper presented at the Conference on Problems of Building and Estimation of Large Econometric Models, Blazejewko, version in Polish published in 1980 in *Przeglad Statystyczny*.

Charemza, W. and Gronicki, M. (1988) *Plans and Disequilibria in Centrally Planned Economies*, North-Holland, Amsterdam.

Charemza, W. and Kiraly, J. (1988) Plans and exogeneity: The genetic-teleological dispute revisited, Discussion Paper No. 71, Department of Economics, University of Leicester.

Colijn, L. (1985) Some proposed methodologies to quantify the influence of macroeconomic disequilibrium on the size of the second economy in Poland, European University Institute, Florence (*mimeo*).

Czerwinski, Z. and Jurek, W. (1984) Mierzenie stopnia nierownowagi na rynku konsumenta (proba estymacji ekonometrycznej), *Ekonomista*, 723–39.

Davidson, J., Hendry, D., Srba, F. and Yeo, S. (1978), Econometric modelling of the aggregate time-series relationship between consumers expenditure and income in the United Kingdom, *Economic Journal*, **88**, 661–92.

Deaton, A. S. (1978) Involuntary saving through unanticipated inflation, *American Economic Review*, **66**, 341–67.

Dlouhý, V. (1981) Nerovnovazne modely a odhad parametru poptavkove funkce, *Ekonomicko-Matematicky Obzor*, **17**, 74–85.

Dyba, K. (1973) Odhad parametru poptavkove funkce na nerovnovaznem trhu, *Politicka Ekonomie*, **21**, 1123–32.

Engle, R. F., Hendry, D. F. and Richard, J.-F. (1983) Exogeneity', *Econometrica*, **51**, 277–304.

Fair, R. C. and Jaffee, D. M. (1972) Methods of estimation for markets in disequilibrium, *Econometrica*, **40**, 497–514.

Freund, J. E. and Walpole, R. E. (1987) *Mathematical Statistics*, 4th Edn, Prentice-Hall, London.

Galasi, R. and Sziraczki, G. (1985) *Labour Market and Second Economy in Hungary* (edited volume), Campus Verlag, Frankfurt.

Gapinski, J. H. (1982) *Macroeconomic Theory: statics, dynamics and policy*, McGraw-Hill, Auckland.

Goldfeld, S. M. and Quandt, R. E. (1975), Estimation in a disequilibrium model and the value of information, *Journal of Econometrics*, **3**, 325–48.

Gomulka, S. (1986) *Growth, Innovation and Reform in Eastern Europe*, Wheatsheaf, Brighton.

Granger, C. W. J. (1969) Investigating causal relations by econometric models and cross-spectral methods, *Econometrica*, **37**, 424–38.

Green, D. W. and Higgins, C. I. (1977) *SOVMOD* I: *A macroeconometric model of the Soviet Union*, Academic Press, New York.

Gronicki, M. and Szreder, M. (1981) Bayesian analysis of a simple model of market in disequilibrium: Repressed inflation case, University of Gdansk (*mimeo*).

Gronicki, M. and Szreder M. (1986) Bayesian analysis of Polish TV market, Paper presented at the Conference on Problems of Building and Estimation of Large Econometric Models, Szczyrk.

Grossman, G. (1977) The 'second economy' of the USSR, *Problems of Communism*, **26**, 25–40.

Guilkey, D. K. and Salemi, M. K. (1982) Small sample properties of three tests for Granger causal ordering in bivariate stochastic system, *Review of Economics and Statistics*, **64**, 668–80.

Harvey, A. C. (1981) *The Econometric Analysis of Time Series*, Philip Allan, Oxford.

Herer, W. and Sadowski, W. (1981) Nawis inflacjny, parts 1 and 2, *Zycie Gospodarcze* Nos **32**, 33.

Houthakker, H. S. and Taylor, L. (1970) *Consumer Demand in the United States*, Harvard UP, Cambridge, Mass.

Howard, D. H. (1976) The disequilibrium model in a controlled economy: An empirical test of the Barro-Grossman model, *American Economic Review*, **66**, 871–9.

Hulyak, K. (1982) Disequilibrium models for consumption goods in Hungary, Paper presented at the Conference of the Applied Econometrics Association, Budapest.

Hulyak, K. (1985) An econometric disequilibrium macromodel for Hungary, Paper preserved at the World Congress of the Econometric Society, Cambridge, Mass.

Ito, T. (1980) Methods of estimation for multi-market disequilibrium models, *Econometrica*, **48**, 7–26.

Katsenelinboigen, A. (1977) Coloured markets in the Soviet Union, *Soviet Studies*, **29**, 62–85.

Kemme, D. (1984) Discriminating among alternative production functions in Polish industry, *Empirical Economics*, **9**, 59–66.

Kemme, D. and Winiecki, J. (1984) Disequilibrium in centrally planned economies, University of North Carolina, Greensboro (*mimeo*).

Klacek, J. and Nesporova, A. (1985) Economic growth and efficiency in six socialist countries, Paper presented at the Conference on Models and Forecasts 85, Bratislava.

Klaus, V. and Rudlovcak, V. (1982a) Nerovnovazne modely v ekonometrii, *Ekonomicko-Matematicky Obzor*, **18**, 216–24.

Klaus, V. and Rudlovcak, V. (1982b) Savings function as an inverse problem of disequilibrium consumption modelling: A case study of Czechoslovakia, paper presented at the Conference of the Association of Applied Econometrics, Budapest.

Kivet, J. F. (1986) On the rigour of some misspecification tests for modelling dynamic relationship, *Review of Economic Studies*, **53**, 241–62.

Kolupa, M. (1961) O podstawowych twierdreniach dla problemu eliminacji obciazen parametrow funkcji popytu, *Przeglad Statystyczny*, **8**, 395–400.

Kolupa, M. (1965) *Badanie Popytu w Warunkach Niedostatecznej Podazy*, PWE, Warsaw.

Kornai, J. (1971) *Anti-Equilibrium. An economic systems theory and the tasks of research*, North-Holland, Amsterdam.

Kornai, J. (1980) *Economics of Shortage*, Vols 1 and 2, North-Holland, Amsterdam.

Kornai, J. (1982) *Growth, Shortage and Efficiency*, Blackwell, Oxford.

Krzysztofiak, M. (1982) Wzrost gospodarczy Polski w latach 1950–1980 na tle europejskich krajow RWPG, *Ekonomista*, 55–538.

Laffont, J.-J. and Garcia, R. (1977) Disequilibrium econometrics for business loans, *Econometrica*, **45**, 1187–204.

Libura, U. (1979) *Konsumpcja a Wykorzystanie Mozliwosci Produkcyjnych w Gospodarce Socjalistycznej*, PWN, Warsaw.

Maddala, G. S. (1986) Disequilibrium, self-selection and switching models, in *Handbook of Econometrics*, Vol. III, (Z. Griliches and M. D. Intriligator, eds) North-Holland, Amsterdam.

Maddala, G. S. and F. D. Nelson (1975) Maximum likelihood methods for models of markets in disequilibrium, *Econometrica*, **42**, 1013–30.

Michalak, K. and Starzynska, W. (1979) Application of disequilibrium indicators in econometric models of some foodstuffs markets, Paper presented at the Conference on Models and Forecasts, Polanica.

Nasilowski, M. (1980) Empirical testing of the Cobb-Douglas production function on the example of five socialist countries, *Acta Oeconomica Polona*, **1**,

Nuti, D. M. (1986) Hidden and repressed inflation in Soviet-type economies, *Contributions to Political Economy*, **5**, 37–82.

Pawlowski, Z. (1959) Problem nieobciazonosci parametrow funkcji popytu a przypadki przejsciowej niedostatecznej podazy dobr na rynku, *Przeglad Statystyczny*, 7, 41–50.

Pickersgill, J. (1980) Recent evidence on Soviet households saving behaviour, *Review of Economics and Statistics.*

Pindak, F. (1983), Inflation under central planning, *Jahrbuch der Wirtschaft Osteuropas*, **10**, 93–131.

Podkaminer, L. (1986) Measurement of macroeconomic disequilibrium in centrally planned economies: A critical review of Professor Portes' results, paper presented at the IIASA Meeting, Applied General Equilibrium Modelling, Laxenburg.

Portes, R. (1986) The theory and measurement of macroeconomic disequilibrium in centrally planned economies, Discussion Paper No. 91, Centre for Economic Policy Research, London.

Portes, R., Quandt, R. E., Winter, D. and Yeo, S. (1983) Macroeconomic planning and disequilibrium: Estimates for Poland, 1955–1980, Birkbeck College Discussion Paper No. 139, London.

Portes, R., Quandt, R. E., Winter, D. and Yeo, S. (1984), Planning the consumption goods market: Preliminary estimates for Poland, 1955–80, in *Contemporary Economic Modelling* (eds. P. Malgrange and P.-A. Muet), Blackwell, Oxford.

Portes, R., Quandt, R. E., Winter, D. and Yeo, S. (1987) Macroeconomic planning and disequilibrium: Estimates for Poland, 1955–1980, *Econometrica*, **55**, 19–41.

Portes, R. and Santorum, A. (1987) Money and consumption goods market in China, *Journal of Comparative Economics*, **11**, 354–71.

Portes, R. and Winter, D. (1978) The demand for money and for consumption goods in centrally planned economies, *Review of Economics and Statistics*, **60**, 8–18.

Portes, R. and Winter, D. (1980), Disequilibrium estimates for consumption goods in centrally planned economies, *Review of Economic Studies*, **47**, 37–59.

Romanski, J. and Welfe, W. (1986) Simulation minimodel W-6 of the Polis economy under disequilibrium, Paper presented at the Conference on Problems of Building and Estimation of Large Econometric Models, Szczyrk.

Spanos, A. (1982) Equilibrium, identities, behavioural and measurement equations in economic modelling, Birkbeck College Discussion Paper No. 110, London.

Starzec, K. (1983) L'economie Polonaise vue a travers des circuits paralleles, (*mimeo*), Paris.

Starzynska, W. (1986) On problems of modelling of food market in disequilibrium (case of Poland), Paper presented at the Conference on Problems of Building and Estimation of Large Econometric Models, Szczyrk.

Starzynska, W. (forthcoming), Rola cen w budowie i weryfikacji ekonometrycznych modeli rynkow nierow nowadze, in *Konsumpcja i Rynek w Warunkach Kryzysu*, University of Lodz.

Viktorinova, B. (1986) Specifikacia a odhad parametra funkci dopytu a ponuky v nerovnovaznom modeli, *Ekonomicko-Matematicky Obzor*, **22**, 144–59.

Welfe, A. (1984), Analiza popytu w warunkach nierownowagi, *Ekonomista*, 1045–64.

Welfe, A. and Welfe, W. (1986) Modele i prognozy rynku w warunkach nierownowagi, *Ekonomista*, 279–314.

Welfe, W. (1973) A medium-term econometric model of the Polish economy, prace IEiS UL, No. 2, series D, Lodz.

Welfe, W. (1976) The unbalanced econometric macromodels, in *Computing Equilibria: How and why*, (J. Los and M. W. Los, eds.) North-Holland, Amsterdam.

Welfe, W. (1978) Disequilibria in econometric macromodels of centrally planned economies, Paper presented at the Conference on Colloques sur Structures Economiques et Econometrie, Lyons.

Welfe, W. (1985) Econometric macromodels of unbalanced growth, prace IEiS UL No. 29, Series D, Lodz.

Winiecki, J. (1985) Portes ante Portas: A critique of the revisionist interpretation of inflation under central planning, *Comparative Economic Studies*, **27**, 25–52.

Wisniewski, M. (1985) Zrodla i rozmiary drugiego obiegu gospodarczego, *Ekonomista*, **1986**, 913–40.

Zwass, A. (1979) *Money, Banking and Credit in the Soviet Union and Eastern Europe*, Macmillan, London.

12

Savings and consumption in the centrally planned economy: a disequilibrium approach

ALEKSANDER WELFE

12.1 INTRODUCTION

The analysis of consumer demand and savings constitutes one of the most developed areas of economics and applied econometrics. Empirical research has been conducted by scientists of the calibre of H. Schultz, H. Wold, R. Stone, H. Theil, F. Modigliani and many others. Classical analysis, however, assumes that consumer demand is fulfilled and realized in the market, in terms of purchases. Thus the unspent part of current income is saved voluntarily and comprises the increase in consumer savings.

The first articles where authors assumed the possibility of the existence of excess demand were published in England during the Second World War (Kalecki, 1941; Rothbarth, 1941; Scitovsky, 1942; Neisser, 1943). These pioneering works were influenced by the market situation at that time and by the rationing system introduced in England during the war years.

The Polish experience in the 50s similarly inspired Pawlowski (1960, 1961). Research on the problem of bias of demand function estimators in a state of disequilibrium was further continued by Kolupa (1961, 1965). But apart from two works by Clower (1965) and Leijonhufvud (1968), which went practically unnoticed for quite a long time, disequilibrium problems were then forgotten. In the early 70s a few very important articles were published which shed new light both on the problem of disequilibrium model specification (Barro and Grossman, 1971; Fair

International Studies in Economic Modelling Series.
Models of Disequilibrium and Shortage in Centrally Planned Economies.
Edited by C. Davis and W. Charemza.
Published in 1989 by Chapman and Hall Ltd, 11 New Fetter Lane, London EC4P 4EE.

and Jaffee, 1972) and its estimation (Maddala and Nelson, 1974; Amemiya, 1974). At the same time Kornai (1971) published a monograph in which he provided a theoretical basis for the analysis of centrally planned economies in a state of disequilibrium. The first attempts to incorporate disequilibrium and imbalances into econometric macromodels were done by W. Welfe (1973, 1976). In the last decade many works have been published which explicate both the theory of disequilibrium models and econometric techniques.[1]

The main idea developed in this chapter is that shortages (excess demand) result in less money spent than planned by consumers. As a result, forced savings remain in their pockets. These savings are partly added to current income to spend in the next period, and hence increase demand even more. Consequently, classical budget constraint has to be modified to capture this phenomenon. The final aim of this chapter is to construct a consistent system of demand equations with the savings function and its estimation. The empirical investigation is based on a sample from 1961 to 1985, covering the period of the Polish crisis and consumer market breakdown.

12.2 BASIC ASSUMPTIONS OF THE MODEL

First, it should be stressed that there are two sources of household income: personal income and consumer credits. The former includes wages, individual (private) farmers' incomes, pensions and other transfers from social funds, non-salary income (of artists, lawyers, writers etc.), income of private entrepreneurs and others. Consumer credits are of secondary importance in Poland and comprise less than 2% of total incomes. Usually they are granted for privileged social groups (for example, young couples) or for special purposes, mainly housing construction and apartment renovations. Hence we have neglected credits both in our theoretical and in our empirical research.

Second, we believe that there is no trade-off between the consumer goods market and the labour market in centrally planned economies. The excess demand results in a decrease in productivity, because workers spend more time standing in lines (sometimes during working hours), are more often on sick-leave (fully paid), etc. Consequently a decrease in productivity takes place, just as in market economies. However, this effect is not accompanied by a decrease in wages because employees continue to get the same amount of money. On the contrary, they can make even more, because managers of socialized enterprises are willing to pay more for overtime, to compensate for the decline in output. In opposition to Barro and Grossman's (1971) model, people accept more overtime work (for example, on Saturdays and Sundays) and need the extra income, which they expect will enable them to buy scarce goods on the free (or black) market, where prices are much higher than on the socialized market.

[1]The most complete bibliography of disequilibrium and rationing models is published by Quandt (1986).

Third, our analysis of households' behaviour in a state of disequilibrium concerns short and medium time periods. Consequently, we assume that market shortages do not change consumers' preferences permanently. This means that after regaining equilibrium, consumers should buy the same quantities of goods as before market breakdown, if their incomes and relative prices remain unchanged.

The above hypothesis has significant consequences both for further research, and for the interpretation of final results. Here disequilibrium is measured as a deviation from the normal situation, closest to equilibrium. A similar concept was introduced by Kornai (1980), who calls it normal disequilibrium or normal path (see also Kornai, 1982).

Fourth, the existence of disequilibrium causes consumers to transfer their demand from markets characterized by insufficient supply to remaining markets. In the simplest case one may assume that the direction and magnitude of transfers is determined by elasticities of substitution. Because the elasticities of substitution among many commodities are close to zero, the transfers of demand do not induce balancing between demand and supply. In other words, there may simultaneously exist markets in deep disequilibrium and those which are balanced or have excess supply. Common observation of the crisis years in Poland justifies this statement. Some markets, mainly luxury durable goods (e.g. wool carpets), because of a rapid price increase and supply increases, had been balanced relatively quickly after a short period of breakdown in 1981, and in recent years have been characterized by excess supply. At the same time, the apparel and footwear market was and still is unbalanced. Thus historical experience denies the hypothesis that households transfer the amount of money they were unable to spend (because of shortages) until they run out of all possibilities of spending it.

Socialized trade in centrally planned economies brings to the market usually more than 90% of the goods purchased by consumers. In some special markets (e.g., agricultural products, some goods manufactured by craftsmen), apart from socialized trade, free trade exists, and there prices play an equilibrating role. It is an unusual phenomenon that even in this case most consumers prefer to wait, and do not transfer their demand from the socialized to the free market, even when the quality and attractiveness of goods is higher in free trade. This results mainly from consumer expectations concerning the improvement of the market situation, and from the hope of occasional good fortune (e.g. a sudden delivery of scarce goods from the wholesale trade), for which they want ready cash.

Therefore, the use of the macrofunction of consumption for the estimation of excess demand may lead to significantly biased results (see, for example, Portes *et al.* 1982; Charemza and Gronicki, 1984). In making a reasonable compromise, on account of the size of the model, one should distinguish homogeneous aggregates from a disequilibrium point of view.

Fifth, as a consequence of accepting the previous assumptions, we can state the following: as a result of insufficient supply of goods, at the end of the period, and after all demand transfers, unspent money still remains in consumers' pockets. We call this forced savings. Consumers can keep these savings in cash, or deposit all or some in banks. The forced character of these savings is manifest in the fact that they would have been spent on goods if commodities (of which the supply was insufficient) had appeared in the market. Thus forced savings are understood as the financial equivalent of positive final excess demand, hence:

$$\mathrm{FSP}_t = \sum_{l=1}^{L} \mathrm{DE}_t^l P_t^l, \ \mathrm{DE}_t^l > 0 \tag{12.1}$$

where

FSP = forced savings, stock, end of the period
DE = excess demand, final, after demand transfers
P = prices, indexes
$l = 1, \ldots, L$ denotes commodity market
$t = 1, \ldots, T$
Further, we adopt the convention that P at the end of a symbol denotes current prices.

Fourth, we assume that excess demand is a deterministic function of variables called indicators of disequilibrium (or their linear or nonlinear combinations)

$$\mathrm{DE}_t^l = f_{\mathrm{E}}(\mathbf{x}_{\mathrm{E}_t}^l, \boldsymbol{\alpha}_{\mathrm{E}}^l) \tag{12.2}$$

With the demand equation

$$D_t^l = f_{\mathrm{D}}(\mathbf{x}_{D_t}^l, \boldsymbol{\alpha}_{\mathrm{D}}^l) + \xi_D^l \tag{12.3}$$

it gives us

$$Q_t^l = f_{\mathrm{D}}(\mathbf{x}_{D_t}^l, \boldsymbol{\alpha}_{\mathrm{D}}^l) - f_{\mathrm{E}}(\mathbf{x}_{\mathrm{E}_t}^l, \boldsymbol{\alpha}_{\mathrm{E}}^l) \cdot U_{D_t}^l + \xi_{D_t}^l, \tag{12.4}$$

subject to

$$Q_t^l = \min\{D_t^l, S_t^l\} \tag{12.5}$$

We use the following symbols:

$\mathbf{x}_{\mathrm{D}}, \mathbf{x}_{\mathrm{E}}$ vectors of explanatory variables in demand and excess demand function, respectively
$\boldsymbol{\alpha}_{\mathrm{D}}, \boldsymbol{\alpha}_{\mathrm{E}}$ vectors of parameters
D demand
S supply
Q transacted quantity, the amount purchased by households
ξ disturbance term, $\xi_{D_t}^l : N(0, \sigma^2)$,
$U_{D_t} = 1$ if $D_t > S_t$ and $U_{D_t} = 0$ if $D_t \leqslant S_t$.

12.3 CONSUMER SAVINGS AND BUDGET CONDITION IN A STATE OF DISEQUILIBRIUM

The stock of consumer savings (GSP) measured at the end of the period consists of two components: the stock of voluntary savings (VSP) and forced savings (FSP):

$$\text{GSP}_t = \text{VSP}_t + \text{FSP}_t \tag{12.6}$$

Obviously, in conditions of equilibrium, the second component in the above identity equals zero and the entire savings have a voluntary character.

Let us assume that because of insufficient supply of goods, and after all possible transfers of demand, consumers have accumulated forced savings at the end of the previous period, $\text{FSP}_{t-1} > 0$. In the current period they can neutralize a fraction of them, whereas the rest will enrich the purchasing power of households:

$$\text{FSP}_{t-1} = \eta\text{FSP}_{t-1} + \kappa\text{FSP}_{t-1} \tag{12.7}$$

where

η is the coefficient of neutralization of forced savings
$\eta \in [0, 1]$
κ is the coefficient of activation of forced savings for the purchase of goods, $\kappa = 1 - \eta$.

The neutralization of forced savings means that consumers will treat them (or part of them) as voluntary savings, planning to utilize them for the purchase of goods in the future, or keeping them in banks as a reserve. When households expect shortages to be temporary, and that price increases will not depreciate their savings, they neutralize the entire FSP_{t-1}-fund, which formally means $\eta = 1$. However, one may point out two main reasons why households keep forced savings. First, they constitute financial security for postponed demand, which occurs with intensity particularly in the durable goods market. Second, a significant role is played by consumers' expectations concerning changes in disequilibrium: when they expect an increase in excess demand one may expect that consumers will try to spend most of their forced savings on goods (thus, $\kappa \cong 1$). It should be particularly emphasized that the neutralization process is not constant over time. However, we treat it as such here, and thus further estimated values of the η and κ parameters express mean tendencies.

Theoretically, it is possible that the stock of voluntary savings may decrease, as a result of several factors. In the past it happened mainly due to market panic resulting from unstable political situations. People were buying primarily food products which could be kept for a long time (e.g. sugar, salt, and flour) and also necessities (textiles, footwear). A direct reason for money withdrawal from banks might be inflation as well, if combined with relatively low interest rates. A decrease in real income with a short-term consumer policy of protecting their standard of living may be the third reason for a voluntary savings decrease.[2]

Thus, the total amount of money which constitutes the effective budget constraint (so-called activated income, YXP) consists of three elements: current income (YP), part of forced savings (κFSP) and voluntary savings (ζVSP)

$$\text{YXP}_t = \text{YP}_t + \zeta\text{VSP}_{t-1} + \kappa\text{FSP}_{t-1} \tag{12.8}$$

[2]This is specially important if we consider the market period, or special consumer groups, e.g. pensioners, large families.

where ζ is the coefficient of utilization of voluntary savings for the purchase of goods, which we call the coefficient of dissavings, $\zeta \in [0, 1]$.

Let us note that processes of neutralization of forced savings and dissavings processes exclude each other. If the consumer decides to incorporate part of forced savings into voluntary savings, he or she will simultaneously not use the latter for the purchase of goods. Then the result in Equation 12.8 can be rewritten as

$$\eta \neq 0 \rightarrow \zeta = 0, \quad YXP_t = YP_t + \kappa FSP_{t-1}$$

or (12.9)

$$\zeta \neq 0 \rightarrow \eta = 0, \quad YXP_t = YP_t + \zeta VSP_{t-1} + FSP_{t-1}$$

For simplicity, henceforth we will be using formula 12.8, bearing in mind the relationship between the η and ζ parameters and its consequences.

To describe the whole process of household saving, let us assume that they have a normal propensity to save, which is directly independent of market situations. This hypothesis has been partly justified by research based on questionnaires (Strzelecka, 1984). It proves that even in the case of deep disequilibrium, consumers save a part of their income voluntarily. Savings which are the result of a normal propensity to save we will call normal savings. These are accumulated for two reasons: as a financial reserve for the future, and for the purchase of goods – mainly durables – whose expense exceeds current income. In the simplest case, one can assume that the increase in normal savings (ΔNSP) is a function of current income (YP), its increment (ΔYP), and the real interest rate paid by banks (TO):[3]

$$\Delta NSP_t = f(YP_t, \Delta YP_t, TO_t) \tag{12.10}$$

where Δ denotes the first difference.

Summing up, the stock of global savings (GSP) equals the stock measured at the end of the previous period, minus a portion of voluntary (ζVSP) and forced savings (κFSP) to be used for the purchase of goods, plus the difference between activated income (YXP) and the value of purchases (QP)

$$GSP_t = GSP_{t-1} - \zeta VSP_{t-1} - \kappa FSP_{t-1} + YXP_t - QP_t \tag{12.11}$$

Similarly, one can write the process of forming the stock of voluntary savings

$$VSP_t = VSP_{t-1} + \eta FSP_{t-1} - \zeta VSP_{t-1} + \Delta NSP_t \tag{12.12}$$

Note that in conditions of equilibrium the second and third components in the above formula equal zero. Hence the increase in voluntary savings is identical to the increase in normal savings.

Households having activated income (YXP) at their disposal plan to devote it to the purchase of goods and to increasing the stock of voluntary savings. This can be written as:

$$YXP_t = DP_t + \Delta NSP_t \tag{12.13}$$

[3]It can be proven that this function is a result of aggregation of the function describing the increase in bank savings and the stock of cash savings (A. Welfe, 1987b). For simplicity we omit the disturbance term, assuming that all classical conditions are met.

where

$$\mathrm{DP}_t = \sum_{l=1}^{L} D_t^l P_t^l$$

The above identity defines the allocation of activated income in *ex ante* categories. In conditions of disequilibrium the value of purchase is less than the planned demand, $\mathrm{QP}_t < \mathrm{DP}_t$, which results in forced savings. A comparison of Equations 12.6, 12.7, 12.11 and 12.12 allows us to show the *ex post* allocation of the YXP-fund:

$$\mathrm{YXP}_t = \mathrm{QP}_t + \Delta\mathrm{NSP}_t + \mathrm{FSP}_t \tag{12.14}$$

In equilibrium the last component in the above formula is equal to zero, of course. Also, activated income (YXP) is reduced under this condition to current income, which implies that Equation 12.14 has the classical shape

$$\mathrm{YP}_t = \mathrm{QP}_t + \Delta\mathrm{NSP}_t \tag{12.15}$$

12.4 THE SYSTEM OF EXPENDITURE FUNCTIONS WITH THE SAVINGS FUNCTION

The main component of total consumer demand (D^l) is the notional demand (DN^l) (this term was first used by Clower, 1965) which in equilibrium corresponds to the conventional understanding of consumer demand. For simplicity we assume that the notional demand is a function of two factors: real current income (Y) and relative prices (PC^l). Assume that at the end of the previous period consumers accumulated forced savings (FS_{t-1}), part of which they decided to use for the purchase of goods ($\kappa\mathrm{FS}_{t-1}$). Then a second component of total demand occurs – postponed demand. Assuming the linear form of function 12.3 it gives us:

$$D_t^l = \beta_0^l + \beta_1^l Y_t - \beta_2^l \mathrm{PC}_t^l + \beta_3^l \kappa \mathrm{FS}_{t-1} + \xi_{D_t}^l \tag{12.16}$$

where FS is the real value of forced savings (to be defined further).

The coefficients of allocation of forced savings β_3^l in the above function express two effects at the same time. One is the process of postponing demand. The second is transfers of demand of a special kind: the portion of forced savings left after fulfilling postponed demand will be treated by consumers as extra income, to be used only for the purchase of goods available in the market, and is thus allocated similarly to current income.

Function 12.16 does not cover explicitly an additional component of total demand: anticipated demand (A. Welfe, 1985), which is the result of consumer expectations. Anticipated demand raises purchases of goods above the level set by normal and postponed demand. Consumers, in wanting to avoid the consequences of price increases (or disequilibrium), are willing to purchase goods and stock them. Their expectations are based mostly on observed price changes. Thus, if the increase in relative prices brings a large increase in demand due to an anticipatory effect, the decrease in demand resulting from the price increase may be compensated for. Hence, if function 12.16 does not include anticipated demand, the estimated coefficient associated with relative prices may have a positive sign (see also Tobin, 1952).

The function of expenditures (Equation 12.4) also includes, besides total demand, excess demand, which is directly unobservable. It has been defined as a deterministic function of disequilibrium indicators.

Indicators of disequilibrium may be classified into two different categories. The first includes variables whose changes one believes to be the result of disequilibrium changes. The most well-known indicator of this type is the Fair–Jaffee (1972) price indicator, having its origin in Walrasian theory of general equilibrium. It has been widely used and modified by other authors (Laffont and Garcia, 1977; Bowden, 1978; Fair and Kelejian, 1977). The indicators belonging to the second group are based on a comparison of available characteristics determining potential demand and supply. Disequilibrium is defined here as the result of functioning factors included in the indicator. To construct this kind of indicator (A. Welfe, 1984), let us transform the identity defining excess demand, $\mathrm{DE}_t = D_t - S_t$, as follows:

$$\mathrm{DE}_t = S_t\left(\frac{D_t}{S_t} - \frac{D_{t-n-1}}{S_{t-n-1}}\right) + S_t\frac{\mathrm{DE}_{t-n-1}}{S_{t-n-1}} \tag{12.17}$$

If we assume that in the initial period $t-n-1$, disequilibrium does not exist (thus $\mathrm{DE}_{t-n-1} = 0$), the second component in the above formula equals zero, of course. The acceptance of a more general assumption, that excess demand in the initial period has a normal value, yields

$$\mathrm{DE}_t = \boldsymbol{\alpha}_{\mathrm{E}} S_t\left(\frac{D_t}{S_t} - \frac{D_{t-n-1}}{S_{t-n-1}}\right), \text{ for } \frac{D_t}{S_t} > 1 + \varepsilon \tag{12.18}$$

where ε is the normal value of relative excess demand

$$\varepsilon = \frac{\mathrm{DE}_{t-n-1}}{S_{t-n-1}}$$

The true values of demand and supply are unknown. However, it is possible to find variables which determine the potential limit for demand and supply. In the case of demand, it may be the volume of current income, or current income increased by forced savings (according to formula 12.8), if their value is known approximately. As concerns supply, one may use sales in socialized trade. They can be treated as instrumental variables.

The practical way of finding ε is not easy, however. It seems that the best solution of this problem is to accept that the closest level to equilibrium is indicated by the minimum of the ratio of determinants, respectively demand ($\tilde{D}$) and supply ($\tilde{S}$):

$$\mathrm{DE}_t^l = \tilde{\boldsymbol{\alpha}}_{\mathrm{E}} I_t^l \tag{12.19}$$

where[4]

$$I_t^l = \tilde{S}_t^l\left(\frac{\tilde{D}_t}{\tilde{S}_t^l} - \min\left\{\frac{\tilde{D}_1}{\tilde{S}_1^l}, \ldots, \frac{\tilde{D}_T}{\tilde{S}_T^l}\right\}\right)$$

Consequently, the function of consumer expenditures can be written as follows:

$$Q_t^l = \beta_0^l + \beta_1^l Y_t - \beta_2^l \mathrm{PC}_t^l + \beta_3^l \kappa \mathrm{FS}_{t-1} - \tilde{\alpha}_{\mathrm{E}}^l I_t^l + \xi_{D_t}^l \tag{12.20}$$

Constructing the savings function, we assume that the dissavings process does not occur in macroscale, which is further justified by empirical results. According to Equations 12.6 and 12.12 and assuming the linear form of the normal savings function defined by Equation 12.10, we obtain

$$\Delta \mathrm{GSP}_t = \psi_0 + \psi_1 \mathrm{YP}_t + \psi_2 \Delta \mathrm{YP}_t + \eta \mathrm{FSP}_{t-1} + \psi_3 \Delta \mathrm{IP}_t + \xi_{G_t} \tag{12.21}$$

where IP is the indicator of global disequilibrium.

We dropped out the TO variable, because the interest rate paid by banks was constant over many years in Poland, irrespective of price changes. Thus TO is identical for these years with the cost-of-living index.

Forced savings have been defined in formula 12.21 – analogously to excess demand (see Equation 12.19) – as a deterministic function of the global disequilibrium indicator, thus

$$\mathrm{FSP}_t = \psi_3 \mathrm{IP}_t \tag{12.22}$$

By directly substituting Equation 12.19 into 12.1, we obtain

$$\mathrm{IP}_t = \sum_{l=1}^{L} \tilde{\alpha}_{\mathrm{E}}^l I_t^l P_t^l \tag{12.23}$$

Taking into account the assumption expressed by Equation 12.1 and Equations 12.22 and 12.23 one would expect that $\psi_3 = 1$. However, because of possible errors in measurement we need to keep parameter ψ_3 in place, and treat it as the measure of approximations used.

The value of forced savings in constant prices (FS) in Equations 12.20 and 12.16 which determines the existence of postponed demand remains undefined. To recalculate it from current prices we suggest using a price index defined in the following way

$$\mathrm{PFS} = \sum_{l=1}^{L} \beta_3^l P_t^l \Big/ \sum_{l=1}^{L} \beta_3^l \tag{12.24}$$

which reflects the allocation of forced savings among different goods.

12.5 METHOD OF ESTIMATION USED

Notice that Equation 12.20 represents a set of L equations defining consumer expenditures on different goods. For identification purposes, the parameters of the

[4]This is the reversal of the trend-through-peaks method (Klein and Summers, 1966).

complete system should satisfy $L + 2$ nonlinear conditions (A. Welfe, 1987a). Moreover, combining Equations 12.23 and 12.24 with 12.20 and 12.21 makes it necessary to estimate a large number of parameters, which however depends partly on the number of goods distinguished.

In connection with what was said earlier, an iterative method is proposed. In its first step, the starting values of parameters $\tilde{\alpha}_E^1, \ldots, \tilde{\alpha}_E^L, \beta_3^1, \ldots, \beta_3^L, \eta$ and ψ_3 are set. Later, the values of assistant variables are calculated, which are defined below. Next, the estimation of the system is made by restricted ordinary least squares (OLS). The estimated values of coefficients $\tilde{\alpha}_E^l, \beta_3^l, \eta$ and ψ_3 constitute the starting point for the second step of the procedure. For the $m + 1$ iteration it can be written as follows:

$$Q_t^{l(m+1)} = \beta_0^{l(m+1)} + \beta_1^{l(m+1)}Y_t - \beta_2^{l(m+1)}PC_t^l + \beta_3^{l(m+1)}AFS_{t-1}^{(m)} - \tilde{\alpha}_E^{l(m+1)}I_t^l + \xi_{D_t}^{l(m+1)} \tag{12.25a}$$

$$\Delta GSP_t^{(m+1)} = \psi_0^{(m+1)} + \psi_1^{(m+1)}YP_t + \psi_2^{(m+1)}\Delta YP_t + \eta^{(m+1)}FSP_{t-1}^{(m)} + \psi_3^{(m+1)}\Delta IP_t^{(m)} + \xi_{G_t}^{(m+1)} \tag{12.25b}$$

$$IP_t^{(m)} = \sum_{l=1}^{L} \tilde{\alpha}_E^{l(m)} I_t^l P_t^l \tag{12.25c}$$

$$FSP_{t-1}^{(m)} = \psi_3^{(m)} IP_{t-1}^{(m)} \tag{12.25d}$$

$$AFS_{t-1}^{(m)} = (1 - \eta^{(m)})(FSP_{t-1}^{(m)}/PFS_t^{(m)}) \tag{12.25e}$$

$$PFS_t^{(m)} = \sum_{l=1}^{L} \beta_3^{l(m)} P_t^l \Bigg/ \sum_{l=1}^{L} \beta_3^{l(m)} \tag{12.25f}$$

Additionally, two conditions are imposed on the parameters of the above system:

$$\sum_{l=1}^{L} \beta_1^l + \psi_1 = 1 \tag{12.26a}$$

$$\sum_{l=1}^{L} \beta_3^l = 1 \tag{12.26b}$$

The first condition assumes that current income will be allocated between goods and savings (normal), the second that forced savings are meant for purchasing goods.

The procedure is continued until the differences between values of parameters $\tilde{\alpha}_E^l, \beta_3^l, \eta$ and ψ_3 obtained in succeeding iterations are smaller than the chosen threshold (here 0.0001). Convergence of the system was reached in empirical investigations in all experiments performed (around 150). These experiments differed in the starting values of the parameters and, slightly, in the specification of some equations (to be discussed in section 12.6). The number of iterations depended mostly on the starting values and varied between 10 and 40 iterations.

Restrictions imposed a priori on parameters may be treated as additional observations; thus the disturbance covariance matrix is singular. The restricted least squares estimator is unbiased if the restrictions are correct and then it is the best within the class of unbiased estimators that are a linear function of sample and nonsample information.

12.6 EMPIRICAL RESULTS

Empirical investigations were made based on statistical data on Poland and covering the period 1961–85. We distinguished six groups of goods: food, alcohol and alcoholic beverages, textiles with apparel and shoes, durables, other nondurables, and services. Together with the savings function, the system consists of seven functions. However, expenditures on individual residential structures, cigarettes and tobacco, and minor payments of consumption character were not considered. One may assume, with high probability, that demand was satisfied in these markets. Heterogeneity of this aggregate as well as some technical limitations associated with the size of the model caused us to omit additional equation(s). Condition 12.26a, however, has had to be modified, and in changed form yields summation of the parameters to 0.95, because omitted components constitute about 5% of total consumer expenditures.

Preliminary results pointed out the necessity of making two further modifications. First, into the equations explaining expenditures on durables we add the interactive variable changing the propensity to consume in the years 1972–75. The same variable was introduced symmetrically into the savings equation. This is due to the fact that in these years a very high acceleration of real income took place, which was treated by consumers as transitory income and saved rather than spent on goods. Second, parameters associated with relative prices in equations explaining expenditures on food, textiles with apparel and shoes, and durables, were set a priori, because their previously obtained estimates had been statistically insignificant and had incorrect signs. This was probably due to the fact that we had neglected anticipated demand in the specification of these equations. The process of anticipation of prices by consumers was especially strong in the eighties, particularly in food, textiles, shoes and durables markets. The results obtained in the last step of the iterative procedure discussed in section 12.5 are given in Table 12.1.

The values of the propensity to save do not raise any objections, and are similar to those obtained by other authors (see, for example, Tomaszewicz and W. Welfe, 1978). Special attention should be devoted to estimates of parameters associated with variable AFS. They express the combined effect of postponing and transferring demand. This explains why the highest value has the coefficient β_3^l in the equation for alcohol and alcoholic beverages (0.534). In years characterized by deep disequilibrium, in this market, consumers could thus spend their remaining money in the easiest way.

The estimate of the coefficient of neutralization of forced savings η equals 0.695 and is difficult to appreciate because one cannot find it in the work of other researchers. Notice that this value expresses the mean tendencies. However, it seems to be realistic and does not contradict common opinion. It can serve as the first approximation and be compared with other results in the future.

Consistent with the definition of forced savings (see Equation 12.1) and identities 12.22 and 12.23, the expected estimation of parameter ψ_3 should be close to unity. The empirical result fully justifies our expectations, which indirectly confirms the hypothesis ($\hat{\psi}_3 = 0.911$). One should also underline the significance of estimates of parameters of the system, and high R^2-coefficient values.

Using formula 12.19 the value of excess demand was calculated. In Fig. 12.1 the trajectories of expenditures (solid line) and demand (broken line) for three groups of goods are presented. The results are generally in accordance with those previously obtained (A. Welfe, 1985, 1986), in spite of enlarging the statistical sample by the years 1984–85. It seems that they confirm common observation and do not contradict public opinion.

We would like to focus on the last decade, because it is easier to judge that period's disequilibrium. The middle seventies are widely considered to have been the period when the consumer market was in its best condition. Excess demand for textiles, apparel and shoes increased substantially in 1978, reaching almost 18% of total sales (see Fig. 12.1(b)). The next year brought an increase in disequilibrium in the durable goods market (excess demand was 7.5%), and 1980 saw a further deterioration in the situation. A real breakdown took place in 1981. The state authorities decided to introduce a rationing system which covered food products (14% of excess demand), alcoholic beverages (42% of excess demand), gasoline and some necessities (footwear, baby clothes). Total excess demand in this year reached the level of 16% of total supply (sales). According to our estimates, the situation

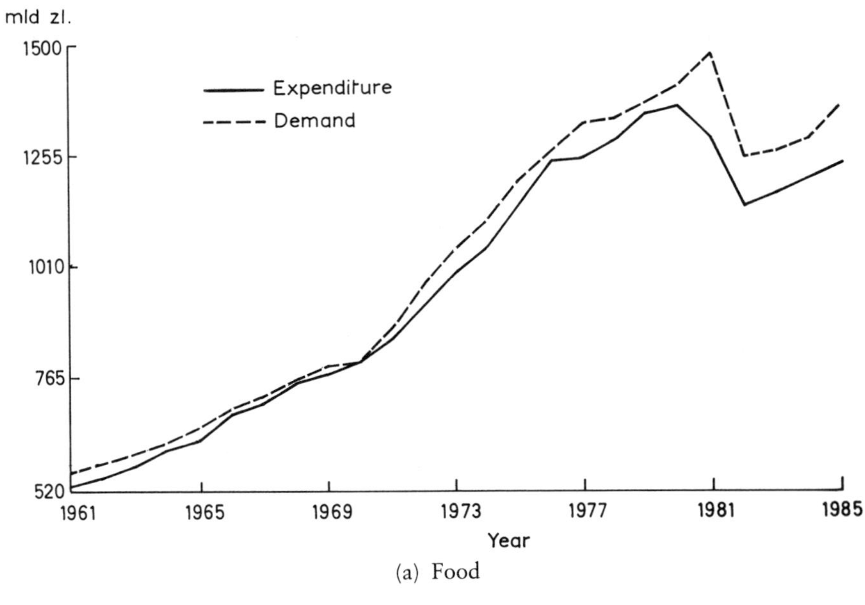

(a) Food

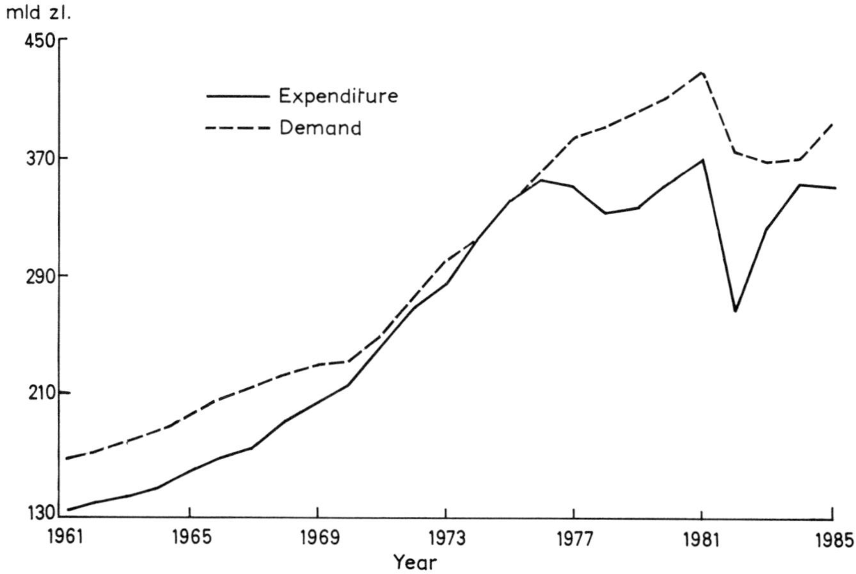

(b) Textiles, clothing and shoes

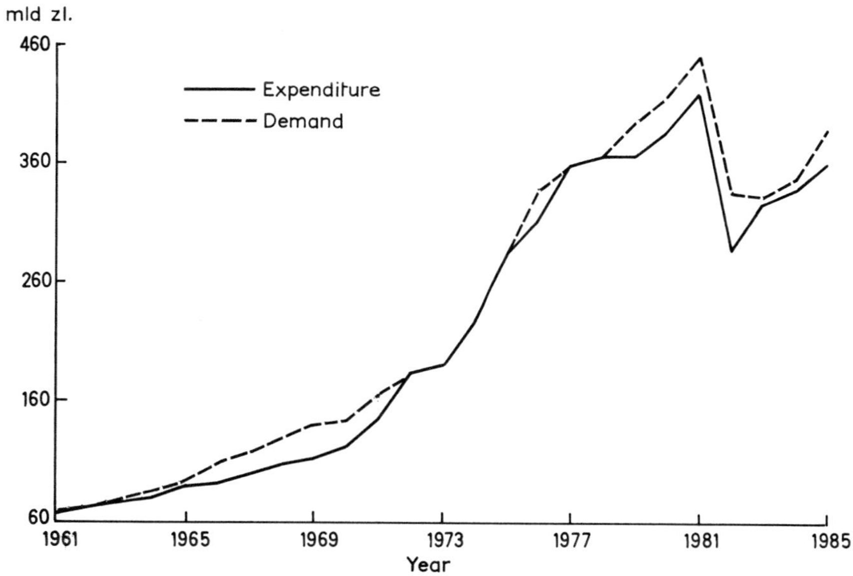

(c) Durable goods

Fig. 12.1 Expenditures and demand for (a) food; (b) textiles, clothing and shoes; (c) durable goods in Poland.

Table 12.1 Estimates of expenditure functions and savings function sample 1961–85

Endogenous variables	*Exogenous variables*									
	Constant	Y_t	$Y_t \cdot U7275$	PC_t^1	AFS_{t-1}	I_t^1	ΔY_t	FSP_{t-1}	ΔIP_t	R^2
Food	232.228	0.307	—	−40.0*	0.132	−2.386	—	—	—	0.999
	(6.01)	(80.27)		—	(0.91)	(20.01)				
Alcoholic beverages	33.468	0.157	—	−145.846	0.534	−0.014	—	—	—	0.979
	(0.30)	(28.00)		(5.92)	(1.40)	(14.70)				
Textiles, clothing and shoes	99.221	0.082	—	−20.0*	0.179	−0.801	—	—	—	0.997
	(20.16)	(70.99)			(2.03)	(20.50)				
Durable goods	−20.762	0.119	−0.014	−50.0*	0.073	−0.002	—	—	—	0.993
	(1.05)	(50.50)	(7.90)		(2.71)	(8.00)				
Other nondurable goods	−49.213	0.227	—	−56.311	0.048	−0.018	—	—	—	0.998
	(3.00)	(73.09)		(2.80)	(4.00)	(37.20)				
Services	22.523	0.032	—	−12.669	0.035	−1.759	—	—	—	0.971
	(5.02)	(20.01)		(1.01)	(2.30)	(4.91)				
Savings	−27.746	0.035	0.03	—	—	—	0.049	0.695	0.911	0.989
	(2.10)	(2.90)	(5.60)				(3.70)	(3.40)	(8.01)	

*The value of the parameter was set a priori. *T*-values are in parentheses.

was even more dramatic in 1982 because of extremely high excess demand for textiles, apparel and shoes (close to 40%) and durables (over 17%). Disequilibrium in the food market (see Fig. 12.1(a)) was still fairly high (almost 10%). After the improvement in 1984, 1985 came to an end with a significant increase in excess demand in all markets excluding alcohol.

There are various extensions of the analysis which will be explored in future work. One can try to incorporate into the demand function anticipated demand, which calls for the formulation of households' expectations. It may be very interesting to estimate the amount of initial excess demand and demand transfers as well. It would also be very important from a practical point of view, because it would allow the calculation of price (or supply) increases necessary to balance the consumer market. Finally, the same structure of expenditure–savings system could be tested with different disequilibrium indicators.

ACKNOWLEDGEMENTS

I would like to take this opportunity to thank Professor W. Welfe for very stimulating and clarifying discussions. I wish also to express my gratitude to Mary LaRue for her language editorial work and typing. The Polish Centrally Co-ordinated Program 10.9 and Project LINK are gratefully acknowledged.

REFERENCES AND FURTHER READING

Amemiya, T. (1974) A note on a Fair and Jaffee model. *Econometrica*, **42** (3), 759–69.

Barro, R. J. and Grossman, H. J. (1971) A general disequilibrium model of income and employment. *American Economic Review*, **61**, 82–93.

Bowden, R. J. (1978) *The Econometrics of Disequilibrium*, North-Holland, Amsterdam.

Charemza, W. and Gronicki, M. (1984) *Ekonometryczna analiza nierownowagi gospodarczej Polski*, PWN, Warsaw.

Clower, R. W. (1965) The Keynesian counter-revolution: a theoretical appraisal, in *The Theory of Interest Rates* (eds F. H. Hahn and F. Brechling), Macmillan, London.

Fair, R. C. and Jaffee, D. M. (1972) Methods of estimation for markets in disequilibrium. *Econometrica*, **40** (3), 497–514.

Fair, R. C. and Kelejian, H. H. (1977) Methods of estimation for markets in disequilibrium: a further study. *Econometrica*, **42** (1), 177–90.

Kalecki, M. (1941) General Rationing, *Oxford Bulletin of Statistics: Studies in War Economics*, Blackwell, Oxford.

Klein, L. R. and Summers, R. (1966) *The Wharton Index of Capacity Utilization*, University of Pennsylvania, Philadelphia.

Kolupa, M. (1961) O podstawowych twierdzeniach dla problemu eliminacji obciazen parametrow funkcji popytu. *Przeglad Statystyczny*, 8 (4), 395–400.

Kolupa, M. (1965) *Badanie popytu w warunkach niedostatecznej podazy*, PWE, Warsaw.

Kornai, J. (1971) *Anti-Equilibrium*, North-Holland, Amsterdam.

Kornai, J. (1980), *Economics of Shortage* North-Holland, Amsterdam.

Kornai, J. (1982) *Growth, Shortage and Efficiency. A Macroeconomic Model of the Socialist Economy*, Basil Blackwell, Oxford.

Laffont, J. J. and Garcia, R. (1977) Disequilibrium econometrics for business loans. *Econometrica*, **45** (5), 1187–204.

Leijonhufvud, A. (1968) *On Keynesian Economics and the Economics of Keynes*, Oxford University Press, Oxford.

Maddala, G. S. and Nelson, F. D. (1974) Maximum likelihood methods for models of markets in disequilibrium. *Econometrica*, **42** (6), 1013–30.

Neisser, H. F. (1943) Theoretical aspects of rationing. *Quarterly Journal of Economics*, **57**, 378–97.

Pawlowski, Z. (1960) Problem nieobciazonosci parametrow funkcji popytu a przypadki przejsciowej niedostatecznej podazy dobr na rynku. *Przeglad Statystyczny*, 7 (1), 41–50.

Pawlowski, Z. (1961) *Ekonometryczne metody badania popytu konsumpcyjnego*, PWN, Warsaw.

Portes, R., Quandt, R. E., Winter, D. and Yeo, S. (1982) Macroeconomic planning and disequilibrium: some preliminary results for Poland, 1955–1980, Discussion Paper, No. 126, Birkbeck College, London.

Quandt, R. E. (1986) *Bibliography of Quantity Rationing and Disequilibrium Models*, Princeton University, New Jersey.

Rothbarth, E. (1941) The measurement of change in real income under conditions of rationing. *Review of Economic Studies*, **8**, 100–7.

Scitovsky, T. (1942) The political economy of consumers' rationing. *Review of Economic Statistics*, **24**, 114–24.

Strzelecka, M. (1984) Zmiany w konsumpcji gospodarstw domowych w latach 1980–1983, Instytut Rynku Wewnetrznego i Konsumpcji, Warsaw.

Tobin, J. (1952) A survey of the theory of rationing. *Econometrica*, **20**, (4), 521–53.

Tomaszewicz, A. and Welfe, W. (1978) Popyt konsumpcyjny jako funkcja dochodow osobistych gospodarstw domowych, in *Modele konsumpcji. Ekonometryczne modele rynku* (ed. W. Welfe), PWE, Warsaw.

Welfe, A. (1984) Analiza popytu w warunkach nierownowagi. *Ekonomista*, No. 5, pp. 1045–65.

Welfe, A. (1985) *The System of Demand Equations in a State of Disequilibrium*, International Institute for Applied Systems Analysis, CP-85-35, Laxenburg.

Welfe, A. (1986) Zgodny uklad rownan popytu w warunkach nierownowagi. *Przeglad Statystyczny*, **33**(3), pp. 258–82.

Welfe, A. (1987a) Estymacja zgodnego ukladu rownan popytu w warunkach nierownowagi. *Przeglad Statystyczny*, **34** (2), pp. 133–43.

Welfe, A. (1987b) Procesy oszczedzania gospodarstw domowych w warunkach nierownowagi. *Finanse*, No. 25–32.

Welfe, W. (1973) A medium-term econometric model of the Polish economy. *Prace IEiS UL*, No. 2, Lodz.

Welfe, W. (1976) The unbalanced econometric macromodels, in *Computing Equilibria: How and Why* (eds J. Los and M. W. Los), North-Holland, Amsterdam.

13

A disequilibrium approach to modelling foreign trade in centrally planned economies

JOSEF C. BRADA and ARTHUR E. KING

13.1 THE THEORY

It is rare in any area of economics that the theoretical framework for the analysis of a major topic should spring in virtually complete form from the mind of one individual and that it should then remain almost unchanged for nearly three decades. Yet this is precisely the case in the analysis of the response of the centrally planned economy (CPE) to changes in foreign trade flows. The analytical framework developed by Franklyn D. Holzman in the 1950s has remained the dominant, if not the only, framework for the analysis of foreign trade adjustment mechanisms in CPEs.[1]

13.1.1 The traditonal approach

Holzman's analysis considers the classical CPE. Such an economy has an inconvertible currency and practices price-equilization, by which domestic suppliers or users of tradables transact their sales and purchases at arbitrarily fixed domestic prices. Thus neither changes in the (arbitrary) exchange rate nor in the terms of trade are felt by agents in the domestic economy. Domestic production, consumption and investment decisions as well as decisions regarding the volume and composition of foreign trade are made by central planners. Planners act so as to maintain full employment of labour and other resources, possibly by maintaining excess demand in factor and product markets through over-full-employment planning.

[1]Holzman's work on the subject is conveniently collected in Holzman (1974, 1987).

International Studies in Economic Modelling Series.
Models of Disequilibrium and Shortage in Centrally Planned Economies.
Edited by C. Davis and W. Charemza.
Published in 1989 by Chapman and Hall Ltd, 11 New Fetter Lane, London EC4P 4EE.

Within this framework, Holzman demonstrated, many of the traditional foreign trade adjustment mechanisms found in market economies do not exist in the CPE. For example, an increase in imports would not serve to reduce domestic production through the foreign trade multiplier because planners would maintain full employment. Similarly a change in the terms of trade would not affect the volume of trade or the domestic level of production and its pattern unless planners altered them. Nevertheless, the CPE was not immune to the effects of changes in trade flows; rather these changes made themselves felt through real rather than market effects. The most important of these real effects was dubbed by Holzman the foreign trade bottleneck multiplier. Any change in the volume of imports or, to a lesser degree, of exports, would disrupt the inter-industry flows of inputs. Some of these inputs would then be in short supply and at full employment could not be replaced through domestic production without reducing the production of other commodities. Moreover, in the case of imports, some commodities either could not be replaced by means of domestic production or could only be replaced at prohibitively high resource cost. Thus, the decline in availability of certain imports would create bottlenecks forcing the reduction of production in sectors requiring importables as inputs, and such reductions in these sectors would then force downstream industries to cut back their production as well. As a consequence, domestic production would decline through a multiplier process even though full employment of resources was maintained by the planners. Of course, the effect operates in the other direction as well. Increases in imports break bottlenecks, and thus small increases in imports can have a multiplier effect that yields large increases in domestic production.

Because planners maintain full employment amidst fixed wages and prices, there is also the potential for a monetary effect of trade fluctuations on domestic activity. The development of this line of reasoning clearly anticipates the modern disequilibrium literature associated with Barro and Grossman.[2] If net imports of consumer goods decline, then, unless planners act to change the consumption–investment mix, the supply of consumer goods must fall. With full employment and fixed prices and wages, excess demand for consumer goods either appears or is exacerbated and work incentives suffer, leading to a further decline in production. Holzman himself believed this to be a minor problem that would '... have only a minor effect on the real economic variables' (1974, p. 114). While this may have been true in the case of the Soviet Union in the 1950s, when consumer goods imports were virtually non-existent, it may have been an important factor for CPEs in the 1970s and 1980s when consumption goods imports were much greater.

13.1.2 The modern theory: formalization and extension

Recent writing on foreign trade adjustment mechanisms in CPEs has either sought to formalize and integrate Holzman's analysis by means of macroeconomic models that build on the disequilibrium framework of Barro and Grossman (1974, 1976) or

[2]Here Holzman was anticipated by Ames (1953). Pryor (1963) is also an important contribution.

to extend Holzman's work to reflect the changing relationship between foreign trade flows and domestic economic activity wrought by reforms in the economic systems of the East European CPEs.

One model which casts Holzman's analysis in the framework of the Barro–Grossman disequilibrium model is that of Portes (1980). Portes constructs a macroeconomic model of a CPE that permits, but does assume the existence of a number of possible combinations of excess supply, excess demand and rationing on labour and consumer goods markets. This generality in disequilibrium regimes is obtained at the expense of a good deal of realism and severely limited possibilities for empirical verification (Ames, 1980; Hewett, 1980; Holzman, 1980). Households supply labour in response to real wages, money balances and, if rationed, to the supply of consumer goods. Their demand for goods is also a function of real wages and money balances, as well as of the demand for labour if there is unemployment. Thus the traditional spillover from consumer goods shortages to the supply of labour may occur if households are rationed. Aggregate output is produced by means of a fixed-coefficient production function with domestic value added (capital and labour) and an imported intermediate good as the two inputs. Aggregate output is divided by planners between consumption, exports and government purchases. Both the imported intermediate good and the exported final good are traded at fixed foreign prices.

Portes assumes that planners seek to maximize a welfare function with consumption and government purchases as arguments subject to a balance of payments constraint as well as to constraints imposed by the technology and by resource availability. Planners manipulate real wages and the volume of exports to achieve their objectives. Since imports are a linear function of domestic production, Portes is able to show how changes in the instruments influence the trade balance and the balance between demand and supply in the consumer goods market. Under appropriate conditions this model can yield a response of output to external shocks that is similar to Holzman's foreign trade bottleneck and monetary effects.

A model more in keeping with Holzman's analysis of CPE adjustment mechanisms is that of Brada (1980, 1982). Brada's analysis of disequilibrium is limited to either equilibrium or excess demand on the markets for labour and consumer goods and excess supply of money, all reflecting the conventional view of macroeconomic disequilibrium in CPEs. This simplification of the disequilibrium regime permits a richer modelling of the macroeconomy. As in Portes' model, households supply labour on the basis of wages, money holdings and the availability of consumer goods; disequilibrium in the money market is added as an additional explanatory variable. Consumption demand is a function of wage incomes (real wages times employment), money stocks, and excess cash holdings. Production is specified in a more realistic way with output depending, in the short run, on employment and exports and imports of intermediate goods. Brada also models explicitly the supply and demand for money on the part of households, and portfolio imbalance thus plays a role in household behaviour. Trade flows, and consequently the effect of their alterations on the macroeconomy, are also more complex. There are four trade flows, exports of final goods and of intermediate goods and the imports of each.

Brada shows that an increase in net exports of the intermediate good must have both a real or foreign-bottleneck-multiplier effect, as well as a monetary one which results from spillovers from consumer goods and money market disequilibria to the supply of labour. In the case of an increase in net exports of final goods, output falls not through the bottleneck multiplier but only as the result of spillovers from the goods and money market. This richer description of the effect of the trade flows on the economy also leads to a more realistic view of the policy choices facing planners, who are assumed to maximize a welfare function with domestic output and the balance of trade as arguments. In such a framework planners need not strive for either domestic or external equilibrium. Hence, in the face of deteriorating terms of trade, planners may opt for a combination of a greater external payments deficit and increased excess demand for goods and excess supply of cash domestically. The policy tools that planners are likely to use to manage disequilibria also differ from those proposed by Portes. Among them are changes in the mix of final and intermediate goods exports and imports, reductions in investment, and domestic inflation accompanied by nominal wage increases to wipe out excess cash hoards.

The other line of research on the foreign trade behaviour of the socialist economies does not explicitly employ the disequilibrium framework of spillovers, but will be sketched here because it raises important issues about the verification of any macroeconomic model of disequilibrium in a CPE. Because of foreign trade reforms undertaken in the 1970s, most notably in Hungary, but in other planned economies as well, some of the assumptions of the classical CPE trade model are no longer valid, and Thomas Wolf (1978, 1980a, 1980b), has explored the behaviour of a modified centrally planned economy(MCPE). In the MCPE, central planning is assumed to be circumscribed to allow enterprises to respond to relative prices in seeking maximum profits; some prices remain fixed while others are free to change in response to market prices; a link is established between the world market price and the domestic price of a subset of tradables; and enterprises are allowed to make some foreign trade decisions on the basis of profitability considerations.

Within the framework of such a reformed economic mechanism, Wolf examines the responses of the MCPE to external shocks with a view to uncovering their implications for domestic resource allocation, production, consumption and trade in fixed-price and flexibly priced goods. Wolf shows that certain decisions regarding the exchange rate, prices or commercial policy can lead to perverse responses by the MCPE to external inflation or to chnges in the terms of trade. While Wolf assumes that full employment is maintained, and that disequilibria faced by consumers on markets for certain goods do not lead to spillover effects on the supply of labour, his analysis is important for disequilibrium macromodelling since the testing of an aggregate macromodel of a MCPE would need to capture resource shifts between sectors and resulting exchange payments imbalances in order to separate such relative price effects from the effects of macroeconomic disequilibria on these same variables.

13.1.3 Implications of theory for empirical modelling of centrally planned economies

The clearest message of Portes' theoretical work for modellers is that the best way to model a CPE is to specify supply and demand equations for domestic markets and the spillover effects from one market to another, and to estimate the entire system, letting the data reveal periods of excess demand and supply and therefore the comparative statics of the system at any period of time. Such a procedure appears to be beyond currently available econometric techniques and consequently the use of the insights provided by theory is perforce partial and imperfect. Nevertheless, the models outlined above, if taken seriously, do provide some guidance for the specification of macroeconomic models.

If one accepts the Holzman–Brada framework, then the domestic economy is assumed to be either at equilibrium or characterized by excess demand for consumer goods and labour and excess supply of cash. This means that domestic activity can be viewed as being supply determined. Linkages between trade flows and domestic economic activity center on the supply multiplier and care must be taken to account for trade in final (consumer) goods and intermediate inputs since their macroeconomic effects may be both quantitatively and qualitatively different. Moreover, planners' equilibrating behaviour should be seen as derived from the maximization of a utility function with domestic production and the trade balance as arguments, suggesting that there are tradeoffs between domestic and balance of payments equilibrium.

If one accepts the Portes model of CPE foreign trade, then it is not permissible to specify the model as supply driven. Moreover, with planners maximizing consumption and investment (government purchases) subject to a balance of payments constraint there can be no trade-off between domestic disequilibrium and the trade balance. Rather exports of the final goods and real wages are the tools used to adjust the economy to domestic equilibrium and the payments target. Such requirements impose severe restrictions on macroeconomic modelling although, as we shall see in the next section, some ingenious efforts have been made by Portes and his associates, as well as by others, to overcome these difficulties.

Finally, the implications of Wolf's work on the MCPE would seem to be that any modelling of the recent experience of the reformed economies of East Europe, most notably of Hungary and Poland, must take into account the relative prices of tradables and non-tradables, the exchange rate and the relationship between the domestic and foreign prices of goods with fixed domestic prices and those with flexible prices in determining consumption, production and trade. These requirements impose severe demands on would-be model builders. First of all, a concern with relative prices implies a degree of disaggregation that may be difficult to achieve on the basis of the data available. Moreover, it is difficult to obtain data on trade flows in both domestic and foreign prices. Thus, practically speaking, broad proxies would have to employed if one sought to implement Wolf's insights into macroeconometric models.

13.2 EMPIRICAL MODELS OF CENTRALLY PLANNED FOREIGN TRADE

Despite the conceptual difficulties of modelling disequilibrium, the appeal of macroeconometric models as tools for forecasting and policy analysis has led to the construction of a variety of macroeconometric models for all the centrally planned economies. In building these models, researchers have had to resolve many difficult theoretical and data questions, with foreign trade only one, albeit an important one, of the considerations that must be kept in mind. Because of this generality and the resulting lack of focus on the theory of domestic and external disequilibrium of the large macroeconometric models, other researchers have turned to small, more manageable, econometric models whose primary purpose is to make the connection between domestic disequilibrium and the trade sector transparent and testable. In this section, we survey both of these types of models.

13.2.1 Foreign trade in large econometric models of centrally planned economies

It is not our purpose here to survey all, or even a large proportion of the econometric models of CPEs, in part because there are too many such models and in part because this would duplicate extant surveys.[3] Rather, we examine the way in which trade and disequilibrium are modelled in some typical macroeconometric models. Among the best of these are those of Tomczyk (1987) who sets out to construct a model of CMEA trade based on the linking of national models through foreign trade equations. Tomczyk assumes that disequilibrium exists both within the economy of each CMEA country and in its trade relations with other CMEA countries and with the West. Thus, in cases of excess demand, observed quantities are modelled as being generated by a supply schedule, while in cases of excess supply, observed quantities are modelled as being generated by a demand schedule. Interactions between trade and the domestic economy are carefully specified but cannot be construed as true spillover effects.

The domestic goods market is assumed to be characterized by excess demand and thus industrial and agricultural production are modelled by means of production functions. Non-agricultural production depends on domestic inputs of capital and labour and on imports of fuels and raw materials. The inclusion of imports in the production function is at the same time appealing and methodologically troublesome. It is troublesome because the output measure in the equation is

[3] Among these are Shapiro and Halabuk (1976), Shapiro (1977), King (1979), Kemme (1981).

net material product (NMP) in industry and it is well known that it is not permissible to include intermediate inputs in a production function for net output. At the same time, it is reasonable to expect that, particularly in the 1970s and early 1980s, fuel and raw material availability must have represented an important bottleneck in production, and thus to exclude the effects of fuel and raw material shortages from a model of domestic non-agricultural production would be to miss a major linkage between foreign trade and the domestic sector. In keeping with the excess demand hypotheses, domestic consumption and investment are modelled as functions of domestic production and imports of either consumer goods or machinery, respectively.[4] While the model thus allows for the operation of a foreign-trade bottleneck effect on production, the monetary effect, where disequilibrium on the consumer goods or money market reduces the supply of labour time or effort, is nonexistent because employment or labour supply is an exogenous variable.

Trade in this model, as in virtually all other macroeconometric models of CPEs, is disaggregated into four commodity groups according to the CTN nomenclature, and into trade with CMEA and non-CMEA (developing and developed market) economies. The disaggregation of foreign trade creates a good deal of complexity as does the fact that import and export equations must reflect both domestic factors, including disequilibrium, and possible disequilibria in foreign markets as well. Thus in intra-CMEA trade, fuels and non-food raw materials and food are in excess demand while machinery and equipment as well as consumer manufactures are in excess supply. Exports of the former are therefore modelled as being supply determined. For example, exports of fuels and raw materials are modelled as a function of industrial output, lagged imports of fuels to reflect bilateral balancing in hard goods, and hard currency debt variable to account for the diversion of exports to Western markets. The feedback between trade and the domestic economy thus comes through the industrial output variable since industrial output is influenced by imports of fuels and raw materials, as described above, and by previous imports of machinery nd equipment.[5] Agricultural exports are modelled in analogous fashion, but since the output of agriculture is independent of import flows, there is no interaction between the domestic economy and agricultural exports to CMEA. The exports of the CMEA countries for these goods are pooled and the total is then allocated to individual countries by means of allocation equations that have a number of trade flows as explanatory variables.

Imports of machinery and equipment and of consumer manufactures are demand driven. Machinery imports of each country depend on investment lagged total exports to socialist countries, and current and lagged machinery imports from the West. For consumer manufactures, consumption and consumer manfactures

[4]It is rather curious that here, as in the production function above, imports rather than net imports are taken as the measure of availability or shortage.

[5]The output of fuels and non-agricultural raw materials is considered part of industrial output. Thus, rather paradoxically, the model uses industrial output to reflect both the consumption of fuels and raw materials and their production.

imports from the West replace investment and Western machinery imports as activity variables. Total import demand for CMEA is pooled and then allocated to exporters by means of trade-driven equations.

Exports of fuels and raw materials, machinery and equipment, and consumer manufactures to non-CMEA countries are modelled as being demand determined, with some effort to include supply constraints. Exports of each of the three categories depend on total exports of the category by non-CMEA countries, relative prices of CMEA and non-CMEA suppliers of the commodity, deviations from the normal ratio of non-CMEA fuel and raw materials imports to industrial production and the lagged value of the commodity group's exports to the non-CMEA region. Food exports are also demand determined with deviations from the normal level of the domestic food consumption to production ratio replacing the fuels and raw materials to industrial production variable.

Imports from the non-CMEA region are modelled in a two-step process. Nominal demands depend on the level of domestic production of the commodity group, relative prices in CMEA and world trade, and imports of the commodity group from socialist countries. These demands are then adjusted to bring them into conformity with a hand-currency budget constraint which depends on export earnings and a willingness to borrow to finance the deficit.

The specifications of this model generally reflect the state of the art not only in East Europe, where data availability, institutional support and professional interest have pushed efforts beyond those of Western researchers, but also among Western modellers of CPEs.[6] To the extent that one can uncover causality within a large scale econometric model, the following generalizations can be made about the role of disequilibrium in modelling. First, the principal way in which disequilibrium enters these models is through the process of specification of behavioural equations. That is, model builders assume that excess demand or supply is the ruling regime for a particular market and then model market behaviour as either determined by the supply schedule or by demand. Since prices are inflexible it is unclear whether these choices are made because modellers actually believe in the existence of pervasive disequilibrium or because the modelling techniques based on simultaneous estimation of prices through the interaction of supply and demand is inapplicable. Less methodologically sound is the frequent practice of mixing supply and demand elements to reflect a ruling regime and the influence of periodic shortages or surpluses arising from shocks to the unmodelled side of the market. A second generalization is that spillovers from disequilibrated markets to other

[6] Among the noteworthy East European modelling efforts are those of Ivan Šujan of the Computing Research Centre in Bratislava (see, for example, Šujan 1987a, 1987b) for Czechoslovakia; of György Szakolczai at the Laboratory of Econometrics at the Institute of Economics of the Hungarian Academy of Sciences (see Szakolczai *et al.*, 1987) for Hungary; and the Welfe group at Lodz for Poland. Among Western models are several of the Soviet Union (surveyed in Hildebrandt, 1985: of Czechoslovakia, Brada and King, 1979; Brada, King and Schlagenhauf, 1981; and of Romania, Brada, Jackson and King, 1984).

A noteworthy and unique contribution is Vaňous (1980) who models CMEA trade flows using proxies for domestic disequilibrium as explanatory variables in a model of the foreign trade of CMEA countries. Vaňous, however, views the domestic economy as exogenous and models on interaction between trade and the domestic economy.

markets are modelled only infrequently. Tomczyk, for example, models only the effect of shortages of fuels and raw materials on production; Brada and King (1979) on the other hand, model only spillovers from consumer market disequilibria to the supply of labour effort. This is understandable, if for no other reason than that there are only so many variables that one can include in any equation, and, as we mentioned above in connection with Tomczyk's production function, it is not evident that simple ways of including such spillovers are methodologically defensible. Particularly weak are the links between domestic disequilibrium and the foreign trade sector. Finally none of the available models even attempts to deal with the issues raised by Wolf. Exchange rates and the relationship between domestic and foreign prices are entirely absent from these models.

In defence of these models and their creators, it must be admitted that similar sins are committed by creators of large-scale econometric models of market economies for which the theoretical underpinning is considerably more developed. What is disappointing is that all the models seem to fit well and to track well despite disparate assumptions about how specific sectors should be modelled and about the nature and effects of disequilibrium. That large scale macroeconometric models are not tools for hypothesis testing is nowhere better shown than in the case of models of CPEs.

13.2.2 Testing the relationship between disequilibrium and trade in CPEs

In view of the difficulty of testing hypotheses regarding disequilibrium and its effects on foreign trade, researchers have turned to smaller models of CPEs where the conceptual complexity of excess demand and supply regimes can be properly specified, estimated and tested. A modelling strategy oriented toward hypothesis testing must, of course, be especially well grounded in the underlying macroeconomic theory of disequilibrium in the CPE and must relate specification of the model to this theory in a direct and transparent way that makes it clear how the specification employed can confirm or disconfirm the hypotheses being tested. One of the available efforts in this direction, and clearly the most rigorous, is the work of Burkett, Portes and Winter (1981) who set out to model the effects of domestic disequilibrium on the trade flows of four CPEs, Czechoslovkia, the GDR, Hungary and Poland. We examine first the theoretical basis of their model and then discuss their findings.

The theory underlying the exercise is the model proposed by Portes described in the previous section wherein planners attempt to maximize a welfare function whose arguments are consumption and government expenditures (including investment) subject to a balance of payments target, and constraints imposed by the need to maintain equilibrium in the goods and labour markets. That is, planners *ex ante* plan for equilibrium which *ex post* may or may not be realized as the result of errors in plan construction and exogenous shocks.[7] Two assumptions of the

[7] Why planners would be unwilling to trade off a greater balance of payments surplus (deficit) for greater (smaller) domestic disequilibrium is unclear. Moreover, Portes' formulation of the problem is akin to viewing policy-makers in a market economy as setting policy so as always to achieve full employment and price stability subject to a balance of payments constraint.

Portes model are modified so that export prices are now given by a downward sloping demand curve and the ratio of imports to domestic production is not fixed The specification derived from this model consists of three equations:

$$M_t = a_0 + a_1\mathrm{NMP}_t + a_2\mathrm{PX}_t + a_3\mathrm{PM}_t + a_4TB_{t-1} + a_5\mathrm{ED}_{t-1} + u_{1t} \tag{13.1}$$

$$X_t = b_0 + b_1\mathrm{NMP}_t + b_2\mathrm{PX}_t + b_3\mathrm{PM}_t + b_4\mathrm{TB}_{t-1} + b_5\mathrm{ED}_{t-1} + u_{2t} \tag{13.2}$$

$$\mathrm{PX}t = C_0 + C_1X_t + C_2\mathrm{WT} + C_3\mathrm{PW}_t + u_{3t} \tag{13.3}$$

where

X_t, M_t = exports and imports in year t
NMP_t = net material product in year t
PX_t, PM_t = prices of exports and imports in year t
TB_{t-1} = exports/imports in year $t-1$
ED_{t-1} = (estimated consumers' demand/estimated supply of consumer goods) in year $t-1$
WT_t = weighted average of CMEA and market economy import values in year t
PW_t = weighted average of CMEA and market economy export prices

Two questions that must immediately arise regarding this specification are whether it is consistent with the Portes model and with what other models might it be consistent. On the first point, Burkett, Portes and Winter write:

> The planners are considered to fix the volumes of exports and imports in relation to four different determinants: the general level of activity in the economy, prices, balance of payments constraint, and excess demand pressures within the economy. The general level of activity is measured by net material product (NMP), prices by the appropriate unit value indices (PX for exports, PM for exports). The balance of payments constraint is measured by the last period's trade balance defined in ratio form (TB).
>
> Two quite different variables were used to measure excess demand inside the CPE. Excess demand can occur either in the consumption goods market or in the investment goods market... In the event, our measure of excess demand for investment goods, the growth of gross investment and defense expenditures, never had a significant role and was dropped from the specification at an early stage of estimation.
>
> Our measure of excess demand on the consumer goods market was taken from the disequilibrium estimates given in Portes and Winter (1980)... Our excess demand variable here is the ratio of fitted demand to fitted supply at constant prices (Burkett, Portes and Winter, 1981).

Since the empirical specification is not based on a fixed relationship between domestic production and imports, as it is in the Portes model, Equation 13.11 lacks any means of determining how planners decide on the substitution of imported inputs for domestic ones, particularly labour, so as to maximize their objectives.[8]

[8]This is precisely the issue raised by Wolf in his models of planned economies' trade behaviour.

Moreover, imports now consist of final goods as well as of intermediates. Thus while greater output, *ceteris paribus*, may require greater imports of inputs such as fuel and raw materials, it should also reduce the need for imports of final goods. If exports are a balancing item for domestic goods markets, then it is unclear why, *ex ante*, they are not derived from the balance of payments target and the desired volume of imports rather than as specified in Equation 13.2. That is, if planned exports and imports are assumed to bear a fixed relationship determined by the payments target then they must be modelled by a system of simultaneous equations not through two entirely independent equations. The use of the lagged ratio of exports to imports in Equations 13.1 and 13.2 does not reflect an *ex ante* trade plan that ensures fulfilment of external payments target as posited by Portes. In fact the specification is more consistent with a trade plan that does *not* attempt to meet a payments target but rather simply sets the level of exports and imports independently of each other and only influences these independent decisions by some consideration of the payments target.

The modelling of the payments target is also unsatisfactory, since it is assumed to be given by the lagged ratio of exports to imports. Thus, issues of debt service, quite serious for some of the countries in this study, are assumed to be irrelevant. More important, since the past year's trade plan may not have been met due to internal or external shocks and planning errors, it is not clear why the realized trade balance of the past year rather than the planned target or the difference between plan and realization should be of relevance to planners.

Finally, the export and import equations reflect a rather curious and imperfect specification of plan realization. Recall that planners construct a plan that should meet the payments constraint and achieve domestic equilibrium. During the course of plan execution, errors and shocks crop up, forcing planners to adjust. In Equations 13.1 and 13.2, the only shocks and errors permitted to influence trade flows are those originating in the domestic economy; shocks or errors originating in the international trade sector have no effect on trade decisions. For example, a failure to realize import plans due to a Western embargo or to Soviet rationing of oil deliveries would have no effect on the volume of export realized during the planning period.[9] What then of the payments target? Surely shocks to export flows should influence the decisions regarding the implementation of import plans, but in this model they do not.

As Burkett, Portes and Winter point out, disequilibrium in the market for investment and defence goods appears not to influence trade decisions, although this may well be due to the rather curious proxy adopted for disequilibrium in this market. The disequilibrium measure for the consumer goods market is more satisfactory, but here too there are some shortcomings. First, it is unclear why it should be last year's consumer goods market disequilibrium that leads to

[9] Ideally such shocks should be seen as having two types of effects, one operating within the trade sector through the balance of payments target and another working through the domestic economy, influencing output and domestic disequilibrium and through them the trade sector. In fact neither set of effects is modelled.

adjustments in the current year's trade flows. True, such a specification is a nice convenience enabling one to avoid the need to deal with interactions between trade flows and the domestic economy, but it is wrong to assert that trade cannot be adjusted to respond to the current year's planning errors and shocks. Indeed, if this year's plan is constructed to be balanced *ex ante* then it makes little sense to adjust it to deal with disequilibria that occurred in the past year and that have in any case been eliminated in the construction of the current year's plan. Indeed, as specified, the trade model is not at all a model where an *ex ante* consistent plan must be adjusted for disequilibria and external shocks that arise in the course of implementation. There are, in fact, no shocks, no errors and no contemporaneous disequilibria, thus in fact there is absolutely nothing in the model that prevents the *ex ante* plan from being realized.

The measure of disequilibrium used is also inconsistent with received economic theory and with the Portes model. Burkett, Portes and Winter use measures of consumer market disequilibria estimated by Portes and Winter (1980) who estimate a savings equation and a consumer goods supply equation on the assumption that any annual observations may be the result of excess supply, excess demand or market equilibrium. The Houthakker–Taylor savings function used by Portes and Winter is clearly at odds with the theory of the planned economy as expounded by Portes (1980) in constructing the model on which Burkett, Portes and Winter purport to base their specification. Portes (1980, p. 99) writes: 'There is no justification for consciously planning excess supply on the labour market (given that planners attach utility to output but not to household leisure) . . .'. Consequently one can assume that, in the course of plan implementation, shocks and errors will result in labour market equilibrium with perhaps some intermittent, but not systematic, excess. The Houthakker–Taylor savings function, on the other hand, is derived on the basis of the existence of persistent involuntary unemployment. Thus theory and quantification once again diverge.

The specification of goods supply is also suspect. Portes and Winter (1977) select a specification for the supply function by regressing observed consumption on a series of variables purporting to explain planners' decisions about the supply of consumption goods. Among these supply determining variables are the trend values of consumption and household assets, deviations from trends of consumption to NMP and investment plus defence to NMP ratios, and deviations in agricultural production. These are variables that planners might consider as relevant if they indeed did desire to equilibrate the consumer goods market, but by selecting them Portes and Winter must recognize that the behavioural hypothesis behind their test for market disequilibrium is that planners strive to eliminate it.[10] Moreover, since the explanatory variables are chosen because they provide a good fit for the entire sample, it is not surprising that in the disequilibrium estimation a large proportion

[10]A test for equilibrium on the consumer goods market that is free of this problem would replace planners' behaviour with the actual supply of goods as reflected by domestic production and net imports of such goods.

of the observations have a high probability of falling on the supply curve. Burkett, Portes and Winter use the measures of disequilibrium on the market for consumer goods to estimate the parameters of Equations 13.1–13.3 for Czechoslovakia (1955–77), the GDR (1957–77), Hungary (1958–77) and Poland (1958–75). The plausibility of their estimates and their implications for our understanding of the role of disequilibrium in CPE trade behaviour are difficult to judge because the authors interpret them within the framework of Portes' model when, in fact, the specification does not reflect that model. In the event, in the import demand equation the coefficient for excess demand on the domestic market has the theoreticaly correct positive sign in the case of Czechoslovakia and the GDR; for Hungary and Poland the coefficient is not significantly different from zero. For the export supply equation, the coefficient of the lagged excess demand variable has the correct negative sign at a borderline level of significance in the case of the GDR and Hungary but it is not significantly different from zero in the case of Poland. For Czechoslovakia the coefficient is significant but positive. If one accepts the specification of the model, then there is some evidence that domestic disequilibrium has an impact on foreign trade flows; whether a better specification would strengthen or disconfirm this finding remains for future reseach.

The Burkett, Portes and Winter effort is thus marred by the failure of the specification to reflect the theoretical underpinning the researchers ascribe to it, by some peculiarities of the specification that are independent of the theory and by the possibility that the estimates of domestic disequilibrium are flawed. Despite these shortcomings, this effort at modelling and hypothesis testing is valuable in that it clearly points the way for future work by at least arguing for the need for a sound theoretical foundation to disequilibrium modelling, for the testing of hypotheses regarding the consequences of disequilibrium and for adopting a multi-country approach to the modelling effort.

Some extensions of the work of Burkett, Portes and Winter have been undertaken by Dlouhý and Dyba (1985) and by Dlouhý in his contribution to this volume (Chapter 7). The former paper is noteworthy for attempting to test the hypothesis that Czechoslovak exports to non-socialist countries may be demand-constrained. The exercise is hampered by the use of the Burkett, Portes and Winter model and the rather *ad hoc* modification of this model to deal with the fact that only trade flows with non-socialist countries rather than total trade are modelled. Nevertheless the authors are able to estimate export supply and demand equations if certain parameters are constrained a priori. This raises the possibility that both the domestic and foreign markets need to be modelled as disequilibrium systems. However, in the Dlouhý and Dyba model the estimates of domestic disequilibrium are dropped from the list of explanatory variables so any such interaction is precluded.

In his contribution to this volume, Dlouhý (Chapter 7) does attempt such a linkage, but he does it in terms of an equilibrium model. The appeal to equilibrium on the consumer goods market enables him to utilize a consumption function to determine the level of consumption. If such equilibrium exists, it is worth noting, then consumption could just as well be based on a planners' supply function of the type specified by Portes and Winter (1977). This possibility leads to the following question: is the behaviour of planners at all consistent with any resource or foreign trade constraints on planners behaviour? Clearly total output must be allocated to consumption, investment and net exports. Consumption demand is based on lagged and current disposable income and lagged savings, and planners are assumed to satisfy this demand. Investment depends on lagged output, investment, change in investment and net exports. Trade with socialist countries is exogenous while with capitalist countries its determinants resemble those of the Burkett, Portes and Winter model. The interactions between the three sources of demand are weak if not non-existent. Consequently, the uses of final output are over-determined and there is no mechanism that ensures that aggregate demand and supply will be equilibrated. To assume that some or all markets are in equilibrium eliminates some of the complexities of disequilibrium estimation, but it simultaneously imposes other requirements on the specification that must not be overlooked.

Thus, at this stage we know little about the realities of CPE foreign trade responses to disequilibrium on the domestic market or on the foreign exchange market. We do know from the extant work on the subject that it is possible that disequilibria may play a role in influencing economic behaviour of consumers and of planners. More attention must be paid to constructing theoretical models of open CPEs that will allow for the testing of hypotheses regarding the existence and consequences of various disequilibria. Similarly, econometric models must be specified in much closer conformity to the theoretical models; otherwise, as we have seen, the estimates we obtain cannot be used to distinguish between rival hypotheses.

13.3 AN ALTERNATIVE METHOD OF ESTIMATING MODELS OF ECONOMIES IN DISEQUILIBRIUM

The maximum likelihood techniques for estimating disequilibria on CPE markets utilized in the studies cited above have the attractive feature of being able to provide simultaneous estimates of the supply and demand curves. At the same time, for purposes of macroeconomic modelling and hypothesis testing they are limited in being able to deal with only one market at a time.[11]

An alternative approach to modelling disequilibrium is to go back to the Barro–Grossman framework and make better use of their theoretical insights regarding rationing and spillovers. That is, if agents find themselves rationed on one market then theory suggests that there will be spillovers to other markets as rationed agents adjust their behaviour not only on the rationed market but on

[11] In theory the maximum likelihood technique can be used to estimate parameters for a multi-market model (see Portes *et al.*, 1987) but we are still far away from modelling an entire macroeconomic model in this way.

others as well. To take the example given by Portes (1980) in his specification of consumers' labour supply as discussed above, labour supply depends on wages and money stock if consumers are not rationed on the goods market; but if they are then the supply of consumer goods, or, put another way, the degree to which they are rationed, influences labour supply. Thus one may estimate supply and demand curves on one market to determine whether it is in equilibrium or one can attempt to uncover the spillover effects from that market to other markets and thereby infer the extent and nature of disequilibrium on the first market.

While this latter methodology is less direct it is appealing because of its potential for overcoming the econometric obstacles that plague existing estimation techniques. Indeed, the first attempt at applying the Barro—Grossman model to a planned economy by David Howard (1976) was based precisely on the concept of spillovers from the consumer goods market to the labour market in the Soviet Union. Howard, however, was forced to assume the existence of excess demand on the goods market and to represent it by proxies. The estimation technique employed here enables us to treat disequilibrium as an unobserved variable whose nature and extent we infer from the behaviour of economic agents and whose plausibility can be verified by economic theory. The investigation of unobservable phenomena through the study of their effects on observable ones is unusual in economics but it is common in the physical sciences. Thus, for example, the existence of the outermost planets was inferred from their gravitational influence on the more visible planets long before astronomers confirmed the existence of these outer planets through direct observation. Similarly, although at the other end of the physical scale, the properties of subatomic particles are gleaned from the effects of their collisions with larger and more visible atoms rather than through direct observation which is, in any case, impossible. We treat disequilibrium in precisely the same way, as an unobservable phenomenon to study through its effects on observable outcomes.

13.3.1 A model of Soviet grain and meat production and trade

The activity that we chose for modelling is Soviet production and trade in grain and meat. There are several reasons for this choice. The first is that there are clear spillover effects by which disequilibrium in one market should influence agents' behaviour in other markets. Moreover, grain and meat production and consumption are sufficiently important to planners to warrant serious concern over the maintenance of equilibrium, but sufficiently small to avoid the need to model feedbacks from meat and grain consumption and trade on domestic macroeconomic variables. Finally, the data on production, consumption and trade in grain and meat are quite reliable and, because grain and meat are relatively homogeneous products, physical quantities can be employed without the concern for the types of aggregation problems that would be involved in, say, the case of consumer manufactures.

Sometime in the early 1960s, the leadership of the Soviet Union chose to increase the production of meat and other animal products. This policy was to be implemented by expanding animal herds and by refusing to allow fluctuations in feed availability resulting from poor harvests to affect the growth of herds. This is not to say that animal herds were to be totally insulated from fluctuations in the harvest. Rather, planners followed a more complex strategy whose objective was to stabilize the growth of meat and animal products. Thus, for example, herds were reduced through distress slaughtering in 1975, which led to a short-term upsurge in meat production but caused a subsequent decline that was offset in part by imports of meat. In 1979–82 herds were maintained on lower rations, thus reducing meat production in the short run because there was no distress slaughter of livestock, and meat imports were stepped up to supplement domestic production.

Since planners have a flexible way of dealing with their objective of providing a given volume of meat for consumption and since the

> ... most important factor hampering the growth of livestock production ... is the feed situation – a chronic shortage of sufficient energy feeds and an imbalance among major feed components, including a serious protein deficit (Severin, 1987)

it is evident that decisions on herd size, grain imports and meat imports must be modelled in a simultaneous fashion. Moreover, decisions on the volume of imports may be influenced by considerations of a foreign exchange constraint reflecting current and future exchange earnings, competing demands for imports of machinery and equipment, and the availability of hard currency credits.

In addition to these biological relationships between animal and crop production, we assume that planners have some perception of disequilibrium on the markets for grain and meat and that they respond to these perceptions in their decisions on supply of and trade in meat and grain. This is consistent with, but goes somewhat beyond, planners' perceptions in the model used by Burkett, Portes and Winter where domestic disequilibrium influenced only trade decisions. It bears emphasizing that, like Burkett, Portes and Winter, in examining foreign trade behaviour we obtain inferences not about meat and grain market disequilibrium as experienced by consumers, because consumers do not affect foreign trade flows, but rather the perceptions of market disequilibrium on the part of planners since their perceptions do influence trade flows. On the basis of these assumptions we specify the model as:

$$H = h[H(-1), \mathrm{GTOT}] \tag{13.4}$$

$$M = m[\mathrm{XG}, \mathrm{XM}, H(-1)] \tag{13.5}$$

$$\mathrm{NIG} = f[\mathrm{XF}, \mathrm{XG}, \mathrm{NPW}] \tag{13.6}$$

$$\mathrm{NIM} = g[\mathrm{XF}, \mathrm{XM}, \mathrm{NPM},] \tag{13.7}$$

$$\mathrm{XG} = X_3[\mathrm{INCOM/POP}, H(-1), \mathrm{GTOT}] \tag{13.8}$$

$$\mathrm{XM} = X_2[\mathrm{INCOM/POP}, M(-1), \mathrm{NIM}(-1)] \tag{13.9}$$

$$\mathrm{XF} = X_1[\mathrm{XF}(-1), E(\mathrm{T}), E(\mathrm{INT}), D, I] \tag{13.10}$$

where

D = Soviet hard currency debt (billion US $)
G = Soviet grain harvest (million metric tons)
GTOT = G(− 1) + NIG (million metric tons)
H = herds (animal units)
I = investment (billion 1970 roubles)
INCOM = consumer incomes (billion 1970 roubles)
E(INT) = expected interest rate on Eurodollar deposits in London (%)
E(T) = expected Soviet barter terms of trade (1966 = 100)
M = Soviet meat production (million metric tons)
NIG = net imports of grain (million metric tons)
POP = Soviet population (million persons)
NPG = price of grain ($/bushel, US Gulf ports) divided by index of Soviet export prices (1966 = 100)
NPM = price of beef (cents/lb – US, all ports) divided by index of Soviet export prices (1966 = 100)
XF = disequilibrium in market for foreign exchange (dimensionless)
XG = disequilibrium in market for grain (dimensionless)
XM = disequilibrium in market for meat (dimensionless)

and the expectations operator is:

$$E(X_t) = X_t\{1.0 + [0.5(X_t - X_{t-1})/X_{t-1} + 0.3(X_{t-1} - X_{t-2})/X_{t-2} + 0.2(X_{t-2} - X_{t-3})/X_{t-3}]\} \qquad (13.11)$$

Equation 13.4 models changes in herd size as depending on the availability of feed, GTOT. GTOT is the sum of last year's grain harvest, which, since it is harvested in the autumn of the previous year, is fed to animals in the current year and the current year's net grain imports, which supplement available domestic supplies in the year in which they are imported. Because most livestock has a relatively fixed need for grain rations and because even ruminants consume a relatively high proportion of their diet in the form of grains, this relationship between feed and herd size is based on relatively immutable biological processes.

The production of meat is modelled in Equation 13.5. The basic variable is, of course, the size of animal herds. Under normal conditions a certain proportion of animals would be slaughtered yielding a normal level of meat production. However, the normal proportion may not be slaughtered if decision-makers observe and respond to disequilibria in the meat and grain markets. In this case the relevant decision-makers are both planners and private individuals who own livestock. If they perceive an excess supply of meat on the market, planners may defer slaughtering animals, preferring to save them for the future. Private owners will perceive relatively lower prices on collective farm markets and may also defer slaughterings. Thus we would expect the coefficient associated with XM to be positive. The expected sign of the coefficient for XG is less certain. If a lack of grain forces distress slaughtering and ample supplies encourage greater efforts at herd formation, the coefficient would be positive. However, as we have seen, the Soviet reaction to grain shortages has been to reduce animal ratios, thus causing a decline in meat production and carrying the herds through poor years. This behaviour suggests that the coefficient should be negative, and thus it cannot be signed a priori.

Net imports of grain, Equation 13.6, reflect the state of the domestic grain market. If there is a shortage domestically, then grain will be imported. There are, of course, constraints on the amount of grain to be imported. One is the opportunity costs of grain imports, measured here by the price of grain divided by the index of Soviet export prices, since exports must ultimately pay for imports. The other explanatory variable is the relationship between the supply of and demand for foreign exchange. If export earnings are plentiful and there are no other pressing import needs, then it is easy for planners to allocate funds for grain imports. On the other hand, if planners face many demands for imported goods then funds must be rationed between competing claims, and grain imports may be reduced. Net imports of meat, Equation 13.7, are modelled in parallel fashion, reflecting the same decision-making process.

The excess demand for foreign exchange, XF, is modelled in Equation 13.10 on the basis of the belief that planners in centrally planned economies face demands for imports that may exceed the availability of foreign exchange and therefore they must ration these demands to fit the amount of foreign exchange available.[12] Equation 13.10 thus contains those elements of supply of and demand for foreign exchange likely to influence planners' perceptions of the balance between claims for and the availability of foreign exchange. First, we believe that any supply–demand imbalances are carried over from one period to the next; unsuccessful claimants for foreign exchange can be expected to continue to press their demands in the future, and thus XF(− 1) enters the equation. The ability to pay for imports depends on the terms of trade, which reflect the relative cost of imports. Planners by their very nature must look to the future and we therefore use expected terms of trade as an explanatory variable. In addition to using the earnings from exports, planners can also borrow foreign exchange, and the amount they are willing to borrow is likely to depend on expected interest rates and the existing stock of foreign debt. Finally, it is evident from Soviet trade statistics that the other major claimant for hard currency imports is the Soviet investment sector, and thus the volume of investment should influence the demand for foreign exchange. We believe that the use of expectational and current variables gives a much more realistic explanation of the foreign exchange constraint faced by the planners than does the lagged trade deficit used by Burkett, Portes and Winter.

The state of the market for meat is given by Equation 13.9. The demand for meat is a function of real per capita incomes. We use per capita rather than total incomes since the demand for meat in the Soviet Union has been driven much more by per capita increases in income than by population growth. The supply of meat is a

[12]Evidence for this belief can be found in Bond and Levine (1979), Hanson (1981) and Hewett (1981).

function of production and net imports. The excess demand for grain, Equation 13.8, is modelled in parallel fashion, save that there are two sources of demand, human consumption and the feed requirement of the livestock sector and the grain supply variable is the sum of lagged production and current imports.

13.3.2 Estimating the model

The model specified above has three variables, XF, XG, and XM for which there are no observations since market disequilibrium is an unobserved variable. Consequently we employ a maximum likelihood technique that permits us to specify the model as if we had data for the unobserved variables and then to estimate all the parameters of the model on the basis of data for the observable variables. Naturally the parameter estimates and data on the observable variables can then be used to calculate estimates of the unobserved variables.

The use of unobservable variables in specifying our model classifies it as a latent variable or causal model.[13] We estimate it by means of the EM (estimation-maximization) algorithm originally formulated by Dempster *et al.* (1977) and then further developed as EMIMIC (estimation of multiple indicator–multiple cause models) by Watson and Engle (1983).[14] Because the estimation procedure has been explained in detail by Engle and Watson (1981) and by Burmeister and Wall (1987) we describe it only briefly.

The structural equations of the econometric model are specified to include the unobservable variables without regard for the availability of data. The structure can be interpreted from a control engineering viewpoint as the state-space form. From the joint distribution of the observable variables a likelihood function for the unobserved variables can be formed. Because it is difficult to maximize this function using usual gradient techniques, the EM algorithm employs a derivative-free procedure. This involves applying a Kalman filter to find minimum mean squared estimates of the conditional mean of the latent variables given the data on the observable variables. These conditional estimates of the unobserved variables and the observed variables are combined in a second stage of the algorithm to maximize the likelihood function by means of an iterative convergence. A major advantage of this procedure is that it yields parameter estimates near the maximum with few computations, although this is achieved at the cost of many computations to then approach the exact maximum.

Following Watson and Engle (1983), we summarize the state-space form of our model by means of two types of equations. The so-called transition equations specify the state or unobserved variables, X_t, as a function of their lagged values to

[13]The latter term is more often used in other social sciences and is related to the LISERL estimators as described in Joreskog and Sorbom (1981). Latent variable models and techniques for parameter estimation are described in Aigner *et al.* (1981).

[14]The basis of the EMIMIC algorithm comes from control engineering. See Mehra (1974) for a comparison of problems in economics and in control engineering and Watson and Engle (1983) for an update and generalization of Mehra.

allow for dynamic behaviour; of observed exogenous and other predetermined variables, $Z1_t$, to allow for the measurement of the unobserved variables; and a stochastic disturbance term V_t:

$$X_t = \boldsymbol{\Phi} X_{t-1} + \boldsymbol{\theta} Z1_t + v_t \tag{13.12}$$

The measurement equations specify the observed dependent variables, Y_t, as dependent upon the unobserved variables X_t; another set of observed predetermined variables $Z2_t$, which may include variables found in $Z1_t$; and a stochastic disturbance term, e_t:

$$Y_t = \boldsymbol{\alpha} X_t + \boldsymbol{\beta} Z2_t + e_t \tag{13.13}$$

where $\boldsymbol{\Phi}$, $\boldsymbol{\theta}$, $\boldsymbol{\alpha}$, and $\boldsymbol{\beta}$ are vectors of parameters to be estimated. The error terms are assumed to be normally and identically distributed with zero means and $V[v_t] = Q$ and $V[e_t] = R$.

To estimate the model the likelihood function must be formed. If $\bar{Y}_t$ is the conditional mean of Y_t given values of Y for all previous periods as well as the values of $Z1_t$ and $Z2_t$, then $\bar{Y}_t$ is a best estimate of Y_t and we define $\eta_t = (Y_t - \bar{Y}_t)$. If we let $V[\eta_t] = H_t$ and $\pi = [\Phi, \theta, \alpha, \beta]$, then the log likelihood function for the state-space form of the model is

$$L(\pi) = f(\eta_t, H_t) \tag{13.14}$$

It is to Equation 13.14 that the EM Kalman filter technique is applied iteratively to estimate $\hat{\pi}$. In the course of maximizing $L(\pi)$ the information matrix can be derived, and its diagonal elements provide the standard errors of $\hat{\pi}$. It should be noted that standard diagnostics for evaluating the quality of individual parameter estimates obtained through full-system parameter estimation are not as yet fully developed. As Watson and Engle (1983, p. 386) indicate '... the estimation procedures are sufficiently complicated and specialized that slight variations on the specification cannot easily be considered and diagnostic testing is almost unknown'. Burmeister and Wall (1987) rely on *t*-ratios, Tchebychev's inequality and Kolmogorov–Smirnov tests to investigate hypotheses of interest to them.

We compiled data for the observable variables over the period 1960–84, and, because of the structure of the expectations variable, the model was estimated over the period 1963–84 by means of the EMIMC algorithm. To assure that the algorithm converged to a global rather than local maximum a variety of starting parameter values and convergence criteria were employed.

13.3.3 Discussion of results

The parameter estimates for the full model are presented in Table 13.1. The most important finding from the estimates for the full model is the insignificance of XF, the foreign exchange constraint, in the equations for meat and grain imports and the fact that none of the variables in the equation for XF are significantly different from zero. This suggests that either shortages of foreign exchange played no role in

Table 13.1 Parameter estimates for Equations 13.4–13.10

$$H = 17.57 + 0.73H(-1) + 0.11 \quad \text{GTOT}$$
$$\quad (2.02) \quad (7.73) \qquad (3.43)$$

$$M = -2.29 - 0.58\text{XG} + 0.53\text{XM} + 0.13H(-1)$$
$$\quad (0.50) \quad (3.85) \quad (1.93) \quad (2.78)$$

$$\text{NIG} = -30.11 + 0.003\text{XF} + 7.30\text{XG} + 555.16 \quad \text{NPW}$$
$$\quad (0.67) \quad (0.05) \quad (3.48) \quad (1.79)$$

$$\text{NIM} = 0.46 + 0.002\text{XF} + 0.09\text{XM} - 46.07 \quad \text{NPM}$$
$$\quad (0.80) \quad (0.14) \quad (2.05) \quad (1.83)$$

$$\text{XG} = 11.95 \quad \text{INCOME/POP} + 0.02H(-1) - 0.05 \quad \text{GTOT}$$
$$\quad (2.56) \qquad (0.33) \quad (2.76)$$

$$\text{XM} = 20.54 \quad \text{INCOME/POP} - 1.25M(-1) + 2.30\ \text{NIM}(-1)$$
$$\quad (1.82) \qquad (1.70) \quad (1.07)$$

$$\text{XF} = 0.10\text{XF}(-1) + 0.04E(\text{T}) + 0.27E(\text{INT}) + 0.5D - 0.07 \quad I$$
$$\quad (0.14) \quad (0.02) \quad (0.05) \quad (0.06) \quad (0.04)$$

t-statistic in parentheses.

influencing the volume of meat and grain imports or that planners perceived no disequilibrium between demands for and supplies of foreign exchange. To test for this possibility we reestimated the model without the variable XF in the equations for NIG and NIM and without the equation for XF. A χ^2 test indicated that the inclusion of XF and Equation 13.10 in the specification did not significantly increase the explanatory power of the model. Consequently, our model of choice consists of Equations 13.4–13.9, the parameter estimates for which are presented in Table 13.2.

Parameter estimates for Equation 13.4, the determination of herd size are significantly different from zero and accord with a priori expectations. Lagged herds and the availability of grain both exert a positive influence on herd size, consistent with the strategic relationship between feed availability and the number of animals that can be maintained. The production of meat, M, depends positively on lagged herds.[15] The greater the excess demand for meat, the greater the proportion of a given herd that will be slaughtered, indicating that both planners and private owners of livestock respond to perceptions of excess demand for meat. Finally, excess demand for grain reduces the supply of meat. This means that the dominant policy regime in the Soviet Union over our sample period has been one of carrying livestock on short rations through periods of feed shortage rather than resorting to distress slaughtering.

[15]The use of lagged herds is due to the fact that data on livestock numbers are given as of the last day of the year. Thus the lagged value is more indicative of availability of animals for slaughter during the year than is the contemporaneous value.

Table 13.2 Parameter estimates for Equations 13.4–13.9

$$H = 17.57 + 0.73H(-1) + 0.11 \quad \text{GTOT}$$
$$\quad (2.02) \quad (7.73) \quad (3.43)$$

$$M = -2.31 - 0.60\text{XG} + 0.55\text{XM} + 0.13 \quad H(-1)$$
$$\quad (0.52) \quad (4.25) \quad (2.90) \quad (3.68)$$

$$\text{NIG} = -30.78 + 7.22\text{XG} + 584.16 \quad \text{NPW}$$
$$\quad (0.69) \quad (3.85) \quad (1.93)$$

$$\text{NIM} = 0.47 + 0.09\text{XM} - 48.02 \quad \text{NPM}$$
$$\quad (1.29) \quad (2.72) \quad (2.48)$$

$$\text{XG} = 12.12 \quad \text{INCOME/POP} + 0.02H(-1) - 0.05 \quad \text{GTOT}$$
$$\quad (2.69) \quad (0.35) \quad (2.90)$$

$$\text{XM} = 19.30 \quad \text{INCOME/POP} - 1.16M(-1) + 2.55 \quad \text{NIM}(-1)$$
$$\quad (2.46) \quad (1.77) \quad (1.54)$$

t-statistic in parentheses.

Net imports of grain, NIG, are modelled by Equation 13.6. The excess demand for grain has a positive influence on imports, reflecting planners' efforts to equilibrate the domestic grain market through imports. The price variable, measuring the opportunity cost of grain in terms of exports is positive, reflecting a price inelastic demand on the part of planners as well as a causal relationship between Soviet grain imports and world market prices. Net imports of meat, on the other hand, respond negatively to the increase in the opportunity cost of meat as measured by NPM and positively to perceived increases in the excess demand for meat on the domestic market. Thus in decisions regarding meat and grain imports the extent of excess demand or excess supply on the domestic market does influence planners' trade decisions.

The excess demand or supply of grain and meat, XG and XM, are given by Equations 13.8 and 13.9. In each case the parameter for per capita income is positive and significant. We used per capita income rather than total income as the demand variable primarily because it has been the increase in Soviet living standards rather than increase in population that has driven the increase in demand for food. In the grain equation the other demand element, the size of livestock herds is not significant. In the same equation the supply element, grain production plus imports, has a negative and significant sign. Thus Equation 13.8 is a well-defined excess demand function. In the excess demand for meat equation the production of meat has the correct sign but the net imports of meat variable has the wrong sign. This may in part be due to the use of lagged meat imports rather than current ones, a specification imposed on us by the estimation procedure.

Overall, the parameter estimates presented in Table 13.2 are consistent in sign with our a priori expectations and with economic theory. Moreover, in most cases parameter estimates are significant at the 5% level, a result that is quite encouraging in view of the fact that the estimation technique employed was devised largely for time series estimation where many more observations were generally available than in structural estimation such as we attempt here.

In view of the fact that EMIMIC is a system estimator, it is worthwhile to examine how well the model performs. Table 13.3 presents the actual and fitted values for the four observable dependent variables and some in-sample prediction diagnostics. The evident close fit between fitted and actual values of the dependent variables is confirmed by the summary diagnostic statistics. For all four variables the correlation between fitted and actual values is quite high, even for time series. This is true even for grain imports, a very volatile series. Our worst fit is for NIM, the net imports of meat, a series where planners have perhaps the greatest discretion in making decisions and also one where constraints on handling and transportation of imports may pose some problems. The model also does an excellent job of finding the turning points in the four series. The Theil inequality coefficient (see

Table 13.3 Actual vs. simulated values of observed dependent variables

	H	*HHAT*	*M*	*MHAT*
1963	123.100	118.648	10.2000	10.3830
1964	110.200	119.884	8.2900	8.7445
1965	113.700	114.557	9.9600	10.2340
1966	121.000	114.304	10.7000	10.7405
1967	124.200	124.576	11.5200	11.5810
1968	122.700	124.053	11.6500	11.8185
1969	121.700	125.108	11.7700	11.9770
1970	122.600	123.635	12.2800	12.4570
1971	130.500	126.823	13.2700	13.3700
1972	134.400	132.919	13.6300	13.8025
1973	134.100	136.498	13.5300	13.5910
1974	138.000	140.513	14.6200	14.7830
1975	141.600	139.882	14.9700	15.2250
1976	136.500	139.145	13.5800	13.7715
1977	138.400	142.675	14.7200	15.0765
1978	143.900	141.955	15.5000	15.5305
1979	147.000	150.115	15.3400	15.6205
1980	148.700	147.911	15.0700	15.4015
1981	149.400	150.760	15.2000	15.5780
1982	150.800	149.237	15.3700	15.5490
1983	153.400	150.974	16.4500	16.6580
1984	156.300	153.807	16.9000	17.0260

In-sample prediction diagnostics

	H vs. HHAT	*M* vs. MHAT
Correlation coefficient	0.965	0.999
(squared)	0.931	0.998
Root mean squared error	3.440	0.228
Theil inequality coefficient	0.013	0.008

Table 13.3 (continued) Actual vs. simulated values of observed dependent variables

	NIG	*NIGHAT*	*NIM*	*NIMHAT*
1963	−7.70000	−7.89592	−0.146000	0.018702
1964	5.70000	6.12231	0.059000	−0.204582
1965	−1.70000	−1.92412	0.230000	0.173708
1966	3.70000	3.15980	0.015000	0.171938
1967	−1.40000	−2.17834	−0.117000	0.084037
1968	−4.10000	−3.83217	−0.017000	0.021330
1969	−6.20000	−5.93509	−0.022000	−0.002514
1970	−5.80000	−5.37220	0.110000	0.083400
1971	−7.20000	−7.08840	0.190000	0.192363
1972	1.40000	1.81553	0.071000	0.076602
1973	21.00000	20.51240	0.054000	0.050807
1974	5.20000	5.02215	0.495000	0.366873
1975	0.40000	1.22726	0.471000	0.436876
1976	25.40000	25.56910	0.321000	0.451563
1977	7.70000	8.29394	0.584000	0.564138
1978	16.60000	16.30330	0.145000	0.578853
1979	12.80000	12.94260	0.577000	0.384069
1980	30.20000	30.45770	0.786000	0.607061
1981	34.90000	34.92100	0.910000	0.755114
1982	45.50000	44.72730	0.907000	0.900805
1983	32.00000	31.84950	0.960000	0.902852
1984	30.50000	30.15110	0.778000	0.916062

In-sample prediction diagnostics

	NIG vs. NIGHAT	NIM vs. NIMHAT
Correlation coefficient	0.9996	0.904
(squared)	0.9993	0.816
Root mean squared error	0.442	0.152
Theil inequality coefficient	0.011	0.158

Pindyck and Rubinfeld, 1981, pp. 364–7) is outstandingly low for all four series, confirming that the model does track well within the sample period. Even for net imports of meat, the series with the highest Theil *U* ratio, 96% of the ratio's value consists of unsystematic error. Consequently, judging the entire model on the basis of its in-sample performance suggest that it performs quite well in terms of the standard statistical criteria.

In view of the robustness of our parameter estimates and of the excellent in-sample fit provided by the model, we have a good deal of confidence that the estimates of disequilibrium generated by the model are also an excellent reflection of Soviet planners' perceptions of disequilibrium on the markets for grain and meat. The estimates of disequilibrium are presented in Table 13.4. For grain, there has been excess demand for the entire sample period, with a secular increase in the absolute value of the perceived gap between supply and demand over time. On the other hand, since the volume of transactions on this market has also grown over the

Table 13.4 Estimates of excess demand/supplies

	$XG = grain$	$XM = meat$
1963	2.32	−2.40
1964	4.37	−4.23
1965	3.19	0.24
1966	3.81	1.01
1967	3.12	0.06
1968	2.95	−0.45
1969	2.70	−0.08
1970	2.71	1.04
1971	2.41	2.16
1972	2.75	1.45
1973	5.02	2.62
1974	3.61	3.32
1975	3.37	2.94
1976	7.04	3.45
1977	4.53	4.29
1978	5.45	5.67
1979	5.08	4.13
1980	7.62	5.77
1981	8.40	6.54
1982	9.81	7.86
1983	8.10	7.68
1984	7.94	7.56

years, the degree of disequilibrium has not grown as rapidly as the data in Table 13.4 indicate. As we mentioned above, our model makes no a priori assumptions about the existence or nature of disequilibrium in domestic markets. Thus in the market for meat both excess supply and excess demand can be found. The excess supply of meat is evident only in the 1960s; thereafter excess demand prevails. The perception of excess supply as revealed by planners' behaviour does not, of course, imply that consumers also perceived that there was an excess supply of meat, although in the light of Podkaminer's (1982) estimates for Poland the possibility should not be dismissed out of hand.

In sum, the results of our modelling effort are quite encouraging. We have proposed an alternative method for specifying a model of multimarket disequilibrium within a planned economy, and we have demonstrated that the estimation of the parameters yields robust results that are consistent with the predictions of theory. We have also shown that for Soviet grain and meat markets since 1963 the degree of excess demand has increased, and, in the case of meat, actually changed from excess supply to excess demand. Our results also show that when planners perceive disequilibrium in important markets they respond to these perceptions through their decisions on domestic production and on international trade flows.

ACKNOWLEDGEMENTS

We are indebted to the National Council for Soviet and East European Research for financial support, to Don E. Schlagenhauf for advice on the implementation of the statistical algorithm employed in this chapter and to Thomas A. Wolf for detailed comments on an earlier draft of this chapter. John DiPippo provided capable research assistance.

REFERENCES

Aigner, D. *et al.* (1984) Latent variable models in econometrics in *Handbook of Econometrics* (eds Z. Griliches and M. Intrilligator), North-Holland, New York.

Ames, E. (1953) The exchange rate in Soviet type economies. *Review of Economic Studies*, **36**, 337–42.

Ames, E. (1980) Counterfactual methods in the study of planned economies: a reconciliation of Tyson and Kenen with Wolf and Portes, in *The Impact of International Economic Disturbances on the Soviet Union and Eastern Europe* (eds E. Neuberger and L. Tyson), Pergamon, New York.

Barro, R. J. and Grossman, H. I. (1974) Suppressed inflation and the supply multiplier. *Review of Economic Studies*, **41**, 87–104.

Barro, R. J. and Grossman, H. I. (1976). *Money, Employment and Inflation*, Cambridge University Press, Cambridge.

Bond, D. L. and Levine, H. S. (1979) Energy and grain in Soviet hard currency trade, in Joint Economic Committee, US Congress, *Soviet Economy in a Time of Change*, US Government Printing Office, Washington.

Brada, J. C. (1980) On the adjustment of centrally planned economies to external economic disturbances: discussion, in *East European Integration and East-West Trade* (eds P. Marer and J. M. Montias), Indiana University Press, Bloomington.

Brada, J. C. (1982) Real and monetary approaches to foreign trade adjustment mechanisms in centrally planned economies: a reconciliation. *European Economic Review*, **19**, 229–44.

Brada, J. C. and King, A. E. (1979) A structural econometric model of Czechoslovakia: 1954–1975. *Empirical Economics*, **4**, 199–219.

Brada, J. C., Jackson, M. R. and King, A. E. (1984) The Romanian balance of payments crisis: an econometric study of its causes and cures, in *New Horizons in East-West Economic Relations* (eds M. R. Jackson and J. D. Woodson, Jr.), East European Quarterly Monographs, Boulder.

Brada, J. C., King A. E. and Schlagenhauf, D. E. (1981) The Optimality of Socialist development strategies. *Journal of Economic Dynamics and Control*, **3**, 1–27.

Burkett, J., Portes, R. and Winter, D. (1981) Macroeconomic adjustment and foreign trade of centrally planned economies, Working Paper No. 736, National Bureau of Economic Research, Cambridge, MA.

Burmeister, E. and Wall, K. (1987) Unobserved rational expectations and the German hyper inflation with endogenous money supply. *International Economic Review*, **28**, 15–32.

Dempster, A., Laird, N. and Rubin, D. (1977) Maximum likelihood from incomplete data via the EM algorithm. *Journal of the Royal Statistical Society* B39, 1–38.

Dlouhý, V. and Dyba, K. (1985) Modelling Czechoslovak foreign trade flows with non-Socialist countries (equilibrium and some disequilibrium estimation). *Czechoslovak Economic Papers*, **23**, 91–110.

Engle, R. and Watson, M. (1981) One factor multivariate time series of model metropolitan wage rates. *Journal of the American Statistical Association*, **76**, 774–81.

Hanson, P. (1981) *Trade and Technology in Soviet-Western Relations*, Macmillan, London.

Hewett, E. A. (1980) A comment on the R. Portes paper, Internal and external balance in a centrally planned economy, in *The Impact of International Economic Disturbances on the Soviet Union and Eastern Europe* (eds E. Neuberger and L. D. Tyson), Pergamon Press, New York.

Hewett, E. A. (1981) The Hungarian economy: lessons of the 1970s and prospects for the 1980s; in Joint Economic Committee, US Congress, *East European Economic Assessment*, US Government Printing Office, Washington.

Hildebrandt, G. G. (ed.) (1985) *Rand Conference on Models of the Soviet Economy*, Rand, Santa Monica.

Holzman, F. D. (1974) *Foreign Trade Under Central Planning*, Harvard University Press, Cambridge, MA.

Holzman, F. D. (1980) A comment on the R. Portes paper, Internal and external balance in a centrally planned economy, in *The Impact of International Economic Disturbances on the Soviet Union and Eastern Europe* (eds E. Neuberger and L. D. Tyson), Pergamon Press, New York.

Holzman, F. D. (1987) *The Economics and Politics of Soviet Bloc Trade and Finance*, Westview, Boulder, Co.

Howard, D. E. (1976) The disequilibrium model in a controlled economy: an empirical test of the Barro–Grossman model. *American Economic Review*, **66**, 871–9.

Joreskog, K. and Sorbom, D. (1981) *LISERL V Users' Guide*, National Educational Resources, Chicago.

Kemme, D. M. (1981) Econometric models of East European economies: a survey of models of Poland. *The ACES Bulletin*, **23**, 51–70.

King, A. E. (1979) Econometric models of Czechoslovakia: a survey. *The ACES Bulletin*, **21**, 71–84.

Mehra, R. (1974) Identification in control and economics: similarities and differences. *Annals of Economic and Social Measurement*, **3**, 21–47.

Pindyck, R. S. and Rubinfeld, D. L. (1981) *Econometric Models and Economic Forecasts*, McGraw-Hill, New York.

Podkaminer, L. (1982) Estimates of disequilibria in Poland's consumer markets, 1965–1978. *Review of Economics* and *Statistics*, **64**, 421–31.

Portes, R. (1980) Internal and external balance in a centrally planned economy, in *The Impact of International Economic Disturbances on the Soviet Union and Eastern Europe* (eds E. Neuberger and L. D. Tyson), Pergamon Press, New York.

Portes, R., Quandt, R. E., Winter, D. and Yeo, S. (1987) Macroeconomic planning and disequilibrium: estimates for Poland, 1955–1980. *Econometrica*, **55**, 19–41.

Portes, R. and Winter, D. (1977) The supply of consumption goods in centrally planned economies. *Journal of Comparative Economics*, **1**, 351–65.

Portes, R. and Winter, D. (1978) The demand for money and for consumption goods in centrally planned economies. *Review Economics and Statistics*, **60**, 8–18.

Portes, R. and Winter, D. (1980) Disequilibrium estimates for consumption goods markets in centrally planned economies. *Review of Economic Studies*, **46**, 137–59.

Pryor, F. L. (1963) *The Communist Foreign Trade System*, MIT Press, Cambridge, MA.

Severin, B. (1988) The USSR: the livestock feed issue, in *Socialist Agriculture in Transition: Organizational Response to Failing Performance* (eds J. C. Brada and K. E. Wadekin), Westview, Boulder, Co.

Shapiro, H. T. (1977) Macroeconometric models of the Soviet Union and Eastern Europe: a tabular survey. *Econometrica*, **45**, 1747–66.

Shapiro, H. T. and Halabuk, L. (1976) Macro-econometric model building in socialist and non-socialist countries: a comparative study. *International Economic Review*, **17**, 529–65.

Šujan, I. (1987a) An analysis of factors contributing to the deceleration in the growth rate of the Czechoslovak economy in the period 1975–1980. *Soviet and Eastern European Foreign Trade*, **23**, 31–53.

Šujan, I. (1987b) Simulations of restoring the external economic equilibrium of Czechoslovakia. *Comparative Economic Studies*, **28**, 61–84.

Szakolczai, G., Bagdy, G. and Windics, J. (1987) The dependence of the Hungarian economy on the world economy: facts and consequences. *Soviet and Eastern European Foreign Trade*, **23**, 54–88.

Tomczyk, P. (1987) *An Econometric Model of the CMEA Trade*, Wydawnictwo Uniwersytetu Lodzkiego, Lodz.

Vaňous, J. (1980) An econometric model of intra-CMEA trade, in *East European Integration and East West Trade* (eds P. Marer and J. M. Montias), Indiana University Press, Bloomington.

Watson, M. and Engle, R. (1983) Alternative algorithms for the estimation of dynamic factor, MIMIC and varying coefficient regression models. *Journal of Econometrics*, **23**, 385–400.

Wolf, T. A. (1978) Exchange-rate adjustments in small market and centrally planned economies. *Journal of Comparative Economics*, **2**, 226–45.

Wolf, T. A. (1980a) External inflation, the balance of trade and resource allocation, in small centrally planned economies, in *The Impact of International Economic Disturbances on the Soviet Union and Eastern Europe* (eds E. Neuberger and L. Tyson), Pergamon Press, New York.

Wolf, T. A. (1980b) On the adjustment of centrally planned economies to external economic disturbances, in *East European Integration and East–West Trade* (eds P. Marer and J. M. Montias), Indiana University Press, Bloomington.

14

Disequilibrium models of investment

IRENA GROSFELD

14.1 INTRODUCTION

Investment processes in the centrally planned economies (CPEs) are characterized by persistent shortages and disequilibria. This assertion is accepted almost unanimously in the literature. Several authors, notably Kornai (1980), go even further and attribute the responsibility for all shortages in the CPE to the state productive sector which is characterized by permanent excess demand, spilling over into the consumption goods market. Such a spillover may or may not exist, but the striking fact is that the discussion of disequilibria and shortages in CPEs usually focuses – and the structure of this volume confirms it – on analysis of the consumption goods market. This is probably due in part to the lack of challenge similar to that offered by Richard Portes, for the study of the consumption goods market. Portes' questioning of the conventional wisdom, according to which there exists a permanent excess demand for consumer goods in the CPEs, provoked a lot of criticism and protests but also stimulated a number of theoretical and empirical studies. Consequently, Western macroeconomics, in its quantity rationing version, and disequilibrium econometrics have been successfully applied in recent years to the study of the consumption goods market. As far as the investment sector is concerned, the permanence of excess demand is accepted by the partisans of the shortage school as well as by those of the disequilibrium school, and there are very few attempts to verify that hypothesis empirically.

An even more important reason why the modelling of shortage and disequilibria in the investment sector is relatively rare is the weakness of the theory of investment and the lack of a well-defined theory of the firm in the CPE.

International Studies in Economic Modelling Series.
Models of Disequilibrium and Shortage in Centrally Planned Economies.
Edited by C. Davis and W. Charemza.
Published in 1989 by Chapman and Hall Ltd, 11 New Fetter Lane, London EC4P 4EE.

I shall start by discussing separately planners' and firms' behaviour in the institutional framework of investment decisions. Next, the reason for the theoretical difficulties in modelling investment in the CPEs will be analysed. In particular I shall argue that an equilibrium may not exist because of the explosive behaviour of demands, and that the literature on disequilibrium theory provides a strong theoretical argument against applying the disequilibrium model to investment processes in the CPEs. In the following section, I shall present some econometric models of investment: those based on the classical disequilibrium approach and those which describe the regulation of investment by signals of internal and external tensions. Finally, a model describing planners' behaviour in terms of a threshold reaction to various constraints will be presented: this model belongs to a wider class of disequilibrium models with two regimes and a switching condition. Comparing different models I shall concentrate on their assumptions and theoretical basis, rather than on computational problems.

14.2 THEORY OF INVESTMENT BEHAVIOUR IN THE CPE

Let us start with a brief description of investment processes in the CPEs. There is no investment market in the CPEs, if by investment market we understand all sales or purchases of investment goods, i.e. all exchanges of investment commodities against money. Investment goods are distributed administratively in physical terms. Money is passive: money expenditures simply follow previously made decisions on real investment. Investment outlays in turn are determined as a result of an interaction between the planners and the firms within the state sector during the process of preparation and implementation of the plan. Consequently any theory of investment determination should take into account firms' behaviour, planners' behaviour and the relationship between the two, rather than treat the state sector as one agent.

14.2.1 Firms' behaviour

The behaviour of firms is the most distinctive feature of the CPE reality and it is determined by its particular institutional framework. Integrated in a hierarchical structure, firms are dependent on and supervised by higher authorities – ministries, industrial associations, etc. Property rights being untradable, the state is actually the owner of the firms: it exercises political control over the economy by imposing its preferences but also by rescuing firms from financial difficulties. Hence, by permanent interventions of all kinds designed to achieve greater compatibility between the macro- and microrationality, the centre creates an institutional environment that is basically unstable. Kornai (1980, 1986) speaks in this context about the paternalistic role of the state and stresses strongly the fact that central intervention in the situation of financial difficulties assures all firms of their survival. His famous term, soft budget constraint, was originally introduced in order to take account of this expectation. Kornai's is unquestionably a very important point (I shall come back to it later), but it misses an important feature of the CPE's institutions. The state intervenes not only to prevent bankruptcy. It intervenes constantly in the working conditions of the firm (changing plans, norms, parameters) and also – what turned out to be particularly pernicious from the motivational point of view – in decisions which the firms could consider as belonging to their own sphere of competence. Although a series of reforms attempted to increase the autonomy of enterprises, the sphere of decisions the centre could make has never been clearly defined. The centre has never renounced the possibility of intervening at any time in the choices of enterprises and of changing – whenever difficulties appear – the rules of the game.

The activity of an enterprise is judged according to a number of criteria which are subject to constant changes. In the traditional Soviet-type system, the maximization of gross production was the principal performance indicator. Later, planners became aware of the shortcomings of a system in which bonuses depended solely on the realization and surpassing of the plan of gross production, and added further indicators designed to direct the enterprises' effort towards the achievement of an objective the planners considered specially important at the moment: reduction in the cost of production, applying new technologies, realization of an assortment plan, quality improvement, etc. The multiplication of these specialized (and often inconsistent) indicators, obliged the firms to choose some tasks at the expense of others, and resulted in a sort of decentralization of decision-making – an effect unexpected and undesired by the planners. Looking for more control, the centre then reacted by adding still more success indicators, more incentives, and more sanctions. To break this vicious circle, the so-called synthetic success indicators (such as profit), taking account of all aspects of enterprise activity, were introduced. But they did not replace the old ones: both types of indicator continued to exist side by side. Profit-making has thus never become the real motive for the firms, even in the reformed CPEs: since differentiation among enterprises has never become acceptable for the state, the authorities attempted to reduce the arising inequalities by withdrawing so-called undue profits from the firms which succeeded too well.[1]

All economic reforms in the CPEs can in fact be viewed as a search for a compromise between the synthetic success indicator and the multiple partial indicators that express the planners' priorities and allow them to control directly all enterprise activity. Each reform proposal declared the necessity of increasing the role of synthetic success indicators and of limiting the number and importance of specialized indicators. After short periods of time, however, the latter always reappeared in one form or another in the incentive structure of the firm. The

[1]For that reason a very interesting model of Goldfeld and Quandt (1987), based on the assumption that the expected profit maximization is the manager's objective, cannot be considered as a model of a real CPE firm. It is rather a model of an idealized market-socialism with a bailout mechanism. It does not mean, however that the profit incentive is incompatible with the soft-budget constraint. But to have a profit incentive operating effectively would require stable economic institutions and a uniform regulation system insensitive to the effect of differentiation of enterprises and of individual incomes.

constantly changing economic institutions have discouraged the firms' efforts and made them take steps designed to assure them a quiet life. One of the simplest ways to realize plan targets and to guarantee some flexibility in the functioning of an enterprise is to obtain an abundance of means of production, and especially of investment. The firms thus have a positive motive to present very ambitious investment demands. But probably still more important is their lack of incentives for not making such demands, since there are no institutional restrictions limiting the firms' claims on investment. Here we return to Kornai's notion of the soft budget constraint: the softness of the budget means that there is no danger of a forced exit. Even if enterprises have to repay credits (which is not always necessary) insolvency is not a real threat: they can always rely on subsidies, price adjustment, tax facilities, and so on. Whereas the demand for inputs in other areas can be limited by storage possibilities and the ratched principle reasoning, in the case of investment the only tactical consideration that can restrain the firms' demand is the strategy of hooking into the plan.[2]

Let us summarize. State intervention in the functioning of the centrally planned economy has two contradictory effects: permanent changes in the regulation system increase firms' uncertainty, while the lack of the risk of bankruptcy creates a feeling of security. Both give the firms an incentive to try to obtain as much investment goods and investment credits as possible; their demand is price inelastic. The metaphor of great investment hunger, often used by East European (particularly Hungarian)[3] economists, describes convincingly the irrepressible investment pressure which is a constant element of the CPE.

All this does not mean that the demand for investment is an infinite demand or that the intensity of the demand does not change. In fact, the demand is determined through a permanent interaction with the planners' behaviour and depends on the planners' attitude towards expansion.

14.2.2 Planners' behaviour

Do not the planners themselves impose some investment pressure on the economy? Until the 1960s, the planners undoubtedly had a tendency to overinvest, and their economic strategy was based on the maximization of the growth rate of investment. But since the 1970s, they appear to be more sensitive to internal and external constraints, and a strategy of moderate growth seems to have progressively replaced a strategy of accelerated growth, although the tendency is not equally visible in all countries.[4] In the absence of other regulatory mechanisms, planners are the only equilibrating element in the CPE. Their decisions concerning the growth rate of investment are not exogenously and arbitrarily imposed on the economy (as suggested by Wiles, 1982). Although some investment projects are unquestionably

[2]The enterprises underestimate the anticipated costs of their investment projects knowing that once accepted and started they will certainly be finished.
[3]See, for example, Báuer (1978), Kornai (1980), Soós (1978).
[4]For example it is much more marked in Hungary than in Poland.

accepted independently of economic considerations, in general the planners' investment behaviour seems to be stable and can be viewed as a rational reaction to various constraints. However, the planners' intervention does not always have a stabilizing effect on the economy; their behaviour varies with and influences the investment cycle.[5] So long as the constraints in the economy are not very hard the planners allow investment outlays to be determined in bargaining between different sectors. Trying to avoid an open explosion of conflicts, the planners accept expansion and postpone restrictive measures: the firms' investment hunger takes over. But when the intensity of shortages attains an unbearable level, making adjustments necessary,[6] the planners succeed in restraining investment pressure, restore their control of the economy, and reduce the growth rate of investment. The alleviation of tensions, however, makes the planners return to their passivity and again allows investment pressure to dominate the investment process.

Various administrative and economic methods have been tried in order to reduce investment demand and to make managers more responsible for their decisions, but in the prevailing institutional framework of the CPEs, all of them proved ineffective. Indeed, the planners' decisions cannot be based on a rational discrimination among investment projects on the basis of marginal evaluation. Because of the inevitable informational decentralization (adequate technical information is available only to firms) and the lack of appropriate efficiency criteria (in the absence of an adequate price system), planners cannot allocate investment credits independently of the demand expressed by the firms in the state productive sector, and they are bound to rely on information exchanges with the firms.

14.2.3 Defining demand for and supply of investment

During the preparation of the plan, central planners send out tentative figures for planned investment expenditures. Enterprises in turn send messages about their investment needs. A bargaining process then takes place and eventually an investment budget and an investment plan are adopted. But the bargaining process does not stop there; it continues during realization of the plan as well. The investment pressure is even stronger then because of the firms' original strategy of hooking into the plan. Several authors affirm that the excess demand for investment is positive but they do not try to define the concept clearly. Others (e.g. Lackó, Chapter 10 in this volume) maintain that it is impossible to measure it. Let us look a little more closely, therefore, at the problem of defining the investment supply and demand and the excess demand for investment which seems to be at the root of the weakness of the theory of investment in the CPE.

[5]In order to avoid a terminological debate on whether cycles in the CPEs exist or not we should rather speak about the succession of phases of acceleration and deceleration in the growth rate of investment.
[6]Some authors (Goldman and Kouba, 1969; Pajestka, 1965) stress the role of the supply-side factors in the stopping of expansion – shortage of investment goods and services, scattering of resources among too many investment projects, the deficit of the balance of trade – others (Olivera, 1960; Bajt, 1971; Grosfeld, 1985, 1986) underline the role of consumption in the crisis and the necessity of taking into account the scarcity of consumers' goods.

A possible solution to the definition problem would be to consider the difference between the firms' proposals and the total investment plan as a measure of excess demand that appears during the preparation of the plan. An alternative would be to look at excess demand in the process of the plan's implementation, and to compare the actual demand for investment, which usually turns out to be higher than predicted (because of hooking into the plan), with the productive capacity of building enterprises, and with the production and import of investment goods.

A crucial difficulty with both these approaches – independently of the unavailability of relevant data – is the specification of the demand function. To be realistic the demand function should take into account the firms' struggle for more investment. If we look for inspiration in recent works in macroeconomic disequilibrium theory, which introduced quantity rationing signals, in addition to prices, into the demand function, it turns out that several such non-Walrasian models are unfortunately based on assumptions that are quite unrealistic with respect to CPE institutions. In Drèze's (1975) model, for example, each agent's demand satisfies perceived constraints in all markets; the agent sends his constrained demand and cannot attempt to explore how binding his constraints are. But this is clearly not appropriate for the CPE firms: all their efforts and capacities of innovation in fact focus on bargaining about constraints and on exploring how binding these constraints really are.

Unlike other theories, Benassy's (1977) model appears to be quite adapted to the CPE situation. This model assumes that effective demand for a given commodity is based only on constraints perceived in other markets, ignoring the constraints on that commodity. In other words, agents can send trade offers violating the constraints they perceive. Whether or not they will do so will depend on the perceived rationing scheme, i.e. the expected relation between the demand and the actual transaction, which depends on observations made by the agents in the previous periods. If the transaction is independent of the effective demand which the agents express, there will be no incentives for them to manipulate the information they send. But if the actual transaction is an increasing function of the agents' excess demand, they will overbid, i.e. express a demand higher than the outcome they expect to obtain. Benassy (1977) showed that if all agents engage in overbidding, this will lead to an explosive behaviour of demands making equilibrium meaningless or non-existent.

The state enterprises clearly overbid during the preparation of the plan. They have no reason to behave honestly and to reveal the truth about their investment needs. On the contrary, being aware that the investment credits they actually receive will be more or less proportional to the demand expressed, they have every incentive to present overambitious demands. Their past experience shows that because the aggregate demand usually exceeds the available investment funds, central planners proceed by cutting the planned investment expenditures in each branch by a certain percentage. Confronted with a sequence of disequilibrium situations, firms have obvious incentives to overbid: they simply try to adjust, looking for a sort of intertemporal optimization. Therefore the excess of the firms' demands over realized (or planned) investment conveys no information about the true magnitude of disequilibria. It makes the modelling of investment in terms of classical disequilibrium macroeconomics very precarious.

14.3 SOME ECONOMETRIC INVESTIGATIONS

There have been some attempts to apply a quantity rationing analysis to the investment sector, taking into account certain specificities of the CPE. The first model to be discussed here is that developed by Charemza and Gronicki (1986). The authors recognize that the modelling of investment is 'a major gap in achieving a fully integrated disequilibrium econometric macromodel of a CPE', and they try to formulate the problem of investment behaviour in such terms as to make the application of a quantity-rationing macrotheory feasible.

This is a legitimate approach. What is somewhat surprising, however, is that the same authors who did not accept Portes' questioning of the all-excess demand hypothesis with respect to the consumer market, and worked extensively with a chronic excess demand hypothesis (Charemza and Gronicki, 1984), now propose a model of investment behaviour in which the hypothesis of excess demand within the state production sector is not accepted – although the authors' own description of the microeconomic features of investment behaviour would suggest it – and is tested against aggregate time-series data.

Consequently the model is formulated as a classical disequilibrium model, with demand function, supply function, and minimum condition, to which the authors add a plan adjustment equation. The formulation follows basically Charemza and Quandt (1982) and Portes *et al.* (1987). There are separate mechanisms generating demand (on the part of enterprises) and supply (on the part of planners). Planners are supposed to maximize investment and to be constrained in their maximization by the critical level of consumption, the balance of payment, the full employment policy, and production capacity. All these constraints are supposed to be taken into consideration by planners in the preparation of the plan; this is not very realistic but the authors are aware of it. The plan itself is seen as an endogenous mechanism for adjustment towards an equilibrium or, in other words, as an instrument of equilibrating the demand and supply.

Enterprises are also seen as economic units looking to maximize their importance and, therefore, their investment, with the maximization being restricted by central rationing. The aggregate investment demand is defined as a sum of investment demand by particular enterprises.

It is in this equation that the main weakness of the model appears. The investment demand is defined there as a function of the difference between the desired level of fixed assets (which is practically determined by the plan of the output) and the level of fixed assets in the previous period. All we know about the behavioural characteristics of the firm makes this relationship difficult to accept. The desired level of fixed assets (and investment demand) is only to a certain degree influenced by the plan of output. The plan can be relatively small and still the desired level of fixed assets can remain very high, simply because the whole motivation of the firm, rooted as it is in the institutional framework of the CPE, cannot be affected by the mere reduction of the plan that the firms inevitably perceive as temporary. The plan of output could be *the* determinant of the investment demand only if the allocation of investment did not conform to a typical manipulable rationing scheme. But it does so conform, and we know that enterprises overbid. Moreover, it is not clear why a dummy variable appears only in the equation for the desired level of fixed assets. If it reflects, as the authors explain, the 'first symptoms of "Polish crisis"', it should be included also on the supply side of the model.

It is also difficult to follow the authors when they admit the possibility that the investment supply can be greater than investment demand 'in the case of temporary harder planning discipline, when enterprises are penalized for inefficient investment and also in the case of huge backlog of investment from previous period' (Charemza and Gronicki, 1986, p. 108). Obviously, we cannot affirm that such a situation is absolutely impossible. But, as I have said already, all attempts to curb the enterprises' investment hunger have until now proved to be ineffective, and it is difficult to imagine what kind of additional constraints put on firms could create an excess supply regime. 'A huge backlog of investment from previous periods' and other constraints can restrain central economic policy-makers from starting new investment and provide an incentive for planners to impose an austerity, but it is not clear why they should dissuade an enterprise with a soft budget constraint from showing an insatiable investment hunger.

An important novelty of the model under discussion is the distinction between the notional and effective increase in fixed assets. This distinction allows the possibility of the freezing of investment and bottleneck effects to be taken into account; in other words, it is admitted that investment can have a negative impact on fixed assets. This is a welcome step towards capturing the reality of investment processes in the CPEs. What is questionable, however, is the model's introduction of possible disturbance through the lagged excess demand in the equation for effective increase in fixed assets. The excess demand of the firm does not necessarily reflect constraints and bottlenecks; it is rather, let us repeat once again, a permanent characteristic of the CPE enterprise. Consequently, the simulated effect on consumption supply of the smoothing of the investment is not convincing. The bottleneck coefficient cannot be set at zero, unless we assume an institutional framework quite different from that prevailing in the CPEs.

The estimated model has been used for simulation of investment excess demand, which appears to be positive in 1960–79, with an annual mean of 23.3% of the quantity observed, and strongly negative in 1980: −28.1%. Although 1980 in Poland was a bad year for econometricians, such a spectacular switch from excess demand to strong excess supply of investment is rather surprising and throws some additional doubts on specification. The authors' interpretation, that the planners succeeded in decreasing excess demand in the second half of the seventies 'by implementing a slightly harder financial discipline for enterprises' and that 'this policy, together with a great backlog in the locked-up investments and with a wide social critique of the over-investment policy, provided to limitation of investment demand and to receiving a negative sign for 1980' (Charemza and Gronicki, 1986, pp. 113–14), does not seem to be satisfactory.

Hulyák (Chapter 9 in this volume) proposed a similar model of the investment goods market integrated in a multimarket disequilibrium macromodel. Her definition of investment demand basically follows that of Charemza and Gronicki (1986). The difference is that Hulyák does not test the hypothesis of excess demand for investment but assumes that the excess demand is positive. She attempts to estimate a changing intensity of excess demand and introduces it into the plan adjustment equation.

In constructing their demand/supply model of investment Charemza, Gronicki and Hulyák intended to fill a gap in the existing disequilibrium macromodel of a CPE. They achieved the objective but some of the models' assumptions are questionable.

They model the investment processes in a classical disequilibrium style, defining demand, supply (which is always (Hulyák) or usually (Charemza and Gronicki) smaller than demand), and adding a plan adjustment mechanism which assumes that planners react to increasing acuteness of shortages (excess demand) in the investment sector. Unquestionably, planners are sensitive to tensions in the investment sphere but excess demand does not seem to play the role of tension indicator. Aware of the firms' bargaining strategy, planners know that investment demand is not a reliable measure of tension. There are other indicators however – e.g. gestation period, investment commitment and freezing of investment – which might give the planners an incentive to adjust. Information coming from outside the investment sector can also influence planners' investment decisions. In fact, several models have been proposed which were designed precisely to identify those disequilibrium indicators appearing significant in planners' reaction function. Their authors do not apply the classical disequilibrium analysis but attempt to endogenize the planners' investment behaviour, which they consider to be an equilibrating element in the CPE, and the key to understanding the fluctuations of the growth rate of investment (see Grosfeld, 1986). Consequently these offers do not model the interaction between enterprises and planners, and they eschew the delicate problem of defining investment supply and demand. Thus, for example, Lackó (1980, Chapter 10 in this volume) shows that in making their investment decisions planners are sensitive to internal and external tension indicators (which are, respectively, total unsatisfied delayed demand and the relative net exports).

The modelling of the planners' behaviour, and particularly that concerning investment, seems to be a necessary step in building a comprehensive model of the CPE. In the absence of any other equilibrating mechanisms in the CPEs, planners are unquestionably a stabilizing factor, reacting rationally to different constraints. Models of this kind implicitly assume, however, that planners react continuously to all information, whatever it is. But we know that the planners' adjustment is not smooth: they do not react early enough to different red signals, thus allowing for an accumulation of imbalances. Their behaviour is characterized by a sort of bounded rationality: it is not always equilibrium oriented: adjustment is not continuous.

We shall now present a model of planners' investment behaviour which takes into account the discontinuous nature of the planners' reaction.

14.4 MODEL OF PLANNERS' BEHAVIOUR WITH A THRESHOLD EFFECT

The hypothesis tested in Grosfeld (1987) is that planners react to disequilibrium signals but only after the imbalances go beyond some threshold. So long as the tensions are weak, planners are inclined to ignore them: exposed to overwhelming investment pressure, engaged in complex bargaining with different pressure groups, and having their own preferences for expansion, they accept imbalances and postpone adjustment.

Thus Grosfeld (1987) is not a classical disequilibrium model, switching between excess-demand and excess-supply regimes but rather belongs to a large class of models switching between two regimes of a different kind: one which is characterized by a routine, conventional planners' behaviour and the other in which active decision-making takes place.

Two equations are specified and estimated in the paper: the investment plan equation and the realized investment equation. The basic idea of the model was to use in each equation two kinds of variable: the variables that represent the usual, mechanical reaction of the planners and the variables that reflect the level of disequilibrium. Standard investment decisions are determined by planned investment (present and past) and past realized investment. The backward looking character of the investment function is akin to that observed by Gács and Lackó (1973) and Portes *et al.* (1987). In the equation for planned investment, the role of disequilibrium indicators is played by the balance of trade (total or with non-socialist countries); in the equation for realized investment that role is played by real wages. In order to account for the threshold effect in planners' behaviour an econometric model inspired by the disequilibrium econometric literature has been used. In that model the dependant variable is determined by one equation or the other depending on a condition which is itself specified in stochastic terms. The final specification can be presented as follows:

$$\mathrm{PINV} = \begin{cases} \alpha_1\mathrm{PINV}_{-2} + \alpha_2\mathrm{INV}_{-1} + u_1 & \text{if } \mathrm{BC}_{-2} \geqslant \alpha_4 + u_1'' \\ \alpha_1'\mathrm{PINV}_{-2} + \alpha_2'\mathrm{INV}_{-1} + \alpha_3'\mathrm{BC}_{-2} + u_1' & \text{if } \mathrm{BC}_{-2} < \alpha_4 + u_1'' \end{cases} \tag{14.1}$$

or

$$\mathrm{PINV} = \begin{cases} \beta_1\mathrm{PINV}_{-2} + \beta_2\mathrm{INV}_{-1} + u_2 & \text{if } \mathrm{BCNS}_{-2} \geqslant \beta_4 + u_2'' \\ \beta_1'\mathrm{PINV}_{-2} + \beta_2'\mathrm{INV}_{-1} + \beta_3'\mathrm{BCNS}_{-2} + u_2' & \text{if } \mathrm{BCNS}_{-2} < \beta_4 + u_2'' \end{cases} \tag{14.2}$$

$$\text{INV} = \begin{cases} \tau_1\text{PINV} + u_3 & \text{if } \text{SR}_{-1} \geqslant \tau_3 + u_3'' \\ \tau_1'\text{PINV} + \tau_2'\text{SR}_{-1} + u_3' & \text{if } \text{SR}_{-1} < \tau_3 + u_3'' \end{cases} \tag{14.3}$$

where all variables are measured in natural logarithms, and

PINV = ratio of planned investment volume to realized investment volume in the previous period,
INV = ratio of realized investment volume in period t to realized investment volume in the previous period,
BC = ratio of the value of total exports to the value of total imports
BCNS = ratio of the value of exports to non-socialist countries to the value of imports from these countries,
SR = ratio of real wages in period t to real wages in period $t-1$,
u_1, u_1', u_1'', u_2, u_2', u_2'', u_3, u_3', u_3'' are error terms assumed to be $N(0, \Sigma)$ and independent.

Disequilibrium indicators enter the equations when their values go beyond a threshold. The idea of threshold reaction appears in Kornai (1980, 1982) and Lackó (1980) but in a slightly different way. Kornai (1982) suggests that planners control the investment process by reacting to three shortage signals: less-than-normal value of consumption, more-than-normal value of investment commitment and above-normal shortage intensity (pp. 46–50). In Lackó (1980) model planners are supposed to be sensitive to internal tension, represented by unsatisfied delayed demand (corresponding to Kornai's investment commitment), and external tension, caused by deterioration in the balance of trade. However, the 'normal shortage, tensions, tolerance limits' are not identified clearly and Kornai goes so far as to say that 'there is no clear rule as to where the tolerance limit lies' (Kornai, 1980, p. 212). Grosfeld (1987) could actually be considered as a test of Kornai's idea of control by norms: a deviation from the normal value of a variable provokes a response on the part of the central planners; the thresholds which make planners change their investment decisions are identified empirically and estimated together with the behavioural equations.

Both disequilibrium indicators perform quite well. The estimated thresholds are believable and consistent with our empirical knowledge. An interesting result is that planners seem to be more tolerant with respect to the deficit in total trade than to that in hard currency: BCNS appears in the planned investment equation when exports and imports with non-socialist countries approach equilibrium (exports/imports in hard currency < 1.095), while BC seems to be considered by planners as a constraint when it shows a deficit of a certain magnitude (total exports/imports < 0.9). The real wage enters the realized investment equation when its growth rate is smaller than 3%. Its coefficient is significant and confirms the planners' sensitivity to the evolution of real wages: when the wage growth is less than acceptable the planners react by reducing the growth rate of investment. The results also confirm the phenomenon which had been observed previously in an unsystematic way. The relationship between the investment plan and realized investment changes according to whether the tensions are acceptable or not: as long as they remain relatively weak the planners appear to be unable to control the investment process and to curb the investment pressure (τ_1 is greater than 1.5). But when the disequilibrium signal is sufficiently important to be taken into consideration, the influence of the plan on realized investment is strongly weakened (for details, see Grosfeld, 1987).

In spite of the good performance of the model, the choice of the variables – disequilibrium indicators – may be criticized and alternative solutions could be proposed. The author herself admits that it might be more appropriate to use balance of payments as an indicator of external disequilibrium, but she argues that this could not be done because of data limitations. Also, excess demand for consumption goods is suggested as a more adequate proxy for consumption pressure. Indeed, the introduction of a measure of excess demand would be the most urgent extension of the model: in the periods of excess demand for consumption goods the real wage variable should not be, and probably is not, considered by planners as a reliable tension indicator. Another variable which might reflect a consumption constraint and, beyond some threshold, result in restrictive investment policies, is a poor agricultural harvest (Mihályi, 1987).

Disequilibrium signals appearing in the model come from outside the investment sector. It might be important to test – as has been suggested by Kornai (1982) and Lackó (1980) – whether some shortage indicators coming from inside the investment sector also influence the planners' investment decisions. Instead of using lagged excess demand of investment – which, as we said, seems to be ill defined because of the overbidding – it would be interesting to verify whether the value of investment commitment going above some threshold plays such a role.

14.5 CONCLUSIONS

The modelling of investment in the CPE depends heavily on the theory of investment we have and on our assumptions about the character of the planning processes. If we see planning as a kind of co-operative game, with a dialogue allowing for a compromise, and assume that we can sensibly define the investment demand of enterprises in the CPE, the quantities demanded by enterprises can be considered as demand and the quantities offered by the planners as supply. Consequently, the Charemza and Gronicki model turns out to be an interesting attempt to apply a quantity-rationing analysis and to take into account certain specificities of the CPE. If, however, we agree that the informational exchange between enterprises and planners, rather than aiming at enriching the informational assets of both sides, is merely an instrument of manipulation of the adversary, i.e. if overbidding is viewed as the main characteristic of investment behaviour and the allocation of investment credits as a manipulable rationing scheme, the use of a conventional econometric disequilibrium macromodel makes no sense. In which case the Grosfeld model seems more appropriate. It is, however, less complex and less ambitious. It does not propose a theory of investment determination but simply explores adjustments aspects and gives a bootstrap theory of planners' behaviour. But perhaps it is more fruitful to start with such a simple explanatory scheme in order to establish the relationships which better represent the behaviour of the real economy.

REFERENCES

Bajt, A. (1971) Investment cycles in European socialist economies: a review article. *Journal of Economic Literature*, **9**, 1.

Báuer, T. (1978) Investment cycles in the planned economy. *Acta Oeconomica*, **21**, 3.

Benassy, J. P. (1977) On quantity signals and the foundations of effective demand theory. *Scandinavian Journal of Economics*, **79**.

Charemza, W. and Gronicki, M. (1984) Simultaneous consumption–labour–money analysis in a disequilibrium framework: Poland 1960–1980, CREES Discussion Paper No. RC/B20, University of Birmingham.

Charemza, W. and Gronicki, M. (1985) Maximum likelihood estimation methods for disequilibrium models in a centrally planned economy, CREES Discussion Paper No. G5, University of Birmingham.

Charemza, W. and Gronicki, M. (1986) A model for investment in Poland: a disequilibrium econometrics approach, *Economic Modelling*, **3**, 106–16.

Charemza, W. Quandt, R. (1982) Models and estimation of disequilibrium for centrally planned economies. *Review of Economic Studies*, **49**.

Dreze, J. (1975) Existence of an equilibrium under price rigidity and quantity rationing. *International Economic Review*, **16**.

Gács, J. and Lackó, M. (1973) A study of planning behaviour on the national economy level. *Economics of Planning*, **13**, 1–2.

Goldfeld, S. M. and Quandt, R. E. (1987) Budget constraints, bailouts and firms under central planning (*mimeo*).

Goldmann, J. and Kouba, K. (1969) *Economic Growth in Czechoslovakia*, IASP, New York.

Grosfeld, I. (1985) La théorie du cycle de Hayek et les économies à planification centralisée. *Revue d'Etudes Comparatives Est–Ouest*, **16**, 4.

Grosfeld, I. (1986) Endogenous planners and the investment cycle in the centrally planned economies. *Comparative Economic Studies*, **28**, 1.

Grosfeld, I. (1987) Modelling of planners' investment behaviour. *Journal of Comparative Economics*, **11**.

Kornai, J. (1980) *Economics of Shortage*, North-Holland, Amsterdam.

Kornai, J. (1982) *Growth, Shortage and Efficiency*, Basil Blackwell, Oxford.

Kornai, J. (1986) Soft budget constraints. *Kyklos*, **39**, 1.

Lackó, M. (1980) Cumulating and easing of tensions. *Acta Oeconomica*, **24**, 3–4.

Mihályi, P. (1987) Why do East European investment booms halt?, Paper presented at the Conference on Fluctuations and Cycles in Socialist Economies, University of Padua, 13–14 March.

Olivera, J. (1960) Cyclical growth under collectivism. *Kyklos*, **13**.

Pajestka, J. (1965) Interpretacja pierwszej fazy rozwoju gospodarczego Polski. *Ekonomista*, **4–5**.

Portes, R. *et al.* (1987) Macroeconomic planning and disequilibrium: estimates for Poland, 1985–80. *Econometrica*, **55**, 1.

Soós, K. A. (1978) Some general problems of the Hungarian investment system. *Acta Oeconomica*, 2, 3.

Wiles, P. (1982) What we still do not know about the Soviet economy. In *The CMEA five-year plans (1981–1985) in a new perspective: Planned and non-planned economies*, NATO, Brussels.

15

Reproduction of shortage in the Hungarian car market

ZSUZSA KAPITÁNY, JÁNOS KORNAI and JUDIT SZABÓ

15.1 INTRODUCTION

The basic question considered in this chapter is: what are the shortage phenomena that appear in the car market, and what are the factors that account for the perpetuation of shortage in this market? Using the general theoretical ideas and analytical apparatus previously developed by one of the authors (Kornai, 1980), we try to discover whether there are any regularities that are theoretically important in the development of demand and supply, in the behaviour of buyer and seller and in the changes that occur in the market situation. Our subject is delimited as follows:

1. Only the Hungarian car market is investigated, although some reference is made to practice in other countries.
2. Only private cars bought by private persons are considered.

In the following, the expression 'car market' is to be understood as carrying the preceding restrictions.

Although the main subject of this chapter is the investigation of the car market, it is our endeavour that it should have a significance beyond the problems of just one sector and should contribute to the general theory of consumer demand and purchase, that is, of the consumer goods market.

International Studies in Economic Modelling Series.
Models of Disequilibrium and Shortage in Centrally Planned Economies.
Edited by C. Davis and W. Charemza.
Published in 1989 by Chapman and Hall Ltd, 11 New Fetter Lane, London EC4P 4EE.

15.2 FORMS OF DISTRIBUTION

People acquire a car through various forms of distribution, through various channels. The percentage shares of the forms of distribution are summarized in Table 15.1. Let us consider each form in turn.

1. The predominant form of distribution of new cars is by queueing for cars sold by the Merkur Company. This is the form investigated primarily in this chapter.
2. Some new cars are sold by rationing, by jumping the queue. The main form here

Table 15.1 The share of the various forms of distribution

	Qualification according to the theoretical forms of distribution	*Share (%)*		
		In total purchase	*In the purchase of new cars*	
		1976	*1976*	*1979*
New cars				
1. Merkur sales following the queue	Queueing	21.6	84.5	82.4
2. Merkur sales according to special quotas	Rationing	3.1	8.2	9.6
3. Merkur sales as lottery prizes	Random distribution	1.0	4.0	3.3
4. Konsumex sales against hard currency	Constrained market transaction or present	0.3	1.4	2.0
5. Private person's import	Market transaction or present	0.5	1.9	2.7
Second-hand cars				
6. Merkur purchase sold to private persons	Constrained market transaction, combined with queueing, rationing or random distribution	6.3	–	–
7. Private person's sale of private person	Constrained market transaction	67.2	–	–

Source: Merkur, Konsumex and OTP Reports, study of the Hungarian Market Research Institute, see Note 1.

Methodological notes: Row 1 consists of those who bought their car against a cheque issued by OTP. The data in row 2 are our own calculations. From the quantity sold by Merkur we deducted the figures in rows 1 and 3 and considered the difference as an approximation of the special quota.

[1]This table is compiled from the following sources: *A lakossági személygépkocsi piac egyes idöszerü problémái* (Some of the present-day problems in the private car market) (manuscript) Hungarian Market Research Institute, Budapest, 1978; *Közlekedési és Hirközlési Évkönyu*, 1979 (Transport and News Communication Yearbook, 1979), Central Statistical Office, Budapest, 1980; *Merkur Adatgyujtemény*, 1979 (Merkur Data Collection, 1979), Merkur Car Trading Co., Budapest (manuscript), 1980.

is that ministries and a few other institutions are granted a special quota which they distribute among their staff and the employees of the sector under their authority. There is no clear and public regulation to determine the share of this special quota or which buyers should benefit from it.

We do not consider it to be our function to comment from an ethical point of view on the form of distribution through rationing by jumping the queue. We do not have enough information at hand to evaluate the criteria of this rationing. As economists and observers of real socio-economic systems we realize that where there is a queue, there must also be people jumping the queue.

3. Another form of distribution is the prize draw (car prize for savings, lottery).
4. At the state-owned foreign trade enterprise Konsumex, cars can be bought for hard currency with or without only a short queueing time.
5. Where a car is imported by a private person from abroad this is, in the final analysis, a market form, since either the car owner or the individual presenting the owner with a car will have effected the purchase under market conditions abroad, as a rule, against hard currency.
6. Some second-hand cars are sold through the state enterprise Merkur. This is done seemingly in market form, yet due to the artificially low prices the buyers of second-hand cars form a queue; sellers state their preferences (rationing); or a decision is made by lot.
7. The overwhelming part of the trade in second-hand cars is from private sellers to private buyers via a market form where prices are formed and transactions concluded through bargaining. However, certain regulations constrain the trade in second-hand cars.[2]

In summary, we can state that the basic form of acquiring a car is to queue for it. Although there is no unregulated free market, the larger part of the trade in second-hand cars does take place – with some administrative constraints – under conditions that can be called a market form.

15.3 SHORTAGE INDICATORS

According to everyday experience, there is no absolute shortage of cars in Hungary. If one wants to buy a car, one can do so, at most one has to wait or to accept forced substitution. There is, however, a relative shortage of cars: supply deviates from the initial demand of the buyers – in quantity; and in composition according to type, quality, age and delivery date. We have tried to measure this deviation in numerical terms, with the aid of shortage indicators.

Data collection has proved difficult. Several indicators exist which are appropriate for describing the market situation and are observable without any great difficulty, yet no regular observation has been arranged.

[2]For example, it has been ruled recently that within three years from the delivery of a new car it can be sold only to Merkur, or to some other specially indicated enterprise or co-operative.

Table 15.2 Computed queueing time (Years)

	Average	*For Trabant*	*For Wartburg*	*For Skoda*	*For Lada*
1966	2.1	2.9	2.6	0.7	–
1967	0.5	1.0	0.6	0.0	–
1968	0.9	1.6	1.4	0.1	–
1969	1.7	4.6	3.1	0.6	–
1970	2.0	4.3	3.3	1.6	–
1971	2.5	4.9	5.6	3.0	–
1972	2.4	4.2	7.6	3.1	1.0
1973	1.6	2.4	4.0	1.6	1.0
1974	1.1	1.1	2.8	1.6	0.8
1975	0.5	0.3	0.8	1.0	0.4
1976	2.3	4.7	4.9	0.4	0.8
1977	2.5	3.7	3.7	0.5	1.8
1978	3.6	5.8	4.8	1.2	3.6
1979	5.3	7.2	5.8	2.3	6.0
1980	3.1	4.0	2.3	1.4	3.4

Source: See note 1; also letter from Merkur about 1980 and 1981.

15.3.1 Queueing time

The most important shortage indicator in the market for new cars is the queueing time. The actual queueing time of buyers could be observed, but these data have not been collected. Therefore, we have to be content with two kinds of indirect approach to the queueing time. One is the promised queueing time; the expectable waiting time which Merkur and the National Savings Bank (OTP) make known to the buyer when joining the queue. The other approach is the calculated queueing time: the quotient of the backlog of orders at the beginning of the year, and of sales transacted in the course of the year. It shows in how many years the turn will come for the last buyer who joined the queue in the previous year, if sales were to stabilize at the present year's level.

In Table 15.2 the time series of the calculated queueing time is given for the average of all models and for a few specially considered models. The queueing time is long even on average, and is especially so for the popular models.

15.3.2 Forced substitution

Forced substitution is one form in which shortage is manifest. In an investigation carried out by the Hungarian Market Research Institute a large number of buyers were asked the following question: 'If you could obtain it immediately, which model would you choose instead of the one you have ordered?' The answers are summarized in Table 15.3. The principal diagonal shows the proportion of those who voluntarily chose the model ordered, at least as far as the available selection is

Table 15.3 Forced substitution ratios among those wanting to buy a car (the present number of those who have ordered a certain model = 100)

Model ordered	*Model preferred in case of immediate delivery*							*Other models not available at present*	*Total*
	Trabant	*Wartburg*	*Skoda*	*Zhiguli*	*Moskvich*	*PF126*	*Dácia*		
Trabant	47	21	1	19	0	5	1	6	100
Wartburg	16	7	7	53	0	1	6	10	100
Skoda	5	10	0	80	0	0	5	0	100
Zhiguli	5	30	7	30	1	3	2	22	100
Moskvich	25	25	0	25	0	0	0	25	100
PF 126	25	11	11	39	0	0	0	14	100
Dácia	0	25	0	50	0	0	0	25	100

Source: Study of the Hungarian Market Research Institute, see note 1.
Methodological note: The investigation took place in 1977; 4120 persons were asked in total; 1406 sent valid answers.

concerned. The elements of the table outside the principal diagonal show the proportion of those who applied forced substitution even at the moment of ordering: the long queueing time for the preferred model compelled them to order another one. The proportion of those making their choice voluntarily is highest for the Trabant – every other buyer in the queue would in fact like to have a Trabant. For the remaining models this ratio is much lower and certain models are chosen exclusively as forced substitutes. The answers confirmed our assumption: forced substitution is present to a high degree in the car market.

15.3.3 Ratio of new to second-hand car prices

Some of the buyers choose, instead of (or besides) queueing, the immediate purchase of a second-hand car. Therefore, one of the indirect indicators of shortage in the car market is the ratio of the prices of new to second-hand cars. In the first column of Table 15.4 we give Danish data.[3] Denmark, like Hungary, does not manufacture cars; nevertheless, the supply of new as well as of second-hand cars is ample. Under such conditions, the price ratio reflects the judgement of the market on the value of the two categories of cars.

In the Hungarian market the constraints on the supply of new cars drive the prices of second-hand cars upwards. Our statistics cover a period in which no rules prohibited the sale of second-hand cars under three years old to private persons. The second column of Table 15.4 shows the relative prices of the Skoda, which is the easiest to buy, and the third column shows those of the Trabant, which is the most popular car; the price ratios for the other models fall between these two extremities. In this period buyers paid a considerable surcharge for a relatively new Trabant in the second-hand car market. Even in the sixth year of use, two-thirds of

Table 15.4 Second-hand car prices as a percentage of new prices

Age of car	*Danish average price* 1	*Hungarian price Skoda S100* 2	*Hungarian price Trabant 601* 3
0–1	70.3	87.8	112.2
1–2	61.5	81.1	104.4
2–3	50.3	73.0	92.8
3–4	40.9	62.2	85.1
4–5	37.1	54.1	75.4
5–6	31.5	43.2	65.8

Source: For column 1 we used the study by N. Keargard (1975), and for column 2 the article 'Mit ér . . . ?', see note 3.
Methodological notes: The Danish data were based on advertisements in 1968. The Hungarian data were based on actual market transactions in the first half of 1977, by taking partial samples.

[3]This table is compiled from data in the following: Keargard (1975); 'Mit ér ma a kocsim?' ('What is my car worth today?') *Autó-Motor*. **30**, (19), 21, 1977; *Vejtransporten*, 1980; *Automobil – importorerness Sammenslutning* , Denmark, 1980.

the new price was paid, while in the Danish car market, on average, the second-hand price falls below one-third of the new price after such long use.

15.3.4 Credit terms

Finally, credit terms are one more indirect shortage indicator. Although in the Hungarian case, credit terms are not highly responsive to short-run changes in the market situation, they do signal lasting changes and are characteristic of the different sales systems. The general rule is that if a credit transaction[4] is made at all, in a buyers' market the seller will grant credit to the buyer, while in a sellers' market the buyer will grant credit to the seller. In most capitalist countries a car can be bought on credit. Latterly, since the difficulties of selling cars have grown, Skoda cars can be bought on credit in Czechoslovakia. In Hungary, however, half of the purchase price, and in Romania the whole of it, has to be paid in advance, when the buyer joins the queue.[5] To summarize, the data confirm clearly that in the past 15 years there has been a chronic and, to a changing degree intensive shortage in the Hungarian car market.

15.4 BEHAVIOUR OF THE BUYER

15.4.1 Demand

In circumstances of chronic shortage we cannot apply a single specific concept of demand, since the buyer's buying intention is interpretable in several ways and evolves over time. Theoretically, in the case of the car market, the following phases can be discerned:

(a) *Hypothetical demand, phase 1*

Buyers are asked the question whether they would like to purchase a car, and if so, which model they would choose if, at a valid Hungarian price, they could obtain any Eastern or Western type for immediate delivery,[6] with a guarantee of parts supply and full and complete service.

[4]In the Soviet Union and in the GDR no advance payment has to be made when joining the queue. It may be that this adds to the length of the queue in these countries.
[5]Our source of information on the Soviet Union, Czechoslovakia and Romania is the article, 'Autóvásárlási körkép a szocialista országokból' ('Panorama of car purchases of the socialist countries'), *Heti Világgazdaság*, 18 July 1981.
[6]Each important Western model has a valid forint price, if only for the purpose of imposing customs duty on privately imported cars. For interpretation of the stipulation of immediate delivery it is assumed that the buyer has no special (so-called optional) wish, but is willing to buy the model available from stock.

(b) *Hypothetical demand, phase 2*

The foregoing question is asked, but narrowed down to the Eastern and Western models, or their present-day descendants, which have been on sale in Hungary during the last 15 years – against forints or hard currencies, or which have been imported privately in any considerable quantity.

(c) *Hypothetical demand, phase 3*

The question is further narrowed down to the models that are currently sold by Merkur against forints, with queueing. Yet buyers are asked to assume that they do not have to queue for any of the models. Thus buyers will consider only the model and the price.

(d) *Initial demand, phase 4*

It is not only the qualitative properties and the price of the model but also the expected queueing time that influences the buyers' decisions on which queue to join. Contrary to the *hypothetical* demands of phases 1–3, this initial demand need not be discovered by questioning but is revealed the act of joining the queue. What is more, the buyers confirm the seriousness of their intentions by depositing a 50% advance payment.

(e) *Revised demand, phase 5*

As a result of several factors, buyers may modify their initial demand and leave the queue. Perhaps they join another queue, either because they have changed their mind, or because the seller has persuaded them to buy another model.[7]

(f) *Actual purchase, phase 6*

To our knowledge, no questioning has yet been done concerning phases 1 and 2 of the hypothetical demand, and regarding phase 3, only the Hungarian Market Research Institute has made some initial steps. (Reference was made to this in respect of Table 15.3.) Data on the initial demand and on the actual purchase are available in the records of Merkur and of the OTP, or can be concluded from them, but so far they have not been systematically examined, or their trends methodically analysed. Information on revised demand is again sporadic.

In speaking of demand, in the present chapter we always mean initial demand, even when this is not specified. Some problems still remain, which have to be solved in order to make our definitions unambiguous. One important distinction is

[7] For example, in December 1978 the queue for the Lada 1600 was opened. Over 14 000 of those waiting for a Lada 1500 went to join the new queue, since in this way their turn came sooner. (See 'Mennyit kell autóra várni?') ('How long to wait for a car?') *Magyar nemzet*, 24 December 1978.

whether we want to measure a stock indicator or a flow indicator. Several definitions of practical consideration can be given for both. In the first column of Table 15.5 a stock indicator is shown: the stock of accumulated and valid orders. The flow indicator of the third column is called the additional buying intention.[8] It can be stated on the basis of the table that the demand for cars fluctuates significantly.

15.4.2 The decision variables of buyers

When they consider purchasing a car, buyers can decide the following:

1. Through what form of distribution can they and, in the case of choice, do they wish to acquire a car? For most buyers, if they want a new car, there is only one form: that of queueing.

Table 15.5 Time series of the demand for cars

	Demand stock: backlog of unfilled orders at the beginning of the year		*Demand flow: additional buying intention*	
Year	*Units*	*Chain index*	*Units*	*Chain index*
1966	33302	—	−2733	
1967	10800	32.4	33159	↑
1968	18501	171.3	40176	121.2
1969	34029	183.9	65413	162.8
1970	74619	219.3	76316	116.7
1971	108164	145.0	59692	78.2
1972	116837	108.0	49019	82.1
1973	106051	90.8	58368	119.1
1974	85912	81.0	47616	81.6
1975	44210	51.5	208836	438.6
1976	163532	369.9	124589	59.7
1977	206791	126.5	238687	191.6
1978	352586	170.5	285315	119.5
1979	530772	150.5	−91880	↓
1980	323547	61.0	41605	↑
1981	259471	80.2		–

Source: column 1 is based upon Merkur Adatgyüjtemény (see note 1), and letter from Merkur; column 2 is based on the former, as well as Közlekedési és Hirközlési Évkönyv (see note 1) and information from OTP and Konsumex. The chain indices in column 4 show the change compared with the previous year by an upward or downward arrow instead of an index, if the additional buying intention changed from a negative number to a positive one, or vice versa, that is, when a quotient cannot be computed.

[8] The additional buying intention is composed of two main items: (1) on a market where there is a queue, the number of those joining the queue, (2) the additional buying intention showing on the market where there is no queue. It can be assumed that the latter is realized within the given year. Therefore, it is approached by summing the following data on purchase: purchase through rationing estimated by comparing the data of Merkur and of the OTP, sale through Konsumex, and private imports by the population.

2. If they have chosen to queue, they can decide for which model to queue. Within this range, they can have a few special choices such as colour and certain accessories.
3. When queueing, they can decide to leave the queue and join another – assuming that the seller agrees to this – or cancel their order.
4. If, finally, their turn comes, they can decide whether to take delivery of the car immediately, or to postpone it, within the term allowed by the seller.

Buyers have no say in the price, which is determined by the seller. Our buyers are therefore price-takers. The situation is different when a second-hand car is purchased from a private person, when bargaining is possible.

15.4.3 Factors explaining the buyer's behaviour

In Table 15.6 the variables explaining the buyer's behaviour are surveyed and classified. In the following parts of this chapter we shall try to throw some light on their effects, in so far as this is possible by relying on experience and deductive reasoning.[9] Some of the explanatory variables are also described numerically in this

Table 15.6 The explanatory variables of the buyer's behaviour

Effect group	*Serial number*	*Denomination*
Standard effects		
The buyer's financial resources	1	The buyer's income
	2	The buyer's savings
	3	The expected selling price of the buyers car at the time of the exchange
Price effect	4	New car price
		Second-hand car price
	5	Fuel price
	6	Other costs of running (spare parts, service)
Cross price effects	7	Public transport tariffs
	8	Housing costs
	9	Other costs of living
Non-standard effects		
Supply effects	10	Limitation of the range of models to be ordered
	11	Expected queueing time
Administrative effects	12	Administrative constraint on the re-sale of cars

[9]Some statements in the Table and in the adjoining analysis concern only the total demand for cars, or total purchases of cars, while others can be understood as concerning the demand and purchase variables for either the total of all models or for each model separately.

chapter. These numerical descriptions are, however, not to be considered more than illustrations.

In Table 15.6 standard and non-standard effects are discerned. First of all we must clarify this distinction. Standard consumption theory, elaborated by Gossen, Engel, Pareto, Hicks and Samuelson, has developed parallel with neo-classical economics and is intertwined with Walrasian equilibrium theory. The part of standard consumption theory that is concerned with demand functions is accepted by some Hungarian economists who otherwise reject its theoretical bases.[10] Standard consumption theory was inspired by the examination of a buyers' market, and of the behaviour of the buyer having a hard budget constraint and paying an effective price. In such a market the fundamental explanatory variables of demand are price and income – although other factors may also be important. In a buyers' market demand can be satisfied more or less without change; intention and realization coincide. Therefore, the direct observation of demand can be replaced by the observation of actual purchases.

In this chapter the attribute 'standard' refers to interrelations which exist between demand on the one hand and price and the financial resources or income of the buyer on the other. In a sellers' market the standard explanatory variables also have a significant influence, assuming that the buyer's budget constraint is hard and the price effective.[11] In the case of cars purchased by private persons both conditions are fulfilled; therefore, the standard factors exert their effect. Besides these, however, there are other, nonstandard factors with considerable influence, characteristic specifically of a sellers' market; these are mainly supply and administrative constraints. Let us first take in turn the standard factors.

(a) *Factor* (1), *income, and factor* (2), *savings*

In the estimation of demand functions the customary procedures consider the buyers' actual purchases as dependent variables.[12] This procedure is justified if the buyers are not constrained by short supply in realizing their intention. In such a case it can be assumed that buying intention and actual purchase coincide. This assumption is unjustified, however, if there is chronic shortage in the market. This is because in the latter case the direct determinant of the actual purchase is supply. Therefore, what is determined in this case by the customary process is not a demand function but a purchase function.

In some markets direct observation of demand is possible even in the case of shortage: this is so in the Hungarian car market, in which, as we have seen, initial demand is recorded. It is this indicator – more exactly, the additional buying intention – that we have analysed.

[10] See for example Hoch, Kovács and Ördög (1980).

[11] For an explanation of the concepts of hard budget constraint and effective price see Kornai (1980).

[12] This is the practice also followed by most Hungarian research on the econometric estimation of demand interrelations. See for example Baranyai *et al.* (1972), and Hulyák, Muszély and Szakolczai (1980) and see note 10, also discussing the car market.

It is evident that even under conditions of chronic shortage the fundamental Engel relationship is effective; income strongly influences the long-term trend of demand. The actual purchase function, however, does not coincide with the demand function. Purchase runs closely parallel with income. There are two indirect effects to be observed. First, in the long run neither the buyer's behaviour manifest in buying, nor the seller's behaviour manifest in selling diverge greatly from demand. When the queueing time has elapsed the customer usually buys the car (only a few orders are cancelled). The supply directly determining the actual purchase cannot diverge far from the long-term trend of demand, either; supply follows demand more or less, apart from the shortage zone. Second, the parallel trends of incomes and car purchases can be further explained by the fact that both change because of the effect of identical factors. The macro-level fluctuations in the tightness of restrictions simultaneously influence income on the buyer's side and car imports on the seller's side; actual purchases will be determined by the latter, that is, by supply.

In a buyers' market, the following chain of causation asserts itself:

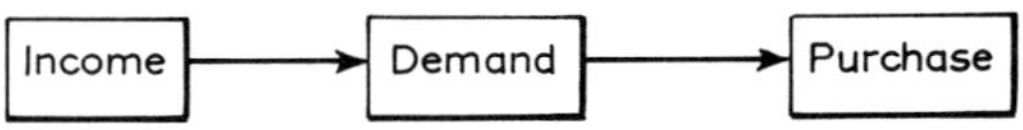

As opposed to this, in a sellers' market the chain of causation is the following:

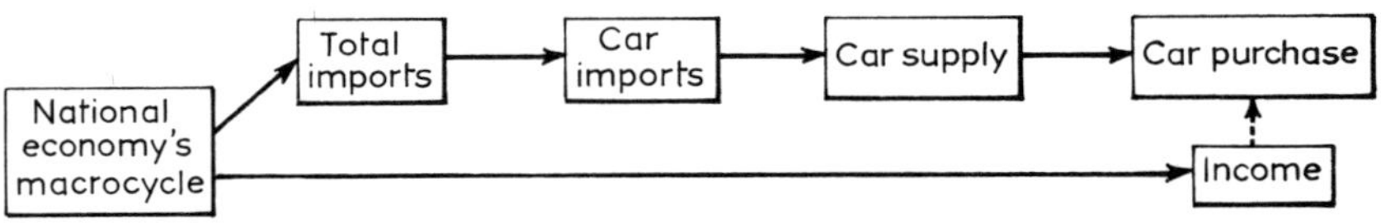

The vertical dotted arrow indicates that income and car purchases are moving closely together, because of a common cause, but it does not refer to any important direct causal link: close correlation, of course, does not necessarily represent causality. Demand is considerably influenced by the savings flow. Further investigation is required into how the savings flow itself depends on other factors, including income, supply on other markets, cross price effects, and so on.

Finally, one more remark on the role of income. The customary approach answers first of all the question, what is the effect of earlier and of simultaneous incomes on demand and purchases. It is common knowledge, however, that demand is also influenced by expectations for the future. This is to be revealed by some other methods, for example, by interviews. A halt in the rise in real wages and expectations of stagnation and recession may be expected to slow down the rise in demand for cars.

(b) *Factor* (3), *value of the car already owned*

Those who possess a car already will get a new one by exchanging it; the selling price of the old car is an important financial resource for buying the new one. Their financial situation is dependent largely on the trend of prices in the second-hand car market, on the administrative rulings that constrain selling, and so on. Thus it would be appropriate to divide car buyers into two large groups: old car owners and new comers. Their market behaviour is different (as is their driving).

(c) *Factors* (4), (5), (6), *own price effects*

The price of new cars had been by and large stable until 1979. In that year a sudden rise occurred. This also meant a considerable increase in the obligatory advance, which has to be considered as a component of the rise in prices. The rise obviously depressed the demand for cars. It played a role in the sudden fall of stock demand in 1979. We cannot find out, however, to what extent this factor was responsible for the fall and to what extent it was caused by other factors which exerted their effect more or less at the same time – the halt in the rise in real incomes, less optimistic expectations, the administrative constraints on the resale of cars, etc. Separation of the various effects in an econometric way seems impossible now because most of them appeared in one single package in 1979.

Demand for second-hand cars is obviously a diminishing function of the price of second-hand cars. It is also one of the standard relationships that the prices of new and second-hand cars have a cross effect on demand: the relative rise in prices of the former will augment the demand for the latter and vice versa.

(d) *Factor* (7), *public transport tariffs*

The present level of public transport tariffs has been stable for years and, compared with the rising costs of motoring, is very low. This acts to reduce demand for cars, while occasional increases in tariffs would boost the demand for cars.

(e) *Factor* (8), *housing costs*

There is no clear relationship between housing costs and the demand for cars. Common microeconomic considerations suggest the existence of a substitution relationship. A number of households can choose between spending their money on improving their housing conditions, or on buying a car. In Hungarian conditions, however, there are a great number of households for which the problem is present in another form. The two items of consumption, costs of housing and costs of a car, diverge substantially. This is confirmed by the studies of Lackó (1975) and Simon (1979), which say that fluctuations in housing costs are reflected in fluctuations in savings and not in changes in other items of consumption, including the purchase of cars.

(f) *Factor* (9), *other costs of living*

There is no clear relationship here, either. We do not know in which way more expensive consumption affects the demand for cars. Those who became accustomed to a car will hardly give it up, but will tend to use their car less.

The nonstandard effects occur as a consequence of the sellers' market character of the economy – and of the car market within it – as well as of the administrative intervention in the market process.

(g) *Factor* (10), *limitation on the assortment offered for sale*

This is an effect of fundamental importance, which draws a sharp line between a buyers' market and a sellers' market. Of course, the buyer can always choose only among the alternatives offered by the seller. The question is, to what extent the seller adjusts the set of supply alternatives to the revealed demand of the buyer. If, say, the buyer were offered products A, B, C, D, E, etc. in 1971 and products B and D were no longer supplied in 1972, this directly affected demand in 1972. For example, the Western cars imported and sold in the 1960s and early 1970s created a demand for these models. Later, however, Western car imports stopped. Another example is the supply of the Trabant and the Wartburg. Bookings for Wartburgs ceased between 1972 and 1974, and for Trabants between 1972 and 1975. The majority of buyers were not prepared to accept forced substitution, but preferred waiting. Thus, when booking started again, postponed demand came out in a burst. This instance shows clearly that the supply offered by the sellers, both foreign and Hungarian, is a major influence on buyers' demand.

(h) *Factor* (11), *queueing time*

The effect of the expected queueing time on the buyer's demand is not unambiguous. On the one hand, the rising queueing time repels the buyer. Some buyers are impatient and prefer to combine forced substitution with a shorter queueing time, switching to less popular models if the queue for the model they prefer is too long.

There is also, however, a contrary effect: the rising queueing time attracts the buyer to the queue.[13] If the queue is very long, it is worth booking in advance for a second purchase. It is also worth joining the queue for the purpose of resale, since in this way a speculative premium can be achieved. It is our impression that this accounted for the sudden rise in demand in the years 1977 and 1978, which was then put to an end by the 50% advance and the prohibition of early resale of second-hand cars.

(i) *Factor* (12), *administrative constraints on resale*

Where there is shortage and queueing, places in the queue will almost inevitably be offered for sale. As in the case of jumping the queue, we do not seek to pass a moral judgement: as economists we have to establish that this is a concomitant of the phenomenon of queueing.

[13] The relationship between shortage and buyer behaviour is examined more fully in Kornai (1980).

Administrative orders banned the sale of cars less than three years old among private persons. (Even before that the sale of cars under one year old between private persons had been specially taxed.) Since the ownership of a car has to be registered with the police, this order is enforceable. In the long run, the administrative constraint influences queueing intentions, that is, demand.

Now that we have come to the end of our list of explanatory variables, we wish to stress that a one-sided emphasis of either the standard or the non-standard effects may lead to false conclusions. It is only by considering their simultaneous effects that we car understand the buyer's behaviour.

15.5 BEHAVIOUR OF THE SELLER

15.5.1 Supply

In the preceding section the concept of the buyer did not need to be specified, since it is self-evident who the buyer is in the case of car purchases by the population. The situation is different with the seller. In talking about 'seller' in this chapter we mean the personification of a multi-level regulation process and not only the person facing the buyer in the course of the selling transaction. Starting from the bottom, the decision levels are the following:

1. Merkur personnel serving the buyer directly
2. Merkur apparatus, headed by the managers. On a level with them is the Mogürt Foreign Trade Company attending, among other things, to the imports of private motor cars
3. medium-level management organs: the Ministry of Domestic Trade, the Ministry of Foreign Trade, the Ministry of Transport and Communication
4. top-level economic organs: the National Planning Board, the Ministry of Finance, and the Office of Supply and Price-Control
5. top party and government organs making the most important decisions.

In our theoretical approach we summarize the simultaneous effects of the activities of all these levels in the abstract category of 'seller'. Real decisions are formed, of course, through the complicated interactions of these levels.[14] In this chapter, however, we do not consider the division of roles within the decision-making process, but try to describe the common resultant of the interactions, first of all from the point of view of whether there is any regularity in it.

As for the concept of supply, it gives less trouble. Under the conditions of a shortage economy, supply is self-evident and easy to measure. What is at the seller's disposal as real physical supply can be sold almost immediately. Therefore, in what follows, supply will be considered as identical with sales. The time series of car

[14]In this connection, when defining the concept of the seller, we disregard private persons offering second-hand cars for sale.

Table 15.7 Time series of sales to private persons

Year	*(Units)*	*(Chain index)*
1964	6857	–
1965	9957	145.2
1966	15 706	157.7
1967	20 785	132.3
1968	20 474	98.5
1969	19 879	97.1
1970	36 665	184.4
1971	44 033	120.1
1972	49 182	111.7
1973	67 307	136.9
1974	77 942	115.8
1975	81 692	104.8
1976	72 661	88.9
1977	84 388	116.1
1978	96 833	114.7
1979	100 270	103.5
1980	103 632	103.4

Source: See note 1, also letter from Merkur.
Methodological note: The figures include not only sales through queueing, but also those through a special quota and prizes.

sales to private persons is shown in Table 15.7. The long-run trend is of considerable increase but this is uneven and fluctuating.

15.5.2 The seller's decision variables

The seller – that is, the collectivity of organs and enterprises attending to car supply – may decide on the following:

1. import intention. Actual imports depend, of course, not only on the Hungarian import intention but also on the willingness of the exporting party.
2. distribution of actual imports among the various groups of buyers – firms, nonprofit institutions and private individuals – and how much of the latter should be through queueing and how much through the special quotas or other forms of distribution
3. granting the requests of those in the queue to alter their orders, or perhaps inviting them to switch to another model
4. suspending the booking of orders
5. determination of the financial terms of the sale. What should be the price of the new car? How much advance should be paid, or, should cars be sold on credit?
6. administrative constraint on resale among private persons
7. regulation of the sale of second-hand cars by the state-owned trade company.

Apart from a little more or less important internal friction, the centrally directed multi-level regulation process exhibits a homogeneous behaviour, so that the tens of thousands of competing buyers are faced with one single, specially large, quasi-monopolistic seller. The seller is particularly gigantic, because it not only controls the supply, and determines the price, as monopolistic sellers usually do, but it also controls the credit terms and the administrative constraints on the market process.[15]

15.5.3 Factors explaining the sellers' behaviour

In Table 15.6 standard and non-standard effects explaining the buyer's behaviour were distinguished. The explanation of the seller's behaviour must begin with a negative statement: the most important standard factors: the price and the expected enterprise profit have no effect on the seller's supply. This statement cannot be clearly confirmed by past experience, since there was only one considerable price increase – in 1979. This price increase was certainly not followed by a standard micro-market reaction, namely an increase in supply. On the contrary, macro-effects predominated: as the price increase kept back demand, so also was supply kept back under the restriction that covered the whole economy. This is manifest not only in the short-term change in supply, but also in the medium-term targets. Our interviews have shown that the five-year supply targets were gradually lowered. The initial target figure in 1978 was – according to several agreeing opinions – to import 650 000 private cars. In the summer of 1981 the 6th Five-Year Plan envisaged imports of 520 000 cars, but at Merkur and at the Ministry of Domestic Trade several voiced the opinion that only 480 000 should be imported.

Both public opinion and a number of economists reacted by thinking: 'it has become more expensive, but at least there will be more . . . ' It must be noted that this relationship does not assert itself automatically. In our circumstances, for example, it does not work in such a market as the one in which new cars are sold.

Let us now turn our attention to the non-standard effects. All of these go back to the fact that the seller is not a genuinely independent, market-oriented firm. It is basically not led by profit considerations, and does not have to compete for the goodwill of the buyers. The seller is a multi-level regulating organization motivated by a number of other factors, and it is able to impose its will upon the buyer. Without aiming at completeness, we shall consider five factors.

(a) *Transport policy*

With a certain amount of simplification, the usual train of thought runs as follows:

Let private motoring develop, but not too fast. It should not outrun its supplying

[15]The attribute 'quasi' was added because some of the buyers can acquire a car through other channels: if they have foreign currency, or foreign relatives willing to present them with a car, if they have a share in the special quota, or are content to buy a second-hand car on the private market. Yet the overwhelming majority of buyers find themselves facing the monopolistic seller.

facilities, the capacity of the road network, the service network, etc., nor should it displace public transport.

There are a number of rational elements in this train of thought. There is much to support the idea that in a planned economy it should not be solely the spontaneous demand of the population that decides the rate of increase of car ownership and that serious consideration should also be given to the indirect, external consequences of the growth of motoring, which then should be constrained accordingly. These are, however, constant factors, which would lead to a smoothly rising upper limit: they cannot account for the wide fluctuations in supply.

(b) *Effect of trade with the socialist countries*

As we have already mentioned, imports depend not only on the Hungarian import intention, but also on the export willingness of the partner countries. Here very definite constraints are encountered from time to time, which halt or reduce imports of one model or another.

Besides these irregular constraints on the exporters' side, there are some more regular relationships as well. If the trade balance between Hungary and a car-manufacturing socialist country becomes more favourable for Hungary, Hungarian car imports from the country in question will increase suddenly the next year. If the foreign trade balance changes for the worse, car imports will slow down or stop. This relationship does not prevail in every market in every year but it can be observed quite frequently. There is also a general and rather close correlation between car imports from the various socialist countries and total imports from the same countries. In these cases, therefore, car imports depend more or less linearly on total imports from the particular country.

(c) *Effect of trade with the capitalist countries*

Here the export willingness problem does not arise: the capitalist car industry is always ready to supply when Hungary is ready to pay in hard currencies.

An obvious relationship exists between the convertible balance of payments and the stock of debts on the one hand, and car imports from capitalist markets on the other hand. When the balance was not deteriorating to any great extent but, on the contrary, was improving, or an improvement was expected, imports from capitalist countries were stable or even growing. When, however, there was a significant deterioration, or prospects were poor, imports were stopped.

(d) *Effect of the national economy's macrocycle*

The continuous close relationship between private car imports and total imports holds not only for the absolute quantities, but more or less for the rates of growth too. Figure 15.1 shows the chain indices of private car imports and of total

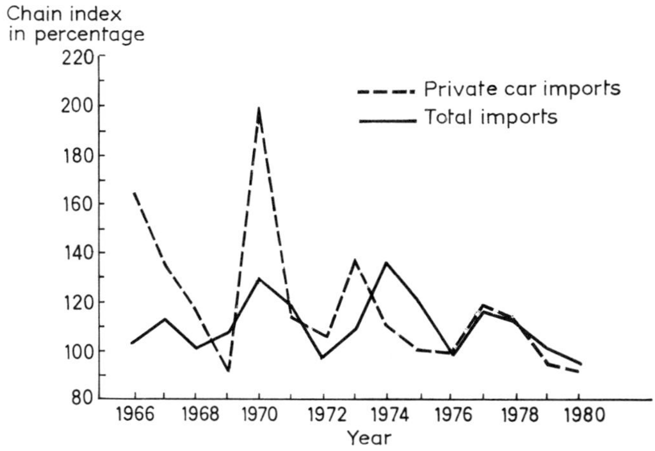

Fig. 15.1 Trends in car imports and total imports in Hungary. Source: see note 16.

imports.[16] Accelerations and decelerations coincide in eleven cases out of fourteen. Therefore, the permanent relationship between economy-wide import decelerations and accelerations on the one hand and car import decelerations and accelerations on the other hand is demonstrable.

(e) *Reaction to shortage signals*

The factors listed so far put an upper limit on the 'zone' of supply currently possible, on the basis of the foreign trade and transport policy constraints. However, there is still quite extensive scope for manouevring by the seller. According to our observation, this manouevring takes place mostly in response to shortage signals.

(f) *Effect of the queuing time*

Figure 15.2 presents the interrelations graphically and also shows the actual queueing time series previously given. The seller observes an upper tolerance limit. Shortage should not be too great. This limit seems to be an average queueing time of two and a half to three years. The motive for observing an upper tolerance limit is to be sought in the political sphere. Economic policy-makers aim to avoid a queueing time which buyers regard as intolerably long. If queueing time exceeds this limit, the 'attraction effect of the queue' will grow and, consequently, the queue will become even longer.

[16] *Statisztikai Évkönyvek* (Statistical Yearbook). Central Statistical Office, Budapest, 1960–69; *Külkereskedelmi Statisztikai Évkönyvek*, 1970–1979 (Foreign Trade Statistical Yearbooks 1970–1979). Central Statistical Office, Budapest, 1971–1980.

Table 15.8 Lower tolerance limit of shortage and supply reaction

Year	*Queueing time*	*Supply chain index*
1966	2.12	157.7
1967	0.52 ⌐	132.3
1968	0.90	→ 98.5
1969	1.70	97.1
1970	2.04	184.4
1973	1.58	136.9
1974	1.10	115.8
1975	0.54 ⌐	104.8
1976	2.25	→ 88.9
1977	2.50	116.1

Source: see Tables 15.2 and 15.7.

When around 1978–79, the upper tolerance limit was greatly exceeded, the seller intervened drastically in order to cut back shortage to the tolerable limit. This was achieved, however, not through increased supply, but by suddenly curtailing demand via price increases, raising of the advance payment, and additional constraints on resale.

Sellers also observe a lower tolerance limit: shortage should not be unduly low. They are specially intent that the sellers' market should not turn into a buyers' market. It seems that this lowest critical value is about six months to one year of waiting. The seller has several motives for acting in this way. The most obvious one is that if the queue were to be too short or, worse still, cease altogether, it would be inconvenient for the seller. Buyers would become more fastidious, and it would be hard or perhaps even impossible to satisfy their wishes because of the suppliers' rigidity. Certain foreign trade considerations of a general character also preclude too short a queue.[17] There is not even enough capacity for storing unsold cars.

The car market hit the lower tolerance limit of shortage twice: in 1967 and in 1974. Contrary to the procedure adopted when the upper tolerance limit was exceeded, the price was not changed now, but – for three years on both occasions – supply was held back. This is demonstrated in Fig. 15.2 and Table 15.8. These were the two cases in which the supply available to the

[17] These arguments were heard, for example, in an interview with a leading member of the car trade. Question: 'In your view, how long should the queue be to be compatible with the Hungarian economic situation?' Answer: 'Two to three years.' Q: 'Is it not possible to strive for a shorter queue?' A: 'We could not work with a queue of less than one year. For that, the whole purchase structure would have to be changed'.

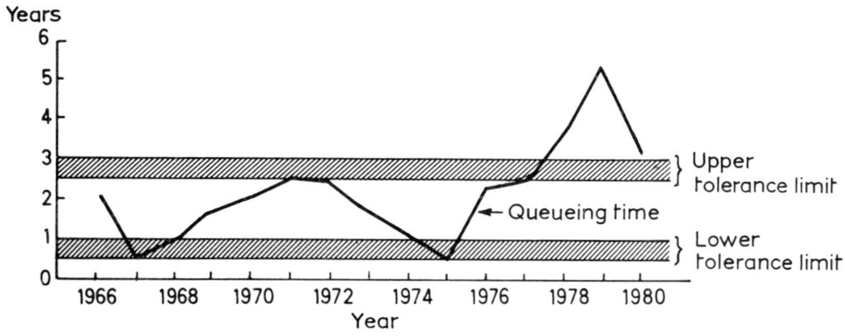

Fig. 15.2 Queueing time and tolerance limit in the Hungarian car market. Source: See Table 15.20.

population did not just grow more slowly, but an absolute reduction took place.[18] As a result of holding back the supply, the average waiting time again rose above the lower tolerance limit.

The sharp turns that follow the surpassing of the tolerance limits largely account for the great fluctuation of supply. The time series also show that a connection between shortage signal and supply (or other decision variables of the seller) exists not only in the proximity of the tolerance limits. It can be demonstrated with the most popular models, Trabant, Wartburg and Lada, that a growing queueing time acts towards increasing supply[19] while a declining queueing time acts towards reducing supply.[20]

The first part of the relationship is almost self-evident: the seller does not want shortage to become too severe. The other part is somewhat perplexing: the seller is opposed to too little shortage.

(a) *Effect of the backlog of unfilled orders*

According to our observation, if the growth of the backlog of unfilled orders speeds up, this will be followed in the next year by a speed-up in sales. This observation is illustrated in Fig. 15.3. As opposed to the queueing time signal, this relationship does not account for the turning-points in sales to private persons, but is effective

[18]The question arises, how far was the slow-down in car imports attributable to the general import restriction, that is, the effect of the macrocycle. We think that the macrocycle played no role in starting the change, since in both cases total imports started to slow down only a year later: in 1968 and in 1975 respectively. The fact, however, that the car market was restricted for three years in both cases and that in the last year of this period supply was reduced absolutely may well have been influenced by the general situation of the economy.

[19]This does not contradict our previous statements. If the queue has exceeded the upper tolerance limit for all models taken together, there will be a reaction not in supply, but in price and in administrative market restrictions. If, however, the queue for certain models has grown within the zone under the tolerance limit, supply will react.

[20]We have examined cases of marked growth or decline in the queueing time for all models of cars. The relationship stated above was found to prevail with the following frequency: Trabant – 5 cases out of 7, Wartburg – 5 cases out of 8, Lada – 3 cases out of 4. Thus, whilst there is not a deterministic rule, there is a perceptible stochastic regularity.

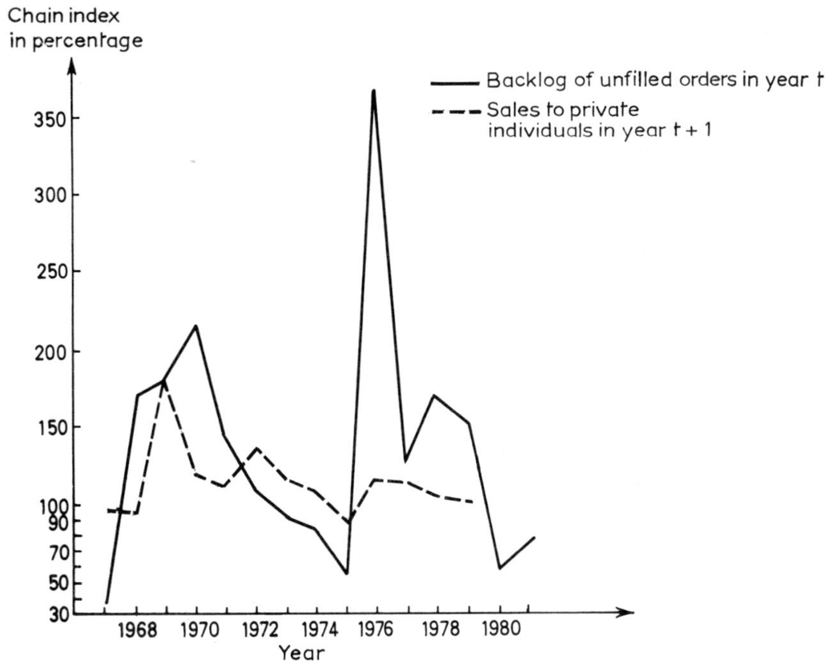

Fig. 15.3 Car sales to private individuals and backlog of unfilled orders in Hungary. Source: Tables 15.5 and 15.7.

rather in the quieter years. The frequency of its effect amounts to two-thirds of the cases (in the years 1969–80). It is thus not particularly strong and is weaker than the effect of the macrocycle and of the foreign trade factors.

Here we want to add a remark on the long-term effects of shortage on car production and supply. In the capitalist car industry there is sharp competition between producers, first of all by offering more up-to-date models with much lower fuel consumption than before. Yet the car industry of the socialist countries has hardly moved in that direction. The East German, Soviet, Czechoslovak and Polish car industries still produce the models of the 1960s – in the case of Trabant the 1950s – with only minor variations. Nor have there been any real signs of a change in this situation since 1973. This causes and will go on causing very serious losses in the socialist countries' energy balance, in which petrol consumption by cars is an ever-increasing item. Besides, with these models the socialist car exporters cannot really break into the Western markets, their low prices notwithstanding.

The citizens of the socialist countries who wish to buy a car would also prefer models with lower fuel consumption, because the price of petrol has increased.[21] On a sellers market, however, the buyers may have wishes, but they cannot force

[21] A few comparative data on fuel consumption (litres per kilometre). Lada 1500: 9–10, Lada 1200: 8–9, Trabant 601: 7–9. Compared with this, the consumption of the Japanese Daihatsu Charade (993 cc) amounts to 6–7, that of the Daihatsu Coure (550 cc) to 7.5. Source: *Autóvásárlók Kézikönyve* (Car Buyers' Manual), Idegenforgalmi és Propaganda Kiadó Vállalat, Budapest, 1979; 'Japán miniautók ('Japanese minicars')', *Heti Világgazdaság*, 13 June 1981.

them upon the seller. Those who control the manufacture of the Trabant and the Lada well know that these ten-to-twenty year old, mass-produced, routine models are sold to a queue stretching 6–10 years ahead, both in their own country and in other socialist countries. Thus they feel no economic pressure towards technical progress, and manufacture of models with much lower fuel consumption.

15.6 THEORETICAL PROPOSITIONS AND CAR MARKET EXPERIENCE

Summarizing our analysis, we can say that the explanatory factors considered 'twitch' the supply, by lengthening or shortening the queue. This is reflected in the buyer's reactions. The level of supply, the length of the queue, changes in prices and other related measures cause demand to fluctuate, which in turn affects the length of the queue and the supply.

We shall now examine some theoretical propositions developed previously by one of us (Kornai, 1980), in the light of our observation of the car market. The propositions were the result of generalization of practical experience and deductive logical reasoning, without reference to statistical data. That is why they were called hypotheses which needed to be tested. Now we can refer to statistics, but only in respect of a single partial market. Although no theory of a wide validity can be proved on this basis, it can be said in any case that the investigation of this field does not disprove but rather supports the theory.

All the propositions refer only to the resource-constrained shortage economy of the socialist system, with both the traditional and the post-reform[22] economic management; therefore, this limitation will not be added to the formulation. The first three propositions are concerned with shortage.

1ST PROPOSITION

Chronic shortage is present in the market in general, and especially in the market of several household consumer goods.

2ND PROPOSITION

The backlog of unfilled orders (that is, the queue) is permanently larger than is justified by organizational factors.

3RD PROPOSITION

A departure from the usual intensity of shortage serves as a feedback signal[23] at all levels of control and planning. In response to this signal the market returns to its usual intensity of shortage.

[22] We mean here the reform of the Hungarian economy put into effect in 1968.
[23] János Kornai, (1980), like a few similar studies, describes the control of the market in the state of chronic shortage in the form of control adjusted to the intertemporal average. We have found that

Our observation of the car market fully confirms these propositions. Particular attention is drawn to the finding which we hold to be decisive in the empirical test of the propositions: the control prevailing on the car market does not allow the queue to shrink too much. As long as this regularity prevails, shortage will subsist.

The indicators of the chronic shortage situation have no internationally valid tolerance limits; the numerical values of the limits may differ from time to time and country to country. In the course of the last 15 years in Hungary, a queueing time of 1 to 3 years has become habitual and accepted on the car market. The seller has become accustomed to adjusting the acquisition routine and sales organization to it. On the other hand, the buyer too has become accustomed to it; it is on this basis that, for the great majority of buyers, the customary forced substitution and queueing propensity rates, the usual scheduling of covering the purchase price, etc. have developed. The habitual intensity of shortage is a self-perpetuating phenomenon: habit, conformism and conservatism are partly a consequence but also a cause of the continuous reproduction of shortage.

In other socialist countries economic management as well as the buyers are accustomed to different degrees of shortage intensity. For example, in the GDR queueing time is 8 to 10 years and there is nothing to indicate that it is likely to change in the future. It depends on a large number of factors, not least on how far the population is willing to accept the established shortage situation and on the degree of consumer dissatisfaction with which the political and economic management can and will 'coexist'. In this respect there are considerable differences between the various countries.

The following three propositions are concerned with the relationship between prices and shortage.

4TH PROPOSITION

Chronic shortage can subsist on the market for a consumer product group, whether the relative price of the latter is low or high.

The price of private cars is relatively high in Hungary. Kravis *et al.* (1975) demonstrate in their book that the relative price of cars in Hungary (compared with the average consumer price level) is about double that in the advanced capitalist countries. In spite of the high price, chronic shortage prevails. Thus the experience gained on the car market supports the 4th proposition. But from this point of view price trends may be even more important.

another form of control prevails on the car market: control reacts mainly to cases of hitting the tolerance limits. The book discusses this form of control in several places too, but in another context. The relation between the two forms of control and, in this connection, a few theoretical and terminological problems of the normal state and of control according to norm will be discussed in an ensuing article. The 1st proposition is developed on pp. 47, 474 and 491–2, the 2nd proposition on p. 140, the 3rd on pp. 59 and 497–8, the 4th on p. 501, the 5th on pp. 338 and 497 and the 6th on pp. 497–8 and 556–7. The 3rd proposition has been formulated in such a way that we are not compelled here to touch upon these theoretical and terminological questions, which require a more detailed explanation.

5TH PROPOSITION

Neither the central planning of consumption, nor the behaviour of firms shows any general automatic tendency to react to the raising of the selling price by increasing supply.

6TH PROPOSITION

A rise in prices does not automatically eliminate shortage of any consumer article.

Our observation of the car market supports these two propositions. The rise in prices has undoubtedly curtailed household demand, but has not led to increased supply. The growth of demand and supply both slow down, so that the usual intensity of shortage persists.

The confirmation of the 5th and 6th propositions is highly illuminating from the point of view of practical economic policy. No rise in prices will eliminate shortage unless supply reactions change. This leads us to the last question which we wish to discuss here.

Will the state of the car market change from a sellers' market to a buyers' market? We are trying to answer the question not in the normative, but in the predictive sense. It would not be impossible for shortage in the car market to cease.[24] There are some sections of the Hungarian market where there is no shortage, or at most sporadic frictional shortages: for example, energy supply, including fuel to the population, and most foods, etc. And yet we are of the opinion that it is improbable, though not impossible, that the car market will change into a buyers' market under the existing institutional conditions.

A transition of this kind would absorb substantial resources at the actual time of the transition. A car would have to be imported without delay from either the socialist or the capitalist market for every member of the queue who maintained their order in spite of the much shorter delivery time.[25] It is true, of course, that this would be only an advance use of resources to be used anyway, since, after elimination of the queue, only new additional demand would have to be satisfied continuously. And now, when a stagnation of incomes is expected, the elimination of the queue would not absorb too many resources, even temporarily. Probably, however, the planners and managers in charge of car imports and sales would not be pleased to see such a change, for the reasons explained earlier. But even if they

[24]It is worth looking again at the Czechoslovak car industry, in which deteriorating exportability and restricted home market demand have led to sales difficulties. In Czechoslovakia, a Skoda can be bought without queueing and credits are even granted for car purchase (see note 5). Of course, there is a great difference between the Czechoslovak and the Hungarian problem. In Czechoslovakia car manufacturing capacity exists but demand has fallen. In Hungary, however, all cars are imported. It depends exclusively on imports, whether supply maintains or eliminates the queue, at a given volume of demand. Car imports are, in a certain sense, a question of free decision. Therefore, in attemtping a prediction we are trying to answer the question, how will the behavioural regularities and decision routines established by the existing mechanism affect this free decision.

[25]A change to a buyers' market and a guarantee of continued good supply would cause many people to withdraw their order, the motive for which may have been overinsurance or speculation. The fictive part of demand would be sure to fall.

came out in favour of effecting the change, the central allocators of resources could not be expected to bring themselves to take such a step. There is chronic shortage in other sectors too, some of them regarded as more important than private car transport. In certain fields shortage is a lot more intensive and, in its social effect, more depressing, too. Why should queueing be eliminated just for private cars? Directly or indirectly, the tendency of the even distribution of shortage would come to prevail. Of course, the arguments which refer to foreign trade considerations and the difficulties of the international balance of payments would be emphasized. Finally, however, the question would not be decided exclusively, perhaps not even primarily, by arguments. Very heavy pressure would be put on the decision-makers to concentrate their efforts on other tasks rather than on transforming one specific partial field from a sellers' to a buyers' market. The final result will be the continuous restoration of shortage in the car market for the foreseeable future.[26] Our prediction is, of course, conditional on there being no essential change in the institutional framework of the economy.

ACKNOWLEDGEMENTS

We express our thanks to the directors and officers of the car trading company Merkur, the Ministry of Domestic Trade, the Hungarian Market Research Institute and the National Savings Bank (OTP) who helped us by revealing problems and supplying data. However, this chapter reflects our own opinion exclusively. Reprinted with permission from *Soviet Studies* (1984), **36**, 236–56 c.

REFERENCES

Baranyai, I. Csahók, I. Rácz, A. Salamin, P. and Schnell, L. (1972) *Fogyasztási cikkekt és szolgáltatások jövedelemrugalmassága*, Central Statistical Office, Budapest (Statisztikai Idöszaki Közlemények, Vol. **246**).

Hoch, R. Kovács, I. and Ördög, M. (1980) *Fogyasztás és jövedelem* Institute of Economics, Hungarian Academy of Sciences, Budapest (manuscript)

Hulyák, K., Muszély, G. and Szakolczai Gy. (1980) *A fogyasztás ökonometriai modellezésének eredményei* Országo Anyag- és Árhivatal és Számitógépalkalmazási Kutató Intézet, Budapest (manuscript).

[26]Collection of materials for this chapter ended in late 1981. The time that has elapsed since then has confirmed our prediction. The central party newspaper, *Népszabadság*, published an interview on 26 July, 1983 with the general manager of the Merkur automobile trading company. (The interview was by Csaba Egerszegi.) It turned out from this – in the words of the general manager – that 'owing to the development of the balance of trade' in 1983 imports would be 16% less than the original quota. 'Because of our purchasing possibilities the average waiting time for a car has lengthened ... And not only the quantity but also the composition of imports developed unfavourably'. According to the interview the two most popular models are the Skoda and the Lada. Skoda imports diminished by more than 40% and delivery of the Lada is not promised until 1988.

Keargard, N. (1975) *Eftersporgsel og varighed for varige goder*, Kobenhavns Universitets Okonomiske Institut, Kobenihavn.

Kornai, J. (1980) *Economics of Shortage*, North-Holland, Amsterdam.

Kravis, I. B., Kenessey, Z., Heston, A. and Summers, R. (1975) *A System of International Comparisons of Gross Product and Purchasing Power*, Johns Hopkins UP, Baltimore.

Lackó, M. (1975) Consumer savings and the supply situation. *Acta Oeconomica*, **15** (3–4), 365–83.

Simon, A. (1979) An econometric study of the consumption and savings of the population. *Matecon*, **15** (4), 67–88.

APPENDIX

ZSUZSA KAPITANY

Although we finished collecting the material for Chapter 15 at the end of 1981, we have continued to observe the operation of the car market. The experience of recent years has shown that our statements apply over a longer period as well as confirming the correctness of our forecast: there has been a chronic shortage in the Hungarian car market even in the eighties. The backlog of unfilled orders, that is, the queue, regularly exceeds double the annual sales. The computed queueing time – our most important indicator of shortage – is rather long even in the eighties, as regards both average cars and popular types.

The unchanged behaviour of the seller can be seen well from column 1 of Table 15.A1: the seller continues to observe the upper tolerance limit already at work in the seventies, that is, that the shortage should not be too great, and that the queueing time should not be longer than 2–3 years. But the seller also cares that the shortage should not be too mild, either. While in the seventies the Trabant was one of the most popular types and exhibited the longest queue, in recent years it can be obtained almost immediately. But the queueing time of two highly popular types, the Skoda and Lada, increased significantly, thus the average queueing time has remained invariable, 2–3 years.

Table 15.A1 Computed queueing time (years)

	Average	*For Trabant*	*For Wartburg*	*For Skoda*	*For Lada*
1981	2.7	3.4	2.6	0.4	5.4
1982	2.6	3.1	3.0	0.3	5.1
1983	2.4	2.5	1.8	0.5	4.2
1984	2.2	1.8	1.0	3.6	3.7
1985	2.0	1.1	0.6	3.2	4.0

Source: Letter from Merkur.

In spite of the fact that two-stroke cars have recently lost much of their popularity, and demand has shifted towards the bigger, more expensive and relatively more up-to-date types (Lada, Skoda), the supply has not reacted to this and the relative purchase proportions of imports have not changed. Thus, for example, while between 1981–85 almost 50% of all buyers were already waiting for a Lada car, their share in imports was merely 27%.

Forced substitution by a considerable number of buyers has remained an invariable phenomenon of the market: they either place an order for a different type from that originally intended, or modify their order under way. If one type of car is not much in demand but can be obtained immediately, while the shortage is chronic in the market for the other types, some of the buyers shift to the market of the type not sought for. In Hungary these types of cars are those that not so long ago were very popular, the East German two-stroke cars (Trabant, Wartburg). For these types of cars in spite of a significant fall in demand in recent years the import purchase proportions and the volume of supply (that is the sales) have remained unchanged. Many of the buyers of two-stroke cars buy this type of necessity, since they want a car immediately (e.g. a significant share of the newly established private taxi cab stock is a Wartburg type). This form of forced substitution is presented in Table 15.A2 which shows the orders and sales data of the two-stroke and relatively cheap Trabant type in recent years.

The size and direction of forced substitutions are also indicated in Table 15.A3, comprising the modification of orders with Merkur in the first quarter of 1986, as a percentage of all modifications. Estimating the changes in orders for the whole year, their number is about 10% of all annual orders. The seven types to be obtained from Merkur have been classified into three groups according to queueing time:

I Types to be obtained immediately (Trabant, Wartburg)
II Types to be obtained after an average waiting time of 1–3 years (Zastawa, Dacia, PF 126)
III Types to be obtained after long queueing (5 years) (Skoda, Lada).

Table 15.A2 Sales, demand stock, demand flow for Trabant (units)

Year	*Sales*	*Backlog of unfilled orders at the beginning of the year*	*Orders: number of those joining the queue*
1981	23 483	79 249	12 222
1982	22 040	67 988	9 700
1983	22 584	55 648	7 488
1984	22 316	40 552	7 259
1985	22 582	25 495	1 105
1986	29 019*	4 018	—

Source: Letter from Merkur.
*Planned data for 1986.

Table 15.A3 Forced substitution ratios (among those who changed their order = 100)

Model ordered	*Model reordered*: *Trabant Wartburg I*	*Zastava Dacia PF 126 II*	*Skoda Lada III*
Type I Trabant Wartburg	8	4	2
Type II Zastava Dacia PF 126	9	34	7
Type III Skoda Lada	8	22	6

Source: Letter from Merkur.

The cause of changes of type within the same group is usually the appearance of a new type within the family of type or group, and such modification is made in the interests of quality or up-to-dateness. But the reason for modification within the group may also be that buyers expect to obtain the new type earlier if they modify their order in time. The reason for a change to another type in another group may also be better quality or up-to-dateness. But the experiences of the car market suggest that the main viewpoint for changes to types in another group is faster or immediate access to a car, by accepting a significant difference in price or even lower quality.

In spite of the rise in car and fuel prices well as the stagnation in real incomes

Table 15.A4 Time series of demand for cars and sales

	Demand stock: backlog of unfilled orders at the beginning of the year		*Demand flow: additional buying intention*		*Sales*	
Year	*Units*	*Chain index*	*Units*	*Chain index*	*Units*	*Chain index*
1982	238 041	94.0	68 543	71.0	90 100	95.1
1983	196 347	82.5	98 932	144.3	81 830	90.8
1984	190 265	96.9	114 823	116.1	87 716	107.2
1985	194 561	102.3	142 002	123.7	96 078	109.5
1986	214 827	110.4	—	—	—	—

Source: See Table 15.5 and 15.7 and footnote 8 in the main text.

in recent years, the demand for cars has kept on growing (Table 15.A4). The growing demand may be explained for example by the general shortage prevailing in the consumer markets, or by the low standards of services and mass transportation. But the main explanation of the phenomenon is probably that the car in Hungary is a working and property asset. The constraints on the supply of new cars push the price of used cars upwards even in the eighties, thus the car may also be a source of continuous income if it is continually changed in time. With rising prices this income is also growing and because of the chronic shortage of popular types of cars, even at a rate exceeding that of the price rise for new cars.

The financial situation (wealth) of individual buyers and their related expectations have a decisive impact on the demand for cars. While the less wealthy and, in addition, the buyer who saves slowly delay the purchase of a new car and exchange it even less frequently than warranted technically, the type of buyer with adequate pecuniary resources accelerates the exchange process. The wealthy type practically always uses a new car, changing the car every three years, selling the previous one at the price of a new one or even at a higher price. The allocation mechanisms established in the car market further increase the chances for income, thus providing significant surplus income for some car-owners.

Owing to the technical state and average age of the stock of cars, the scrappings and replacements that can no longer be delayed will expectably further increase the demand in the rest of the eighties. In 1985, more than 7% of private cars were older than 15 years, and 60% of the stock was older than 6 years. The average age of the stock is 8.2 years and the expected lifetime 16 years. In the last five years the share of new cars has diminished within the stock, while that of cars older than 9–10 years has grown significantly. This disadvantageous age distribution of the stock of cars indicates an abnormal slowing down in the change and replacement of stock and the latter is menaced by dangerous ageing and technical deterioration.

16

Modelling parallel markets in centrally planned economies: the case of the automobile market in Poland

WOJCIECH CHAREMZA, MIROSLAW GRONICKI and RICHARD E. QUANDT

16.1 INTRODUCTION

Most empirical work dealing with socialist economies appears to be oriented towards macroeconomic problems or at least to the study of highly aggregated time series. Examples are provided by Howard (1976), Lacko (1975), Podkaminer (1982), Portes and Winter (1980), Portes *et al.* (1983, 1984a, b, 1985), Welfe (1983), and Charemza and Gronicki (1988). Only relatively infrequently has a particular market been the target of detailed empirical investigation; a case in point is the study of the Hungarian car market by Kapitany, Kornai and Szabo (1984 and Chapter 15 of this volume).

In the present chapter we examine the complicated structure of a consumer durables market in which shortages occur and which generate parallel mechanisms of exchange. Such an undertaking differs markedly from analogous endeavours in the context of free-market economies. Some of the more salient differences are the following.

1. In the free-market context it is commonly assumed that either equilibrium is attained or if not, a continual approach to equilibrium occurs because of a partial adjustment to a discrepancy between desired and actual stocks. In socialist economies prices are controlled and change only infrequently and it is

International Studies in Economic Modelling Series.
Models of Disequilibrium and Shortage in Centrally Planned Economies.
Edited by C. Davis and W. Charemza.
Published in 1989 by Chapman and Hall Ltd, 11 New Fetter Lane, London EC4P 4EE.

commonly argued that socialist economies invariably exhibit excess demand (Kornai, 1980; Pickersgill, 1980; Winiecki, 1982; Balicki, 1983), without any tendency to approach equilibrium. Although this proposition is debatable for aggregate demand and supply (Portes and Winter, 1980; Portes *et al.*, 1983, 1986, 1985), we consider the assumption of permanent excess demand to be reasonable for particular consumer durables.

2. The notions of permanent income and rational expectations play a prominent role in free-market approaches to characterizing durables markets (Bernanke, 1984). Whether the change in desired stocks of durables such as cars is proportional to permanent income in socialist economies is certainly debatable. It is also not clear whether the rational expectations view is entirely reasonable in an environment in which central planners can abruptly alter both the lifetime prospects and the transitory component of income.
3. If it is true that there is permanent excess demand, we must ask what theory of consumer behaviour is compatible with that state of affairs. According to the Dreze (1975) concept, consumers maximize utility subject to the budget constraint and *all* quantity constraints simultaneously; hence their effective demands ought not to exceed their allotted rations. It is also not clear whether consumers can be thought of as adjusting their labour supply between work and leisure (and between the normal economy and the underground economy – Charemza and Gronicki, 1988) in response to perceived rationing in commodity markets. We must therefore investigate, at least schematically, what kind of utility function is compatible with a reasonable definition of permanent excess demand.
4. In comparison with free market economies, the data problems are enormous. Even for the car market, for which data seem to be reatively abundant, no data (such as registration or insurance records) are available from which a quarterly series on the stock of cars could be constructed and so no aggregate equations such as those of Chow (1957) can be estimated. There are no panel data and so procedures such as those of Bernanke (1984) are not possible. In certain submarkets one cannot even ascertain the quantity of cars transacted. In these circumstances, model formulation and estimation will be difficult.

In section 16.2 we introduce a simple model of the consumer that is compatible with the stylized facts of the market; this model is applied in the computation of queue-length in the econometric work. In section 16.3 we discuss the basic features of the automobile market. In section 16.4 we formally specify the model. Section 16.5 is devoted to problems of estimation and section 16.6 contains a discussion of results. Section 16.7 contains some brief conclusions.

16.2 A MODEL OF THE CONSUMER

It is a fact of life in socialist economies that certain consumers' durables such as cars are, on the whole, not readily available. Consumers have to queue and this is said to be compatible with the normal state of the market (Kornai and Weibull, 1978; Katz and Owen, 1984). An important feature of such a situation is that, at any one moment or over any unit period, the number of consumers requiring service (demand) is greater than the number being serviced (supply); moreover, that this situation can represent an equilibrium in the sense that there are no forces tending to change the queue. We illustrate this with a very simple model, related to that of Lindsay and Feigenbaum (1984), and lacking the elaborateness of Katz and Owen (1984). The role of the model is to provide an interpretation of excess demand for the case of durable goods and to provide the context for the computation of queue length and waiting time in section 16.6.

We posit the simple utility function

$$V(x, y) = U(x) + \gamma e^{-\beta w} y \tag{16.1}$$

where x is a composite good with price normalized to unity, $U(x)$ is concave, $y = 1$ if the consumer enters the queue for the durable good at the present time and zero otherwise, w is the amount of time the consumer has to wait for the delivery of the durable good. The parameter γ measures the utility contribution of the durable good relative to the composite good and may assume different values for different consumers.[1] The waiting time transforms future car services into present utility and β acts as the discount rate (for simplicity we assume that β is identical for all consumers). Both γ and β are assumed to be positive. The additive separability of the two types of goods appears reasonable but otherwise the utility function is chosen to express the effect of queueing as conveniently as possible. Given an income M, the budget constraint is

$$\begin{aligned} &x = M \text{ if the consumer does not enter the durable good queue} \\ &x = M - p \text{ otherwise} \end{aligned} \tag{16.2}$$

where p is the price of the durable good paid irreversibly at the time of entry into the queue.[2] The expected utility of the consumer is

$$\begin{aligned} &U(M) \text{ if the consumer does not enter the queue} \\ &U(M - p) + \gamma E(e^{-\beta w}) \text{ otherwise} \end{aligned}$$

The consumer will enter the queue if

$$U(M - p) + \gamma E(e^{-\beta w}) > U(M) \tag{16.3}$$

Assume that (1) we have a continuum of consumers, each characterized by a particular value of $\gamma > 0$, with density $h(\gamma)$, and (2) arrivals in the queue and service in the queue are independent with arrival and service rates λ and μ respectively ($\mu > \lambda$) and that the interarrival and interservice times are exponentially dis-

[1] We would gain some generality at the cost of a substantial increase in complexity if we write the utility function as $U(x) + \gamma f(w)$, with $f' < 0, f'' > 0$.

[2] This is a reasonable simplification of reality, even though in practice only a fraction of the price need be paid upon entering the queue. However, allowing for a sequence of payments would unnecessarily complicate the model.

tributed (which preserves the Markov property of the process). In the present context the arrival rate is the rate at which customers arrive at the queue and the service rate is the rate at which customers in the queue obtain the durable good. Then the density function of waiting time w is

$$f(w) = (\mu - \lambda)e^{-(\mu - \lambda)w} \tag{16.4}$$

It follows immediately that[3]

$$E(e^{-\beta w}) = \frac{\delta}{\delta + 1} \tag{16.5}$$

where $\delta = (\mu - \lambda)/\beta$. Then Equation 16.3 becomes

$$\gamma > \frac{\delta + 1}{\delta}[U(M) - U(M - p)] \equiv \gamma_0 \tag{16.6}$$

and the fraction of consumers entering the queue is

$$P = \int_{\gamma_0}^{\infty} h(\gamma)\, d\gamma = 1 - H(\gamma_0) \tag{16.7}$$

where H is the cumulative distribution function of h. If N is the total number of consumers, the arrival rate λ is NP. The arrival rate depends on P and in turn determines P as the outcome of optimal decisions by consumers; an equilibrium exists if the mapping $P \rightarrow P$ possesses a fixed point. The key notion here is that one of the parameters of the queueing, the arrival rate, is endogenous. The arrival rate (as well as the service rate, and the parameters of the utility function) determines the expected waiting time and that influences the results of utility maximization, which in turn determines a new value of the arrival rate. Consumers are in equilibrium if no further adjustment is necessary; i.e. if the expected waiting time assumed in the utility maximization exercise yields an arrival rate which implies that same expected waiting time. It is simple to show that

$$P = 1 - H\left\{\frac{(\mu - NP)/\beta + 1}{(\mu - NP)/\beta}[U(M) - U(M - p)]\right\} \tag{16.8}$$

has a unique solution for P in the $0 < P < \mu/N$ interval.[4] Then the expected waiting time and queue length corresponding to this solution value of P are themselves equilibrium values.

[3]We take the expectation unconditionally (i.e. not conditioned by queue-length) since we are more interested in the case in which this variable is not observed. This is because order books for cars in Poland have never been published to our knowledge and we conjecture that they have not even been aggregated. For a more detailed but similar development see Katz and Owen (1985).

[4]The following fixed-point argument establishes the existence of a unique equilibium. Consider the two sides of Equation 16.8 as functions of P. The left-hand side is a 45° line through the origin. The right-hand side has value $1 - H[(\mu/\beta + 1)(U(M) - U(M - p))/(\mu/\beta)]$ at the origin and has a negative derivative with respect to P. Hence a unique intersection exists in the $0 < P < 1$ interval. Furthermore, the limit of the right-hand side as $P \rightarrow \mu/N$ is 0; hence μ/N must exceed the solution value. This then also assures that $\lambda < \mu$ at equilibrium as required.

From Equation 16.8 we immediately obtain several comparative statics results. Denote by P_M, P_p, P_δ the partial derivatives of P with respect to M, p and δ respectively, and let ΔU denote $U(M) - U(M-p)$. Differentiating Equation 16.8, we obtain

$$P_M = (\beta\delta^2 + H'N\Delta U)^{-1}\beta\delta(\delta+1)H'[U'(M-p) - U'(M)]$$
$$P_p = -(\beta\delta^2 + H'N\Delta U)^{-1}\beta\delta(\delta+1)H'U'(M-p)$$
$$P_\delta = (\beta\delta^2 + H'N\Delta U)^{-1}\beta H'\Delta U$$

H', ΔU and U' are positive and $U'(M-p) - U'(M)$ is positive (by virtue of the concavity of U). Hence $P_M > 0$, $P_p < 0$ and $P_\delta > 0$. From the latter it follows that an increase in the service rate increases the willingness to enter the queue, whereas an increase in β, i.e. the rate at which utility decays with waiting time, reduces the willingness to enter the queue. But no matter how the parameters of the problem change, in equilibrium there will always be a queue, even though more unfavourable circumstances reduce the willingness of consumers to enter the queue.

In the light of the above it is convenient to interpret excess demand in the aggregate time series sense to consist of the difference between the quantity demanded by consumers in the queue on the average in a given time interval and the quantity made available in that time interval. Although in a flow sense there is no excess demand in the queueing model, since the arrival rate must be less than the service rate, the amount of goods consumers are willing to purchase in any one time interval may be much greater than the amount of goods delivered in that interval. Consumers optimize on the basis of their income, prices and waiting time and an equilibrium emerges in which the waiting time is endogenous and is jointly determined by the consumer's decision to enter the queue. Although consumers' expectation of long waits reduces their probability of entering, in equilibrium there is no reason for the queue to disappear. Disequilibrium is thus somewhat of a misnomer: the lack of equality between demand and supply is an equilibrium property of the system.

16.3 AN EXAMPLE: THE AUTOMOBILE MARKET

The markets for cars and dollars (and other Western currencies) are the best-known examples of parallel markets in centrally planned economies, especially in Poland. These markets are large and exert a strong influence on smaller legal and illegal fields of private activity. They are also relatively well organized and statistical data about them, however incomplete, are more plentiful than about other markets. The Polish car market is described in detail by Krasinski *et al.* (1980) and by Starzec (1983).

The complicated structure of the car market stems from the fact that various systems exist for conveying new cars from the seller (the state) to consumers. The institutional arrangements provide for different prices and methods of payment in each of the resulting submarkets.

A major distinction is between cars that are on the market for the first time and second-hand cars (though note that this distinction is not the same as that between new and used cars, since some new cars are purchased for speculative reasons and are immediately resold). The supply of cars sold for the first time is predominantly from the state. In the period under investigation (1973-IV to 1982-IV) the private importation of new and used cars was negligible and amounted to 4.6% of deliveries in 1974 and 3.6% in 1979 (Krasinski *et al.*, 1980). The rest of the supply came from the state and consists of cars manufactured in Poland, as well as imported.[5] The cars sold by the state are actually sold by state enterprises: primarily by Polmozbyt, which sells cars for Polish currency (zlotys) and by Pewex, which sells cars (and other consumption goods) for US dollars and other hard currencies.[6] In the 1974–79 period the percentage of total state deliveries accounted for by dollar sales ranged from 16 to 23% (Krasinski *et al.*, 1980).

The official zloty price is paid by consumers to Polmozbyt in instalments. A fraction is due upon entering the queue, another fraction during the waiting period and the balance upon delivery. Delivery of a car occurs after a long period in the queue, which anecdotal evidence reports as measured in years. The price is extremely stable: for the Fiat 125p, for example, it was changed only three times in the 1974–82 period. This zloty price is invariably the lowest price at which a car can be purchased.[7]

The cars sold by Pewex were usually available without appreciable queueing and were often of higher quality. Although it is convenient here to pretend that all these sales were against dollars, they were in fact against dollars and some other hard currencies as well as against dollar vouchers. (For a similar practice in Hungary, see Kapitany, Kornai and Szabo, 1984.) The legal aspect of these transactions is not straightforward. Polish citizens were allowed to import and possess Western currencies but could legally sell them only at official banks at an official price. Dollar vouchers, on the other hand, could be obtained at banks for dollars at par and could also be freely traded at presumably market clearing prices and also used to purchase cars in Pewex. The free market for dollar vouchers and the black market for dollars are closely related and in this chapter they are regarded as a single market. The prices of dollar vouchers and black-market prices for dollars were nearly identical and from the point of view of potential car buyers it made no

[5] In 1972–82 the cars produced domestically were mainly Fiat 125p, Fiat 126p, Syrena and Polonez (since 1978). The import of new cars accounted for 10–30% of total deliveries and consisted mainly of Skoda (from Czechoslovakia), Trabant (from the GDR), and Zaporozec (from the USSR).

[6] Besides Pewex there exist some other enterprises authorized to sell cars for dollars but they are of minor importance.

[7] The basic price was subject to some other types of variation. Sometimes it was combined with a lottery for a place in the queue. At other times (1979—80) an express price was introduced which was substantially higher than the basic price but in effect eliminated queueing. There also existed a combined dollar—zloty price as well as a system of coupons to avoid queueing that was available to privileged groups of consumers (Krasinski *et al.*, 1980).

difference whether they paid Pewex in dollars or dollar vouchers. If not imported from abroad, either of these instruments had to be purchased privately. In general, hard currencies played a significant role in private Polish economic activity. It is estimated that US dollars in circulation in Poland ranged from $1.5 to 6 billion (Mojkowski, 1984) and various estimates put the monthly volume of black-market hard-currency transactions alone at $2.5–4.2 million.[8]

A third market for cars is the second-hand market in which cars are sold by private firms specializing in the sale of second-hand cars, by advertising and principally through car exchanges organized in large cities as open-air markets.[9] There is reason to believe that prices in the second-hand market (in zlotys, dollars or dollar vouchers) were market clearing.

To summarize: there are in effect three markets for cars in Poland. The first is the official zloty market for new cars that does not have a market clearing price. The second is the official dollar market for new cars (Pewex) that involves shorter queues and higher prices. Finally, there is the second-hand market in which queues are negligible. In the latter two, the price presumably does clear the market.

16.4 THE BASIC MODEL

An ideal model would specify demand and supply functions for all three submarkets and the stock of cars would play an important role in this specification at various points. Unfortunately, such a model would not be practical because key data required by it are not available. For the period investigated, 1973-IV to 1982-IV, we have found no quarterly data on the stock of cars, nor on the quantities of cars traded in the Pewex market and in the second-hand market.[10] In general, one would have to consider this lack a serious disadvantage in attempting to estimate a car model. However, in Poland it appears reasonable to argue that, because of permanent shortage, the stock of cars has never reached a level at which it would exert an appreciable effect on the demand. We therefore content ourselves with an inferior specification in which some of the structural equations are already partially reduced; i.e. they stand between the original structural equations that would emerge from agents' optimizing behaviour and the reduced form equations.

[8]Starzec (1983) suggests that dollars are brought to Poland by tourists, diplomats, Polish citizens temporarily abroad (seamen, etc.); to this we may add undetermined amounts of remittances from relatives in Western countries. It should be noted that there is some disagreement about what the facts are: Simon (1982) estimates private hoards of US dollars in 1980 as no greater than 300 million. However, all sources seem to stress that hard currencies represent a substantial part of the entire stock of money in private hands.

[9]As indicated earlier, second-hand cars are not necessarily used cars in that a new car purchased from the state could be immediately resold at a profit in the second-hand market.

[10]Although an annual series on the stock of cars does exist, we did not feel confident about attempting a quarterly interpolation of the series.

16.4.1 Notation

The variables in the econometric model are as follows:[11]

D = demand for new cars in the zloty market
BM\$ = black-market price of US dollars
RP = retail price index
PF = zloty price of second-hand cars
INC = nominal personal income
DUM = the success illusion dummy; equals 1 for 1978-I to 1979-I and 0 otherwise
PO = the official zloty price in the new car market
PCD = the official dollar price of new cars in the Pewex market
CAR = quantity of cars (supply) made available to the zloty-denominated new car market
$E(\ )$ = expectation of ( )
ξ_1, ξ_2, ξ_3 = normally distributed error terms, independent of one another.

We generally omit the time subscript for simplicity except for the subscripts +1 and −1 to denote leads and lags.

16.4.2 Equations

The jointly determined variables of the model are D, BM\$/RP and PF/RP. We write the three equations by normalizing on these variables in turn as

$$D = \alpha_0 + \alpha_1(\text{INC/RP}) + \alpha_2[E(\text{BM\$}_{+1}/\text{RP}) - \text{BM\$/RP}] + \alpha_3\text{DUM} + \alpha_4(\text{PF/RP} - \text{PO/RP}) + \xi_1 \tag{16.9}$$

$$\text{PCD}(\text{BM\$/RP}) = \beta_0 + \beta_1(D - \text{CAR}) + \beta_2(\text{PO/RP}) + \xi_2 \tag{16.10}$$

$$\text{PF/RP} = \varepsilon_0 + \varepsilon_1(\text{PF/RP})_{-1} + \varepsilon_2(\text{BM\$/RP} - (\text{BM\$/RP})_{-1}) + \varepsilon_3(\text{PO/RP}) + \xi_3 \tag{16.11}$$

The first of these equations is the demand function for new cars in the zloty market and is a proper structural equation. In view of the previous discussion, it is the demand (per unit period) of all those who receive cars from the state and who are willing to pay zlotys for them, plus those who are still in the queue. A priori we expect $\alpha_1 > 0$ (an increase in real personal income increases car demand), $\alpha_4 > 0$ (the relative price effect; the greater is the free market price relative to the official new car price, the greater the demand for new cars) and $\alpha_3 > 0$. This latter is attached to a dummy variable which takes values of 1 during certain quarters (1978-I to 1979-I) of the Gierek government when the illusion of success for the Polish economy appeared to be widespread.[12] The coefficient α_2 measures the effect of an expected

[11] For a more detailed definition of variables see Appendix A.
[12] The substantial investments made by the Gierek government in the first half of the 1970s had backfired for a variety of reasons, including bad harvests, and led to increasing foreign trade deficits and attempts by the government to raise food prices. In 1978, a significant propaganda effort was undertaken to justify past policies by pointing to the increase in consumer goods. See Montias (1982).

real increase in the price of dollars. Dollars are not only the way to buy a Pewex car but are also a generalized store of value and investment instrument for consumers. The increase in the expected value of dollars may lead to some substitution away from Pewex cars to zloty cars, and may in particular increase the demand for zloty cars for speculative purposes. At the same time, the expectation of more expensive dollars in the future, will tend to increase the current demand for dollars and, *ceteris paribus*, reduce the current demand for new cars. On balance, we expect the latter effect to predominate and thus α_2 is expected to be negative.[13]

The observed queue length should, *ceteris paribus*, be included in the demand equation. As we indicated before, the queue length is unfortunately not observable. We attempted a compromise solution by including as a variable the queue length that is implicitly given by the model itself. As was the case with some other attempted extensions of the model, this placed too heavy a burden on it and rendered most coefficients insignificant upon estimation.

The second equation of the model can be thought of as resulting from equating demand and supply of cars in the Pewex market. Let this demand be given by $D_x = D_x(\text{PCD}(\text{BM\$/RP}), \text{PO/RP}, D - \text{CAR}, \text{INC/RP})$, expressing that demand depends on own price (recalculated in zlotys through the black-market dollar price), the competing product's price, on excess or unsatisfied demand in the competing market and on income. It is our hope that the income variable will capture the motivations to demand dollars that are in addition to those emanating from the car market. It is much harder to describe the supply function. Kapitany, Kornai and Szabo (1984) argue that supply is not affected by price and profitability but by the planners' general social goals, some macrovariables such as trade balance, and by excess demand for the cars. We therefore take supply to be a linear function of excess demand in the zloty new-car market plus a random error. Equating demand and supply yields an equation that explains the dollar price of Pewex cars. Income already affects D_x through $D - \text{CAR}$ and we delete it from D_x in the main runs to reduce the number of parameters to be estimated. (In one run, however, we retain INC/RP in the equation.) The dependent variable explained by Equation 16.2 is PCD(BM\$/RP), i.e. the implicit zloty price of Pewex cars (in this, however, PCD is exogenous and fixed by the authorities and only BM\$/RP is jointly determined). A priori $\beta_1 > 0$ and $\beta_2 > 0$, since the latter measures the substitution effect of higher prices in the official zloty market, whereas the former measures the spillover effect from that market (i.e. the effect of the excess demand in that market).

The third equation of the model is a partially reduced structural equation that explains PF/RP, the market-clearing price in the second-hand car market. The equation may again be thought of as resulting from equating demand and supply in that market. Supply in this market is likely to depend on recent past prices and we include $(\text{PF/RP})_{-1}$ as an explanatory variable. If supply responds negatively to

[13]Note that $E(\text{BM\$}_{+1}/\text{RP})$ is an expectation of a future value and in conformity with frequent practice is taken to depend only on past, hence predetermined, values.

lagged price and positively to current price and if demand has an inverse relation to current price, the autoregressive coefficient ε_1 will be positive. The coefficient ε_3 is also expected to be positive. The case of ε_2 is more difficult. It measures the effect of a recent increase in the real price of black-market dollars.[14] This effect is not fully unambiguous. On the one hand, to the extent that dollars are an important liquid asset, an increase in the value of dollars generates a positive real balance effect. On the other hand, if the demand for dollars is inelastic, an increase in its value generates a negative income effect. A further consideration also suggests a negative effect: in critical times of uncertainty, dollars are a particularly good store of value (easily and safely stored, etc.) whereas cars are a particularly unsuitable asset (bulky, subject to damage, and potentially unusable because of gasoline restrictions and other reasons). Hence general increases in the value of dollars may be related to lower free market prices. On balance we posit a negative relationship here.

There are a number of alternatives one could consider. Specifically, the level of personal income could reasonably be included in each equation, since the demand for each type of car is likely to depend on it. One might well include in the equations the quantities of other durable goods in short supply (e.g. housing) in order to measure the spillovers due to rationing in these markets. The excess demand in the zloty new-car market may spill over into the market for second-hand cars (for a similar suggestion see Kapitany, Kornai and Szabo, 1984). Some of these other possibilities are briefly discussed in section 16.5. However, we are already estimating 15 parameters (12 α's, β's, ε's and 3 error variances) in the basic model with 35 effective data points. Any substantial increase in the number of parameters is likely to create difficulties and we have attempted to stick to as parsimonious a formulation as possible, even if not entirely correct from the theoretical point of view.

The model is a disequilibrium model in that demand for new cars is not assumed to be equal to supply. In fact, we assume that we observe the supply (CAR) and that in every period

$$D \geqq \text{CAR} \tag{16.12}$$

and that, as usual, the demand D is not observed by the econometrician. This makes for a disequilibrium model that is slightly different from the usual ones. We now turn to the problems of estimation.

16.5 METHODS OF ESTIMATION

16.5.1 Maximum likelihood

The generally preferred method of estimation for disequilibrium models is maximum likelihood. The joint density (for observable and unobservable vari-

[14]Unlike Equation 16.9 where the black-market price of dollars was intended to capture in part a speculative effect, here we are intending to capture a current effect; hence the use of a current difference rather than a leading difference.

ables) is obtained from the structural equations and the unobservable variable is integrated out over the regions corresponding to the excess demand and excess supply regimes. If no a priori sample separation is given, the density is the sum of the two integrals. Simplifying for the moment by restricting attention to the standard simple disequilibrium model in which there exists only a demand function, supply function and min condition, the density of the transacted quantity can be written as

$$h(Y) = g_1(Y)(1 - G_2(Y)) + g_2(Y)(1 - G_1(Y)) \tag{16.13}$$

where g_1 and g_2 are densities and G_1 and G_2 are the cumulative distributions of demand D (supply) conditional on supply S (demand) and evaluated at $Y = \min(D, S)$ (Hartley, 1974). In the present case we made the a priori assumption that only excess demand occurs. (For the derivation, see Appendix B.) This is equivalent to the assumption that the term in Equation 16.13 corresponding to excess supply (say $g_2(1 - G_1)$) is zero, which is identical to the assumption that the conditional probability $Pr\{S > D | Y\} = 0$. The likelihood function is then the product over the observations of terms such as Equation 16.13.

16.5.2 Estimation of the condensed model

An alternative approach is to condense the model by solving Equation 16.9 for D and substituting this in Equation 16.10. This yields an ordinary nonlinear simultaneous equations system and may be estimated by the full-information maximum likelihood method (FIML). It can be shown (Quandt, 1985; Portes, Quandt and Yeo, 1985) that the density function of the random variables BM\$/RP and PF/RP from this condensed model is identical with that part of the density function derived in Appendix B that corresponds to the term in Equation 16.B3 that does not involve the cumulative normal $1 - \Phi(\ )$. Note that in the condensed model the variable D is replaced for all its data points, which follows from the assumption that all observations lie on the supply curve and that the probability that CAR exceeds D is equal to (or close to) one. This indicates that the condensed method is based on a stronger a priori assumption than the disequilibrium FIML method.

16.5.3 Concluding comments

It is to be noted that in general it is not obvious that the parameters of the function representing the long side of the market are estimable. Sufficient conditions for estimability are not known, although some necessary conditions for estimability are derived in Quandt (1985). These conditions are satisfied for the parameters of the equations. Even then, the variance of the equation may not be identified; this can be seen most clearly by inspection from the condensed model. For this reason, in the condensed procedure we only estimate a combined variance for the first equation. In the disequilibrium maximum likelihood method we arbitrarily fix the value σ_1^2 at 2.0.

16.6 EMPIRICAL RESULTS

The principal results are displayed in Table 16.1. Columns are designated by model number and estimation method. *M* designates the disequilibrium maximum likelihood procedure and *C* the condensed procedure. Model 1 is the basic one discussed in section 16.3 with the proviso that the expectation variable $E(\text{BM\$}_{+1})$ is replaced by $\text{BM\$}_{+1}$, i.e. perfect foresight is assumed. In this model we obtain the real expected future price of dollars by deflating by RP. In a variant, we deflate by RP_{+1}, which implies that the real value of dollars is foreseen perfectly. The results of this variant are extremely similar to those of Model 1 and are not reported explicitly. In Model 2, the variable INC/RP in the demand for new cars is replaced by DEP/RP, where DEP represents the end-of-quarter deposits of households in savings institutions. In Model 3 we modify Model 1 by including the spillover

Table 16.1 Results of estimation*

	Model				
Coefficients	*1/M*	*2/M*	*3/M*	*4/M*	*1/C*
α_0	23.117	37.651	31.297	23.393	− 5.273
	(0.666)	(1.004)	(0.939)	(0.167)	(− 0.001)
α_1	4.415	− 1.533	1.112	4.711	4.453
	(0.598)	(− 0.288)	(0.117)	(0.636)	(0.604)
α_2	− 11.849	− 12.312	− 14.827	− 10.007	− 11.843
	(− 2.189)	(− 1.943)	(− 2.179)	(− 1.908)	(−2.187)
α_3	18.047	19.484	18.515	17.731	18.040
	(3.515)	(3.143)	(3.187)	(3.433)	(3.515)
α_4	164.245	189.116	187.319	160.821	164.200
	(4.301)	(3.630)	(3.272)	(4.262)	(4.294)
β_0	− 3.190	− 3.435	− 3.289	− 3.059	− 2.530
	(− 3.183)	(− 3.285)	(− 3.669)	(− 0.929)	(− 0.021)
β_1	0.023	0.021	0.021	0.023	0.023
	(4.887)	(4.128)	(3.697)	(4.837)	(4.879)
β_2	4.086	4.319	4.174	3.931	4.085
	(5.449)	(5.514)	(5.401)	(5.369)	(5.448)
ε_0	− 0.330	− 0.354	− 1.434	− 0.757	− 0.330
	(− 0.451)	(− 0.463)	(− 0.589)	(− 1.305)	(− 0.451)
ε_1	0.962	0.969	2.178	0.825	0.962
	(2.986)	(2.898)	(1.011)	(4.012)	(2.984)
ε_2	− 0.504	− 0.527	− 0.796	− 0.318	− 0.504
	(− 1.291)	(− 1.260)	(− 0.753)	(− 1.692)	(− 1.289)
ε_3	0.480	0.501	0.323	0.435	0.481
	(0.833)	(0.830)	(0.329)	(1.180)	(0.832)
ε_4	—	—	0.006	0.262	—
	—	—	(0.673)	(2.349)	—
Log *L*	41.157	41.055	42.679	44.998	41.157

* *t*-ratios are given in parentheses below the estimates.

term $\varepsilon_4(y_1 - \text{CAR})$ in the equation explaining the price of second hand cars. In Model 4 we include in this equation the term $\varepsilon_4(\text{INC/RP})$.

The parameter values have the expected pattern of signs in Model 1/*M*. Personal income is not significant in Equation 16.9, although the other coefficients in that equation and in Equation 16.10 are significant. The only significant coefficient in the second-hand car price equation is that of the lagged price although all coefficients at least have the expected sign. In particular we note that the spillover from the normal (zloty) new-car market to the Pewex car market (β_1) is positive and significant, that the effect of an expected dollar appreciation (α_2) has a significant negative effect on the demand for new zloty cars, and that a recent past appreciation of the dollar (ε_2) has a nonsignificant negative effect on the free market price. The insignificance of this coefficient may well reflect the conflicting influences that the right-hand-side variable was expected to have on the free market price.

In Model 3 we also include the spillover from the new zloty market in Equation 16.11. The effect is a nonsignificant increase in the loglikelihood and a numerically very small measured positive spillover effect; we thus have no reason for preferring model 3/*M*.

Model 1/*M* and 1/*C* are directly comparable. The likelihood function values are essentially identical and all coefficients except two of the constant terms are essentially identical. This underscores the similarity of the respective likelihood functions and is compatible with the fact that the estimates in 1/*M* imply a value near 1 for $1 - \phi(\ )$ in the density function (Equation 16.B3). Model 1/C does a worse job, however, of estimating the constant terms and the resulting excess demand predictions are less plausible.

We also estimated some variants of Models 2/*M*. Since they use different variables, they are not nested in 1/*M* and likelihood value comparisons are not particularly meaningful. Their coefficient estimates are broadly similar to those of 1/*M* without any particular reason for preferring them to 1/*M*.

We estimated a model similar to model 3/*M* from which, however, ε_3 was excluded; it did not differ from Model 3/*M* materially. We also estimated a model in which the perfect foresight of $\text{BM\$}_{+1}$ was replaced by an ARMA (4, 4) prediction: this left the estimates of Equations 16.10 and 16.11 largely unchanged but made the coefficients of the new car demand completely insignificant.

We finally examined whether it is a serious misspecification not to include INC/RP in Equations 16.10 and 16.11. Its inclusion in Equation 16.10 yielded a *t*-value of 0.08 and an increase in the likelihood only in the 5th place. However, its inclusion in Equation 16.11 (Model 4/*M*) with coefficient ε_4 yielded a significant coefficient and the likelihood ratio comparison with Model 1/*M* is also significant at the 0.01 level. We conclude that model 4/*M* appears to be the most reasonable one. Our further discussion is based on this model and its comparison to Model 1/*M*.

Setting the error terms equal to zero, the model can be solved simultaneously for the endogenous variables and thus it can be used to make a within-sample prediction of demand *D*. Then $\hat{D} - \text{CAR}$ can be interpreted as the excess demand.

Table 16.2 Excess demands and waiting times

		CAR	Excess demand 1/M	Excess demand 4/M	Mean waiting time 1/M	Mean waiting time 4/M	Mean waiting time upper 90% confidence limit for 4/M
1974	I	21.3	16.7	12.5	1.83	1.63	11.34
	II	21.8	17.0	9.5	1.83	1.48	9.97
	III	19.7	19.9	15.2	2.06	1.83	10.90
	IV	20.2	14.9	15.3	1.79	1.80	10.00
1975	I	19.3	15.6	13.9	1.86	1.78	11.43
	II	26.0	18.3	12.9	1.74	1.54	8.22
	III	24.4	20.6	19.7	1.89	1.85	9.33
	IV	41.3	16.8	17.4	1.43	1.45	5.62
1976	I	33.3	29.6	29.5	1.92	1.91	7.77
	II	37.4	32.5	28.1	1.90	1.78	6.94
	III	34.2	31.0	33.0	1.94	1.99	7.87
	IV	41.1	33.2	37.1	1.83	1.92	6.79
1977	I	51.6	31.9	30.5	1.64	1.61	5.20
	II	51.5	39.4	41.0	1.78	1.82	5.56
	III	48.5	36.3	38.1	1.77	1.81	5.63
	IV	41.1	35.0	42.1	1.88	2.05	6.82
1978	I	65.6	31.5	30.1	1.50	1.47	4.42
	II	52.5	43.1	44.8	1.80	1.87	5.63
	III	47.1	52.7	55.5	2.14	2.20	6.04
	IV	58.3	40.4	48.2	1.71	1.84	5.12
1979	I	66.2	40.5	38.5	1.63	1.60	4.32
	II	51.6	44.8	47.6	1.89	1.94	5.67
	III	49.6	58.6	60.1	2.20	2.23	6.39
	IV	50.5	54.7	62.6	2.10	2.26	6.07
1980	I	49.8	53.7	57.4	2.10	2.17	5.91
	II	40.1	72.0	73.8	2.82	2.87	8.00
	III	29.0	83.4	84.1	3.91	3.94	10.77
	IV	45.7	85.6	92.3	2.90	3.04	7.54
1981	I	48.6	34.4	48.0	1.73	2.01	5.57
	II	37.8	64.1	79.6	2.72	3.14	9.08
	III	25.6	90.7	93.3	4.58	4.68	12.34
	IV	31.3	131.4	126.1	5.23	6.20	11.12
1982	I	28.2	112.6	102.9	5.03	4.68	12.31
	II	35.1	138.1	115.9	4.96	4.33	9.52
	III	34.4	63.0	48.1	2.86	2.43	7.85

Both CAR and this predicted excess demand are displayed in Table 16.2.[15] Returning to our queueing interpretation of section 16.2, we attempt to compute the mean waiting time for new cars. For this purpose we assume as an approximation that the total queue length is the sum of the excess demand and the number of customers actually serviced; i.e. CAR. Denote the sum of these by Q (queue) and further assume that in any quarter the queue length is the expected queue length. Now $E(Q) = \rho/(1-\rho)$ where $\rho = \lambda/\mu$ is defined as the traffic intensity; equating $\rho/(1-\rho)$ to Q allows us to compute ρ, and since we know the service rate (CAR), we can compute the arrival rate λ. Finally, we obtain the mean waiting time as Q/λ (Phillips, Ravindrau and Solberg, 1976). These estimated mean waiting times are displayed in the next to last column of Table 16.2. Models 1/*M* and 4/*M* tell rather similar stories. During the early period, i.e. from 1974-I to some time in 1979, excess demand as a percentage of car deliveries ranged from 50.1 to 124.0. At the same time, the mean waiting time was remarkably steady, ranging from $1\frac{1}{2}$ quarters to just over two quarters. This is substantially shorter than anecdotal evidence would have it. It is clear that this computation gives only a point estimate for the mean waiting time and it is of interest to compute an upper boundary of an interval estimate (see below). One also notes that in 1980–82 there begins a sharp increase in the mean waiting time, relieved only temporarily in 1981–I, and reaching the $4\frac{1}{2}$ to 6 quarter range in the 1981-III to 1982-II period, followed in the last quarter for which we can make a forecast by a sharp drop in both excess demand and mean waiting time. This final drop may well be due to the existence of gasoline rationing. In general, the sharp increase in waiting times in 1980–82 appears to be correlated with the flagging performance of the Polish economy in this critical period.

To obtain an upper confidence limit for the mean waiting time, we resorted to stochastic simulations of Model 4/*M*. The estimated coefficients of the equations were taken to be the mean value and the estimated asymptotic covariance matrix of the estimates was taken to be the covariance matrix of a normal distribution. Coefficients were generated by drawing from this distribution. Equation errors ξ_1, ξ_2, ξ_3, distributed normally with zero mean and variances equal to those estimated from the model, were then added and the mean waiting time for each time period was obtained from the solution of the model. This experiment was replicated 100 times. For each quarter we then determined the 90th percentile of the distribution of waiting times. These are displayed in the last column of Table 16.2. The interpretation of these figures is the same as that of conventional confidence intervals: for example, using a 0.1 level of significance, we cannot reject the claim that in 1982-I the mean waiting time was 12 quarters. There are two particularly noteworthy things about this last column of Table 16.2. First, many of the mean waiting times are quite compatible with the anecdotal evidence of 2–3 years, and none is less than a year. Secondly, whereas the point estimates show relatively

[15] The predicted values of BM$/RP and PF/RP track the actual values tolerably if not very well: for these two variables the R^2's obtained by regressing predicted on actual values are 0.51 and 0.43 respectively.

homogeneous behaviour in the 1970s, the last column shows waiting times in 1974 and early 1975 about as high as in the critical 1981–82 period.

We finally examine the implied price and income elasticities of the endogeneous variables from 4/*M*. Evaluating the elasticities of demand for new cars with respect to (real) income from Equation 16.9 yields income elasticities ranging from 0.28 at the beginning of the period to 0.06 near the end. The corresponding price elasticities range from − 4.64 to − 0.65.[16] Alternatively, one may obtain the elasticities from the reduced form multipliers. These income elasticities range from 0.03 to 0.12 and the corresponding price elasticities from − 3.25 to − 0.54. The key observation is that the income elasticity of demand for new cars is very small, whereas the price elasticity is quite substantial, in fact, greater than unity in absolute value for almost half the period. If we consider the variant of the model in which income is included in the second equation, the income elasticity rises slightly, but still remains low.

The stochastic simulations can also be used to obtain confidence limits for these elasticities. Since the point estimates of the income elasticities are very low, we are interested in the upper 90% confidence limit which ranges from a low of 0.138 to a high of 0.694 with no noticeable trend except for generally low values in the last 5 quarters. In the case of the price elasticities, the point estimates are increasing over the period. Here we are interested in the lower 10% confidence limit which varies from − 1.098 to − 5.859. The stochastic simulations therefore are compatible with a small but non-negligible and almost trendless income elasticity and a sizeable and slightly increasing price elasticity.

16.7 CONCLUDING REMARKS

It seems that parallel markets in centrally-planned economies can be subjected to econometric analysis in the future. However, the methods employed may be slightly different from the traditions of applied econometrics. First of all, data are difficult to obtain and in many cases are subject to severe approximations. Consequently, the results have to be treated with additional caution. Secondly, estimation techniques are also different from those applied for models of markets in a more flexible economy. The all-excess-demand hypothesis and permanent unobservability of some variables require non-trivial estimation algorithms. In practice almost every model requires its own specific method of estimation. Problems of specification and estimation are therefore closely related. Unlike the case of a genuine market, in a centrally-planned economy price is not sufficiently flexible to indicate the direction and the strength of the excess demand changes. The estimation problem becomes more complicated, since one cannot use prices as disequilibrium indicators. Finally, the specification of the model must be

[16]These elasticities are obtained for each year by using the actual values of the predetermined variables for that year and computing the predicted value of $[\partial y_1/\partial(\mathrm{INC/RP})][(\mathrm{INC/RP})/y_1]$, etc.

sufficiently complex as well as sufficiently flexible so as to

1. accommodate the problems of data availability
2. cope with the problem of exhibiting the relation between the first and second (and in some cases third) markets.

Our results indicate that disequilibrium econometrics can provide insights into problems of this kind. The results on particular car markets are largely in agreement with a priori expectations as well as with some anecdotal evidence. In particular, the estimated fall in excess demand and waiting time at the end of 1982 seems to be a good forecaster of events in the out of sample years 1983–85, in which the free market car price declined (mainly because of reduction in demand) and waiting time for the officially delivered cars was substantially reduced.

APPENDIX A. SOURCES AND CONSTRUCTION OF DATA

The following are the principal sources of data:

BLS *Bulletin of Labour Statistics*, International Labour Office, 1974–82
BS *Biuletyn Statystyczny GUS*, Warsaw, 1974–82
CTPM Official catalogues of Polmozbyt, 1974–82
CTPX Official catalogues of Pewex, 1974–82
PCY *Pick's Currency Yearbook*, Pick's Publication, 1975–79
PT *Przeglad Techniczny*, Warsaw, 1983
QRP *Quarterly Review for Poland*, Economist Intelligence Unit, London, 1982
RS *Rocznik Statystyczny GUS*, Warsaw, 1974–82
RSF *Rocznik Statystyczny Finansow*, Warsaw, 1978, 1982
VE *Veto*, Warsaw, 1982–83
ZG *Zycie Gospodarcze*, Warsaw, 1974–82

All the variables were constructed from raw data. The methods of construction required numerous adjustments to yield consistent data series, since in several cases the same source of raw data was not available for the entire period. We now list the principal methods of construction and difficulties with the various data series.

BM$ Black-market price of dollars computed as a moving average from end of period figures in *PCY* for 1973:4 through 1979:4. For 1980:1 to the end the data came from several sources, principally from *PT*, No. 12, 1983, p. 26. The annual data in the latter were converted to quarterly by adjustments based on *QRP* and *VE*.

CAR Deliveries of new cars to official zloty market. The sources are *BS* and *RS*. In cases of disagreement we took the more recently published figures. In some cases of overlapping annual and quarterly data, we regarded the former as more accurate and adjusted the quarterly data to agree with the annual totals.

INC Households' total personal income in billions of zlotys. The annual figures in *RS* were adjusted to yield quarterly figures on the basis of the monthly figures in *BS* for the principal components.

PCD Index of the official Fiat 125p price (in US dollars) from *CTPX* (1974 = 100.0).

PF Index of freemarket price of cars (1974 = 100.0). This was obtained from the quarterly average prices for seven different types and ages of cars reported in *ZG* by computing their principal components. Only the first component accounting for 98.9% of total variance was used.

PO Index of the official zloty price for the Fiat 125p (1974 = 100.0) from *CTPM*.

RP Index of retail prices. Published quarterly in *BS* up to the end of 1975. After that the annual price index from *RS* was adjusted to provide quarterly interpolations. This relied on quarterly retail sales and on quarterly household expenditures.

DEP Household's deposits in savings institutions. Sources are *RSF*, *BS*, *RS*.

APPENDIX B. DERIVATION OF THE LIKELIHOOD FUNCTION

The derivation becomes more compact if we introduce the following notation: $y_1 = D$, $y_2 = \text{BM\$/RP}$, $y_3 = \text{PF/RP}$, $z_1 = \alpha_0 + \alpha_1(\text{INC/RP}) + \alpha_2 E(\text{BM\$}_{+1}/\text{RP}) + \alpha_3\text{DUM} - \alpha_4(\text{PO/RP})$, $z_2 = \beta_0 - \beta_1\text{CAR} + \beta_2(\text{PO/RP})$, $z_3 = \varepsilon_0 + \varepsilon_1(\text{PF/RP})_{-1} - \varepsilon_2(\text{BM\$/RP})_{-1} + \varepsilon_3(\text{PO/RP})$. The equation system then is

$$\begin{aligned} y_1 + \alpha_2 y_2 - \alpha_4 y_3 &= z_1 + \xi_1 \\ -\beta_1 y_1 + \text{PCD} y_2 &= z_2 + \xi_2 \\ -\varepsilon_2 y_2 + y_3 &= z_3 + \xi_3 \end{aligned}$$

Assume now that the error terms ξ_1, ξ_2, ξ_3 are normally distributed with zero means and covariances and with variances $\sigma_1^2, \sigma_2^2, \sigma_3^2$. The joint density of the endogenous variables for period t then is

$$f(y_{1t}, y_{2t}, y_{3t}) = \frac{|\text{PCD}_t - \beta_1(\varepsilon_2\alpha_4 - \alpha_2)|}{(2\pi)^{3/2}\sigma_1\sigma_2\sigma_3} \exp\left\{ -\frac{1}{2}\left[\frac{(y_{1t} + \alpha_2 y_{2t} - \alpha_4 y_{3t} - z_{1t})^2}{\sigma_1^2} + \frac{(-\beta_1 y_{1t} + \text{PCD}_t y_{2t} - z_{2t})^2}{\sigma_2^2} + \frac{(-\varepsilon_2 y_{2t} + y_{3t} - z_{3t})^2}{\sigma_1^2} \right]\right\} \quad (16.\text{B1})$$

On the assumption that demand always exceeds supply (CAR), the pdf of the observable random variables is

$$h(y_{2t}, y_{3t}) = \int_{\text{CAR}_t}^{\infty} f(y_{1t}, y_{2t}, y_{3t})\, dy_{1t} \quad (16.\text{B2})$$

Define

$$w_{1t} = \alpha_2 y_{2t} - \alpha_4 y_{3t} - z_{1t}$$

$$w_{2t} = \mathrm{PCD}_t y_{2t} - z_{2t}$$

$$A_t = \frac{(-\varepsilon_2 y_{2t} + y_{3t} - z_{3t})^2}{\sigma_3^2}$$

$$B_{1t} = \frac{\sigma_2^2 w_{1t} - \sigma_1^2 \beta_1 w_{2t}}{\sigma_2^2 + \beta_1^2 \sigma_1^2}$$

$$B_{2t} = \frac{\sigma_2^2 w_{1t}^2 + \sigma_1^2 w_{2t}^2}{\sigma_2^2 + \beta_1^2 \sigma_1^2}$$

By completing the square on y_{1t} and integrating, we obtain

$$h(y_{2t}, y_{3t}) = \frac{|\mathrm{PCD}_t - \beta_1(\varepsilon_2\alpha_4 - \alpha_2)|}{2\pi\sigma_3(\sigma_2^2 + \beta_1^2\sigma_1^2)^{1/2}} \exp\left\{-\frac{1}{2}\left[\left(\frac{\sigma_2^2 + \beta_1^2\sigma_1^2}{\sigma_1^2\sigma_2^2}\right)(B_{2t} - B_{1t}^2) + A_t\right]\right\}$$
$$\times\left[1 - \Phi\left(\frac{\mathrm{CAR}_t + B_{1t}}{\sigma_1\sigma_2/(\sigma_2^2 + \beta_1^2\sigma_1^2)^{1/2}}\right)\right] \qquad (16.\mathrm{B3})$$

where $\Phi(\ )$ is the cumulative standard normal distribution function. The likelihood function is

$$L = \prod_t h(y_{2t}, y_{3t}) \qquad (16.\mathrm{B4})$$

It can be shown by the same technique that if the spillover term $\varepsilon_4(y_{1t} - \mathrm{CAR}_t)$ is included in Equation 16.11 the density function becomes

$$h(y_{2t}, y_{3t}) = \frac{|\mathrm{PCD}_t(1 + \varepsilon_4\alpha_4) - \beta_1(\varepsilon_2\alpha_4 - \alpha_2)|}{2\pi(\sigma_2^2\sigma_3^2 + \beta_1^2\sigma_1^2\sigma_3^2 + \varepsilon_4^2\sigma_1^2\sigma_2^2)^{1/2}}$$
$$\times \exp\left\{-\frac{1}{2}\left[C_0(C_{2t} - C_{1t}^2)\right]\right\}\left[1 - \Phi\left(\frac{\mathrm{CAR}_t + C_{1t}}{1/C_0^{1/2}}\right)\right]$$

where

$$C_0 = \frac{\sigma_2^2\sigma_3^2 + \beta_1^2\sigma_1^2\sigma_3^2 + \varepsilon_4^2\sigma_1^2\sigma_2^2}{\sigma_1^2\sigma_2^2\sigma_3^2}$$

$$C_{1t} = \frac{\sigma_2^2\sigma_3^2 w_{1t} - \sigma_1^2\sigma_3^2\beta_1 w_{2t} + \sigma_1^2\sigma_2^2\varepsilon_4 w_{3t}}{\sigma_2^2\sigma_3^2 + \beta_1^2\sigma_2^2\sigma_3^2 + \varepsilon_4^2\sigma_1^2\sigma_2^2}$$

$$C_{2t} = \frac{\sigma_2^2\sigma_3^2 w_{1t}^2 + \sigma_1^2\sigma_3^2 w_{2t}^2 + \sigma_1^2\sigma_3^2 w_{3t}^2}{\sigma_2^2\sigma_3^2 + \beta_1^2\sigma_1^2\sigma_3^2 + \varepsilon_4^2\sigma_1^2\sigma_2^2}$$

where w_{1t} and w_{2t} are as before and

$$w_{3t} = \varepsilon_2 y_{2t} + y_{3t} - z_{3t}$$

ACKNOWLEDGEMENTS

We are indebted to Gregory C. Chow and Stephen M. Goldfeld, Barbara G. Katz and David Kemme and to the referees for useful suggestions. Responsibility for any errors is our own. Financial support from the National Science Foundation and from the National Council of East European and Soviet Research is gratefully acknowledged. Reprinted with permission from *European Economic Review* © Elsevier Science Publishers D. V. (North Holland), **32** (1988) 861–83.

REFERENCES AND FURTHER READING

Balicki, W. (1983) Theory of disequilibrium in centrally-planned economies. *Jahrbuch der Wirtschaft Osteuropas*,**10**, 9–39.

Bernanke, B. S. (1984) Permanent income, liquidity, and expenditure on automobiles: evidence from panel data. *Quarterly Journal of Economics*, **99**, 587–614.

Charemza, W. and Gronicki M., (1988) *Plans and Disequilibria in Centrally Planned Economies*, North-Holland, Amsterdam (to be published).

Chow, G. C. (1957) *Demand for Automobiles in the United States*, North-Holland, Amsterdam.

Dreze, J. H. (1975) Existence of an exchange equilibrium under price rigidities. *International Economic Review*, **16**, 301–20.

Hartley, M. (1977) On the calculation of the maximum likelihood estimator for a model of markets in disequilibrium, Discussion Paper No. 409. Economic Research Group, SUNY Buffalo.

Howard, D. H. (1976) The disequilibrium model in a controlled economy: an empirical test of the Barro–Grossman model. *American Economic Review*, **66**, 871–9.

Kapitany, Zs., Kornai. J. and Szabo, J. (1984) Reproduction of shortage on the Hungarian car market. *Soviet Studies*, **36**, 236–56.

Katz, B. G. and Owen, J. (1984) Disequilibrium theory, waiting costs, and saving behavior in centrally planned economies: a queueing-theoretic approach. *Journal of Comparative Economics*, **8**, 301–21

Katz, B. G. and Owen, J. (1985) Rationing by waiting lists: comment, New York University, Graduate School of Business.

Kornai, J. and Weibull, J. W. (1978) The normal state of the market in a shortage economy: a queue model. *Scandinavian Journal of Economics*, **80**, 375–98.

Kornai, J. (1980) *Economics of Shortage*, North-Holland, Amsterdam.

Krasinski, Z., Garczarczyk, J. Kwissa, B. *et al.* (1980) *Rynek Motoryzacyjny w Polsce* (Market for Cars and Their Services in Poland), WKil, Warsaw.

Lacko, M. (1975) Consumer savings and the supply situation. *Acta Oeconomica*, **15**, 365–84.

Lindsay, C. M. and Feigenbaum, B. (1984) Rationing by waiting lists. *American Economic Review*, **74**, 404–17.

Mojkowski, J. (1984) Dollar musi parzyc. *Polityka-Export-Import*, No. 9, September.

Montias, J. M. (1982) Poland: roots of the economic crisis. *The ACES Bulletin*, **24**, 1–19.

Phillips, D. T., Ravindrau, A. and Solberg, J. J. (1976) *Operations Research: Principles and Practice*, Wiley, New York.

Pickersgill, J. (1980) Repressed inflation and price controls in the Soviet household sector, Research Conference on the Second Economy of the USSR, Kennan Institute for Advanced Russian Studies, Washington, DC.

Podkaminer, L. (1982) Estimates of the disequilibrium in Poland's consumer markets 1965–1978. *Review of Economics and Statistics*, **64**, 423–32.

Portes, R., Quandt, R. E., Winter, D. and Yeo, S. (1983) Macroeconomic planning and disequilibrium: estimates for Poland 1955–1980, Discussion Paper No. 139, Birkbeck College, London.

Portes, R. Quandt, R. E., Winter, D. and Yeo S. (1984a) Planning the consumption goods market: preliminary estimates for Poland 1955–1980, in *Contemporary Macroeconomic Modelling* (eds P. Malgrange and R. A. Muet), Blackwell, Oxford.

Portes, R., Quandt, R. E., Winter, D. and Yeo, S. (1984b) Sur la taille des erreurs de plantification. *Annales de l' INSEE*, **55/56**, 245–54.

Portes, R., Quandt, R. E. and Yeo, S. (1985) Tests of the all-excess-demand hypothesis, Financial Research Center, Research Memo No. 59, Princeton University.

Portes, R. and Winter, D. (1980) Disequilibrium estimates for consumption goods markets in centrally planned economies. *Review of Economic Studies*, **47**, 137–59.

Quandt, R. E. (1985) On the identifiability of structural parameters in all-excess-demand disequilibrium models, Financial Research Center, Research Memo No. 56, Princeton University.

Simon, C. (1982) L'economie polonaise en mouvement: reformes, inflation et recherches de l'equilibrie. *Revue d'Etudes Comparatiues Est-Ouest*, **13**, 35–78.

Starzec, K. (1983) L'economie polonaise vue a travers des circuit paralleles (*mimeo*), Paris.

Vanous, J. (1980) Private foreign exchange markets in Eastern Europe and the USSR, Discussion Paper No. 80–20, Department of Economics, University of British Columbia, Vancouver.

Welfe, W. (1983) An aggregate econometric model of consumer goods markets in disequilibrium WA-1, Paper presented at the Economic Society Meetings, Pisa, 28 August–3 September.

Winiecki, J. (1982) Investment cycles and an excess demand inflation in planned economies: sources and processes. *Acta Oeconomica*, **28**, 147–60.

17

Priority and the shortage model: the medical system in the socialist economy

CHRISTOPHER DAVIS

17.1 INTRODUCTION

The model of the socialist economy that has been developed by Janos Kornai and his associates reflects the many insights of these economists into the behaviour and interactions of socialist economic institutions (Kornai, 1980, 1982b, 1986; Kapitany *et al.*, Chapter 15; Kemme, 1987 and Chapter 4; Lackó, Chapter 10). Despite its positive features, the shortage model has gaps in coverage and is imprecise in some definitions and specifications of hypotheses. Several of its problems, especially related to the concept of the soft budget constraint, have been noted by critics such as Gomulka (1985), Hare (Chapter 3) and Nuti(Chapter 5). However, neither they nor Kornai have adequately analysed one of the apparent major determinants of sectoral and institutional behaviour and performance in a shortage economy, namely the central decision-makers' priorities that are expressed during plan formulation and implementation.

The objectives of this chapter are to investigate theoretically and empirically several issues related to the interconnections between the traditional shortage model and central planner's priorities. In order to accomplish this, three topics are examined. First, an assessment is made of the degree of influence the priorities of central authorities exert on sectoral and institutional performance in the socialist economy. Second, the main hypotheses concerning the behaviour of socialist institutions are derived from the shortage model and tested for consistency with observable phenomena. Third, an evaluation is made of whether detected

International Studies in Economic Modelling Series.
Models of Disequilibrium and Shortage in Centrally Planned Economies.
Edited by C. Davis and W. Charemza.
Published in 1989 by Chapman and Hall Ltd, 11 New Fetter Lane, London EC4P 4EE.

inconsistencies in the shortage model can be remedied by incorporating into it the concept of priority.

The theoretical arguments in this chapter are supported by an empirical study of the medical system in the socialist economy, which presents an interesting and significant case for analysis using the shortage model. First, this sector is comprised of economic institutions that have been neglected in Kornai's research: state budget financed facilities (he calls these non-profit institutions). Second, shortage phenomena, such as queues, deficits of goods and rationing schemes, are pervasive in the socialist medical sector due to the interaction of the substantial, rapidly growing demand for medical care and the tight constraints on resource allocations to health. Third, contemporary socialist medical systems are afflicted by serious shortage-related problems: low quality of medical services, sluggish technological progress, inability to control a number of infectious and degenerative diseases, and rising mortality rates (Davis, 1987a, c). Finally, Western health services are also experiencing shortages and rationing (Cooper, 1975; Aaron and Schwartz, 1984). Comparative study of their counterparts in socialist economies can clarify the common causes of these problems and possible solutions.

The analysis begins in section 17.2.1 by defining the concept of priority and nine indicators that can be used to measure its impact on sectoral performance. The hypothesis that the medical system is a low priority sector of the socialist economy is evaluated in section 17.2.2. Attention is shifted from the sectoral to the institutional level in section 17.3, where an attempt is made to extend the shortage model to describe the behaviour of state budget financed medical institutions. Ten hypotheses about socialist medical institution behaviour in a shortage environment are developed from Kornai's general theory and empirically evaluated to determine their validity and consistency with priority-generated phenomena. Finally, section 17.4 presents some conclusions about the significance and impact of priority in the shortage model.

Two introductory observations should be made about the medical institution model presented in this chapter. First, this author argues elsewhere that the medical system in the socialist economy is a component of the health sector, which is comprised of economic institutions that are closely involved in the production of the population's health: households, medical system, medical supply network, medical industry, medical foreign trade, and the central health bureaucracy (Davis, 1979, 1984, 1987c). Within this complex health production process the medical system provides households with preventive and curative services that are produced using a variety of inputs from supply agencies and are primarily financed by the state budget (see Burenkov *et al.*, 1979). Given this, the chapter examines in detail the three institutions that exert the greatest influence on medical system behaviour: central planners in section 17.2.2; consumers in section 17.3.1; and the medical supply network in section 17.3.2.

A second point is that the empirical evaluation of the shortage model is based upon data concerning the Soviet medical system in the period 1965–85 and previous analyses of this author (Davis, 1979, 1983a, 1984, 1985, 1987c). Although this may appear to give the chapter's model a country-specific character, preliminary investigation suggests that it does depict many characteristics of medical system behaviour in the East European socialist economies.

17.2 THE DEFINITION AND MEASUREMENT OF THE PRIORITY OF THE SOCIALIST MEDICAL SYSTEM

In traditional approaches to the study of the socialist economy the issue of priority has been recognized as significant but only minimal attention has been devoted to its formal analysis. Priority has been similarly neglected in shortage models, despite, the fact that in a chronic shortage environment planners have to be more forceful in priority-linked resource allocations during plan formulation and interventions during the plan implementation period. Only a few shortage models incorporate indicators of priority and none treat the priority ranking of the central bureaucracy as an important determinant of institutional behaviour and economic outcomes (Lacko, 1984). Since one of the hypotheses of this chapter is that priority plays a crucial role in influencing sectoral performance in a shortage economy this issue is examined in more detail below.

17.2.1 Definitions of priority in the centrally planned economy

Priority is an expression of the degree of the leadership's commitment to ensure that objectives concerning a sector, programme, or social group are attained irrespective of circumstances in the economy or a market (Ericson, 1987). In the sectoral case, priority can be defined as an instrument of central decision-makers used to ensure that the most important sectors of an economy, according to some ranking, are able to produce the goods and services deemed optimal or sufficient by the authorities. This is accomplished through preferential allocation of resources during plan formulation and protection of the supplies and operating conditions of high priority sectors from disruptions caused by endogenous problems, exogeneous shocks, and plan inconsistencies (Davis, 1988a).

The priority system is used to control developments in a socialist economy that is characterized by considerable complexity in its structure and processes. It follows, therefore, that use must be made of a variety of priority-linked instruments that affect phenomena such as the allocation to a sector of total resources, wages, investment, and supplies of intermediate goods. Given this, the priority of a sector should be evaluated using an array of indicators rather than one measure.

In section 17.2.2 planners' priorities with respect to a sector of the economy (l) are assessed on the basis of nine indicators. Four of these indicators relate priorities expressed during plan formulation. The first is a conventional measure of the weight given to sectoral output in a planner's welfare function. The second reflects Kornai's idea of control by norms. In particular it evaluates the response of planners, expressed in the allocation of resources to sector l, if a performance indicator exceeds an established tolerance limit (Kornai, 1980, chapters 10 and 12). Priority is also manifested in the ranking of wage rates established for a sector relative to the economy average or to measures of labour quality. The fourth indicator evaluates whether the centrally determined financial norms of a sector are adequate relative to the prevailing actual prices of planned inputs.

Additional indicators assess priorities on the basis of the behaviour of authorities during the plan implementation period. One simple measure of priority is the degree to which output plans are fulfilled. This condition should be almost guaranteed for a high priority sector (Ericson, 1988). The sixth indicator reflects the idea that the hardness of budgets, as the concept is defined by Kornai, varies by sector in accordance with priorities (Kornai, 1986). As a general rule, the more important a sector the softer the budget and vice versa. The next two indicators assess whether the authorities have ensured that a sector obtains planned amounts of inputs and investment in the face of the supply disruptions and deficits that are characteristic of a shortage economy. Finally, priority could, in principle, be evaluated using Kornai's shortage intensity function $Z(t)$, defined in Equation 17.13. This would involve measuring the shortage intensity in sector l relative to the average for the whole economy or to the normal value in the sector.

17.2.2 The sectoral priority of the socialist medical system

Priority-linked initial allocations and reallocations of resources ensure that supplies to the most vital sectors of the economy are protected and attainment of their goals is assured. On the other hand, the leadership's priorities generate tight resource constraints on less important sectors during plan formulation. Furthermore, the redistribution of resources in the plan implementation period, which is a zero-sum game, causes a tightening of constraints in less important areas and makes their original plans more difficult to fulfill. For these reasons low priority sectors in a socialist economy is characterized by low weights in the planners' preference function, unresponsiveness of resource allocations to tolerance limit violations, low relative wages and financial norms, underfulfilment of output and input plans, hard budget constraints, and high shortage intensity. In this section an attempt is made to assess the sectoral priority of the socialist medical system using the nine indicators mentioned above and Soviet data for the period 1965–85.

(a) *Weight in planners' preference functions*

One traditional opinion is that priority is expressed in the utility weight (v_l) given to sectoral output (X_l) or some other measure of sectoral performance by central planners during the solution of the optimization problem:

$$\max V = \sum_{l} v_l X_l \quad \text{s.t.} \ A\mathbf{X} \leqslant \mathbf{R} \tag{17.1}$$

where V = the planners' utility from a given output mix, A = the matrix of input-output coefficients, X = the output vector of the X_l, R = vector of available inputs, $\sum_l v_l = 1$. In this case a test of a high priority sector could involve the assessment of whether its weight was greater than the average across sectors, or $1/h$. Conversely, a low priority sector would have a below average weight.

There are numerous difficulties with such an indicator of priority. On the theoretical level it is not clear that the utility weights actually do represent priorities rather than preferences. If the authorities determined that an achievement of a certain level of output in sector 1, X_1^*, was of highest priority, then it is unlikely they would be content with receiving as a substitute a large volume of output from low priority sector n such that there was equivalence in utility values: $v_1 X_1^* = v_n \ddot{X}_n$. In reality, the planners probably have discontinuous preferences represented by a lexicographic ordering (Deaton and Muellbauer, 1980, p. 27; Varian, 1984, p. 114; Ericson, 1988). This means that they have a ranking of outputs and distribute resources so that the most important sector receives inputs needed to achieve its targets before the other sectors, in sequence according to priority, are provided for. If lexicographic orderings exist then utility weights in a social welfare function would be inappropriate indicators of priority.

Another difficulty with the use of utility analysis in evaluating priority is the determination of the structure and weights of the social welfare function shown in Equation 17.1. For example, it is not clear whether in reality planners include medical system output in the objective function or view health allocations as an input cost (i.e. an element of A). If the former is assumed to be the case, then the problem becomes one of finding proxy measures that make explicit the implicit preferences of central decision-makers. Basu has shown that this task is problematic both in theory and practice (Basu, 1980).

If a welfare approach to priority definition is employed despite these obstacles, then it is possible to develop related indicators. It could be argued that although the utility weights are unobservable, the shares of national income according to output or expenditure adjust over time to reflect these valuations. Therefore one could gain insights into the relative values of decision makers by comparing a centrally planned economy's shares of national income generated by a sector or utilized for a purpose to an international standard. For example, if an end use share was significantly above the average for other nations then one might be able to infer that the activity in question had a high priority. This type of relationship is, often asserted to exist when a country spends a substantial share of national income on defence in peacetime. Conversely, end-use shares below the international average suggest that the activity has a low priority. This relationship can be expressed as follows

$$\frac{X_h(t)}{G(t)} < \frac{\hat{X}_h(t)}{\hat{G}(t)} \quad \text{implies low priority} \tag{17.2}$$

where $X_h(t)$ = an expenditure or output component of national income associated with sector h, $G(t)$ = a national income measure (e.g. NMP, GNP), and a circumflex signifies an international standard.

Table 17.1 Resource allocation responsiveness to tolerance limit violation in the health sector

	Social cost indicator crude death rate		*State budget health expenditure*			*Health share of net material product utilized (4)*
Year	*Deaths/1000*	*Percentage above minimum rate (1)*	*Millions roubles*	*Proportionate change (2)*	*Share of total budget (3)*	
1965	7.3	–	6 669	–	6.6	4.2
1970	8.2	12.3	9 284	+0.39	6.0	4.1
1975	9.3	27.4	11 470	+0.24	5.3	4.0
1980	10.3	41.1	14 821	+0.29	5.0	4.1
1985	10.6	45.2	17 611	+0.19	4.6	3.9

Notes: (1) $C(t)/C(t_0)100$, where t_0 represents the period when the social indicator was at a minimal value (equivalent to p).
(2) $b(t+1) - b(t)/b(t)$
(3) $b(t)/B(t)$
(4) $X_h(t)/G(t)$
Sources: Narodnoe (1975), pp. 43, 742, 745, 565–66, 744.
Narodnoe (1985), pp. 31, 559, 565, 411.

In the case of the Soviet medical system two national income share measures of priority can be utilized. First, the health expenditure share of net material product can be compared to those of other socialist countries. The Soviet share for the period 1965–85 is shown in the final column of Table 17.1 to be in the range 3.9–4.1%. According to Kaser (1976), in the early 1970s the Soviet health share was the sixth lowest of the seven European CMEA countries (Kaser, 1976, p. 33). In 1980 the Soviet health share of NMP utilized of 4.1% placed it in the middle of the 2.8–5.1% range for the six East European nations. The specific country percentage shares in decending order were: Poland 5.1; Czechoslovakia 4.9; Bulgaria 4.3; Hungary 3.6; GDR 3.3; and Romania 2.8 (Lazarcik, 1987 and the 1980 statistical yearbooks of the respective countries). From this evidence one could infer that in 1970 the medical system in the USSR had a low priority relative to the standards of other socialist countries. During the seventies the health share of NMP declined in the USSR. However, the priorities of the East European health services apparently diminished more significantly, with the result that the relative standing of the Soviet medical system improved.

Comparisons of the health share of reconstructed Soviet GNP with Western standards also suggest a low priority. In 1980 the USSR devoted 2.4% of GNP to health whereas in most OECD countries shares were in the 6.0–11.0 range (USSR, 1982). Although the validity of such East–West contrasts is questionable because of the difficulties in determining appropriate relative prices and exchange rates in non-market economies, it should be kept in mind that the determinants of health expenditure in a socialist economy, such as wage rates and budget norms, are themselves reflective of national priorities (see sections 17.2.2(c) and (d)). Given this, GNP shares may be reasonably accurate guides to the priority of sectors.

(b) *Resource allocation responsiveness to tolerance limit violations*

Kornai believes that one cause of the reproduction of shortage and investment tension in a socialist economy is that the planners' allocation to a sector of resources, especially investment, has a cyclical pattern of postponement-release that is governed by responses to performance indicators, such as social costs, exceeding predetermined tolerance limits (Kornai, 1980, Chapter 10). This policy of putting out fires entails neglecting a sector while conditions remain within tolerance limits, even if they are deteriorating, but allocating sufficient resources to it when limits are violated to bring social costs back into the acceptable range.

Although Kornai suggests that this is a universal rule, in reality its validity appears to vary between sectors in accordance with their priority. In high priority sectors there is an immediate, if not an anticipatory, resource allocation response to tolerance limit violation. But in a low priority area no such reaction is guaranteed and social costs can remain above tolerance limits for extended periods.

Resource allocation responsiveness to developments in the Soviet medical system during 1965–85 did not conform to Kornai's general rule. Instead it suggested that the authorities were prepared to tolerate high social costs in the health area and that the medical system had a low priority. This can be seen in Table 17.1, which contrasts trends in social cost and resource indicators.

The social cost indicator ($C(t)$) selected is the crude death rate and the tolerance limit (ρ) is set at the minimum achieved level ($C(t_0)$) in 1965 of 7.3 deaths per 1000. Column 1 shows that this death rate rose from 7.3 in 1965 to 10.6 in 1985, or by 45.2%. It should be noted that this trend was representative of mortality developments in the Soviet Union. After 1965 most adult age-specific death rates rose significantly, infant mortality increased by 36% from 1971 to 1976, and life expectancy declined (Davis and Feshbach, 1980; Dutton, 1979). Also, the expectation that the crude death rate should remain stable or decline is not an unreasonable one, since this was the experience of most developed countries in the world.

The other columns describe trends in resource allocation response. State budget expenditures on health ($b(t)$) did increase, as shown in column 3. However, a better measure of responsiveness would be the proportionate change in budget allocations. If behaviour had conformed to Kornai's rule, or this chapter's high priority definition, then budget allocations would have exhibited increasing proportionate change for the periods $t = 1, 2, \ldots n$ from the time the tolerance limit was initially violated in period $t = 0 (C(0) > \rho)$ until the social cost returned to within the tolerance limit in period n:

$$\frac{b(t+1) - b(t)}{b(t)} > \frac{b(1) - b(0)}{b(0)} \tag{17.3}$$

Instead, there was variability over time with proportionate budget changes from 1970 to 1975 and 1980 to 1985 being lower than in preceding periods despite continuing increases in the crude death rate.

The health share of the total state budget ($B(t)$) is the next priority indicator. The high priority sector response to a tolerance limit violation would have been an increase in the budget share. In the USSR, however, the health share declined from 6.6% in 1965 to 4.6% in 1985 which indicated that the following low priority relationship obtained in a period of social cost tolerance limit violation:

$$\frac{b(t+1)}{B(t+1)} < \frac{b(t)}{B(t)} \tag{17.4}$$

A similar, if less pronounced, trend is evident with respect to the health expenditure ($X_h(t)$) percentage of NMP utilized ($G(t)$). It decreased from 4.2% in 1965 to 3.9% in 1985. This can be expressed more formally as:

$$\frac{X_h(t+1)}{G(t+1)} < \frac{X_h(t)}{G(t)} \tag{17.5}$$

Given the increases in most mortality rates and the decline of life expectancy, this reduction in the share of national income devoted to health provided evidence of a resource allocation responsiveness to medical system problems that was characteristic of a low priority sector.

(c) *Relative wage rates*

In a centrally planned economy wage rates are determined by the planners rather than market forces. Although labour quality factors and market conditions influence wage setting, the priority of the sector or program that employs labour is of considerable importance. As a general rule high priority sectors have generous wages whereas wage rates in low priority sectors are below the economy average.

In the case of the Soviet medical system the average wage of health personnel (W_h) has been consistently below that of the national average ($\bar{W}$). In 1984, for example, W_h was 131.4 roubles per month versus $\bar{W}$ of 184.8 roubles per month. Furthermore, the ratio ($W_h/\bar{W}$) deteriorated over time from 0.75 in 1970 to 0.71 in 1984.

Sectoral priority could also be assessed by contrasting the relative standings of wages and labour quality. If the average sectoral educational level was to be used as a labour quality proxy then an indication of low priority in the medical sector would be a ratio of health to average wage rate that was disproportionately low with respect to the ratio of the medical system educational standard to that of the whole economy:

$$\frac{W_h}{\bar{W}} < \frac{E_h}{\bar{E}} \tag{17.6}$$

The data from the USSR for 1975 monthly wage rates and shares of employees with higher or specialized secondary education are as follows (Narodnoe, 1975, pp. 532–33, 546–47, 552):

$$\frac{W_h}{\bar{W}} = \frac{102.3}{145.8} = 0.70 < 2.27 = \frac{0.50}{0.22} = \frac{E_h}{\bar{E}}$$

In other words, medical system wages were 30% below average but educational standards of the health labour force were 2.3 times higher than the national average. Such a discrepancy could exist only in a low priority sector.

(d) *Adequacy of financial norms*

The budgets of institutions in socialist economies are calculated using centrally determined financial norms F_i, where i represents the $1, \ldots n$ input categories, that express allowable expenditure for a planned quantity of inputs $\bar{a}_i$, (henceforth a bar over a character denotes a planned value). There is, however, variability across sectors in the adequacy of these norms with respect to actual input prices ($\tilde{p}$) that reflects relative priorities (henceforth a tilde over a character denotes an actual value). Norms are usually generous in high priority sectors and generate initial budgets sufficient to cover the costs of planned acquisitions of inputs. In a low priority sector the inverse relationship can exist. Due to stingy financial norms the initial budget of sector h (b_h^0) can be inadequate to cover the actual costs of the projected supply plan (b_h^*):

$$b_h^0 = \sum_i F_i \bar{a}_i < \sum_i \tilde{p}_i \bar{a}_i = b_h^* \tag{17.7}$$

The socialist medical system appears to be a low priority sector in the sense of Equation 17.7. Judging from available Soviet data the financial norms governing the calculation of medical institutions' budgets have been low relative to their needs and the prices of inputs. The norms that determined cost estimates for new construction have been outdated and insufficient. For example, the Minister of Health USSR reported in 1987 that hospitals were being built in Tadzhikistan with a normed capital construction cost of 5000 roubles per bed, which contrasted unfavourably with the 40 to 80 thousand roubles spent per bed in other socialist countries (Na perelome, 1987). The capital repair norms have been miserly as well, with the standard of 37.5 kopeks per cubic metre being only one-quarter of what was needed to compensate for depreciation (Babanovskii, 1976, p. 81). One additional example is that the expenditure norms for medicines and dressings until recently were about one rouble per patient day in a hospital bed and 4 kopeks per outpatient polyclinic consultation, which were values several times lower than necessary (Zagalskii, 1986).

(e) *Fulfilment of the output plan*

One of the primary reasons for the existence of the priority system is to guarantee the fulfilment of the initial output plans of the most important branches of the economy (Ericson, 1987). But given limitations on resources this commitment means that in low priority sector h actual output is often lower than original plan targets:

$$\tilde{X}_h(t) < \bar{X}_h(t) \tag{17.8}$$

Assessment of the priority of the medical system on the basis of fulfilment of initial output plans is made difficult by ambiguity in the definition of output and lack of relevant data. If output is defined in terms of the health status of the population, then the rising mortality rates indicate a rather severe underachievement. A second set of output measures relate to the provision of the n different medical services that make up the output vector $\mathbf{X}_h = (x_{h1}, X_{h2}, \ldots X_{hn})$. In recent years the production of services measured by summary indicators such as outpatient consultations, hospital bed-days and hospitalizations appears to have satisfied plan targets (Davis, 1987c). However, a quantity drive evidently existed in the medical system that involved substantial quantity-for-quality substitution (see section 17.3.1(b). As a result, the quantity of basic medical services provided usually exceeded targets whereas their quality was often below planned standards. If one distinguishes between quantity (superscript Q) and quality measures (superscript 1) then in a low priority sector the situation that frequently obtains is:

$$\tilde{X}_h^Q(t) \geqslant \bar{X}_h^Q(t) \quad \text{but} \quad \tilde{X}_h^q(t) < \bar{X}_h^q(t) \tag{17.9}$$

A third approach to measuring medical system output is to use proxy indicators based on inputs of labour (*L*) and capital (*K*). Although this appears to be a questionable procedure, it has provided the foundation for Soviet health planning, which uses numbers of hospital beds, polyclinic capacities, and doctor positions as key indicators (Popov, 1976). In the USSR the input-based targets of medical system output are routinely overfulfilled, but again this achievement depends upon quantity-for-quality substitution. For example, it has been a long-standing Soviet practice to reach hospital bed targets by crowding extra beds into existing wards or corridors, even if this involves the violation of sanitary norms of floor space per bed or of norms that specify the amount of supporting equipment or staff per bed (Davis, 1979, 1983a). In 1987 Dr Chazov, the Minister of Health USSR, complained that although the minimum sanitary norm of ward floor space per hospital bed was 7 square metres, the actual average amount in the medical system was 4.2 (Na perelome, 1987).

Although the evaluation of the priority of the medical system on the basis of output indicators makes sense intuitively, it is difficult to implement empirically. This is due to measurement problems and the ability of most sectors in a shortage economy to achieve targets through quantity-for-quality substitution. Despite this, the examination of trends in the three categories of output indicators suggests that medical system performance resembles that of a low priority sector.

(f) *Hardness of budget constraints*

In the *Economics of Shortage*, Kornai creates the impression that the budget constraints of all institutions producing goods and services in a socialist economy are relatively soft (Kornai, 1980). However, in a later work he acknowledges that priority influences the degree of budget hardness:

> 'In sectors enjoying priority in central plans, the constraint is softer. In sectors permanently neglected, it is hard' (Kornai, 1982a, p. 508).

The budgetary condition of a sector in a shortage economy can be expressed more formally using the symbols introduced in Equation 17.7 for the initial budget b_h^0 and defining actual *ex post* budget expenditure (b_h^e) as the sum of the products of actual prices ($\tilde{p}_i$) and actual inputs ($\tilde{a}_i$). In the case of a low priority sector *h* the usual situation would be:

$$b_h^e = \sum_i \tilde{p}_i \tilde{a}_i \leqslant b_h^o = \sum_i F_i \bar{a}_i \tag{17.10}$$

In contrast, a high priority sector would be characterized by actual expenditure routinely greater than planned (the soft budget constraint condition).

The actual situation with respect to the budget constraint of the medical system is discussed in more detail below in section 17.3.2(a). To summarize, the conclusion from the empirical study of the Soviet case is that the budget constraint of the medical system was relatively hard by the standards prevailing in the economy as a whole, which is what one would expect for a low priority sector. In the spectrum ranging from hard household budgets to soft heavy industry budgets the medical system was closer to the former than the latter. Financial authorities enforced strict budget discipline and health administrators were not even permitted to move funds from budget articles with a surplus to others approaching spending limits. Although there was some softness in the capital budget, this was more reflective of adjustments in funding needed to compensate for the unrealistically low initial budget generated by stingy financial norms and to facilitate fulfilment of *ex ante* construction plans expressed in physical terms than of the central authorities' tolerance of excessive spending by subordinate institutions.

(g) *Fulfilment of supply plans*

Another major function of priorities is to protect the initial planned supplies of intermediate goods destined for important sectors of the economy from disruptions caused by endogenous and exogenous factors (Gregory and Stuart, 1981, p. 180). This means that for high priority sectors the authorities want to ensure nonnegativity in the vector of deviations of actual from originally planned physical quantities of inputs, i.e. $(\tilde{a}_i - \bar{a}_i) > 0$. If input quantities are standardized across sectors and the quantity deviations are expressed in value terms by multiplying them by respective prices then the summed value of a high priority sector would be greater than the average for all n sectors. In contrast, a low priority sector h in a socialist economy would not be so protected and therefore would experience chronic shortages of inputs, i.e. $(\tilde{a}_{ih} - \bar{a}_{ih} < 0)$. As a result the following rule would apply:

$$\sum_i p_{ih}(\tilde{a}_{ih} - \bar{a}_{ih}) < \frac{1}{n}\sum_l\left(\sum_i p_{il}(\tilde{a}_{il} - \bar{a}_{il})\right) \tag{17.11}$$

The analysis of the supply of intermediate goods to the medical system in section 17.3.2(c) indicates that this sector was plagued by chronic shortages. Despite the modest levels of demands for inputs generated by the spartan financial norms and modest activity level targets, supply plans of the medical system were regularly underfulfilled. Although there is insufficient data to estimate the hypothesized relationship shown in Equation 17.11, the available evidence indicates that there was recurrent negativity in the deviations of actual from planned supplies $(\tilde{a}_{ih} - \bar{a}_{ih})$ and suggests that the supply disruptions affecting the medical system were more severe than the all-economy average.

(h) *Fulfilment of investment plans*

Numerous studies have shown that a state of investment tension frequently exists in the socialist economy due to planners' errors and firms' behaviour (Kornai, 1980; Lackó, 1984). This means that the demand for capital goods and construction services considerably exceeds their supply. One method used by central planners to deal with this situation is to apply priorities in the investment reallocation process during plan implementation. The original investment budgets and planned supplies of capital goods of high priority sectors are protected, if not augmented, so they are usually able to fulfil or overfulfil their taut plans. This means that the ratio of actual investment ($\tilde{I}(t)$) to initially planned investment ($\bar{I}(t)$) would be greater than or equal to one and would be above the average for all $l = 1, 2, \ldots h, \ldots n$ sectors of the economy. However, in order to protect investment plans of the institutions deemed important by the authorities it is necessary to siphon off capital goods and services from low priority sectors. In consequence, the investment plans in low priority sector h are chronically underfulfilled and its ratio of actual to planned investment is below the average for the whole economy:

$$\frac{\tilde{I}_h(t)}{\bar{I}_h(t)} < \frac{1}{n}\sum_l \left(\frac{\tilde{I}_l(t)}{\bar{I}_l(t)}\right) \tag{17.12}$$

In assessing the investment plans of the Soviet medical system one has to recognize that due to the sluggish resource allocation responsiveness of planners to tolerance limit violations in the health area (see section 17.2.2(b)) the capital plan targets in physical terms tend to be modest relative to needs. Furthermore, the stingy financial norms used in capital budget estimation generate initial plans with insufficient investment funds to cover the costs of projected capital construction, repair, and equipment acquisition.

Despite the modest nature of health investment plans, their implementation in the USSR was hampered by a routine neglect of contractual obligations by construction organizations, erratic supplies of capital goods from industry, and failure of higher authorities to make compensatory interventions (Davis, 1983a, 1987c). Construction firms knew that when shortages in their labour or material supplies developed and forced a reduction in activity, they should cut back on non-productive medical facilities rather than those being built for more important branches. This resulted in chronic underfulfilment of construction plans and long delays in the completion of medical facilities (Utverzhdat, 1986). For example, a 1987 report by the Ministry of Health stated that:

> Even in the past year preliminary figures indicate that 83% of State capital investment for the whole country was carried out, in the RSFSR – 80%, in the Moldavian SSR – 79%, Tadzhik SSR – 72%. Worst of all, in Kirgizia, Azerbaidzhan, Georgia and several other republics hospitals and polyclinics are built using outdated designs and only 2.5 to 7–8 thousand roubles are spent per hospital bed when the budget estimate norm for the twelfth five-year plan is 18.4 thousand roubles (Na perelome, 1987).

Although it is not possible to estimate the ratios shown in Equation 17.12, available evidence does suggest that the record of investment plan fulfilment in the medical system was worse than the mean for the economy as a whole (see section 17.3.2(b)).

(i) *Degree of shortage intensity*

In his book *Growth, Shortage and Efficiency* Kornai claims that one of the most important determinants of the behaviour of the socialist economy is the degree of shortage intensity (Kornai, 1982b). He argues that the shortage intensity of the entire economy can be represented by a macroindex $Z(t)$, which is an increasing function of observable indicators of shortage and fluctuates around a normal value, $Z^*(t)$ (see Hare, Chapter 3).

If one accepts that $Z(t)$ can be defined and measured, then it should be viewed as the all-economy average. But one would expect the normal state of shortage to vary by sector in accordance with priorities. For example, the strategic machine-building and metal-working industry would have a low shortage intensity indicator, whereas the textile industry would be in the inverse situation. This suggests that the macroindex of shortage intensity should be expressed as the average of the l sectoral intensities and that the index for a low priority sector h would be characterized by:

$$Z_h(t) > Z(t) = \frac{1}{n}\sum_l Z_l(t) \tag{17.13}$$

There are numerous unresolved problems associated with utilizing Kornai's shortage intensity index in empirical investigations, such as definition of appropriate indicators and collection of relevant data. Nevertheless, an attempt was made to evaluate summarily the prevalence of shortage related phenomena in the medical system on the basis of the detailed findings in section 17.3. This indicated that there were chronic and pervasive shortages of all types of inputs to the medical system: labour, building space, machinery and equipment, instruments, medicine and even basic commodities such as bed linen (see sections 17.3.2(b) and (c)). On the output side there were widespread deficits of most medical services that required specialist labour, equipment, or high-quality inputs (see section 17.3.1(a)). Shortage intensity was also reflected in the rising Soviet mortality rates, because they indicated a situation of social costs exceeding tolerance limits due to insufficient allocations of resources. This high intensity of shortage phenomena helps to explain the March 1987 statement of the Minister of Health USSR that in the medical system:

> unresolved problems, difficulties and deficiencies have accumulated to a greater degree than in any other sphere of activity of our society (Na perelome, 1987).

The evaluation of the characteristics and performance of the Soviet medical system on the basis of the nine indicators presented above demonstrates that by all criteria it was a low priority sector of the economy during 1970–85. It also shows that planners' priorities and associated resource allocation rules exert important external influences on medical institution performance. In fact, this analysis suggests that many features of the shortage model may be reflections of central priorities rather than outcomes of endogenous processes and that certain of

Kornai's concepts, such as the soft budget constraint, are not universal across sectors of the economy.

17.3 SOCIALIST MEDICAL INSTITUTION BEHAVIOUR AND THE SHORTAGE MODEL

In Kornai's model of the socialist economy the behaviour of producing institutions (e.g. firms) is the primary cause of the initiation and reproduction of the shortage phenomena that dominate micro- and macroeconomic performance. The emphasis placed on producers in the explanation of economic phenomena is one of the features that distinguishes the shortage model from more traditional ones, which typically attribute at least equivalent importance to planners and consumers. Given this, section 17.3 examines the main assumptions underlying the producing institution's behaviour in the shortage model and the hypotheses derived from them in the case of the socialist medical facility.

The shortage model of the socialist firm appears to be based on three major assumptions. First, the markets for the output of the firm are characterized by chronic shortages. In other words, the initial demands of buyers routinely and substantially exceed the firm's supply of goods and services. Second, firms have a quantity drive, which means they consistently strive to expand output on the basis of extensive growth. Third, the budget constraint of the firm is soft, so there is no effective financial restriction on its demand for inputs.

Kornai derives from these assumptions numerous hypotheses about the operating conditions and behavioural problems of socialist firms. On the output side the firm is supposed to produce goods and services of low quality and/or of obsolete standards and to maintain meagre inventories of saleable goods. Four other hypotheses relate to production processes within the firm: production technology is below international standards; the pace of technological innovation is sluggish; bottlenecks are regularly encountered in production; and output composition is routinely altered in an unplanned manner due to forced substitution of inputs. Kornai also develops three hypotheses concerning the input side of the firm: chronic shortages exist in markets for inputs; there is a high degree of investment tension; and the firm tries to maintain maximum inventories of all categories of inputs.

In this section an attempt is made to evaluate the validity of these assumptions and hypotheses on the basis of theoretical criteria and empirical evidence. The theoretical component of the assessment consists of comparing Kornai's hypotheses concerning medical institution behaviour with those derived in section 17.2 from concepts of priority. The empirical analysis involves the comparison of shortage model hypotheses with evidence about the actual performance of state budget financed medical facilities in the USSR.

In section 17.3.1 topics related to the output behaviour of a socialist medical facility are examined, such as the existence of the seller's market, quantity drive, and quality of output. Hypotheses concerning input behaviour are evaluated in section 17.3.2: hardness of budget constraints; investment tension; and conditions in intermediate goods markets. Section 17.3.3 assesses the production process within the medical institution, and especially the issues of technological level, technological progress, production bottlenecks, and forced substitution. To avoid duplication in coverage, no examination is made here of the shortage model hypotheses concerning inventories, slack, and friction of adjustment that Hare discusses in Chapter 3.

17.3.1 The output behaviour of the socialist medical institution

(a) *Conditions in the market for outputs*

Kornai believes that in a shortage economy the demand for a firm's output consistently exceeds supply (Kornai, 1980). This generates chronic shortages for buyers and suction in the market for the products of selling institutions. The existence of sellers' market is an important cause of institutional behaviour in the shortage model.

This section analyses conditions in the market for medical services in the USSR. The main issues examined are the backlog of untreated illness in the early 1970s, developments in demand during 1970–85, the degree of shortage intensity in the medical care market in the 1980s, and the effects of shortages on household and medical facility behaviour.

In the USSR the provision of medical services always has been insufficient to cope with the illnesses of the population (Davis, 1979, 1983b, 1987b). In the 1970s Soviet authorities tried to assess the scale of the problem by conducting large scale sample surveys of illness rates (M). The analysts then examined patients' records in medical facilities to determine how many cases of existing illness per 1000 population had been presented to doctors for treatment (T). Calculations were made, on an annual average basis, of the rates of untreated illness (U):

$$U = M - T \tag{17.14}$$

Estimates also were made of morbidity iceberg coefficients, S, which measured the ratio of treated illness to total morbidity:

$$S = \frac{T}{M} \tag{17.15}$$

A review of the six Soviet studies by this author indicated that in urban areas $U = 598$ cases of illness per 1000 population and $S = 0.64$, whereas in rural areas $U = 980$ cases of illness per 1000 and $S = 0.27$ (Davis, 1979). In other words, at that time over two-thirds of the cases of illness in the countryside did not receive medical treatment by doctors. Although many patients obtained lower level care from middle medical staff (nurses and feldshers) the Soviet authorities considered this to be an unacceptable situation.

The demand for medical care is a function of a variety of variables, such as need (N), disposable income (D), educational standard (E), capacity of the health system (H), and the price of medical services (P) (Acton, 1975; Cullis and West, 1979):

$$d = f(N, D, E, H, P) \tag{17.16}$$

In this relationship demand increases in response to growth in N, D, E and H but falls in reaction to price rises:

$$\frac{\partial d}{\partial N} > 0, \quad \frac{\partial d}{\partial D} > 0, \quad \frac{\partial d}{\partial E} > 0, \quad \frac{\partial d}{\partial H} > 0, \quad \frac{\partial d}{\partial P} < 0$$

Recent developments in these variables in the USSR have generated substantial growth in the demand for medical services. The Soviet population's need for medical care increased significantly during 1970–85 because of a growth in illness generated by trends in demography, consumption, and the health environment (Davis, 1979, 1987c; Feshbach, 1983). Trends in virtually all of the other variables were conducive to the growth in demand. For example, from 1970 to 1985 the index of real per capita income rose by 62%, the number per 10 000 population with higher and middle education increased from 483 to 701, and the numbers of doctors went up from 668 thousand to 1170 thousand (Narodnoe, 1985). Finally, no money prices were introduced for officially provided medical services.

According to Kornai, if the supply of a good or service has not reached the population's satiation level and the price of the product is not an effective one, then

> there will necessarily be a shortage of the product... In the case of free or almost free distribution the shortage is certainly more intense than it would be in the case of an effective price (Kornai, 1980, p. 416).

These preconditions appear to have been valid with respect to medical services, which were officially distributed free-of-charge and did not satiate the population's needs. In consequence one would expect to observe chronic shortages in the medical care market.

In order to measure shortage it is necessary to have an operational definition of this concept as well as relevant data. The conventional approach would be to define shortage (z) as the excess of the population's notional demand for medical care (d) over actual supply (s):

$$z = d - s > 0 \tag{17.17}$$

However, Kornai argues that this representation is misleading (even in a single market) in a chronic shortage environment because forced substitution effects complicate the definition of demand and distort its measurement (Kornai, 1980, Chapter 5; Kemme, Chapter 4). He recommends measuring the existence of shortage through the use of a shortage intensity function Z:

$$Z = f(z_1, z_2, \ldots z_n) \tag{17.18}$$

where the z_i are indicators of shortage-related phenomena such as forced substitution, length of queues, waiting time, crowding, and production disruption (Kornai, 1980, p. 151). Since the shortage model is being examined in this chapter it is appropriate to adopt Kornai's multivariate approach.

There was ample evidence of shortage phenomena in the Soviet medical system during the 1980s (see Davis, 1983a, 1987a). For example, an authoritative 1986 *Pravda* article reported that:

> People write that many hospitals are overcrowded, patients lie in corridors, there are not enough medical personnel or many medicines. In polyclinics there are queues, it is difficult to see specialists or to obtain diagnosis (Chernyak, 1986).

Another shortage indicator mentioned by Kornai is that of the intensity of criticisms by consumers (or 'voice as a signal'). Over the years dissatisfied Soviet citizens have submitted numerous complaints about deficiencies in the medical system to party, state and local government authorities as well as to the media. According to *Pravda* 66 000 letters of complaint were sent to the Ministry of Health in 1985, a total that was 'significantly more than in the previous year' (Chernyak, 1986). It is possible, however, that the upsurge was more the result of the new *glasnost* policy of General Secretary Gorbachev than of a deterioration in medical care.

The degree of shortage in the medical services market is reflected as well in measurements of second economy activity. Nuti (Chapter 5), Charemza (Chapter 11) and numerous studies surveyed in Davis, 1988b correctly argue that the performance of official markets in disequilibrium in the socialist economies should not be studied in isolation from related unofficial markets (Nuti, Chapter 5; Charemza, Chapter 10). As shortages intensify in the former, transactions in the second economy grow. In the case of the Soviet medical care market second economy activity increased in the 1970s and early eighties. It became more common for patients to provide medical staff with gifts and bribes in order to avoid queues and obtain specialized, high quality medical services that were difficult to receive at the zero money price in the official medical system.

It appears that the growth in the output of medical services during 1970–85 did not significantly increase the morbidity iceberg coefficient (S) discussed above. In February 1987 a leading doctor, S. N. Federov, stated that the Soviet medical system treated only '30% of those people who need it' (Sluzhba, 1987). The maintenance of a large backlog of untreated illness can be interpreted as further evidence of a shortage of medical services.

Examination of medical institution performance reveals numerous cases of supply deficiencies in the medical care market (Davis, 1983a, 1987c). In many rural districts and cities there were inadequate numbers of first-contact general doctors, insufficient average amounts of doctors' consultation time, and deficits of specialist outpatient doctors and diagnostic units (Pochemu, 1979). Due to supply shortages in polyclinics many patients were provided services in a tardy manner or incorrectly referred on to inpatient facilities. Hospitals could not accept all referrals, so they had to regulate demand through waiting lists. Despite this, excessive numbers were admitted to hospitals which resulted in overcrowding, long waits in bed for diagnosis or treatment, and heavy demands placed on hospital specialist departments.

All this evidence indicates that during 1970–85 the demands of Soviet consumers for medical services considerably exceeded available supplies. This situation resulted in chronic shortages in the medical care market and exerted sustained, severe pressure on the medical institutions to increase their output of medical services.

(b) *The quantity drive*

In the *Economics of Shortage* (1980) Kornai explains how the existence of sellers' markets and taut central plans over an extended period affects the motivation of managers of firms and non-profit institutions and instills in them a behavioural pattern summarized by the term the quantity drive (Kornai, 1980; Hare, Chapter 3). This describes the tendency of administrators of institutions in sector h to strive for continuous increases in the quantity of output ($X_h^Q(t)$) in order to satisfy queueing customers and demanding superiors, subject to the maintenance of minimum quality standards and the observance of resource constraints. The increases in output are characteristically achieved through extensive development, which means that the proportional growth in inputs is at least as great as that of outputs.

A review of the Soviet literature indicates that the quantity drive has played a major role in medical system development (Davis, 1979, 1987b, c). It was induced and has been sustained by the chronic shortages in the market for medical care and a health strategy that called for rapid growth in the volume of basic medical services using simple facilities, cheap labour, and modest material inputs.

The medical system quantity drive in the period 1970–85 is illustrated in Table 17.2. It shows that four important medical system outputs increased in the 22–63% range. This growth in output was not generated primarily through improved productivity, however. Instead, reliance was placed on increasing production through the expansion of inputs: doctors by 75%; middle medical personnel by 49%; and hospital beds by 35%. Section 17.3.1(c) argues that these upward trends in the quantities of output and inputs were not matched by similar advances in quality standards of services.

Contemporary Soviet authorities appear to have reached the conclusion that the quantity drive in the medical system has caused distortions in the resource allocation pattern and has contributed to the ineffectiveness of medical institutions in coping with challenges posed by trends in illness. In February 1986 Minister of Health Burenkov stated that:

> The health service as a branch of the national economy today cannot go forward only on the basis of growth in the number of doctors, hospital beds and outpatient visits (Vashe, 1986).

Table 17.2 The quantity drive in the Soviet medical system, 1970–85

Indicator	*Year* 1970	1975	1980	1985	*1985 as % of 1970*
Outputs					
Outpatient visits (millions)	1938.4	2296.9	2750.5	3168.4	163
Hospital bed-days (millions)	828.3	962.9	1063.7	1162.1	140
Hospitalizations (millions)	52.2	57.3	62.7	69.6	133
Preventive screenings (millions)	101.3	106.9	112.5	123.2	122
Inputs					
Doctors (thousands)	668.4	834.1	997.1	1170.4	175
Middle medical personnel (thousands)	2123.0	2515.1	2814.3	3158.9	149
Hospital beds (thousands)	2663.3	3009.2	3324.2	3607.7	135

Sources: (1) Narodnoe (1980, pp. 496–97)
(2) Narodnoe (1985, pp. 539–43)
(3) Davis (1987c)

Instead, it should adopt a course of intensive development based on scientific–technological progress. Over the succeeding year public complaints about the medical system and its failure to engage in reconstruction (*perestroika*) grew and a new Minister of Health was appointed. This resulted in stronger criticism of the quantity drive and related planning procedures. At a February 1987 Trade Union Congress, a leading Soviet doctor blamed many of the serious health problems in the USSR, such as low life expectancy, on 'the unending, extensive growth of the health service' (Sluzhba, 1987). A month later the new Minister of Health, E. I. Chazov, identified the 'extensive tendency in health service development' as a negative phenomenon requiring remedial attention and said that:

> The pursuit of growing quantities of hospital beds and numbers of outpatient visits has led to a grave situation . . . caring about these numbers, they forgot about quality (Na perelome, 1987).

(c) *Quality of output*

Over the past several decades health systems in Western countries have improved significantly the quality of medical services they produce in the areas of diagnosis, treatment, and care. These advances have been generated by several factors. First, developments in medical science, technology and techniques have increased potential capabilities of medical systems. Second, medical care has become more resource intensive and expensive. A given type of medical service is now produced by skilled professionals using capital equipment, pharmaceuticals and other goods that are of higher standards than those in past periods. Third, managers of Western medical institutions prefer to improve the quality of medical service output than to expand its quantity because this enables them to maximize revenue, prestige, and career advancement (Cullis and West, 1979).

The shortage model implies that medical institutions in a resource constrained economy will exhibit a contrasting behavioural pattern and will sacrifice the quality of output in their drive to expand its quantity. In the case of the medical facility in sector *h* this means that the actual quality of its output ($\tilde{X}^q_h(t)$) will often fall below planned values ($\bar{X}^q_h(t)$) or international standards ($\hat{X}^q_h(t)$). Kornai believes that the low quality level of products of suppliers is one of the most consistent and serious consequences of chronic shortage in a market (Kornai, 1980, p. 7).

In the USSR developments in the quality of medical care during 1970–85 were in conformance with hypotheses derived from Kornai's theories (Davis, 1983a, 1987a; Feshbach, 1983). Although there were variations between regions, medical subsystems and facilities, as a general rule the average quality of medical care was low relative to contemporary Western standards and the rate of improvement was slower. Diagnoses in the USSR often were superficial and inaccurate, many medical treatments were unsophisticated and administered tardily, the risk of infection from hospital surgery was high, and the care of hospital patients by nurses and orderlies was poor (Chernyak, 1986; Na perelome, 1987; Sluzhba, 1987; Trudnye, 1987; Utverzhdat, 1986).

The problems of medical care quality in the USSR were caused in part by behavioural patterns of health sector personnel. Central planners paid little attention to quality measures of output and allocated insufficient funds to the medical system. The low medical system wages did not attract the top students to medical schools, which in any case offered training of a mediocre standard, nor did they provide adequate financial incentives for staff either to maintain skills or to perform at high levels.

Shortages and low technical standards of many material inputs to the health service further constrained the quality of medical care. A substantial number of medical buildings were old, not designed for the practice of contemporary medicine, and not properly maintained. Deficiencies in medical technology supply, discussed below in section 17.3.3(a), impeded diagnosis and treatment in most medical facilities. Deficits of sophisticated, effective drugs and diagnostic chemicals also had an adverse impact on medical service quality. Finally, Soviet medical facilities did not provide high standards of hotel amenities, such as good food and clean linen due to the stingy financial norms governing input acquisitions and erratic supplies from the rest of the economy.

17.3.2 The input behaviour of socialist medical institutions

(a) *The budget constraint*

One concept of Kornai that has attracted much attention is that of the soft budget constraint (Kornai, 1980, 1986a). It has been critically evaluated by Hare in Chapter 3 and by Gomulka elsewhere so its features will not be reviewed again (Gomulka, 1985). Instead, attention is focused on the specific issue of the softness of the budget constraint of medical institutions.

If one derived a hypothesis about the nature of the budget constraint of a socialist medical institution from the experience of Western countries then it would postulate a relatively soft budget. For example US medical facilities have been able to pass on most costs associated with their quality drive to the government or insurance agencies and therefore have not experienced binding budget constraints. The discussion in section 17.2 suggested, however, that the situation should be different in the socialist economy because the central authorities give the medical system a low priority in the resource allocation process and impose on it relatively severe budgetary limits. In fact, the budget constraints of medical facilities could be super hard, or flexible downward, in a shortage economy in the sense that they could grow tighter in the course of the plan period as the authorities respond to inconsistencies in plans and bottlenecks in production by reallocating resources and financial entitlements.

Generalization about the budget hardness or softness of medical institutions is complicated by aggregation problems. The typical budget of a state-financed medical facility is subdivided into expenditure articles that are defined by categories of inputs (e.g. labour, medicine, food, capital) (Babanovskii, 1976). Managers are usually unable to transfer funds from one article where there is underspending to another where the budget constraint has become binding. This creates difficulties because market conditions and central provision of supplemental funds vary according to input. Furthermore, the institution could have a soft budget constraint for certain articles and hard ones for others. As is evident, the representation of the budget constraint by a summary indicator of hardness may therefore distort the reality of the budget process.

Examination of empirical evidence from the USSR during 1970–85 indicates that medical institutions had rather tight initial budgets due to their low priority ranking and to the utilization of low wage rates and stingy financial norms in budget calculations. Another contributing factor was that budgets were planned on the basis of medical institution activity levels that were closer to those implied by normed capacities than to the higher ones elicited by the suction of chronic shortages in the medical care market.

Budget discipline in the Soviet medical system was fairly strict. The Ministry of Finance demanded that allocated funds be spent in the sums and patterns outlined in original budgets. Furthermore, demand responsiveness to input prices was relatively high, at least in certain budget categories. There were recurrent reports in

Table 17.3 Actual versus planned expenditure by health budget article in the USSR, 1971–74

	Ratios of actual to planned expenditure					
Budget article	*1970*	*1971*	*1972*	*1973*	*1974*	*Average 1970–74*
No. 3: Administration	1.06	1.07	1.05	1.04	1.04	1.05
No. 10: Medicinal dressings	0.94	0.97	0.98	1.00	1.00	0.98
No. 12: Inventory/equipment	1.25	1.40	1.25	1.40	1.35	1.33
No. 16: Capital repairs	1.44	1.44	1.36	1.41	1.43	1.42
No. 14: Light inventory/ uniforms	0.92	0.93	0.85	0.90	—	0.90

Source: Babanovskii (1976, pp. 59, 65, 77, 80, 83).

the Soviet press of medical facilities not purchasing expensive newer medicines and supplies in wholesale pharmacy markets because of insufficient funds. This phenomenon conflicts with Kornai's hypothesis that 'the normal price-responsiveness of the traditional socialist firm on the input side is weak for instantaneous and short-term adjustments' (Kornai, 1980, p. 336).

Budget constraints of Soviet medical institutions did not have the same degree of hardness in all input categories. Table 17.3 shows ratios of actual expenditures to planned ones for five budget articles. In two cases, articles 10 and 14, there was consistent underspending. Section 17.3.2(c) below suggests that in the case of medicines this was due to the medical system encountering real resource limits before budget constraints became binding. The three other cases in Table 17.3 indicate that there was some consistent overspending as well. For example, during 1970–74 actual expenditure on capital repairs was on average 42 percent higher than planned. However, this was partially due to the fact that institutions had to spend more than budgeted to obtain the original planned volume of capital goods and services because the financial norms that determined budgets were set so low. Also, in 1970, articles 12 and 16 made up, respectively, 2.3% and 3.1% of the total Ministry of Health budget so the aggregate impact of the overspending was marginal. Nevertheless, this pattern of differential budget article hardness appears to be consistent with Kornai's hypothesis that the budgets of institutions are softer for investment goods than for current supplies.

(b) *Investment*

Kornai believes that the quantity drive of socialist firms is accompanied by a strong desire to expand productive capacity and a related investment hunger that is not constrained by the soft budget (Kornai, 1980; Hare, Chapter 3). As a result of the firms' virtually instable demand for investment and the paternalism of central authorities, *ex ante* investment plans are usually unrealistically ambitious. During plan implementation it emerges that costs of approved programs have been underestimated, new capital projects are added, and actual supplies of capital goods and services turn out to be less than projected. This causes investment tension throughout the economy and chronic shortages in capital markets. In consequence, the firm as buyer experiences difficulties in obtaining machinery, equipment and construction services and its investment plans tend to be underfulfilled.

On the basis of Kornai's writings one would expect that medical facilities in a resource-constrained economy would exhibit the same investment-related behaviour as other economic institutions (Kornai, 1980, Chapter 9; Lacko, 1984). However, the material presented in previous sections suggests that some modification of this hypothesis is necessary. It probably would be the case that an expansion drive and investment hunger would exist because of medical managers' desire to increase capacity and capabilities in order to cope better with demand pressure in the market for medical services. But the stern paternalism, low priority, and relatively hard budget constraint affecting the medical system would make it difficult to obtain initial approval of projects, investment funding, and allocations of rationed capital goods. In consequence there would be a restrained, rather than insatiable demand.

This line of argument might lead one to conclude that medical system investment plans would be more feasible and investment tensions would be lower than in other sectors of the economy. Several factors militate against this favourable outcome, however. First, the stingy financial norms applied in health capital budget formulation are not accurate reflections of the actual costs of investment work. As attempts are made to implement plans for a given volume of capital acquisitions or construction, the costs rise beyond budget limits. Second, the low priority status of health means that the central authorities are prepared to tolerate a higher-than-average intensity of shortage in this sector (see sections 17.2.2(b) and 17.2.2(i)). When investment tensions increase throughout the economy the bureaucracy will either initiate or tolerate a siphoning of capital goods and services away from the medical system to high priority areas. This will amplify investment tension in the medical system with the consequences of underfulfilment of investment plans and postponement of capital project completion.

The propositions outlined above concerning investment tensions in the medical system capital market of a socialist economy were evaluated using Soviet data. In the USSR during 1970—85 there were numerous forces that exerted upward pressure on the demand for investment. Previous sections of this chapter have demonstrated the existence in the USSR of shortages in the medical care market, the quantity drive, and an extensive growth strategy that generated increased output through increments in labour and capital. Another contributing factor was the deficient state of the existing capital stock. The number of medical buildings was insufficient to provide planned or actual volumes of services at established standards and the quality of facilities was poor due to design problems, use of old buildings, and lack of repairs. In 1987 the Minister of Health USSR made the following comment about the state of medical buildings:

> But what can one say, when in the Tadzhik SSR more than 50% of hospitals are located in spaces and buildings that have been one hundred percent depreciated. Only 35% of district hospitals in the country have hot water supply, and in 27 there is no sewerage, in 17 – no water supply (Na perelome, 1987).

Equivalent problems existed in the area of medical technology (see section 17.3.3(a)).

Although these factors contributed to the expansion drive and investment hunger in the Soviet medical system, its budget constraint was relatively hard. In consequence the Ministry of Health had to struggle to obtain approval of investment project funding and allocations of capital goods and services. The low capital construction and repair norms used in financial planning meant that investment budget constraints were reached earlier than anticipated. Medical facilities were able to obtain some extra investment funding from local soviets and economic enterprises, but these supplements were usually insufficient to guarantee that the planned volume of work was carried out in a timely manner (see section 17.2.2(h)).

In sum, available evidence indicates that the medical system in the USSR had the expansion drive and investment hunger postulated by Kornai, even though its budget constraint was relatively hard and the demand for capital goods and services was kept under control during plan formulation. A high degree of investment tension existed because of priority-linked intersectoral siphoning and failures of construction organizations to carry out planned investment work.

(c) *Conditions in the markets for intermediate inputs*

Kornai argues that the quantity drive, hoarding tendency, and soft budget constraint cause socialist firms' demands for intermediate inputs to exceed possible supplies, thereby generating chronic shortages (Kornai, 1980, Chapter 4). As a result of this situation, the firm as a buyer must engage in shortage-related behaviour such as searching, queueing and forced substitution.

One might expect the circumstances and the behaviour of a socialist medical institution in intermediate goods markets to be different from that of an industrial firm with a soft budget constraint. Although medical facilities exhibit a quantity drive, their low priority status results in relatively hard budget constraints, modest supply acquisition plans, and difficulties in building up inventories of inputs. Despite the moderate nature of medical institution demands, however, disruptions and shortages in intermediate goods markets are likely to be caused by suppliers underfulfilling agreed delivery plans to a low priority institution and by the siphoning off of commonly used goods by firms with a soft budget constraint.

In assessing the validity of these revised shortage model hypotheses for the Soviet medical system during 1970–85 attention is restricted to the demand for and supply of pharmaceuticals, which are crucial inputs into the production of medical services. Available evidence suggests that the demands of medical institutions for pharmaceutical products were restrained by the central authorities in several ways (Davis, 1983a, 1984, 1985, 1987c). First, the medical facility activity levels that were used in budget calculations corresponded to official capacity utilization norms, not to the higher actual usage rates elicited by shortages in the medical care market. Second, the financial norms that linked permissible budget expenditure on medicines to activity indicators (e.g. outpatient visits, hospital bed-days) were low relative to prevailing prices (see section 17.2.2(d) and Zagalskii, 1985). Through this manipulation of methodology the authorities were able to formulate tight budgets governing medical system expenditures on pharmaceuticals and modest plans for medicine acquisition.

Despite the understated nature of medical system demand for pharmaceuticals, the Soviet medical industry chronically underfulfilled its supply contracts with respect to quantities and assortment of medicaments. There were numerous articles that documented these shortfalls in the seventies (Tak pochemu, 1976; Melnichenko, 1976; Chelovek, 1978; O lekarstvakh, 1979). *Pravda* reported that in 1979 the health service did not receive full, planned amounts of 226 medicines and 209 items of medical equipment and that some orders were only 58–60% satisfied (Industriya, 1980). The supply plan underfulfilment problem continued into the eighties according to numerous sources. For example, a 1985 article stated that:

> In the Russian Federation there is a very difficult situation with respect to the provision of medical establishments and the population with medicaments and medical goods. Enterprises are fulfilling the orders of the health service on average by 70–75%. For a series of the most important and widely used medicines, the fulfilment is even lower: antibiotics 50–70%; fermented preparations 30–70%; preparations for treatment of cardiovascular illness 30–70% (Zagalskii, 1985).

Pharmaceutical supply problems were exacerbated by the process of siphoning. However, this operated in a somewhat more complex manner than Kornai suggests. First, higher quality medicaments were siphoned off from the public medical system by closed medical subsystems (e.g. those of the Ministry of Railroads) that had softer budget constraints. Second, there was siphoning from the medical system wholesale market to the medical goods retail market. This was due to the facts that there was considerable pent-up consumer demand for medicaments and that pharmacies paid greater attention to retail sales, which generated profits, than to wholesale trade, which did not (Na perelome, 1987).

As a result of these market conditions there were widespread shortages of inputs of pharmaceuticals to the medical system (Davis, 1983a, 1987c). For example, a 1986 article reported that:

> For a series of preparations requirements are not fully satisfied. There is insufficient medicine for treatment of the digestive organs, long-lasting forms of insulin, types of medicines for infants, and polysynthetic antibiotics.

> Deliveries are made irregularly, and sometimes fall through completely. In the past year, for example, pharmacies in the Gorkovskaya oblast did not receive 40% of their cardio-vascular preparations (Uluchshat, 1986).

This finding is consistent with Kornai's hypothesis. On the other hand, the stocks of medicines held by Soviet medical facilities appear to have been kept low due to regular above-plan output of medical services and below-plan receipts of pharmaceutical goods. This situation represents a deviation from the predicted input inventory behaviour of firms. As in the case of investment, the shortages in intermediate input markets appear to be more a result of the low priority of medical institutions and siphoning than of soft budget constraints and almost insatiable demands.

17.3.3 Production within the socialist medical institution

(a) *Technological level and innovation*

Over the past several decades the organization, effectiveness and costs of Western health systems have been significantly influenced by the substantial progress made in biomedical sciences, clinical techniques and medical technology (Fuchs, 1974; Cullis and West, 1979; Aaron and Schwartz, 1984). Major breakthroughs in theoretical and applied sciences have clarified the causes and processes of diseases and have identified potential cures. Western pharmaceutical and medical equipment companies have made effective use of new discoveries and have been responsible for an acceleration in technological innovation in processes and products. Advances in science, technology and techniques have diffused rapidly between and within Western countries due to: the existence of relatively open channels for information and trade flows; the pull of growing demands by patients and medical staff for the best possible medical care; and the push given by the innovation and marketing activities of the medical industry.

Some analysts believe that rapid technological progress is a universal phenomena and is causing a convergence in the organizational features and performance characteristics of the medical systems in all countries. The shortage model, however, would suggest that this is not the case. Kornai argues that one of the most adverse effects of chronic shortage on economic performance is the inhibition of the producer's incentive to engage in quality-enhancing technological innovation (Kornai, 1959, 1971, 1980 and section 17.3.1(c)). If this is correct, then one would hypothesize that a socialist medical system operating in a shortage environment would have a relatively low technological level and a slow pace of technological innovation.

The technological level of Soviet medical institutions can be defined as the set of techniques used to produce medical services. Technological innovation refers to the upgrading of the set of techniques through the improvement of existing elements or addition of new ones (process innovation) and to the introduction of new products (product innovation) (Berliner, 1976, Chapter 1). In order to understand the state of and developments in medical technology it is necessary to examine consumers' demand, medical industry supply and the internal behaviour of medical facilities.

In the case of demand, consumers had a minimal impact on technological innovation in medical institutions. Among the reasons for this were: lack of knowledge about available techniques, especially those in the West; satisfaction at receiving medical care based on relatively outdated technologies given the shortage environment; willingness to accept the explanations and decisions of medical authorities in the areas of diagnosis and treatment; and inability to influence production decisions through price signals in the market for medical care (Davis, 1987a).

Technological development in the medical system was impeded on the supply side by deficiencies in medical technology domestic distribution, production and foreign trade (Davis, 1984, 1985, 1987a, 1987c). Soviet sources indicate that the agency which managed wholesale trade in medical equipment and instruments, *Soyuzmedtekhnika*, had many shortcomings, such as inaccurate projections of demand, inadequate storage facilities, inappropriate distribution of products between regions and institutions, and poor after-sales repair service and spare parts supply (Utverzhdat, 1986). The Soviet medical industry contributed to problems in technology supply as well. During 1970–85 industrial enterprises chronically underfulfilled plans for technology deliveries to the health system. At a January 1987 meeting of the Ministry of Health USSR Collegium it was stated that the medical system's orders for medical equipment were only 75–80% satisfied (Trudnye, 1987). The insufficient production of goods by the Soviet medical industry was caused, in turn, by problems associated with its own low priority status in a shortage economy: low technological level, inadequate innovation, and erratic supplies from other enterprises (Davis, 1987c). Foreign trade organizations also contributed to difficulties in the supply of medical technology in that they did not import sufficient quantities of machinery, equipment, instruments, patents, and licenses to alleviate domestic shortages.

Given the conditions in output and input markets, Soviet medical institutions had neither the incentives nor the means to promote technological progress. Section 17.3.1(b) argued that medical facility managers primarily attempted to raise the quantity of services provided in order to reduce shortages in the medical care market rather than, as in the West, to increase quality on the basis of new technologies. Furthermore, there was no pressure for doctors to engage in resource intensive defensive medicine. Staff were not worried about consumer complaints and concentrated on working competently with existing technology, even if it was outdated. Among other factors that inhibited technological development in the medical system were tight constraints on capital acquisition budgets and insufficient supplies of medical machinery and equipment.

As a consequence of these factors the technological level of the Soviet medical system was low by international standards. In 1987 the Ministry of Health provided a confirmation of this point:

> If in the projected costs of our hospitals the share in value terms of medical and technological equipment comprises on average 15%, then in the establishments of other member countries of the CMEA this share is 40% (Na perelome, 1987).

Most medical establishments in the USSR were short of machinery and equipment of all kinds, ranging from sophisticated technologies for diagnosis, such as computer aided tomography and other advanced X-ray equipment, to basic technologies such as stethoscopes, thermometers, and syringes. Few disposable medical products of glass, plastic or paper were available either. Existing equipment was often utilized in rooms that were designed for other functions and did not meet modern hygienic standards. Much of the machinery was obsolete due to low replacement rates, which in turn were a function of the stingy budget norms mentioned above. In addition, maintenance of existing medical technology was made difficult by shortages of engineering staff and spare parts.

The evidence of deficiencies in technological levels and innovation in Soviet medical institutions is consistent with related hypotheses of Kornai's. However, other enterprises and branches of the economy in the USSR achieved technological levels and rates of progress that were relatively closer to Western standards. It may therefore be the case that the acute technological problems of medical institutions were the result of low priority status amplifying the general processes outlined in the shortage model.

(b) *Production bottlenecks and forced substitution*

The firm in a resource-constrained economy often encounters difficulties in acquiring all the inputs called for by its production plan (Kornai, 1980, chapter 2). As a result, certain input constraints become binding earlier than scheduled, production bottlenecks develop, and the enterprise has to engage in alteration in the composition of output, reduction in the volume or quality of output, and forced substitution of inputs. Kornai suggests that similar problems will develop in budget-financed institutions such as medical facilities because of chronic shortages in their input markets.

The experiences of Soviet medical institutions during 1970–85 appear to have been consistent with Kornai's hypotheses concerning production. Bottlenecks in medical service production developed frequently due to disruptions in the planned supply of labour, capital, and intermediate goods. In the case of labour there were growing shortages of middle medical personnel relative to doctors; by the late 1970s there was a deficit of 600–900 thousand nurses and feldshers. This often undermined the work of medical teams because of the unavailability of experienced and skilled nurses. In polyclinics and hospitals production bottlenecks often existed in diagnostic units, laboratories, operating theatres, and treatment departments. These were caused by unanticipated shortages of staff, machinery, anesthetics, drugs, or other commodities. For example, in the 1970s the insufficient supply of film by industry seriously impeded X-ray diagnosis of patients in medical facilities.

Forced substitution was a regular response to shortages (Davis, 1983a). In urban medical facilities, where there were reasonably full complements of doctors but deficits of middle medical personnel, doctors performed the work of their missing and less-skilled subordinates. The situation in rural medical institutions was different because many doctor positions were chronically unfilled. This obliged feldshers to provide patients with medical care instead of doctors. Throughout the medical system labour was substituted for capital due to shortages or malfunctioning of technology. Deficiencies in the supply of office equipment meant that medical staff had to expend excessive amounts of time on paperwork. In numerous cases doctors were forced to use old-fashioned and time-consuming diagnostic and treatment techniques because of shortages of technology. Another form of substitution was of an inferior input for a planned one. In the case of medicines this behaviour was caused by either underfulfilment of supply plans by industry or the need to stay within hard budget constraints by using a cheaper but less effective product.

17.4 CONCLUSIONS

In the shortage model analytical attention is shifted from its traditional focus on central planners to subordinate institutions and endogenous processes in the economy. Kornai believes that the behaviour of firms is the primary cause of the reproduction of the shortage phenomena that characterize the socialist economy. The central authorites are assumed to be actors of secondary importance, to have a paternalistic attitude to subordinates, and to tolerate the firms' quantity drive, investment hunger, and slack budget discipline. Available evidence suggests, however, that the role of central planners is greater and more varied than that implied by the original shortage model. In particular, it appears that the priorities of decision-makers concerning resource allocation are major determinants of sectoral and institutional behaviour in the socialist economy.

The examination of priority in section 17.2.1 showed that the central authorities' priorities are expressed during both plan formulation and implementation in a variety of ways. Among these are: the structure of or weights in preference functions; responsiveness to violations of sectoral performance norms; wage rates; financial norms; the fulfilment of initial output, investment and supply plans; hardness of budget constraints; and degree of sectoral shortage intensity.

The role and impact of sectoral priority was then empirically evaluated on the basis of a case study of the Soviet medical system in the period 1965–85. One finding in section 17.2.2 was that by almost all criteria the medical system had a low priority ranking in the USSR. A second was that some features of medical institution behaviour consistent with Kornai's theory could also be explained as a by-product of this low priority. In contrast, other actual sectoral characteristics generated by low priority status were in conflict with those of the shortage model. This made it necessary to examine more closely the validity of the shortage model of the firm.

In section 17.3 ten hypotheses associated with Kornai's model of the firm were evaluated using Soviet data about the characteristics and behaviour of medical institutions. It was found that almost all were consistent with observable phenomena: shortages in output markets, quantity drive, low quality output, technological problems, forced substitution, production bottlenecks, investment tension, and shortages in input markets. A major exception appeared to exist with respect to the budget constraint. Soviet medical institutions had relatively hard budget constraints rather than soft ones. Given the importance attributed to the concept of the soft budget constraint in Kornai's work this inconsistency raised questions about the validity of the original formulation of the shortage model.

Although it is difficult to generalize from one case study, the evidence from the examination of the medical system suggests that central planners play a more active role than postulated in Kornai's model through their priority-linked resource allocations. It also appears that the production sphere of a shortage economy should be partitioned into low, medium and high priority sectors. Low priority sectors are characterized by a high shortage intensity and a relatively hard budget constraint. In contrast, a high priority sector is minimally afflicted by shortages and has a soft budget constraint (Davis, 1988a). Contrary to the implications of *Economics of Shortage* the soft budget constraint is not a universal feature of the shortage economy. One should not expect to observe both pervasive shortages and a soft budget constraint in the same sector. On the other hand, shortage-related phenomena such as investment tension and deficits in supplies exist even in sectors with hard budget constraints because of the impact of siphoning of resources from low to high priority sectors. In sum, the incorporation of the concept of priority in the shortage model would both enrich its description of institutional behaviour in the socialist economy and eliminate certain inconsistencies evident in the Kornai initial formulation.

ACKNOWLEDGEMENTS

I would like to thank the following scholars for their comments on earlier drafts of this chapter: Martin Cave, Wojciech Charemza, Gregory Hildebrandt, János Kornai, John Litwack and Mark Schaffer. The preparation of this chapter was supported by grants from the Economic and Social Research Council (B00232171) and the PEW Charitable Trust.

REFERENCES AND FURTHER READING

Aaron, H. J. and Schwartz, W. B. (1984) *The Painful Prescription: Rationing Hospital Care*, The Brookings Institution, Washington DC.

Acton, J. P. (1975) Non-monetary factors in the demand for medical services: some empirical evidence. *Journal of Political Economy*, **83** (3).

Babanovskii, I. V. (1976) *Voprosy Finansirovaniya Zdravookhraneniya v SSSR*, Meditsina, Moscow.

Basu, K. (1980) *Revealed Preference of Government*, Cambridge University Press, Cambridge.

Berliner, J. (1976) *The Innovation Decision in Soviet Industry*, MIT Press, Cambridge.

Burenkov, S. P., Golovteev, V. V. and Korchagin, V. P. (1979) *Sotsialisticheskoe Zdravookhranenie: Zadachi, Resursy, Perspektivy Razvitiya*, Meditsina Moscow.

'Chelovek obratilsya v apteku' (1978) *Pravda*, 15 September.

Chernyak, A. (1986) Zdorovye dorozhe bogatstva. *Pravda*, 15 October.

Cooper, M. H. (1975) *Rationing Health Care*, Croom Helm, London.

Cullis, J. and West, P. A. (1979) *The Economics of Health: An Introduction*, Martin Robertson, Oxford.

Davis, C. (1979) The economics of the Soviet health system, Dissertation submitted to Cambridge University, for the PhD in Economics, Cambridge.

Davis, C. (1983a) The economics of the Soviet health system, in *Soviet Economy in the 1980s: Problems and Prospects*, US Congress Joint Economic Committee USGPO, Washington DC.

Davis, C. (1983b) Economic problems of the Soviet health service: 1917–1930. *Soviet Studies*, **35**, 343–61.

Davis, C. (1984) *The Medical and Pharmaceutical Sectors of the Soviet Economy*, Wharton Econometric Forecasting Associates, Special Report, Washington DC.

Davis, C. (1985) *Opportunities in the Soviet Pharmaceutical Market*, Scrip Country Report London.

Davis, C. (1987a) The Soviet health system: a national health service in a socialist society, in M. Field (ed.) *Success and Crisis in National Health Systems: A Cross National Approach*. Tavistock Press, London (to be published).

Davis, C. (1987b) The Soviet medical system during 1928–32: development strategy, resource constraints and health plans, in S. Gross Solomon and J. F. Hutchinson (eds) *Social Medicine in Revolutionary Russia*. Indiana University Press, Bloomington (to be published).

Davis, C. (1987c) Developments in the health sector of the Soviet economy, 1970–90, in *Gorbachev's Economic Plans* US Congress, Joint Economic Committee USGPO, Washington DC.

Davis, C. (1988a) The high priority defense industry in the Soviet shortage economy, in M. Raven and C. Wolf (eds) *The Defense Sector in the Soviet Economy*, Institute of International Studies, Berkeley (to be published).

Davis, C. (1988b) The second economy in disequilibrium and shortage models of centrally planned economies. *Berkeley-Duke Occasional Papers on the Second Economy in the USSR*, No. **12** (July).

Davis, C. and Feshbach, M. (1980) *Rising Infant Mortality in the USSR in the 1970s*, Bureau of the Census Report, Series P-95, No. 74, Washington DC.

Deaton, A. and Muellbauer, J. (1980) *Economics and Consumer Behaviour*, Cambridge University Press, Cambridge.

Dutton, J. (1979) Changes in Soviet mortality patterns, 1959–77. *Population and Development Review*, **5**, 267–91.

Ericson, R. C. (1988) Priority, duality and penetration in the Soviet command economy, *The RAND Corporation*, N–2643, Santa Monica.

Feshbach, M. (1983) Issues in health problems, in *Soviet Economy in the 1980s: Problems and Prospects* US Congress, Joint Economic Committee, USGPO, Washington DC.

Fuchs, V. R. (1974) *Who Shall Live?: Health, Economics and Social Choice*, Basic Books, New York.

Gomulka, S. (1985) Kornai's soft budget constraint and the shortage phenomenon: a criticism and restatement. *Economics of Planning*, **19** (1).

Gregory, P. R. and Stuart, R. C. (1981) *Soviet Economic Structure and Performance*, Harper and Row, New York.

'Industriya zdorovye' (1980) *Pravda*, 9 February.
Kaser, M. (1976) *Health Care in the Soviet Union and Eastern Europe*, Croom Helm, London.
Kemme, D. M. (1987) Thoughts on the chronic shortage model, Paper presented at the Conference on Modelling of Disequilibrium and Shortage in Centrally Planned Economies, Birmingham, 6–8 April.
Kornai, J. (1959) *Overcentralisation of Economic Administration*, Oxford University Press, Oxford.
Kornai, J. (1971) *Anti-Equilibrium*, North-Holland, Amsterdam.
Kornai, J. (1980) *Economics of Shortage*, North-Holland, Amsterdam.
Kornai, J. (1982a) Adjustment to price and quantity signals in a socialist economy. *Economie appliquée*, **35**, (3).
Kornai, J. (1982b) *Growth, Shortage and Efficiency*, Basil Blackwell, Oxford.
Kornai, J. (1986) The soft budget constraint. *Kyklos*, **39** (1).
Lacko, M. (1984) Behavioral rules in the distribution of sectoral investments in Hungary, 1951–1980. *Journal of Comparative Economics*, **8**, 290–300.
Lazarcik, G. (1987) Personal communication to the author about East European NMP utilized and health expenditure statistics on 22 October.
Melnichenko, A. K. (1976) Pyatiletka meditsinskoi promyshlennosti. *Izvestiya*, 1 August.
'Na perelome' (1987) *Meditsinskaya Gazeta*, 25 March.
Narodnoe Khozyaistvo SSSR v 1965 g...v 1985 g, (1966–1986) Finansy i Statistika, Moscow.
'O lekarstvakh' (1979) *Izvestiya*, 23 June.
'Pochemu ochered' v polikliniku' (1979) *Sovetskaya Rossiya*, 21 March.
Popov, G. A. (1976) *Ekonomika i Planirovanie Zdravookhraneniya*, Izdatelstvo Moskovskogo Universiteta, Moscow.
'Sluzhba zdorov'ya trebuyet rekonstruktsii' (1987) *Meditsinskaya Gazeta*, 27 February.
'Tak pochemu raster "defektura"?' (1976) *Literaturnaya Gazeta*, 5 May.
'Trudnye shagi perestroika' (1987) *Meditsinskaya Gazeta*, 23 January.
'Uluchshat' lekarstvennoe obespechenie' (1986) *Meditsinskaya Gazeta*, 14 March.
USSR: Measures of Economic Growth and Development, 1950–80, (1982) US Congress, Joint Economic Committee, USGPO, Washington DC.
'Utverzhdat delom' (1986) *Meditsinskaya Gazeta*, 15 October.
Varian, H. (1984) *Microeconomic Analysis* (2nd edn), W. W. Norton and Co., London.
'Vashe mnenie, tovarishch ministr?' (1986) *Meditsinskaya Gazeta*, 5 February.
Zagalskii, L. (1985) Tabletki dlya Zolushki. *Meditsinskaya Gazeta*, **29** November.
Zagalskii, L. (1986) Tabletki pod podushkoi. *Meditsinskaya Gazeta*, **16** April.

18

Conclusions and future prospects

WOJCIECH CHAREMZA
and
CHRISTOPHER DAVIS

The diversity of current work and dynamism in the field of quantitative modelling of disequilibrium and shortage in centrally planned economies are reflected in the contributions made in this volume, which clarify theoretical concepts, resolve disputes between modelling schools, amplify existing disequilibrium and shortage models, improve estimation techniques, and uncover sources of data. Summaries are presented of the main findings related to shortage models in section 18.1 and disequilibrium models in section 18.2. This is followed by a reassessment of the contrasts and similarities between the disequilibrium and shortage models in light of the new research. Finally, section 18.4 identifies a number of topics that have been inadequately examined in past disequilibrium and shortage model studies but merit further investigation because of their importance. These include the issue of aggregation, cross-section disequilibrium analysis, the role of plans and expectations in models, the second economy in a CPE, data development, and extension of the coverage of disequilibrium and shortage models to include neglected countries and sectors.

18.1 THE SHORTAGE MODEL

Numerous contributions are made in the book to the theoretical development and application of the shortage model. It is generally agreed that Kornai's original ideas expressed in the *Economics of Shortage* and subsequent works are interesting and insightful, but difficult to employ in rigorous empirical modelling. Attempts

International Studies in Economic Modelling Series.
Models of Disequilibrium and Shortage in Centrally Planned Economies.
Edited by C. Davis and W. Charemza.
Published in 1989 by Chapman and Hall Ltd, 11 New Fetter Lane, London EC4P 4EE.

therefore are made to remedy this deficiency by clarifying concepts, identifying and introducing into models measurable shortage-related variables, and developing appropriate data sources.

Hare (Chapter 3) shows that there has been an evolution of Kornai's concept of economic imbalance and related indicators. In *Anti-Equilibrium* Kornai used the notion of 'suction' to describe conditions in a single market with buyers queueing for sellers. However, it proved difficult to measure suction even in simple cases. Kornai introduced the alternative concept of 'shortage' in *Economics of Shortage*, which referred to a multidimensional state of imbalance that he believed should be described by a variety of indicators (e.g. length of queues, search time, forced substitution, forced saving) not just the conventional univariate measure of excess demand. This multivariate approach to shortage measurement, employed by Davis (Chapter 17) to describe conditions in the Soviet medical system, can be helpful but usually involves some arbitrariness in the choice of shortage vector elements and in cross-section and temporal comparisons of shortages. Perhaps in response to these problems Kornai adopted an aggregated 'macro index of shortage' in *Growth, Shortage and Efficiency* that reflects the movement of its component partial shortage indicators and is an important explanatory variable in his dynamic model of the socialist economy. Lackó (Chapter 10) surveys several models of investment and foreign trade that are based upon single shortage variables.

Kornai's ideas concerning the soft budget constraint firms are reviewed by Hare (Chapter 3), Kemme (Chapter 4), and Lackó (Chapter 10). Hare argues that the soft budget constraint concept is not adequately modelled in a formal sense or fully supported by empirical evidence and summarizes criticisms of it by Gomulka and Soós. He believes that the importance of the soft budget constraint in explaining the reproduction of shortage has been overstated and that more attention should be paid to factors such as lack of competition, price inflexibility, and high adjustment friction. Davis (Chapter 17) makes the related points that not all institutions or sectors of the socialist economy have soft budget constraints and that sufficient conditions of the reproduction of shortage are the existence of a soft budget constraint in at least one sector (usually that possessing a high priority) and siphoning of resources from hard to soft budget constraint sectors.

Another controversial notion of Kornai's is that of the 'almost insatiable demand' of firms for inputs, which is generated by the quantity drive and the soft budget constraint. Lackó (Chapter 10) accepts the idea of boundlessly growing demand whereas both Hare (Chapter 3) and Nuti (Chapter 5) dispute it on theoretical grounds. Davis (Chapter 17) argues on the basis of empirical findings that low priority institutions in the production sphere of a socialist economy can have hard budget constraints and consequently have finite demands for labour, capital, and intermediate goods.

Hare (Chapter 3) and Lackó (Chapter 10) suggest that there are substantial difficulties in elaborating the macrolevel shortage model theoretically and that controversies surround the issue of aggregation of shortage-related variables. Given this, at the present stage of its development the shortage model might be most beneficially utilized empirically in microeconomic investigations of single markets and of the behaviour of specific institutions. Examples of these are the studies by Kapitány, Kornai, and Sźabo (Chapter 15) of the automobile market in Hungary and by Davis (Chapter 17) of the service-producing medical institution in the Soviet economy.

The findings of contributors indicate that the role of planners in the socialist economy is more significant than implied in the traditional shortage model. Hare (Chapter 3) stresses the importance in shortage model theory of the vertical planning hierarchy in guiding the economy through 'control by norms' and in explaining the normal intensity of shortage. Kemme (Chapter 4) discusses in several places the responsibility of central planners for the chronic shortages in consumer and producer commodity markets in CPEs. According to Lackó (Chapter 10), the actions of planners are important determinants of the intensity of shortages in investment and foreign trade. Kapitány, Kornai and Sźabo (Chapter 15) point out that planners determine the supply of automobiles, which in turn affects the degree of shortage in that market. Finally, Davis (Chapter 17) claims that the central authorities' priorities with respect to end-uses of national income, sectors, and programmes exert strong influence on microeconomic conditions. As a general rule, there is an inverse relationship between the priority ranking and intensity of shortage in a sector.

Hare (Chapter 3) criticizes Kornai's shortage model investigations for their inadequate empirical support of theoretical ideas and arguments. Lackó (Chapter 10) admits this has been a problem, but shows that there has been progress in developing data on shortage-related variables in models of investment, foreign trade, and consumption and in carrying out econometric estimation and hypothesis testing. The studies of Kapitány, Kornai and Sźabo (Chapter 15) and Davis (Chapter 17) demonstrate that data can be found to evaluate relationships specified in microeconomic shortage models. Despite this, it is noticeable that the data collection and processing, estimation, and hypothesis testing techniques of the shortage modellers are less advanced than those of analysts working with disequilibrium models.

The contributions in this volume suggest that there has been a considerable evolution in the shortage model from its initial formulation by Kornai. Various theoretical concepts have been added, dropped, or modified in attempts to clarify the causes and consequences of shortage in CPEs. In the process shortage models have acquired greater diversity, while remaining linked to the original work of Kornai.

18.2 THE DISEQUILIBRIUM MODEL

The Introduction in Chapter 1 explained that disequilibrium econometric modellers can be roughly divided into the known excess demand (or chronic shortage) group, which accepts the existence of permanent positive excess demand in markets of centrally planned economies, and the testable excess demand group (headed by Richard Portes), which believes that despite microeconomic evidence of shortage phenomena such as queueing and rationing one should carry out statistical testing of disequilibrium hypotheses. The contributions to the volume have helped to resolve numerous points of contention and to clarify the nature of the remaining disputes between these alternative approaches. As a result, disequilibrium modellers are leaving entrenched, extreme positions to seek compromises with each other.

A review of the arguments indicates that the known excess demand and Portes factions are at present still divided by their differing degrees of belief in current statistical methods of verification of the excess demand hypothesis. Indirect or explicit discussions of this issue are presented by Portes (Chapter 2), Nuti (Chapter 5), Quandt (Chapter 6), and Charemza (Chapter 11). It has become clear that the Portes school partisans are not arguing that they believe excess supply regimes were frequently present in Eastern European consumption markets in the seventies. Instead they are saying that if a given statistical test indicates the presence of excess supply, then they are ready to accept this result. In contrast, their opponents strongly believe that chronic excess demand exists in CPEs and argue that the Portes group econometric models used to generate contradictory findings are imperfectly specified and estimated.

The theoretical and econometric arguments on both sides essentially concern the problem of model selection. Put simply, the problem at present is that if one analyst considers that the results of model A are more consistent with a priori expectations concerning disequilibrium than those generated by model B, then model A is selected. Clearly this is not the optimal approach to empirical econometric modelling. Current subjective practices indicate that there is an important deficiency in the statistical methodology of disequilibrium modelling, namely the absence of criteria or techniques that would permit the objective selection of the superior model from a set of comparable ones.

One difficulty in remedying this situation is that the methods used in traditional econometrics to select the best among non-nested models cannot be directly adapted to the case of disequilibrium modelling in CPEs because the crucial dependent variable (whichever is the greater of demand or supply) may not be observable. However, it should be possible to develop suitable disequilibrium model selection procedures through adoption of model encompassing techniques and unification of statistical data.

The convergence in the work of disequilibrium modellers is apparent if one reviews developments in disequilibrium indicator models, which traditionally reflected the chronic excess demand hypothesis and were constructed over the past decade primarily by East European scholars (see the reviews of this type of model in the chapters by Dlouhý (7), W. Welfe (8), Hulyak (9), Charemza (11) and A. Welfe (12)). In the initial studies there was virtually no testing of hypotheses about the sign of excess demand, apart from the t-Student test applied to estimated coefficients of disequilibrium indicator variables. The model builder usually determined not only the sign of excess demand but also its magnitude through the selection of disequilibrium indicators. Although this practice initially was forced on the analysts by the deficiencies in computer hardware and software in Eastern Europe, academic conservatism maintained traditional approaches to disequilibrium indicator modelling even after computational constraints were eased.

Substantial recent developments in disequilibrium indicator modelling are reflected in the chapters of this book. First, analysts have taken advantage of new estimation methods, reviewed by Quandt (Chapter 6), that allow them to relax constraints on model specification. This is a by-product not only of the spread of the computer revolution to Eastern Europe, but also of closer international scientific cooperation. For example, at the current time it is possible for a scholar in Budapest to estimate disequilibrium models using advanced computers with sophisticated estimation techniques in the USA or Great Britain. Second, the authors express scepticism about the value of traditional deterministic univariate disequilibrium indicators and introduce multivariate forms. A. Welfe (Chapter 12) uses a multivariate deterministic disequilibrium indicator model in his investigation of consumption in Poland whereas Charemza (Chapter 11) and Charemza, Gronicki and Quandt (Chapter 116) use stochastic multivariate disequilibrium indicators.

Other chapters of the book explore alternative methods of constructing disequilibrium models of centrally planned economies. Grosfeld (Chapter 14) has developed a novel way of applying discrete choice modelling in her work on investment. Brada and King (Chapter 13) have introduced a latent variable, or state-space, model of foreign trade which is estimated using Kalman filter techniques. One advantage of this approach is that it does not require that all the variables of the model be observable and allows gaps in data and changes in data frequency. At present there are still some theoretical and empirical problems which have to be solved before Kalman filtering can be used to estimate larger and more complicated state-space models, but these are likely to be overcome in the future.

Significant progress is achieved in explicitly portraying the second economy in disequilibrium models. In the past, attempts have been made to analyse theoretically legal and illegal private markets and activities in the socialist economy and their equilibrating roles, but little has been done to model them econometrically. Nuti (Chapter 5) outlines a monetary model of the official and unofficial markets in a CPE, with excess demand in the former and flexible, equilibrating prices in the latter. Charemza (Chapter 11) develops a disequilibrium econometric model of the consumption sphere in Poland that analyses the spillover effects of shortages in official retail markets on the second economy hard currency and consumer commodity markets. The implications for consumer behaviour of the existence of the second economy are analysed in a similar way by Charemza, Gronicki, and Quandt (Chapter 16) in their model of the Polish automobile market.

Advances are made as well in developing appropriate estimation techniques for disequilibrium models of CPEs. Quandt (Chapter 6) makes several fundamental contributions. Of special importance are his rigorous formulation of econometric theory as it relates to disequilibrium modelling of socialist economies and his technique for estimating Portes-type testable models in the known excess demand situation. Other novel approaches to estimation are adopted by Grosfeld (Chapter 14), who uses a switching regression technique in her study of investment, and by Brada and King (Chapter 13), who introduce Kalman filter estimation of a disequilibrium state-space model of foreign trade in a CPE. Finally, Charemza (Chapter 11) shows how a standard package full-information maximum likelihood method can be used to estimate a multi-market disequilibrium model.

Dlouhý (Chapter 7), W. Welfe (Chapter 8), and Hulyak (Chapter 9) critically review the existing macroeconomic disequilibrium models for, respectively, Czechoslovakia, Poland, and Hungary and in each case make significant original contributions. Their surveys and others in the volume make it clear that the modelling effort to date has focused on these three countries. In contrast, there is minimal coverage in disequilibrium models of the important and distinctive economies of the USSR, GDR, Bulgaria, and Romania.

18.3 A REASSESSMENT OF THE RELATIONSHIP BETWEEN SHORTAGE AND DISEQUILIBRIUM MODELS

The findings in the chapters of this volume lead to re-evaluation of some conventional opinions concerning the contrasts between disequilibrium and shortage models of CPEs. First of all, there is considerable heterogeneity within the two modelling schools. Until recently the common assessment has been that shortage models are highly homogeneous and faithful to the original formulations of Kornai, whereas the disequilibrium modellers are divided into the testable and known excess demand camps and disagree fundamentally with each other. However, it is argued in section 18.1 that shortage models have become more diverse and in section 18.2 that many apparent conflicts between the disequilibrium modellers have been resolved. Secondly, it appears that recent developments in the theoretical and empirical work of the two schools have made their differences less acute than suggested in Table 1.1 of the Introduction.

In the case of the concept of aggregate demand, there has clearly been a convergence of opinion between the two groups. The shortage modellers have recognized the need for aggregate indicators that summarize the effects and trends of shortage. This is evident from the contribution by Lackó in Chapter 10. On the other hand, there is growing dissatisfaction within the disequilibrium school with the use of the Keynesian concept of aggregate demand in a CPE context (see Kemme, Chapter 4). The Burkett study discussed by Charemza (Chapter 11) attempts to bridge the gap between the insufficiently aggregated shortage models and the possibly over-aggregated disequilibrium models.

In the past, analysts using the shortage and the known excess demand (disequilibrium indicators) models were prepared to accept the existence of permanent shortages whereas those using the Portes-type models were not. At present, though, there is a more general recognition of the need to test rigorously hypotheses concerning the nature of the disequilibrium regime. As Quandt (Chapter 6) has shown, there also new are ways of dealing with problems of disequilibrium model identification and estimation in a chronic excess demand environment. So there are now fewer disagreements between analysts on this issue.

The treatment of prices by the two different types of modellers has converged as well. Lackó (Chapter 10) shows that shortage models can use prices as important market signal variables that supplement traditional quantity indicators. On the other hand, current disequilibrium models recognize that the character and influence of prices is variable. Depending upon circumstances prices can be regarded as fixed (for example, in official retail markets), as subject to long-run inflationary feedback (see W. Welfe, Chapter 8), or as freely negotiated (in the case of the second economy discussed by Nuti (Chapter 5) and Charemza (Chapter 11)).

Both disequilibrium and shortage models are paying more attention to the impact of planning on economic processes and are portraying the role of plans in a similar manner. For example, some shortage models described by Lackó (Chapter 10) use plans as explicit econometric variables, as does the Charemza–Gronicki disequilibrium model reviewed by W. Welfe (Chapter 8). Plans in these models can either act as policy instruments or determine targets.

Compromises between the two schools can be perceived with regard to the contentious issue of the existence of the soft budget constraint. Davis (Chapter 17) argues that the soft budget constraint is not a necessary condition for the reproduction of shortages if one considers a CPE disaggregated into sectors with differing priorities. In the disequilibrium modelling camp, Quandt (Chapter 6) discusses a disequilibrium model for a firm under central planning with a bailout mechanism, which can be viewed as the equivalent of a soft budget constraint condition.

18.4 THE AGENDA OF FUTURE RESEARCH

Despite the numerous advances made in the disequilibrium and shortage modelling of centrally planned economies, many old issues remain unresolved and several important new ones have emerged in the chapters of this book. This section surveys these problems, which probably will constitute the agenda of future research by disequilibrium and shortage modellers.

The first problem is the traditional one of aggregation in macroeconomic disequilibrium econometric models. As Hare (Chapter 3) points out, Kornai strongly criticized the Barro–Grossman–Portes models for their treatment of aggregation of variables, especially their conceptualization and measurement of excess demand. New arguments in support of Kornai are advanced by Kemme (Chapter 4). The summary message of this critique is that disequilibrium econometric macromodels should be abandoned in favour of disaggregated ones.

Whatever the merits of this criticism, however, analysts will continue to build the aggregated disequilibrium models for centrally planned economies because of the strong demand for concise, simple descriptions of disequilibrium. Politicians, managers, scholars, and journalists would like to know, even approximately, the magnitude of imbalance in a sector or in the entire economy and the extent to which it has changed over time. (This situation is similar to that of the GNP indicator. Despite the well known deficiencies of this measure of economic output and growth it remains in use.) Given the demands for summary disequilibrium indicators and their continuing supply, the objectives of analysts should be to improve their information content and to reduce their aggregation bias.

There are two possible ways of dealing with the aggregation problem. First, one could initially build large, disaggregated models, compute unbiased equilibrium prices and forced saving, and then aggregate *ex post* (see A. Welfe in Chapter 12). Second, one could elaborate a macromodel that was designed in advance to reduce aggregation bias and then estimate it. However, both approaches have disadvantages. The first one requires detailed statistical data because the model would need several spill over and resource constraint variables. In the case of stochastic disequilibrium indicator models, the information requirements could make estimation intractable even with modern computers. An additional problem is that such a model would have so many associated hypotheses that it would be virtually impossible to test all of them. The second approach is theoretically more difficult. In recent papers connected with the *ex ante* aggregation problem in disequilibrium models the required assumptions are particularly stringent: e.g. that the spillover effects on all micromarkets are identical (on the other hand, the present disaggregated disequilibrium models do not allow for any spillovers at all). In considering the options, it appears that the *ex ante* modelling approach offers greater potential. It probably will prove easier to solve the theoretical aggregation problems in small models than to formulate and test the hypotheses of disaggregated disequilibrium indicator models.

Another topic for future research is the development of cross-section disequilibrium models of centrally planned economies. Cross-section analysis would help to resolve aggregation problems by providing additional information on prices that could be used for *ex post* aggregation. It also would assist in bridging the gap between the shortage model and disequilibrium methodologies, in that the study of disaggregated economic units would not require special assumptions concerning aggregation and common research techniques could be employed. It is clear from recent research that analysts from both schools are conducting compatible investigations. Examples of this are the shortage model studies of housing in Budapest mentioned by Lackó (Chapter 10) and the Goldfeld and Quandt formal disequilibrium model of the firm with a bailout (or soft budget constraint) condition described by Quandt (Chapter 6).

The third issue that deserves attention is the use of plan and expectation variables in disequilibrium and shortage models. For over a decade plans have been explicitly incorporated in disequilibrium econometric models of centrally planned economies. Nevertheless, their roles in initiating and reproducing imbalances and shortages are still unclear. There are numerous different hypotheses about this, ranging from regarding planners' errors as the primary cause of disequilibrium, through identifying plans as expressions of the rational expectations of planners, to minimizing the responsibility of plans for disequilibrium phenomena. Similar uncertainty surrounds the treatment of planners in the shortage model. For example, Hare (Chapter 3) points out that the normal state of shortage is generated by the agents' expectations about future shortages, which in turn are influenced by the plans and policies of central authorities. Analysts vary in their assessments of the character of plans. They can be influential or neutral, endogenous or exogenous, time invariant or flexible, and expectation or decision oriented. These alternatives have not been thoroughly examined and convincingly treated in either disequilibrium or shortage models.

It is highly likely that a major topic of future research will be disequilibrium and shortage model analysis of a CPE with parallel state and private (legal and illegal) sectors. Almost all chapters of the book mention the need to study the second economy. The comprehensive survey by Davis of *The Second Economy in Disequilibrium and Shortage Models of Centrally Planned Economies*, mentioned in the Introduction, indicates that considerable theoretical and applied work has already been done on this topic but many unexplored or unresolved issues remain. In general, more research is needed on the interconnections between a state sector characterized by fixed prices, shortages (excess demand), and significant frictions of supply adjustments and second economy markets with flexible prices and supply responses. Among the more specific issues requiring investigation are the spillover effects on private markets of imbalances in official retail markets, the demand-responsiveness of second economy supply, the siphoning of commodities and labour from the first economy, and the operations of second economy barter and bribes markets. Charemza (Chapter 11) has already built a disequilibrium model of the second economy and shows that it is possible to evaluate indirectly the degree of imbalance in the state sector using data on legal and illegal private activities. It is to be anticipated that more work of this type will be carried out in the future.

A fifth issue on the research agenda is data collection and processing. Naturally, this is connected to all others. The disaggregated disequilibrium models and cross-section studies require detailed and consistent sets of data. Any analysis of the role of plans in the socialist economy must be based on time series of plan variables. In order to analyse second economy activities, new sources of data need to be developed. Finally, improved statistical information is needed to extend empirical analysis into areas that have previously only been examined theoretically, such as modelling of labour supply responses to excess demand in official consumption markets.

Some pessimistic modellers may think that little additional, suitable data are available, but in reality prospects are bright. Although desired supplemental data usually cannot be found in official statistical yearbooks, they are published elsewhere in relative abundance and of reasonable quality. The problem is not that data do not exist, but rather that they are disguised or presented in obscure publications.

The statistical offices of many East European countries often provide household survey data on money budgets and occasionally on time budgets. Some cross-section data for firms are also available. For example, detailed information about the economic performance of 500 state enterprises in Poland has been published in a monthly business magazine. Data on official plans can be found for most of the CMEA countries either in government publications or, more easily, in the United Nations annual *Economic Survey for Europe*. Obtaining data about the second economy clearly is a more serious problem. Nevertheless, Charemza (Chapter 11) shows that time series exist of the black market rates of exchange of foreign into domestic currencies. Through careful and imaginative work additional information can be found about shortage phenomena (e.g. queueing and waiting lists), labour supply, bribery, theft of state property, and so on.

It is true that much important data are often genuinely unavailable. In this situation the common practices have been either to ignore the associated topics, however significant, or to make use of poor quality, biased official data. These appear to be questionable solutions to a difficult problem. For instance, if one attempts to model money demand in a CPE without taking into account households' transaction and asset demands for foreign currencies because those data are not available, the result will be a model that is subject to systematic errors in measurement. Perhaps a more appropriate response would be to develope approximate estimates of the missing data and include obviously important variables in the model. As an example, one could use the number of foreign tourists entering the country under study as a proxy for changes in private domestic stocks of foreign currency.

It is hoped that in the future attempts will be made to organize more systematic surveys to fill gaps in the existing data sets. Some promising steps in this direction have already been made, such as the survey of households' housing conditions and rents made in Budapest in 1981, which generated several empirical shortage model papers on housing.

The final problem that should be rectified is the imbalance in the coverage of disequilibrium and shortage modelling of centrally planned economies with respect to countries and sectors of the economy. Most disequilibrium econometric models have been elaborated for Poland, only a few for Czechoslovakia and Hungary, and virtually none for Bulgaria, the GDR, Romania, and the USSR. This undoubtedly results in a distorted image of disequilibrium in Eastern Europe. Poland has unique features such as its significant private agriculture and trade sectors and liberal regulations concerning households' rights to own hard currency assets. Other CMEA countries differ substantially in their economic systems and in the roles played by central planning agencies. Given this, conclusions from a Polish disequilibrium model analysis are unlikely to be valid for other socialist economies. The extension of disequilibrium modelling to neglected countries probably would produce some interesting, and perhaps unexpected, results.

In the case of shortage models, an even more pronounced geographical bias is evident. Virtually all empirical studies are focused on Hungary. An exception to this rule is the analysis by Davis (Chapter 17) of the medical sector in the USSR.

Disequilibrium and shortage models are also unbalanced in their coverage of sectors or institutions. In the case of the shortage model, most theoretical and empirical studies analyse household behaviour as consumers and as suppliers of labour, retail markets, and the activities of an aggregated industry sector. Shortage modellers have neglected agriculture, domestic trade organizations, planners, state service-producing institutions, foreign trade, the second economy, and financial processes.

As Charemza (Chapter 11) indicates, disequilibrium modelling efforts have concentrated on consumption and retail markets. Perhaps this is due to the belief that the neoclassical theory of consumer behaviour, based on maximization of the household's utility function, can be relatively easily adapted to the conditions of a planned economy and can generate well defined consumer demand functions. In contrast, only modest advances have been made in modelling the labour market, investment, producer goods markets, foreign trade, the service sector, agriculture, and the behaviour of central economic authorities.

An obvious deficiency in the work to date is the lack of a formal disequilibrium theory of the firm in a CPE. It is difficult to uncover a clear, meaningful definition of the socialist firm, let alone a thorough theoretical analysis of industrial enterprise objectives, constraints, and behaviour. This contributes to the inadequacy of the treatment in disequilibrium model of the supply side of consumption markets. The supply functions used in econometric models tend to be based on intuition and *ad hoc* assumptions rather than on convincing theory. Perhaps this is not a major problem for disequilibrium indicator modellers, who can regard the supply of consumer goods as exogenously determined. But in Portes-type testable models the supply function plays a much more important role. Adherents to this approach therefore need to develop better models of the supply-side behaviour of firms.

The problems of defining investment demand and supply, and consequently, of formulating appropriate disequilibrium investment models, have been assessed by Grosfeld (Chapter 14). Future work will be impeded by the fact that the development of alternative disequilibrium estimation techniques based on the second economy is more difficult for investment than for consumption. Different obstacles hinder the formulation and estimation of disequilibrium models of labour market behaviour. In Western labour markets, the usual manifestation of aggregate disequilibrium is unemployment. In the case of the repressed inflation regimes of centrally planned economies, however, unobservable labour shortages prevail. Furthermore, the ideological commitments by socialist states to full employment, and restrictions they place on labour participation decisions, raise questions about the meaningfulness of the labour supply function. One might expect that this would lead to rejection of the Barro–Grossman hypotheses concerning labour supply responses and the supply multiplier. But this does not take into account the possibility of more fully modelling labour supply responses to excess demand in official markets by introducing second economy labour markets. In several CPEs (e.g. Hungary and Poland) workers have considerable freedom to adjust the quality of their efforts and to distribute actual labour time between the private and state sectors. If this behaviour were to be better analysed, then it would be possible eventually to construct and estimate a unified disequilibrium model of the consumption and labour supply behaviour of households in centrally planned economies.

Author index

Subject index